Farm Animals: Health and Management

Farm Animals: Health and Management

Edited by Johann Casini

□SYRAWOOD
PUBLISHING HOUSE

New York

Published by Syrawood Publishing House,
750 Third Avenue, 9th Floor,
New York, NY 10017, USA
www.syrawoodpublishinghouse.com

Farm Animals: Health and Management
Edited by Johann Casini

International Standard Book Number: 978-1-68286-793-8 (Hardback)

Cataloging-in-Publication Data

Farm animals : health and management / edited by Johann Casini.
     p. cm.
Includes bibliographical references and index.
ISBN 978-1-68286-793-8
1. Domestic animals. 2. Livestock. 3. Animal health. 4. Livestock--Management.
I. Casini, Johann.
SF41 .F37 2019
636--dc23

TABLE OF CONTENTS

PREFACE

I am honored to present to you this unique book which encompasses the most up-to-date data in the field. I was extremely pleased to get this opportunity of editing the work of experts from across the globe. I have also written papers in this field and researched the various aspects revolving around the progress of the discipline. I have tried to unify my knowledge along with that of stalwarts from every corner of the world, to produce a text which not only benefits the readers but also facilitates the growth of the field.

Farm animals are domesticated animals that are raised to produce labor and commodities like, eggs, milk, wool, and meat, etc. These include cattle, goats, swine, poultry, etc. To maintain the health of farm animals, good husbandry practices such as hygiene maintenance and proper feeding are extremely important. Diseases like foot-and-mouth disease, classical swine fever, rabies, tuberculosis, etc. affect farm animals. Vaccines and antibiotics are available for the prevention and cure of some of these diseases. To promote animal welfare, animal husbandry involves considerations of aspects of animal reproduction, behavior, longevity, physiology, freedom from immunosuppression and diseases, etc. This book provides significant information of this discipline to help develop a good understanding of farm animals and their health and management. The book is appropriate for students seeking detailed information in this area as well as for experts.

Finally, I would like to thank all the contributing authors for their valuable time and contributions. This book would not have been possible without their efforts. I would also like to thank my friends and family for their constant support.

Editor

Novel SNPs in *IL-17F* and *IL-17A* genes associated with somatic cell count in Chinese Holstein and Inner-Mongolia Sanhe cattle

Tahir Usman[1,2†], Yachun Wang[1†], Chao Liu[1†], Yanghua He[1], Xiao Wang[1], Yichun Dong[1], Hongjun Wu[3], Airong Liu[4] and Ying Yu[1*]

Abstract

Background: Bovine mastitis is the most common and costly disease of lactating cattle worldwide. Apart from milk somatic cell count (SCC) and somatic cell score (SCS), serum cytokines such as interleukin-17 (IL-17) and interleukin-4 (IL-4) may also be potential indicators for bovine mastitis. The present study was designed to investigate the effects of single nucleotide polymorphisms (SNPs) in bovine *IL-17F* and *IL-17A* genes on SCC, SCS and serum cytokines in Chinese Holstein and Inner-Mongolia Sanhe cattle, and to compare the mRNA expression variations of the cows with different genotypes.

Results: A total of 464 lactating cows (337 Holstein and 127 Inner-Mongolia Sanhe cattle) were screened for SNPs identification and the data were analyzed using fixed effects of herd, parity, season and year of calving by general linear model procedure. The results revealed that SNP g.24392436C > T in *IL-17F* and SNP g.24345410A > G in *IL-17A* showed significant effects on SCC and IL-4 in Holstein ($n = 337$) and on IL-17 and IL-4 in Sanhe cattle ($n = 127$). The homozygous GG genotype of SNP g.24345410A > G had significantly higher mRNA expression compared with the heterozygous AG genotype.

Conclusions: The results indicate that *IL-17F* and *IL-17A* could be powerful candidate genes of mastitis resistance and the significant SNPs might be useful genetic markers against mastitis in both dairy and dual purpose cattle.

Keywords: Chinese Holstein, Inner Mongolia Sanhe cattle, *Interleukin 17A*, *Interleukin 17F*, Mastitis susceptibility

Background

Bovine mastitis is characterized by a range of chemical, physical and bacteriological changes in the milk accompanied by pathological changes in the udder tissues [1]. Mastitis is the most costly disease of dairy cattle and reported to cause $2 billion dollars annual losses to the U.S. dairy industry and about $35 billion to the world dairy sector [2, 3]. Genetics and environment are the most important factors that contribute to mastitis development.

Milk somatic cell count (SCC) and somatic cell score (SCS) are the most suitable indirect indexes to evaluate the degree of mastitis because these traits are convenient, inexpensive and easy to record [4]. The genetic evaluation and indirect selection of cattle for lower SCC or SCS may reduce the incidence of susceptibility to mastitis [5, 6]. The cytokines in serum such as interleukin-4 (IL-4), IL-6, IL-17, tumor necrosis factor-α (TNF-α) and interferon-γ (IFN-γ) also act as indirect indexes in inflammatory conditions [7], which suggests that beside SCC and SCS, serum cytokines could be considered as crucial indicators for bovine mastitis. Previous studies indicated that mutations in *IL-17F* and *IL-17A* genes were related with inflammatory conditions e.g. inflammatory bowel disease [7], asthma [8], rheumatoid

* Correspondence: yuying@cau.edu.cn

†Equal contributors

[1]Key Laboratory of Animal Genetics, Breeding and Reproduction, Ministry of Agriculture of China, National Engineering Laboratory for Animal Breeding, College of Animal Science and Technology, China Agricultural University, Beijing 100193, People's Republic of China

Full list of author information is available at the end of the article

arthritis [9], ovarian cancer [10], colon cancer [11] and breast cancer [12]. The role of the two genes in disease process in humans and other model animals indicates that *IL-17F* and *IL-17A* could be potential candidate genes for mastitis resistance in bovine as well.

IL-17 is a bridge of the innate and adaptive immune systems. The IL-17 family contains six members with *IL-17A* as the founding member (which was cloned about two decades ago) [13] and *IL-17F* is the most recently discovered member of the family [14]. From the alignment of the predicted amino acid sequence, it was found that *IL-17F* and *IL-17A* share the strongest homology compared to other members of the family and have similar functions to induce inflammatory response [15]. Both *IL-17F* and *IL-17A* were reported to be associated with increased risk of certain subtypes of gastric cancer [16]. However, the influence on the pathophysiological features of ulcerative colitis is partially different among *IL-17A* and *IL-17F* polymorphisms. A study revealed that *IL-17A*/-197A allele was significantly associated with chronic relapsing phenotype of ulcerative colitis and the -197A/A homozygote was more frequent in steroid dependent cases, while the *IL-17F*/7488 T allele was associated with the chronic continuous phenotype [17]. The regulation of IL-17 directly influences the nature of the cellular recruitment at the site of inflammation and it represents a marker of molecule regulating neutrophil and eosinophil infiltration [18].

It was reported that pure dairy breeds e.g. Holstein are more susceptible to mastitis than dual purpose breeds e.g. Simmental and Sanhe cattle [4]. Sanhe cattle is a precious dual purpose breed of northern (Sanhe area) China. The Sanhe area is located in the Inner Mongolia between longitude 117°15' ~ 124°02' east and latitude 47°05' ~ 51°30' north. Its average temperature in winter ranges between −20 °C to −31 °C and the mean annual temperature is −0.95 °C [5]. However, the genetic difference of Sanhe cattle with Holsteins and the genetic effects on mastitis resistance in this breed is rarely reported till date. The present study was designed to 1) evaluate *IL-17F* and *IL-17A* as candidate genes for association analysis with SCS as well several serum cytokines in Chinese Holstein and Sanhe cattle, and to 2) identify novel potential SNP markers for bovine mastitis resistance which could be used for udder health improvement.

Materials and methods

Cattle population and sample collection

A total of 337 Chinese Holstein cows were randomly selected from three dairy cattle farms located in Northern China (Qiqihar, Tianjin and Shanxi) and 127 Sanhe cattle were randomly collected from Sanhe cattle breeding farm in Hailar, Inner Mongolia (Fig. 1). The cows ranged between parity one and five and were milked thrice a day. The cattle were fed a lactation diet as recommended by the Dairy Association of China for lactating cows.

Blood samples were collected from the jugular vein of all selected cattle in three 9 mL tubes, one for DNA extraction (in EDTA coated tube), the second for RNA extraction (containing TRIzol) and the third for serum isolation by an in-house technician of the farm. The RNA extraction tube was immediately placed at −80 °C to avoid any damage to the RNA. For serum isolation, the blood samples were placed at room temperature 30 min to enable blood coagulation and then centrifuged at 3,000 r/min for 10 min to separate serum. The serum samples were then stored at 4 °C and sent to Beijing Huaying Biological Technology Research Institute within 24 h to detect the concentration of IL-4, IL-6, IL-17, IFN-γ and TNF-α (Sino-UK, China) using radioimmunoassay. Briefly, each serum sample was initially centrifuged for 5 min at 3,000 r/min at 4 °C. Then, 100 μL of the supernatant ("cold" antigen), 100 μL of antibody and 100 μL of radiolabeled antigen (125-I, "hot" antigen) were mixed and then stayed at 4 °C for 24 h. Next, 500 μL of separating solution was added in and mixed. The mixture was stayed at room temperature for 20 min and then centrifuged for 25 min at 3,500 r/min at 4 °C. Finally, the supernatant was removed and the radioactivity of the bound antigen remaining in the precipitate was measured by a gamma counter.

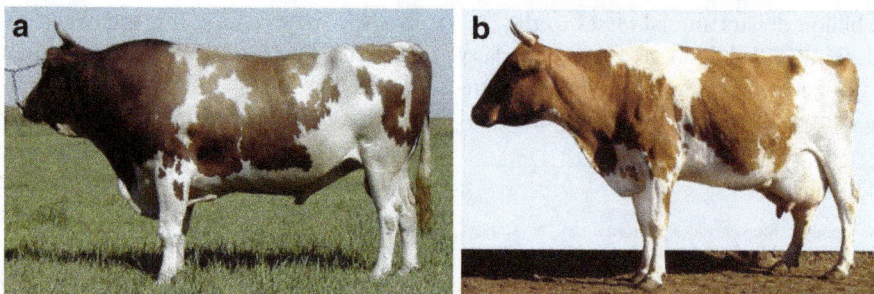

Fig. 1 Sanhe cattle in Inner Mongolia, China. **a** Sanhe bull; **b** Sanhe cow

Fresh milk samples were collected in 50 mL tubes from all the four quarters of each cow's udder and mixed with 0.03 g potassium dichromate, stored at 4 °C and then sent to Beijing Dairy Cattle Centre for somatic cell count (SCC) test within 48 h. SCS were converted from SCC according to Rupp and Biochard [19] using a formula $SCS = \log_2 (SCC/100,000) + 3$.

DNA isolation and SNPs identification
Genomic DNA was extracted from whole blood using Tiangen Blood DNA Kit (Tiangen Biotech Co., China) following the manufacturer's instructions. The quantity and quality of DNA were measured using NanoDrop™ ND-2000c Spectrophotometer (Thermo Scientific, Inc.).

All SNPs in bovine *IL-17F* and *IL-17A* genes were identified by sequencing polymerase chain reaction (PCR) amplicon using a DNA pool constructed with genomic DNA of thirty randomly selected cattle (50 ng/uL per sample). Based on the information of the NCBI and UCSC database, primers for 12 SNPs were designed to amplify the two genes fragments (Table 1). Two SNPs, one in *IL-17F* (g.24392436C > T) and another one in *IL-17A* (g.24345410 A > G), were arbitrarily chosen and screened in the two cattle populations using SNaPshot assay (ABI Multiplex SNaPshot, USA).

RNA isolation, purification and reverse transcription
The Bioteke RNA Isolation Kit (Bioteke, Beijing) was used to extract total RNA from peripheral blood samples of the Sanhe and Holstein cattle. RNase-Free DNaseSet (Qiagen, Germany) was used to purify RNA ensuring that genomic DNA was discarded. Reverse transcription was carried out with High Capacity cDNA Archive Kit (ABI, USA) according to the manufacturer's protocol. PCR primer sets for bovine *IL-17A* and *IL-4* genes were designed by software oligo6.0, considering the golden rules for real-time PCR (RT-PCR) (Table 1). The amplification efficiency of these primer pairs was tested by RT-PCR firstly. The mRNA expression of the two genes was normalized against the housekeeping gene glyceraldehyde-3-phosphatedehydrogenase (*GAPDH*) cDNA in the corresponding samples. The primers sequences of *GAPDH* are also listed in Table 1.

Quantitative RT-PCR
Quantitative real time polymerase chain reaction (qRT-PCR) was carried out to determine the mRNA expression levels of *IL-17A* and *IL-4* genes. The reactions were performed in a total volume of 20 μL containing 2 μL cDNA, 1 μL each primers, 10 μL SYBR Green Master Mix (Roche, Switzerland), 6 μL nuclease-free water using the following amplification condition: 94 °C for 10 min, followed by 44 cycles of 94 °C for 15 s, 60 °C for 10 s, 72 °C for 10 s, and 72 °C for 30 s. Fluorescence signals were collected during 60 °C step. Mean was derived from the two repeats for each sample. Light Cycler 480 RT-PCR system was used to perform amplification, detection and data analyses.

Statistical analyses
The SCC data were classified into three grades: (I) SCC ≤ 200,000 cells/mL; (II) 200,000 cells/mL< SCC < 500,000 cells/mL; and (III) SCC ≥ 500,000 cells/mL. The influence of SCC levels on five serum cytokines and SCS in the two cattle populations was analyzed using the

Table 1 Primer pairs of PCR and real time qRT-PCR used in the present study

SNP	Gene	Upper Primer (5′–3′)	Lower Primer (5′–3′)	PCR type
1	*IL-17F*	GTCATTGGAACATCTCAGGAC	GCAGATCAGCTCAGCTGAAGA	Normal PCR
2	*IL-17F*	ATAATAGTCCTTACATTGACT	GCAAAGCATCAGGAGAGGTTG	Normal PCR
3	*IL-17F*	TGACCTGCTTACTGCCTAAGT	TAAGAGATTTGCTATAGAGTG	Normal PCR
4	*IL-17F*	TGACCTGCTTACTGCCTAAGT	TAAGAGATTTGCTATAGAGTG	Normal PCR
5	*IL-17F*	AACCAAAATGGATTAGCAAGT	GTACAGGGCCAGTTAGTGAA	Normal PCR
6	*IL-17F*	AACCAAAATGGATTAGCAAGT	GTACAGGGCCAGTTAGTGAA	Normal PCR
7	*IL-17F*	AACCAAAATGGATTAGCAAGT	GTACAGGGCCAGTTAGTGAA	Normal PCR
8	*IL-17F*	AACCAAAATGGATTAGCAAGT	GTACAGGGCCAGTTAGTGAA	Normal PCR
9	*IL-17A*	CAGTTCAAGTACACAAATGAGC	GGTGTTTATCCATCCTACATAC	Normal PCR
10	*IL-17A*	TATGAGTATCTGTTTTGCCTAG	CAGTTAGACTTGCTGTCTCTCT	Normal PCR
11	*IL-17A*	TATGAGTATCTGTTTTGCCTAG	CAGTTAGACTTGCTGTCTCTCT	Normal PCR
12	*IL-17A*	TATGAGTATCTGTTTTGCCTAG	CAGTTAGACTTGCTGTCTCTCT	Normal PCR
	IL-17A	AGGGTCAACCTAAACATCGTT	GTACCTCTCAGGGTCCTCATT	Real-time PCR
	IL-4	AGGGTTGGAATTGAGCTTAGG	TGGCTTCATTCACAGAACAGG	Real-time PCR
	GADPH	GCTGCTTTTAATTCTGGC	CTTTCCATTGATGACGAG	Real-time PCR

general linear model procedure of SAS 9.1 using the following model:

$$y_{ij} = \mu + a_i + e_{if}$$

Where y_{ij} represents SCS or serum concentration of cytokine IL-4, IL-6, IL-17, TNF-α and IFN-γ; μ is overall mean; α_i is effect of SCC levels; e is the random error.

Associations of the 12 SNPs with SCS and five serum cytokines were analyzed using the GLM model 2 in the two breeds separately (SAS 9.1):

$$y_{ijkl} = \mu + a_i + \beta_j + \gamma_l + e_{ijkl}$$

Where y_{ijkl} represents each phenotype; μ is overall mean; α_i is the fixed effect of genotype; β_j is the fixed effect of the herd, year and season of birth; γ_l is the fixed effect of parity; e is the random error.

In model 2, the estimated genotype effects were further divided into additive effect (A) and dominant effect (D). The additive effect was the mean deviation of two homozygous genotypes (Formula 1), and the dominant effect was calculated by the deviation of heterozygous genotype from the mean of two homozygous genotypes (Formula 2) [20].

$$A = \frac{AA - BB}{2} \text{ (Formula 1)}$$

$$D = AB - \frac{AA + BB}{2} \text{ (Formula 2)}$$

Where, AA, AB and BB were least square means of genotype AA, AB and BB, respectively.

Student t test was used for qRT-PCR analyses for comparing the difference of mRNA expression level of *IL-17A* and *IL-4* between different genotypes of *IL-17A*.

Results
SNP discovery and genotypes of bovine IL-17 F and IL-17A
In the present study, a total of 12 SNPs (comprising 8 SNPs in *IL-17 F* and 4 SNPs in *IL-17A* gene) were revealed by screening the pooled DNA of 30 randomly selected Chinese Holstein and Sanhe cattle (Table 2). Of the 8 SNPs in *IL-17 F*, one each was located in exon 4 and intron 3 and six were located in 2 kb promoter region, whereas, in *IL-17A*, three SNPs were located in exon 3 and one in 2 kb promoter region. Out of the 12 SNPs, two were then genotyped in a total population of 337 Holstein and 127 Sanhe cattle (Table 3). Allele and genotype frequencies and Chi square test χ^2 results are summarized in Table 3. Chi square test (χ^2) showed that genotype frequencies of all SNPs in the population were in Hardy–Weinberg equilibrium.

Table 2 Information of the 12 SNPs identified in the bovine *IL-17F* and *IL-17A* genes

SNP	Gene	Region	Position	Mutation	SNP ID
1	*IL-17F*	Exon3	BTA23: 24391125	C-T	rs110506339
2	*IL-17F*	Intron2	BTA23: 24392436	C-T	Novel
3	*IL-17F*	2 kb promoter	BTA23: 24398104	A-G	Novel
4	*IL-17F*	2 kb promoter	BTA23: 24398109	A-G	rs109355109
5	*IL-17F*	2 kb promoter	BTA23: 24398704	G-C	Novel
6	*IL-17F*	2 kb promoter	BTA23: 24398855	C-T	rs110523413
7	*IL-17F*	2 kb promoter	BTA23: 24398786	C-G	Novel
8	*IL-17F*	2 kb promoter	BTA23: 24398886	T-G	Novel
9	*IL-17A*	2 kb promoter	BTA23: 24345410	A-G	rs133156805
10	*IL-17A*	Exon3	BTA23: 24350367	A-G	Novel
11	*IL-17A*	Exon3	BTA23: 24350396	C-G	Novel
12	*IL-17A*	Exon3	BTA23: 24350409	A-G	Novel

Effect of three SCC grades on SCS and cytokines of Chinese Holstein and Sanhe cattle
The descriptive statistics for SCC, SCS and each serum cytokine were listed in the Additional file 1: Table S1. The effect of SCC grades on SCS and serum cytokines were analyzed in the two cattle population (Table 4). The results showed that SCC grade had highly significant effect on SCS, IL-17 and IFN-γ in Holstein and on SCS, IL-17 and IL-6 in Sanhe cattle ($P < 0.001$). In Holstein cattle, the grade 3 SCC was significantly associated with higher values of SCS and IL-17 compared with grad I and II. Whereas, in Sanhe cattle, SCC grade II was significantly associated with higher values of cytokine IL-17 and IL-6 compared to grade 1 and 3 ($P < 0.001$).

Effects of the SNPs on mastitis indicator traits
The results of association study are shown in Table 5. SNP (g.24392436C > T) in *IL-17F* showed significant association with SCS in Holstein ($P < 0.05$) and highly significant association with cytokine IL-17 in Sanhe cattle ($P < 0.01$). Whereas, the association of SNP (g.24345410A > G) in *IL-17A* was found significant with cytokine IL-4 in both Holstein and Sanhe cattle ($P < 0.05$). The CC genotype of SNP (g.24392436C > T) was significantly associated with higher SCS compared to the other genotypes in Holstein, whereas, the TT genotype was significantly associated with higher cytokine IL-17 ($P < 0.01$) than genotype CC and CT. The AG genotype of *IL-17A* was significantly associated with higher values of IL-4 than AA genotype in both Holstein and Sanhe cattle ($P < 0.01$).

The additive and dominant effects of the SNPs
To dissect the genotype effects of the significant SNPs, their additive and dominant effects were calculated using

Table 3 Genotype and allele frequencies and Hardy–Weinberg equilibrium test of the 2 SNPs in Chinese Holstein and Innar-Mongolia Sanhe cattle

SNP/Gene	Breed	Genotype frequency			Allele frequency[*]		χ^2 Test (P)
g.24392436C > T		CC	CT	TT	C	T	
IL-17 F	Holstein	0.68 ($n = 228$)	0.27 ($n = 91$)	0.05 ($n = 16$)	0.81	0.19	3.75 ($P > 0.05$)
	Sanhe Cattle	0.61 ($n = 76$)	0.31 ($n = 38$)	0.07 ($n = 9$)	0.77	0.23	1.81 ($P > 0.05$)
g.24345410A > G		AA	AG	GG	A	G	
IL-17A	Holstein	0.15 ($n = 51$)	0.44 ($n = 149$)	0.41 ($n = 137$)	0.37	0.63	0.987 ($P > 0.05$)
	Sanhe Cattle	0.29 ($n = 36$)	0.34 ($n = 42$)	0.36 ($n = 45$)	0.46	0.54	12.08 ($P < 0.05$)

*The wild-type allele are in the left, n = number of cow. $\chi^2_{0.05(1)} = 3.84$

formula 1 and 2, respectively. It was found that both the additive and dominant effects of SNP g.24392436C > T in gene *IL-17F* on cytokine IL-17 were highly significant ($P < 0.001$, Table 6). In addition, the dominant effect of the SNP g.24345410A > G of *IL-17A* was found significant on IL-4 in both Holstein and Sanhe cattle ($P < 0.01$).

Effects of the combination genotypes on mastitis indicator traits

The effects of the combination genotypes are mentioned in Table 7. The results showed that the combination genotype of SNPs in *IL-17F* and *IL-17A* were significantly associated with IL-17 cytokine ($P < 0.01$) and showed a tendency towards significance for association with SCS ($P = 0.054$) in Sanhe cattle.

mRNA expression level in different genotypes of IL-17A and its association with IL-4 gene, cytokine IL-4 and SCS

The results of qRT-PCR for mRNA expression of *IL-17A* and *IL-4* gene showed that mRNA expression of *IL-17A* was significantly higher in genotype GG compared with genotype AG (Fig. 2). Moreover, the AG genotype of *IL-17A* gene showed significantly higher mRNA expression compared with AA and GG genotype with respect to *IL-4* gene. Notably, the AG genotype of *IL-17A* gene was associated with higher values of cytokine IL-4 compared

to the other genotypes in both Holstein and Sanhe cattle ($P < 0.05$).

In order to display the trends of mRNA expression of different genotypes of *IL-17A* gene with the mRNA expression of *IL-4* gene, cytokine IL-4 and SCS values, we draw a line chart as shown in Fig. 3. The mRNA expression of the different genotypes of *IL-17A* gene showed the same trends for mRNA expression of *IL-4* gene, and the values of cytokine IL-4 and SCS. The AG genotype of *IL-17A* had lower mRNA expression and higher values of the other 3 indicators i.e. *IL-4* gene, IL-4 cytokine and SCS.

Discussion

In the present study, a total of 12 SNPs were identified in *IL-17F* and *IL-17A* genes in Holstein and Inner-Mongolia Sanhe cattle, of which 2 SNPs were arbitrarily chosen for genotyping and further screening to evaluate their potential association with mastitis. The two SNPs were found significantly associated with mastitis indicator traits. To the best of our knowledge, this is the first study to examine the associations of polymorphisms in *IL-17A* and *IL-17 F* genes with bovine mastitis and to evaluate these genes as prognostic markers for mastitis in dairy cattle.

Genetic polymorphisms in *IL-17F* and *IL-17A* were reported to be significantly associated with susceptibility

Table 4 Analysis of SCC grade and cytokines of Chinese Holstein and Sanhe cattle

SCC Grades*	Breed	N, %	SCS[%]	IL-4, ng/mL	IL-6, pg/mL	IL-17, pg/mL	IFN-γ, pg/mL	TNF-α, pg/mL
I	Holstein	133 (52.0%)	1.84 ± 0.10^A	1.043 ± 0.032	163.88 ± 3.88	14.80 ± 0.49^B	44.97 ± 1.04^A	1.108 ± 0.029
II		30 (11.7%)	4.73 ± 0.22^B	1.043 ± 0.070	167.99 ± 8.28	16.41 ± 1.05^{AB}	45.92 ± 2.21^A	1.079 ± 0.061
III		93 (36.3%)	7.88 ± 0.12^C	1.090 ± 0.039	154.49 ± 4.61	17.21 ± 0.59^A	38.28 ± 1.25^B	1.063 ± 0.034
P value			<0.01	0.64	0.20	<0.01	<0.01	0.60
I	Sanhe cattle	54 (42.5%)	2.42 ± 0.14^C	0.986 ± 0.023	113.71 ± 3.44^a	12.76 ± 0.75^B	35.67 ± 1.18	1.11 ± 0.034
II		53 (41.7%)	4.60 ± 0.13^B	0.953 ± 0.021	125.67 ± 3.10^a	15.90 ± 0.68^A	37.23 ± 1.06	1.08 ± 0.030
III		20 (15.8%)	6.62 ± 0.22^A	0.949 ± 0.032	110.93 ± 5.04^b	11.90 ± 1.02^B	33.61 ± 1.74	1.19 ± 0.046
P value			<0.01	0.50	<0.01	<0.01	0.19	0.15

Note: *SCC grades: (I) SCC < 200,000 cells/mL; (II) 200,000 cells/mL < SCC < 500,000 cells/mL; and (III) SCC > 500,000/mL. [%]Means with different superscripts within the same column and breed are significantly different at $P < 0.01$ (capital letter) or $P < 0.05$ (small letter)

Table 5 Effects of the SNPs on SCS and cytokines of Chinese Holstein and Sanhe cattle

SNPs	Breed	Genotype, n	SCS[ac]	IL-4, ng/mL	IL-6, pg/mL	IL-17, pg/mL	IFN-γ, pg/mL	TNF-α, pg/mL
IL-17 F	Holstein	CC (228)	5.93 ± 0.47^a	1.105 ± 0.056	174.87 ± 6.80	14.44 ± 0.82	46.58 ± 2.00	1.16 ± 0.05
g.24392436 C > T		CT (91)	4.96 ± 0.55^b	1.153 ± 0.062	179.86 ± 7.56	14.47 ± 0.91	44.44 ± 2.22	1.15 ± 0.06
		TT (16)	4.60 ± 0.92^b	1.182 ± 0.101	175.49 ± 12.29	17.09 ± 1.48	45.53 ± 3.62	1.09 ± 0.09
		P value	<0.05	0.50	0.70	0.17	0.47	0.79
	Sanhe	CC(76)	4.14 ± 0.81	0.91 ± 0.067	107.27 ± 9.80	12.38 ± 2.10^A	37.06 ± 3.36	1.06 ± 0.090
		CT(38)	4.31 ± 0.86	0.93 ± 0.071	109.09 ± 10.46	10.99 ± 2.24^A	36.28 ± 3.59	1.03 ± 0.096
		TT(9)	4.10 ± 1.08	0.92 ± 0.089	118.35 ± 13.06	17.68 ± 2.87^B	37.60 ± 4.48	0.95 ± 0.12
		P value	0.9	0.94	0.62	<0.01	0.87	0.41
IL-17A	Holstein	AA (51)	5.51 ± 0.60	1.026 ± 0.070^B	179.33 ± 8.58	14.05 ± 1.04	43.96 ± 2.52	1.15 ± 0.07
g.24345410 A > G		AG (149)	5.71 ± 0.51	1.190 ± 0.057^A	173.45 ± 7.05	15.12 ± 0.85	46.29 ± 2.07	1.13 ± 0.06
		GG (137)	5.35 ± 0.51	1.105 ± 0.058^a	178.59 ± 7.17	14.55 ± 0.87	46.13 ± 2.11	1.16 ± 0.06
		P value	0.73	<0.05	0.59	0.46	0.55	0.79
	Sanhe	AA(36)	4.65 ± 0.82	0.872 ± 0.067^b	104.20 ± 9.99	11.67 ± 2.24	35.80 ± 3.46	1.04 ± 0.094
		AG(42)	3.77 ± 0.82	0.965 ± 0.066^a	108.42 ± 9.94	11.86 ± 2.23	37.64 ± 3.44	1.08 ± 0.093
		GG(45)	4.19 ± 0.85	0.905 ± 0.069^{ab}	115.65 ± 10.39	13.67 ± 2.34	37.72 ± 3.60	1.03 ± 0.097
		P value	0.14	<0.05	0.09	0.18	0.51	0.56

[ac]Means with different superscripts within the same column and breed are significantly different at P < 0.01 (capital letter) or P < 0.05 (small letter)

of breast cancer in human [12]. In a bovine model of tuberculosis, a higher expression of IL-17 gene was reported to be positively associated with bovine tuberculosis suggesting IL-17 as a potential biomarker for prognosis in bovine tuberculosis [21]. In the present study, we found that the SNP in IL-17F was significantly associated with SCS in Holstein cattle and the effect of the combination genotype on SCS showed a tendency towards significance in Sanhe cattle. The SNPs in IL-17 F and IL-17A were significantly associated with serum

Table 6 The additive and dominant effects of the SNPs on SCS and cytokines in Chinese Holstein and Sanhe cattle

SNPs	Breed	Effect[oo]	SCS	IL-4, ng/mL	IL-6, pg/mL	IL-17, pg/mL	IFN-γ, pg/mL	TNF-α, pg/mL
IL-17 F	Holstein	A	0.66 ± 0.44	-0.04 ± 0.05	-0.30 ± 5.87	-1.32 ± 0.71	0.52 ± 1.73	0.03 ± 0.05
g.24392436 C > T		P value	0.14	0.43	0.96	0.06	0.76	0.56
		D	-0.30 ± 0.59	0.01 ± 0.06	4.68 ± 7.69	-1.30 ± 0.92	-1.60 ± 2.26	0.03 ± 0.06
		P value	0.61	0.88	0.54	0.16	0.48	0.67
	Sanhe	A	0.02 ± 0.35	-0.01 ± 0.03	-5.53 ± 4.31	-2.65 ± 0.98	-0.26 ± 1.47	0.05 ± 0.04
		P value	0.95	0.84	0.20	<0.01	0.85	0.19
		D	0.19 ± 0.49	0.01 ± 0.04	-3.71 ± 6.02	-4.04 ± 1.31	-1.04 ± 2.06	0.03 ± 0.06
		P value	0.69	0.84	0.53	<0.01	0.61	0.62
IL-17A	Holstein	A	0.07 ± 0.27	-0.04 ± 0.03	0.37 ± 3.81	-0.24 ± 0.46	-1.08 ± 1.12	-0.002 ± 0.03
g.24345410 A > G		P value	0.78	0.20	0.92	0.59	0.34	0.96
		D	0.27 ± 0.41	0.12 ± 0.04	-5.51 ± 5.43	0.81 ± 0.66	1.23 ± 1.60	-0.02 ± 0.04
		P value	0.51	<0.01	0.31	0.22	0.44	0.59
	Sanhe	A	0.23 ± 0.22	-0.02 ± 0.02	-5.72 ± 2.68	-1.00 ± 0.60	-0.96 ± 0.92	0.01 ± 0.025
		P value	0.3	0.36	<0.05	0.10	0.3	0.79
		D	-0.65 ± 0.37	0.08 ± 0.03	-1.50 ± 4.55	-0.81 ± 1.01	0.88 ± 1.57	0.01 ± 0.0412
		P value	0.08	<0.01	0.74	0.42	0.57	0.31

Note: [oo]A means additive effect, D means dominant effect

Novel SNPs in IL-17F and IL-17A genes associated with somatic cell count in Chinese Holstein...

7

Table 7 Effects of combination genotypes of IL-17 F and IL-17A on SCS and serum cytokines in Chinese Holstein and Sanhe cattle

Breed	Genotype	n	SCS	IL-4, ng/mL	IL-6, pg/mL	IL-17[æ], pg/mL	IFN-γ, pg/mL	TNF-α, pg/mL
Holstein	CCAA	49	5.71 ± 0.61	1.02 ± 0.07	179.14 ± 8.76	14.1 ± 1.05	44.57 ± 2.56	1.17 ± 0.07
	CCAG	101	6.02 ± 0.56	1.18 ± 0.06	171.31 ± 7.73	15.00 ± 0.93	47.53 ± 2.26	1.11 ± 0.06
	CCGG	78	6.02 ± 0.57	1.07 ± 0.07	176.25 ± 8.29	14.1 ± 0.99	47.28 ± 2.41	1.21 ± 0.06
	CTAG	47	5.43 ± 0.66	1.18 ± 0.07	176.97 ± 9.01	15.08 ± 1.08	44.66 ± 2.64	1.19 ± 0.07
	CTGG	42	4.57 ± 0.69	1.11 ± 0.07	183.34 ± 9.28	14.07 ± 1.08	44.98 ± 2.64	1.12 ± 0.07
	TTGG	16	4.6 ± 0.92	1.18 ± 0.10	175.63 ± 12.38	17.10 ± 1.49	45.79 ± 3.63	1.1 ± 0.09
	P value		0.23	0.11	0.83	0.35	0.71	0.51
Sanhe	CCAA	35	4.83 ± 0.35	0.88 ± 0.02	114.30 ± 4.28	13.82 ± 0.90[C]	34.88 ± 1.41	1.15 ± 0.04
	CCAG	25	3.51 ± 0.39	0.99 ± 0.03	121.22 ± 5.00	15.15 ± 1.05[BC]	37.60 ± 1.65	1.18 ± 0.04
	CCGG	16	4.99 ± 0.51	0.88 ± 0.04	130.81 ± 6.57	17.22 ± 1.39[AB]	37.60 ± 2.17	1.23 ± 0.06
	CTAG	17	5.00 ± 0.49	0.94 ± 0.04	116.96 ± 6.46	12.68 ± 1.32[C]	34.72 ± 2.13	1.19 ± 0.05
	CTGG	23	4.10 ± 0.44	0.94 ± 0.03	125.2 ± 5.82	14.58 ± 1.22[BC]	36.75 ± 1.92	1.10 ± 0.05
	TTGG	9	4.42 ± 0.66	0.93 ± 0.056	131.11 ± 8.37	20.12 ± 1.89[A]	40.37 ± 2.95	1.08 ± 0.08
	P value		0.05	0.11	0.15	<0.01	0.40	0.51

[æ]Means with different superscripts within the same column and breed are significantly different at *P* < 0.01 (capital letter)

cytokine (IL-17 and IL-4) in both Holstein and Sanhe cattle. These cytokines are connected with the innate and adaptive immune system. The results of the present study provided the first evidence that the *IL-17A* promoter polymorphism, whose function is still unclear, is significantly associated with cytokine IL-4 of bovine mastitis. The findings of the study reveal that SNPs in both the *IL-17F* and *IL-17A* genes have similar type of influence on both the Chinese Holstein and Sanhe cattle. Thus, the SNPs that have significant association with mastitis indicator traits in Holstein and Sanhe cattle breed could be considered as important genetic markers

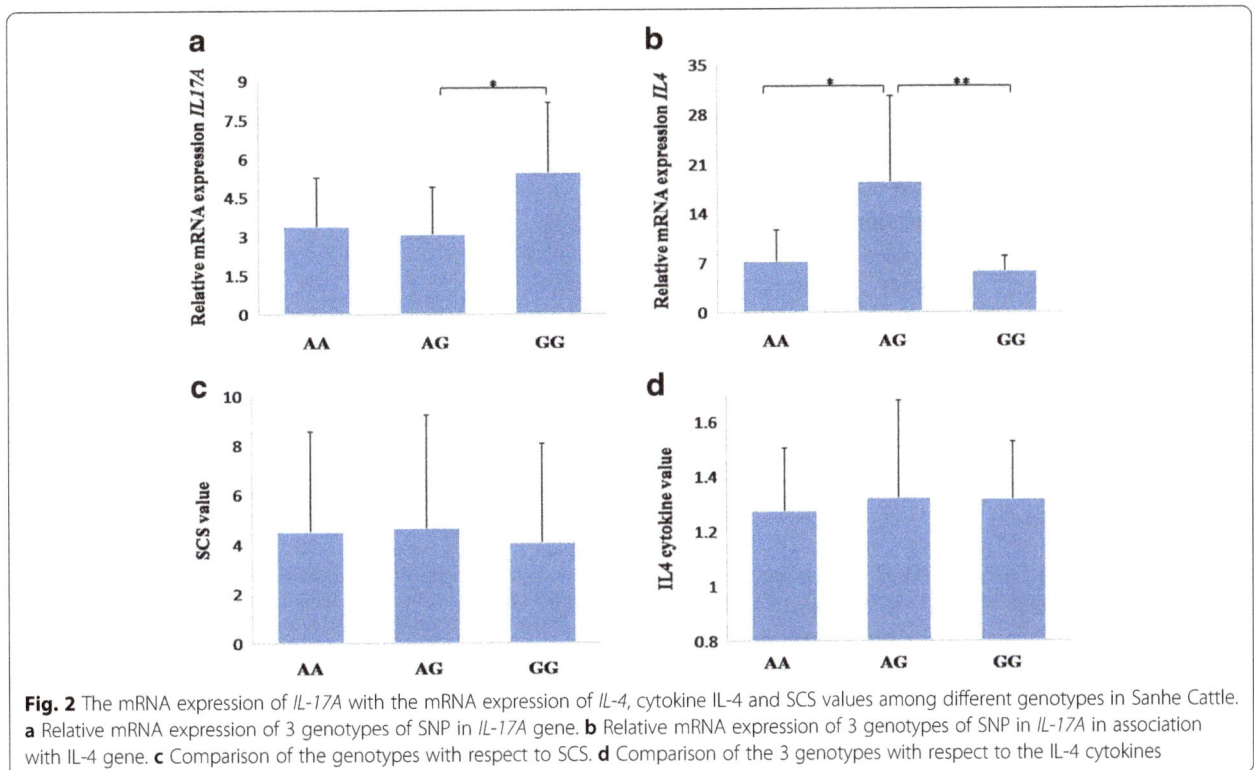

Fig. 2 The mRNA expression of *IL-17A* with the mRNA expression of *IL-4*, cytokine IL-4 and SCS values among different genotypes in Sanhe Cattle. **a** Relative mRNA expression of 3 genotypes of SNP in *IL-17A* gene. **b** Relative mRNA expression of 3 genotypes of SNP in *IL-17A* in association with IL-4 gene. **c** Comparison of the genotypes with respect to SCS. **d** Comparison of the 3 genotypes with respect to the IL-4 cytokines

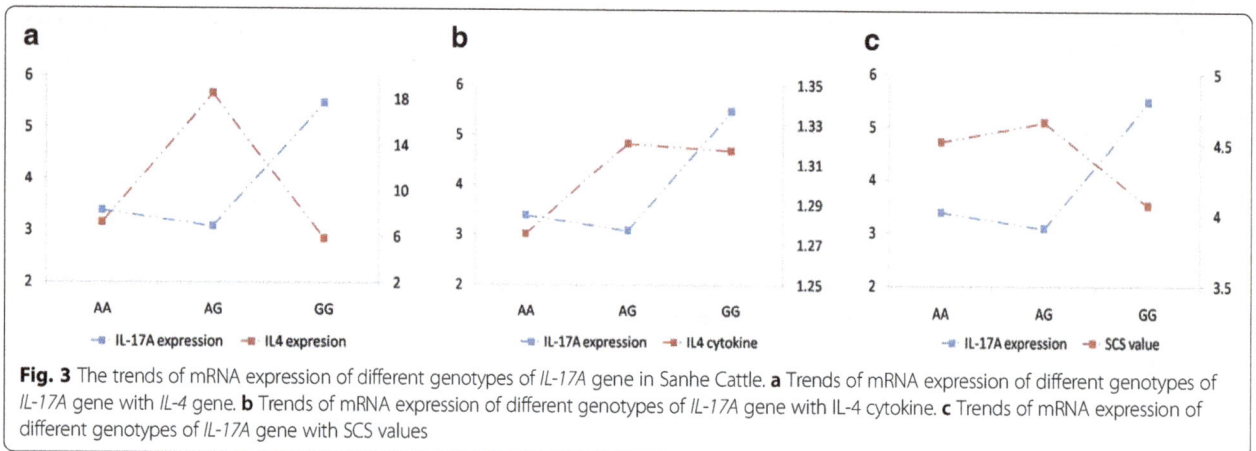

Fig. 3 The trends of mRNA expression of different genotypes of *IL-17A* gene in Sanhe Cattle. **a** Trends of mRNA expression of different genotypes of *IL-17A* gene with *IL-4* gene. **b** Trends of mRNA expression of different genotypes of *IL-17A* gene with IL-4 cytokine. **c** Trends of mRNA expression of different genotypes of *IL-17A* gene with SCS values

in mastitis susceptibility studies in dairy cattle. In addition, although SCS is continuous trait which generally serves as an important indicator for subclinical mastitis, is highly influenced by various environmental factors. Thus, the records of clinical incidence of mastitis should be collected and analyze their association with these two SNPs in *IL17F* and *IL17A*.

The cytokine IL-6 is a major player in hematopoiesis as well as in the immune system. It possesses both pro- and anti-inflammatory properties and is a pleiotropic inflammatory cytokine involved in numerous biological functions including hematopoiesis, inflammation, immune regulation and oncogenesis [22]. It was demonstrated that detection of cytokine IL-6 in milk indicated subclinical mastitis earlier than the detection of elevated SCC [23]. The author concluded that the detection of IL-6 in milk could be a reliable prediction marker for subclinical mastitis. In the present study, SNPs in *IL-17F* and *IL-17A* genes were non-significantly associated with cytokine IL-6, but were significantly associated with SCS, cytokine IL-17 and IL-4. The epigenetic regulation of gene expression by IL-6 can lead to tumor progression by altering the promoter methylation and the genes regulatory pathways [24]. Noticeably, in the present study the grades of SCC were significantly associated with SCS, cytokines IL-6 and IL-17. The significant association of the two SNPs in these genes with serum cytokines and mastitis indicator trait is a positive clue to consider these genes as candidate genes in mastitis resistance studies.

Conclusion

In conclusion, both *IL-17A* and *IL-17F* gene polymorphisms (g.24392436C > T and g.24345410A > G) may provide valuable information for predicting the prognosis of bovine mastitis. However, further studies are recommended to validate both *IL-17* gene expression and early lymphocyte activation as biomarker of immune status. The present study provides preliminary findings of

the relationship between *IL-17A* and *IL-17F* gene with SCC/SCS and cytokines levels. The results infer that *IL-17A* and *IL-17F* genes could be crucial modifiers of inflammatory diseases and the SNPs might be useful markers of genetic resistance against bovine mastitis development in both dairy and dual purpose cattle. We suggest further in-depth research using large population size to evaluate the association of these genes with bovine mastitis.

Abbreviations
SCC: Somatic cell count; SCS: Somatic cell score; DHI: Dairy Herd Improvement; PCR: Polymerase chain reaction; IL-4: Interleukin-4; IL-6: Interleukin-6; IL-17: Interleukin-17; TNF-α: Tumor necrosis factor-α; IFN-γ: Interferon-γ

Acknowledgements
We acknowledge the China Dairy Data Center and Xieerltala Breeding Farm for providing blood samples and DHI data.

Funding
This work was financially supported by the National Natural Science Foundation of China (31272420), the Earmarked Fund for Modern Agro-industry Technology Research System (CARS-37), the Fund for Basic Research from the Ministry of Education of the People's Republic of China (2011JS006), the National Key Technologies R & D Program (2011BAD28B02) and the Program for Changjiang Scholar and Innovation Research Team in University (IRT1191). The funders had no role in study design, data collection and analysis, decision to publish, or preparation of the manuscript.

Authors' contributions
YY designed the experiment and supervised the project. CL and TU carried out the study and wrote the initial draft of the paper. XW, YY, YD, CL, YW, HW and AL collected the samples. CL, XW, HY and DY conducted the lab work. YW, YD and YY performed data analyses and contributed in writing the manuscript. All authors read and approved the paper.

Competing interests
The authors declare that they have no competing interests.

Author details
[1]Key Laboratory of Animal Genetics, Breeding and Reproduction, Ministry of Agriculture of China, National Engineering Laboratory for Animal Breeding, College of Animal Science and Technology, China Agricultural University, Beijing 100193, People's Republic of China. [2]College of Veterinary Sciences and Animal Husbandry, Abdul Wali Khan University Mardan, Mardan 23200, Pakistan. [3]Xieerltala Breeding Farm, Hailaer 021012, Inner Mongolia, China. [4]Agricultural and Animal Husbandry Administration Bureau, Hailaer 021000, Inner Mongolia, China.

References
1. Mansor R, William M, David CB, Andrew B, Amaya A, Justyna SY, et al. A peptidomic approach to biomarker discovery for bovine mastitis. J Proteomics. 2013;85:89–98.
2. Sordillo LM, Streicher KL. Mammary gland immunity and mastitis susceptibility. J Mammary Gland Biol Neoplasia. 2002;7:135–46.
3. Mubarack HM, Doss A, Dhanabalan R, Venkataswamy R. Activity of Some Selected Medicinal Plant Extracts Against Bovine Mastitis Pathogens. J Anim Vet Adv. 2011;10:738–41.
4. Wang XP, Xu SZ, Gao X, Ren HY, Chen JB. Genetic polymorphism of TLR4 gene and correlation with mastitis in cattle. J Genet Genomics. 2007;34:406–12.
5. Yuan Z, Li J, Li J, Zhang J, Gao X, Gao HJ, et al. Investigation on BRCA1 SNPs and its effects on mastitis in Chinese commercial cattle. Gene. 2012;505:190–4.
6. Chu MX, Lai JM, Yu CY, Chang WS, Lin HH, Su YC. Polymorphism of exon 2 of BoLA-DRB3 gene and its relationship with somatic cell score in Beijing Holstein cows. Mol Biol Rep. 2012;39:2909–14.
7. Fujino S, Andoh A, Bamba S, Ogawa A, Hata K, Araki Y, et al. Increased expression of interleukin 17 in inflammatory bowel disease. Gut. 2003;52:65–70.
8. Jin EH, Choi EY, Yang JY, Chung HT, Yang YS. Significant association between IL-17 F promoter region polymorphism and susceptibility to asthma in a Korean population. Int Arch Allergy Immunol. 2011;155:106–10.
9. Paradowska GA, Wojtecka-Lukasik E, Trefler J, Wojciechowska B, Lacki JK, Maslinski S. Association between IL-17 F gene polymorphisms and susceptibility to and severity of rheumatoid arthritis (RA). Scand J Immunol. 2010;72:134–41.
10. Kato T, Furumoto H, Ogura T, Onishi Y, Irahara M, Yamano S, et al. Expression of IL-17 mRNA in ovarian cancer. Biochem Biophys Res Commun. 2001;282:735–8.
11. Tong Z, Yang XO, Yan H, Liu W, Niu X, Shi Y, et al. A protective role by interleukin-17 F in colon tumorigenesis. PLoS One. 2012;7(4):e34959.
12. Wang L, Jiang Y, Zhang Y, Wang Y, Huang S, Wang Z, et al. Association analysis of IL-17A and IL-17 F polymorphisms in Chinese Han women with breast cancer. PLoS One. 2012;7(3):e34400.
13. Yao Z, Painter SL, Fanslow WC, Ulrich D, Macduff BM, Spriggs MK, et al. Human IL-17: a novel cytokine derived from T cellsJ Immunol. 1995;155:5483–6.
14. Starnes T, Robertson MJ, Sledge G, Kelich S, Nakshatri H, Broxmeyer HE, et al. Cutting edge: IL-17 F, a novel cytokine selectively expressed in activated T cells and monocytes, regulates angiogenesis and endothelial cell cytokine production. J Immunol. 2001;167:4137–40.
15. Hayashi R, Tahara T, Shiroeda H, Matsue Y, Minato T, Nomura T, et al. Association of genetic polymorphisms in IL17A and IL17F with gastro-duodenal diseases. J Gastrointestin Liver Dis. 2012;21:243–9.
16. Wu X, Zeng Z, Chen B, Yu J, Xue L, et al. Association between polymorphisms in interleukin-17A and interleukin-17 F genes and risks of gastric cancer. Int J Cancer. 2010;127:86–92.
17. Arisawa T, Tahara T, Shibata T, Nagasaka M, Nakamura M, Kamiya Y, et al. The influence of polymorphisms of interleukin-17A and interleukin-17 F genes on the susceptibility to ulcerative colitis. J Clin Immunol. 2008;28:44–9.
18. Schnyder-Candrian S, Togbe D, Couillin I, Mercier I, Brombacher F, Quesniaux V, et al. Interleukin-17 is a negative regulator of established allergic asthma. J Exp Med. 2006;27:2715–25.
19. Rupp R, Boichard D. Genetic parameters for clinical mastitis, somatic cell score, production, udder type traits, and milking ease in first lactation Holsteins. J Dairy Sci. 1999;82:2198–204.
20. Falconer DS, Mackay TFC. Introduction to Quantitative Genetics (4th ed.). Longman. 1995; ISBN 978–0582243026
21. Blanco FC, Bianco MV, Meikle V, Garbaccio S, Vagnoni L, Forrellad M, et al. Increased IL-17 expression is associated with pathology in a bovine model of tuberculosis. Tuberculosis. 2011;91:57–63.
22. Nishimoto N, Kishimoto T. Interleukin 6: From bench to bedside. Nat Clin Pract Rheumatol. 2006;2:619–25.
23. Sakemi Y, Tamura Y, Hagiwara K. Interleukin-6 in quarter milk as a further prediction marker for bovine subclinical mastitis. J Dairy Res. 2011;78:118–21.
24. Wehbe H, Henson R, Meng F, Mize-Berge J, Patel T. Interleukin-6 contributes to growth in cholangiocarcinoma cells by aberrant promoter methylation and gene expression. Cancer Res. 2006;66:10517–24.

2

Supplemental Smartamine M in higher-energy diets during the prepartal period improves hepatic biomarkers of health and oxidative status in Holstein cows

Mario Vailati-Riboni[1], Johan S. Osorio[1,4], Erminio Trevisi[2], Daniel Luchini[3] and Juan J. Loor[1*]

Abstract

Background: Feeding higher-energy prepartum is a common practice in the dairy industry. However, recent data underscore how it could reduce performance, deepen negative energy balance, and augment inflammation and oxidative stress in fresh cows. We tested the effectiveness of rumen-protected methionine in preventing the negative effect of feeding a higher-energy prepartum. Multiparous Holstein cows were fed a control lower-energy diet (CON, 1.24 Mcal/kg DM; high-straw) during the whole dry period (~50 d), or were switched to a higher-energy (OVE, 1.54 Mcal/kg DM), or OVE plus Smartamine M (OVE + SM; Adisseo NA) during the last 21 d before calving. Afterwards cows received the same lactation diet (1.75 Mcal/kg DM). Smartamine M was top-dressed on the OVE diet (0.07% of DM) from -21 through 30 d in milk (DIM). Liver samples were obtained via percutaneous biopsy at -10, 7 and 21 DIM. Expression of genes associated with energy and lipid metabolism, hepatokines, methionine cycle, antioxidant capacity and inflammation was measured.

Results: Postpartal dry matter intake, milk yield, and energy-corrected milk were higher in CON and OVE + SM compared with OVE. Furthermore, milk protein and fat percentages were greater in OVE + SM compared with CON and OVE. Expression of the gluconeogenic gene *PCK1* and the lipid-metabolism transcription regulator *PPARA* was again greater with CON and OVE + SM compared with OVE. Expression of the lipoprotein synthesis enzyme *MTTP* was lower in OVE + SM than CON or OVE. Similarly, the hepatokine *FGF21*, which correlates with severity of negative energy balance, was increased postpartum only in OVE compared to the other two groups. These results indicate greater liver metabolism and functions to support a greater production in OVE + SM. At 7 DIM, the enzyme *GSR* involved in the synthesis of glutathione tended to be upregulated in OVE than CON-fed cows, suggesting a greater antioxidant demand in overfed cows. Feeding OVE + SM resulted in lower similar expression of *GSR* compared with CON. Expression of the methionine cycle enzymes *SAHH* and *MTR*, both of which help synthesize methionine endogenously, was greater prepartum in OVE + SM compared with both CON and OVE, and at 7 DIM for CON and OVE + SM compared with OVE, suggesting greater Met availability. It is noteworthy that *DNMT3A*, which utilizes S-adenosylmethionine generated in the methionine cycle, was greater in OVE and OVE + SM indicating higher-energy diets might enhance DNA methylation, thus, Met utilization.

(Continued on next page)

* Correspondence: jloor@illinois.edu
[1]Mammalian NutriPhysioGenomics, Department of Animal Sciences and Division of Nutritional Sciences, University of Illinois, Urbana, IL 61801, USA
Full list of author information is available at the end of the article

(Continued from previous page)

Conclusions: Data indicate that supplemental Smartamine M was able to compensate for the negative effect of prepartal energy-overfeeding by alleviating the demand for intracellular antioxidants, thus, contributing to the increase in production. Moreover Smartamine M improved hepatic lipid and glucose metabolism, leading to greater liver function and better overall health.

Keywords: Energy, Methionine, Nutrigenomics, Transition period

Background

The transition period, defined as last 3 weeks prepartum through 3 weeks postpartum, is one of the most important stages of lactation in dairy cattle. Years of strong genetic selection and improvement have allowed modern dairy cows to reach high production performance, both in quantity and quality. However, this has made the transition between late pregnancy to early lactation a significant period of metabolic and immune challenges [1–3]. Because failure to adequately meet these challenges can compromise production, induce metabolic diseases, and increase rates of culling in early lactation [4], the management of the transition cow remains a focal point for dairy producers.

Following the "steaming up" concept of RB Boutflour [5], transition cows during the dry period were first traditionally offered a high fiber/low energy density ration, to then increase the energy density of the ration with a lower fiber content in the last month of gestation (i.e. "close-up" period). This early century practice is still embedded in the modern dairy industry. However, multiple studies have consistently reported negative effects of prepartum energy overfeeding on cow health and productivity. Among these, prepartum hyperglycemia and hyperinsulinemia together with marked postpartum adipose tissue mobilization (i.e., greater blood NEFA concentration) [6–11] have strong negative impact on postpartal health indices [12–15].

Our general hypothesis was that supplementation with rumen-protected methionine (Smartamine M, Adisseo NA) could ameliorate the transition to lactation and the health status of the cows, while controlling and reducing the negative effects of prepartal excess energy. In fact, methionine (Met) itself was able to increase both quantity and quality of production [16, 17], controlling the inflammatory and the oxidative stress status that characterize the transition period [18–20]. These outcomes are partly due to Met's ability to enhance liver function, reducing triacylglycerol accumulation and improving the metabolic capacity of the liver to orchestrate the metabolic transition into lactation [16–20]. Furthermore, Met itself, and several of its metabolites, display an immunonutritional role both in humans [21–24] and in dairy cows [16]. Therefore, in the present study we used serum and plasma biomarkers coupled with targeted hepatic transcriptome analysis from transition cows fed prepartum either a control low energy, a higher-energy, or a higher-energy diet supplemented with rumen-protected Met. Production and immune responses have been published elsewhere [25].

Methods

Experimental design and dietary treatments

All procedures were approved by the Institutional Animal Care and Use Committee (IACUC) of the University of Illinois. Complete details of the experimental design and animal management have been reported previously [25]. Briefly, 65 multiparous Holstein were enrolled and completed the trail remaining healthy throughout the length of the study. All cows were fed ad libitum the same control lower-energy diet (CON; NE_L = 1.24 Mcal/kg DM; no Met supplementation) during the far-off dry period (i.e., -50 to -21 d relative to parturition). Consequently, during the close-up period (i.e. -21 d to calving), cows were randomly allocated to either a higher-energy diet (OVE; NE_L = 1.54 Mcal/kg DM), OVE plus Smartamine M (OVE + SM; Adisseo, NA) or remained on CON. The same basal lactation diet (NEL = 1.75 Mcal/kg DM) was fed to all cows postpartum until d 30 relative to parturition. Smartamine M was top-dressed during the entire experiment over the OVE or lactation diet from -21 through 30 d relative to parturition at a rate of 0.07% of offered DM. For the current study, only a subset of cows were considered for blood biomarker (n = 10 per group) and hepatic gene expression (n = 8 per group) analyses.

Blood sampling and biomarker analysis

Blood was sampled at -26, -21, -10, 7, 14 and 21 d relative to parturition by coccygeal venipuncture using evacuated tubes (BD Vacutainer; BD and Co., Franklin Lakes, NJ) containing either clot activator or lithium heparin for serum and plasma, respectively. Blood was used for determination of (i) metabolic biomarkers: cholesterol, creatinine, growth hormone (GH), insulin-like growth factor 1 (IGF-1), leptin, urea; (ii) liver health biomarkers: albumin, bilirubin, ceruloplasmin, gamma-glutamyl-transpeptidase (GGT), glutamic oxaloacetic transaminase (GOT), haptoglobin, interleukin 6, serum amyloid A (SAA); (iii) and oxidative status biomarkers: β-carotene, glutathione, nitric oxides (NO_x, NO_2, NO_3), paraoxonase, antioxidant capacity (oxygen radical absorbance capacity, ORAC), total reactive oxygen metabolites (ROM), tocopherol.

Concentration of albumin, cholesterol, bilirubin, creatinine, urea, GOT, and GGT were assessed using kits purchased from Instrumentation Laboratory (Lexington, MA) using a clinical auto-analyzer (ILAB 600, Instrumentation Laboratory). Concentrations of ROM were analyzed with the d-ROMs-test, purchased from Diacron (Grosseto, Italy). Concentrations of haptoglobin, ceruloplasmin, paraoxonase and NOx were analyzed using the methods previously described [26–28], adapting the procedures to a clinical auto-analyzer (ILAB 600, Instrumentation Laboratory). SAA and ORAC determinations were performed using the Synergy 2 Multi-Detection Microplate Reader (BioTek Instruments, Inc., Winooski, VT). SAA concentration was assessed with a commercial ELISA immunoassay kit (Tridelta Development Ltd., Maynooth, Co. Kildare, Ireland), while ORAC was determined measuring the fluorescent signal from a probe (fluorescein) that decreases in the presence of radical damage [29]. Quantification of GH, IGF-1, and leptin concentration was as previously described [14]. Bovine IL-6 (Cat. No. ESS0029; Thermo Scientific, Rockford, IL) plasma concentration was determined using commercial ELISA kits, while plasma vitamin A, vitamin E, and β-carotene were extracted with hexane and analyzed by reverse- phase HPLC using an Allsphere ODS-2 column (3 μm, 150 × 4.6 mm; Grace Davison Discovery Sciences, Deerfield, IL), a UV detector set at 325 nm (for vitamin A), 290 nm (for vitamin E), or 460 nm (for β-carotene), and 80:20 methanol:tetrahydrofurane as the mobile phase.

Hepatic gene expression analysis
Liver tissue was harvested via percutaneous biopsy under local anesthesia at -10, 7 and 21 d relative to parturition. Tissue samples were immediately snap frozen in liquid nitrogen and then stored at -80 °C. Complete information about RNA extraction and qPCR procedures can be found in Additional file 1. Briefly, RNA samples were extracted from the frozen tissue and used for cDNA synthesis using established protocols in our laboratory [30]. The qPCR performed was SYBR Green-based, using a 6-point standard curve. Genes selected for transcript profiling are associated with (i) energy metabolism: insulin like growth factor-1 (*IGF1*), pyruvate carboxylase (*PC*), phosphoenolpyruvate carboxykinase 1 (*PCK1*), pyruvate dehydrogenase kinase 4 (*PDK4*); (ii) fatty acid metabolism: acyl-CoA oxidase 1 (*ACOX1*), apolipoprotein B (*APOB*), γ-butyrobetaine hydroxylase 1 (*BBOX1*), carnitine palmitoyltransferase 1A (*CPT1A*), 3-hydroxy-3-methylglutaryl-CoA synthase 2 (*HMGCS2*),

microsomal triglyceride transfer protein (*MTTP*), peroxisome proliferator activated receptor α (*PPARA*), solute carrier family 22 member 5 (*SLC22A5*), trimethyllysine hydroxylase, ε (*TMLHE*); (iii) hepatokines: angiopoietin like 4 (*ANGPTL4*), fibroblast growth factor 21 (*FGF21*); (iv) the methionine cycle: betaine–homocysteine S-methyl transferase (*BHMT*), betaine–homocysteine S-methyl transferase 2 (*BHMT2*), DNA (cytosine-5-)-methyltransferase 1 (*DNMT1*), DNA (cytosine-5-)-methyltransferase 3 α (*DNMT3A*), methionine adenosyltransferase 1A (*MAT1A*), 5-methyltetrahydrofolate-homocysteine methyltransferase (*MTR*), phosphatidylethanolamine N-methyltransferase (*PEMT*), S-adenosyl homocysteine hydrolase (*SAHH*); (v) the antioxidant system: cystathionine-beta-synthase (*CBS*), cysteine sulfinic acid decarboxylase (*CSAD*), cystathionine gammalyase (*CTH*), glutamate-cysteine ligase catalytic subunit (*GCLC*), glutathione peroxidase 1 (*GPX1*), glutathione reductase (*GSR*), glutathione synthetase (*GSS*), superoxide dismutase 1, soluble (*SOD1*), superoxide dismutase 2, mitochondrial (*SOD2*); (vi) and the inflammatory response: ceruloplasmin (*CP*), haptoglobin (*HP*), nuclear factor κB subunit 1 (*NFKB1*), retinoid X receptor α (*RXRA*), serum amyloid A2 (*SAA2*), suppressor of cytokine signaling 2 (*SOCS2*), signal transducer and activator of transcription 3 (*STAT3*), signal transducer and activator of transcription 5B (*STAT5B*). Primer sequences and qPCR performances are reported in Additional file 1.

Statistical analysis
After normalization with the geometric mean of the internal control genes, qPCR data were $\log_2$ transformed prior to statistical analysis to obtain a normal distribution. Statistical analysis was performed with SAS (v9.3). Both datasets (blood and qPCR) were subjected to ANOVA and analyzed using repeated measures ANOVA with PROC MIXED. The statistical model included diet (D; CON, OVE, and OVE + SM), time (T; d -26, -21, -10, 7, 14, and 21 for blood biomarkers, d -10, 7, and 21 for qPCR analysis) and their interaction (D*T) as fixed effect. Cow, nested within treatment, was the random effect. For blood data, data pre-treatment at d-26 relative to parturition, when available, were used as a covariate. The Kenward-Roger statement was used for computing the denominator degrees of freedom, while spatial power was used as the covariance structure. Data were considered significant at a $P \leq 0.05$ using the PDIFF statement in SAS. For ease of interpretation, expression data reported in Table 1 and Fig. 1 are the $\log_2$ back-transformed LSM that resulted from the statistical analysis. Standard errors were also adequately back-transformed.

Table 1 Effect of feeding a control lower-energy diet (CON, 1.24 Mcal/kg DM; high-straw) during the whole dry period (~50 d), a higher-energy (1.54 Mcal/kg DM) diet without (OVE) rumen-protected methionine during the last 21 d before calving, or OVE plus rumen-protected methionine (Smartamine M; OVE + SM; Adisseo NA) from -21 d before calving through the first 30 d postpartum on hepatic gene expression (relative mRNA abundance, $\log_2$ back-transformed LSM) in Holstein cows

	Diet[1]			SE[2]	P-value[3]		
	CON	OVE	OVE + SM		D	T	D*T
Energy metabolism							
IGF1	1.91	1.97	2.41	0.22	0.19	<.0001	0.17
PC	0.24	0.23	0.20	0.02	0.24	<.0001	0.09
PCK1	0.33[a]	0.25[b]	0.31[a]	0.02	0.03	0.10	0.59
PDK4	0.34	0.31	0.78	0.37	0.35	0.03	0.98
Fatty acid oxidation, Lipoprotein and Cholesterol synthesis							
ACOX1	1.34	1.21	1.27	0.07	0.41	0.75	0.61
APOB	1.93	1.67	1.82	0.11	0.22	0.32	0.42
BBOX1	0.39	0.36	0.38	0.02	0.61	0.04	0.96
CPT1A	0.13	0.13	0.14	0.01	0.89	<.0001	0.49
HMGCS2	1.13	1.00	0.93	0.10	0.32	0.22	0.56
MTTP	1.48[a]	1.50[a]	1.28[b]	0.08	0.05	0.02	0.84
PPARA	0.43	0.40	0.46	0.02	0.16	<.0001	0.03
SLC22A5	2.89	2.92	3.08	0.29	0.88	<.0001	0.14
TMLHE	0.42	0.40	0.41	0.02	0.68	0.003	0.39
Hepatokines							
ANGPTL4	0.01	0.01	0.02	0.002	0.34	<.0001	0.03
FGF21	0.27	0.13	0.22	0.07	0.15	0.005	0.0006
Methionine cycle and methylation							
BHMT	1.13	0.96	1.02	0.11	0.54	0.002	0.70
BHMT2	0.38	0.45	0.37	0.04	0.20	0.21	0.57
DNMT1	0.02	0.02	0.02	0.001	0.53	0.03	0.64
DNMT3A	0.99[a]	1.26[b]	1.20[b]	0.09	0.04	0.05	0.91
MAT1A	1.46	1.55	1.52	0.08	0.71	0.09	0.86
MTR	0.05[a]	0.03[b]	0.04[ab]	0.002	0.01	0.31	0.65
PEMT	0.33	0.29	0.30	0.02	0.31	0.07	0.95
SAHH	1.39	1.25	1.39	0.06	0.17	0.0003	0.0009
Antioxidant system							
CBS	1.41	1.66	1.51	0.11	0.27	0.08	0.93
CSAD	0.23	0.26	0.22	0.04	0.64	0.0004	0.36
CTH	0.45	0.46	0.45	0.02	0.93	0.24	0.86
GCLC	0.06	0.06	0.05	0.004	0.80	0.01	0.35
GPX1	1.39	1.37	1.39	0.10	0.99	0.04	0.77
GSR	0.28	0.30	0.26	0.02	0.11	<.0001	0.11
GSS	0.45	0.49	0.46	0.02	0.34	0.78	0.95
SOD1	3.56	3.64	3.40	0.12	0.27	0.07	0.52
SOD2	4.06	4.08	4.10	0.21	0.99	0.92	0.53
Inflammatory response							
CP	2.48	2.13	2.37	0.22	0.49	0.003	0.85
HP	0.18	0.13	0.14	0.09	0.89	0.01	0.48
NFKB1	2.29	2.14	2.48	0.14	0.25	0.04	0.83

Table 1 Effect of feeding a control lower-energy diet (CON, 1.24 Mcal/kg DM; high-straw) during the whole dry period (~50 d), a higher-energy (1.54 Mcal/kg DM) diet without (OVE) rumen-protected methionine during the last 21 d before calving, or OVE plus rumen-protected methionine (Smartamine M; OVE + SM; Adisseo NA) from -21 d before calving through the first 30 d postpartum on hepatic gene expression (relative mRNA abundance, log$_2$ back-transformed LSM) in Holstein cows *(Continued)*

RXRA	0.57	0.56	0.58	0.03	0.86	0.20	0.66
SAA2	0.01	0.01	0.02	0.002	0.68	0.02	0.95
SOCS2	2.09	1.84	2.26	0.24	0.44	0.16	0.60
STAT3	1.38	1.42	1.39	0.11	0.96	0.05	0.43
STAT5B	2.30	2.26	2.45	0.07	0.15	0.002	0.07

[1]Prepartum dietary treatment: CON = control energy, OVE = moderate energy, OVE + SM = OVE supplemented with rumen-protected methionine (Smartamine M, Adisseo Inc.)
[2]SE = greatest standard error of the mean
[3]D = diet, T = time, D*T = diet by time interaction
[a, b]Significant difference among dietary groups (P ≤ 0.05). Differences reported for genes with a significant (P ≤ 0.05) Diet effect

Results
Blood biomarkers
Metabolism
Time affected all metabolic biomarkers (cholesterol, creatinine, GH, IGF1, leptin, urea; T, $P < 0.001$). However, no effect of diet or its interaction with time was detected (D, D*T, $P > 0.05$) (Fig. 2).

Health status
No effects of diet or its interaction with time were significant for haptoglobin or IL-6 concentration (D, D*T, $P > 0.05$). Diet affected albumin concentration (D*T, $P < 0.05$), with greater ($P < 0.05$) postpartum concentrations in OVE + SM compared with both other groups. The overall concentration of total bilirubin, ceruloplasmin, and serum amyloid A tended to be affected by diet (D, $P < 0.10$), with greater levels ($P < 0.05$) in OVE cows compared with CON (total bilirubin), OVE + SM (ceruloplasmin), or both other groups (SAA). Diet also affected GGT and GOT concentration, as OVE + SM had greater ($P < 0.05$) overall GGT concentration (D, $P < 0.10$), especially postpartum (d 14 and 21), compared with OVE, and lower ($P < 0.05$) GOT concentration postpartum (d 7 and 14) (D*T, $P < 0.05$) compared with CON cows. Time affected the concentration of all previous health biomarkers (T, $P < 0.01$).

Antioxidant and oxidative status
No effect of diet was detected for ORAC, total ROM and tocopherol (D, $P > 0.05$). Total NO$_x$ also were not affected by diet, despite the fact that concentrations of both NO$_2$ and NO$_3$ had significant interactions or diet effects (NO$_2$, D*T, $P < 0.10$; NO$_3$, D, $P < 0.05$, D*T, $P < 0.10$). Diet had a strong effect on GSH concentration (D, $P < 0.001$), with greatest concentration ($P < 0.05$) in OVE + SM cows compared with both other groups. When interacting with time, diet tended to affect blood concentration of β-carotene and retinol (D*T, $P < 0.10$). For the first, the response was due to a greater ($P <$

0.05) concentration in OVE + SM cows compared with CON at -21 and -10 d, and to a lower ($P < 0.05$) concentration in OVE compared with CON at 14 d relative to parturition. In the case of retinol, the interaction was due the increasing ($P < 0.05$) concentration postpartum from 7 to 21 d in OVE + SM cows, while in CON and OVE cows the concentration remained constant ($P > 0.05$). This led to a greater ($P < 0.05$) retinol concentration in OVE + SM at 21 d postpartum compare with OVE. Diet also affected paraoxonase concentration, with overall greater level ($P < 0.05$) in CON compared with OVE and OVE + SM (D, $P < 0.05$). This difference was due to greater ($P < 0.05$) concentration in CON cows at -21, -10 and 7 d relative to parturition (D*T, $P < 0.05$).

Gene expression
Energy metabolism
Cows fed the CON or OVE + SM diets had greater PCK1 expression compared with OVE cows (D, $P < 0.05$). Diet also affected the expression of the fatty acid metabolism related genes MTTP (D, $P < 0.05$) and PPARA (D*T, $P < 0.05$). Expression of MTTP was in fact greater ($P < 0.05$) in CON and OVE cows, compared with OVE + SM, while PPARA expression was greater ($P < 0.05$) prepartum (-10 d) for OVE + SM compared with CON and OVE, and lower ($P < 0.05$) early postpartum (7 d) for OVE compared with the other two groups.

Hepatokines and inflammation
Diet alone did not affect genes related to hepatokiens and the inflammatory response (D, $P > 0.05$). However, the hepatokines ANGPTL4 and FGF21 had a significant interaction with time (D*T, $P < 0.05$). For FGF21 this significance was due to a greater ($P < 0.05$) prepartal expression in CON and OVE + SM compared with OVE cows, while for ANPTL4 no differences among dietary groups were detected across the analyzed time points ($P > 0.05$).

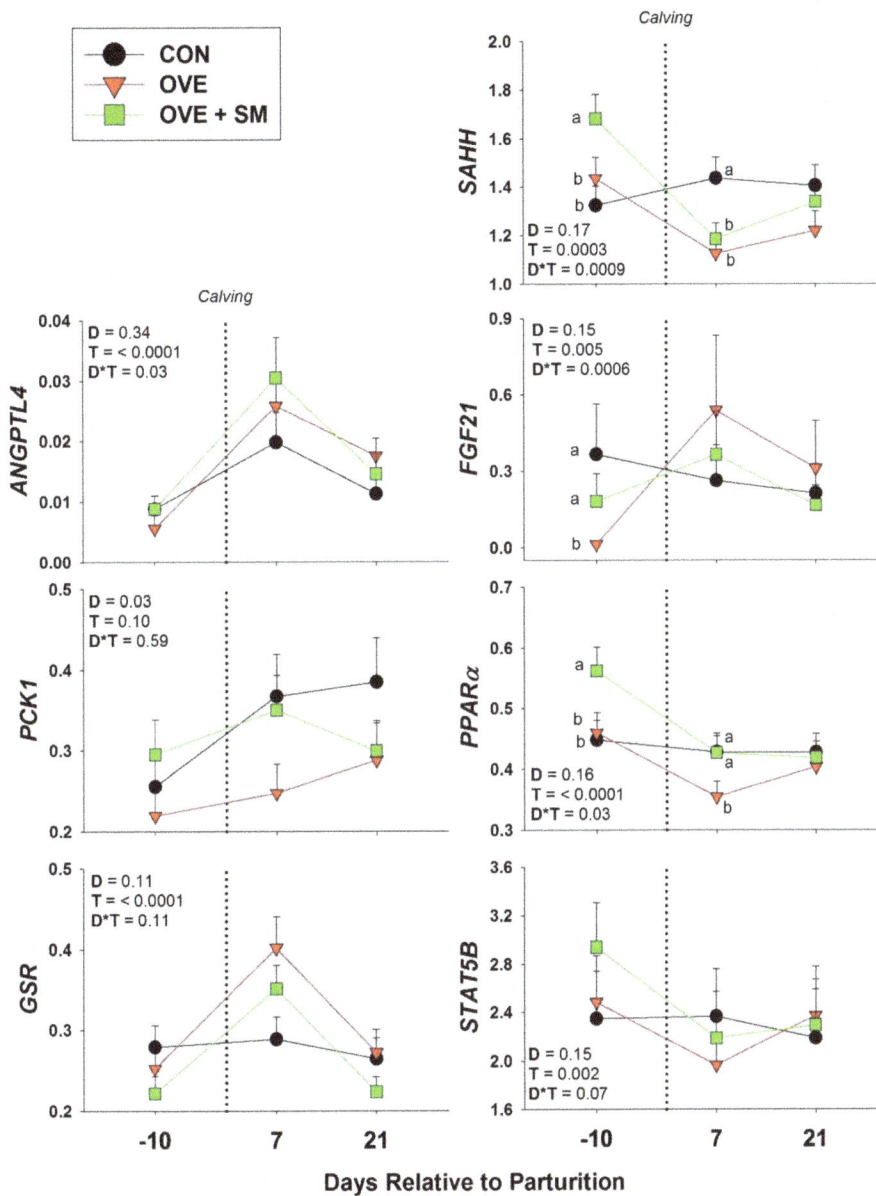

Fig. 1 Effect of feeding a control lower-energy diet (CON, 1.24 Mcal/kg DM; high-straw) during the whole dry period (~50 d), a higher-energy (1.54 Mcal/kg DM) diet without (OVE) rumen-protected methionine during the last 21 d before calving, or OVE plus rumen-protected methionine (Smartamine M; OVE + SM; Adisseo NA) from -21 d before calving through the first 30 d postpartum on hepatic gene expression (log$_2$ back-transformed LSM) in Holstein cows

Methionine cycle and antioxidant system

No genes concerning the antioxidant system were significantly affected by diet, or its interaction with time (D, D*T, $P > 0.05$). However, *MTR* and *DNMT3A*, genes of the methionine cycle, had an overall effect of diet (D, $P < 0.05$). Expression of *MTR* was greater ($P < 0.05$) in CON compared with OVE, with OVE + SM having an intermediate level of expression, while *DNMT3A* expression was greater ($P < 0.05$) in OVE and OVE + SA compared with CON cows. Furthermore, *SAHH* expression was greater (D*T, $P < 0.05$) prepartum in OVE + SM

cows compared with the other dietary groups; whereas, expression was greater ($P < 0.05$) early postpartum (7 d) in CON cows compared with OVE and OVE + SM.

Discussion

Overfeeding dairy cows in the weeks prior parturition (e.g. close up period) has been previously linked with a more pronounced negative energy balance postpartum, due to bigger drops in voluntary dry matter intake (DMI) along with sustained lipid mobilization and possible accumulation of triacylglycerol (TAG) in the liver

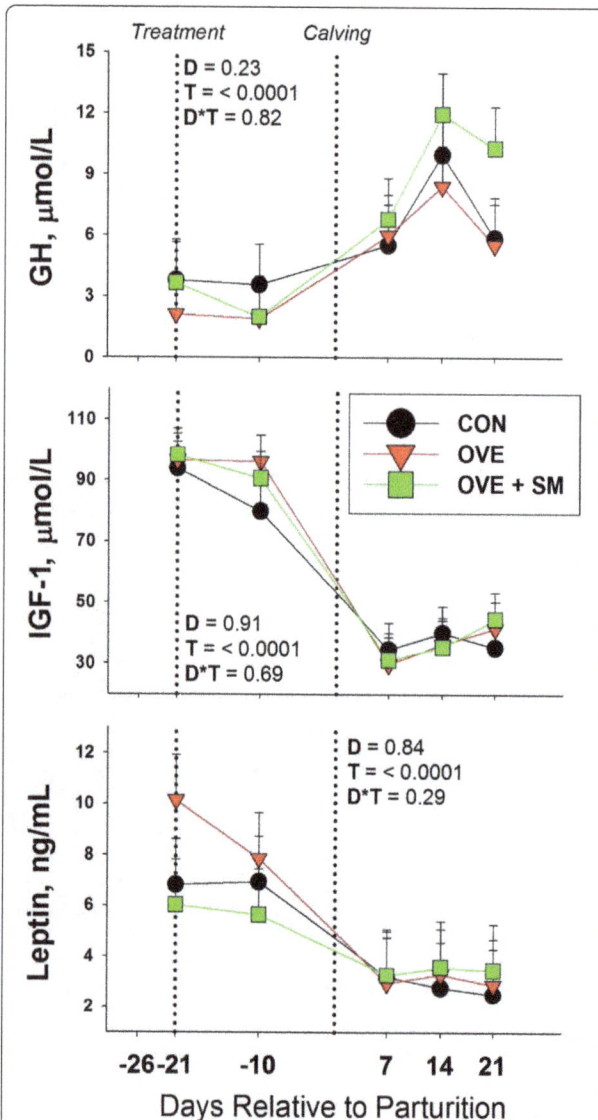

Fig. 2 Effect of feeding a control lower-energy diet (CON, 1.24 Mcal/kg DM; high-straw) during the whole dry period (~50 d), a higher-energy (1.54 Mcal/kg DM) diet without (OVE) rumen-protected methionine during the last 21 d before calving, or OVE plus rumen-protected methionine (Smartamine M; OVE + SM; Adisseo NA) from -21 d before calving through the first 30 d postpartum on endocrine profiles in Holstein cows

SM cows compared with OVE had greater postpartal DMI and better milk production, matching the performance of the control-fed group [25]. Despite the fact that the improved DMI, likely a consequence of the improved health status, could easily explain the improved production performance, other cellular and physiologic also likely were contributing factors.

The hepatic transcriptome revealed how Met supplementation restored *PCK1* expression (an important gluconeogenic gene) to the level of control-fed cows. At least postpartum this could be explained by the higher insulin concentration in OVE + SM [25], as hepatic *PCK1* mRNA expression is directly related to insulin level [32]. The increased insulin concentration also could explain why circulating glucose was lower in OVE + SM cows [25] compared with CON, i.e. overfeeding alone does not affect peripheral insulin resistance [9], and the increased insulin concentration was not followed by changes in GH or IGF1, hence, the improved milk production with OVE + SM also might have resulted from an increase in glucose availability directly channeled to peripheral tissues and the mammary gland. In the latter case it would have contributed to greater lactose production. Peripheral tissues, i.e. adipose and muscle, rely mainly on GLUT4 (an insulin-dependent transporter) for glucose uptake, while the mammary gland uses mainly GLUT1 (usually described as insulin-independent) as the preferred glucose transporter [33]. However, a recent study revealed that insulin increases GLUT1 expression in bovine mammary explants, thus, providing evidence of a functional link between circulating insulin and mammary glucose uptake [34].

Supplementing Met also increased both fat and protein percentage during the first week of lactation [25]. Because biomarkers of muscle catabolism were not affected by diet (e.g. urea and creatinine) and DMI was similar in CON and OVE + SM, we speculate that Met itself, combined with higher circulating insulin, might have been the primary cause of the improved protein percentage. In fact, previous research demonstrated that an increase in amino acid supply (e.g. abomasal casein infusion) could markedly improve milk protein yield, especially when the circulating level of insulin was artificially raised through a clamp [35, 36]. The lower inflammation status and greater liver function around calving in the OVE + SM cows (lower concentrations of albumin and greater bilirubin, ceruloplasmin, GGT, GOT, and SAA) would have guaranteed higher availability of plasma amino acids [37] to the mammary gland for protein synthesis. The increase in fat content, which agrees with several previous studies [16, 38–41], might have been related to cellular pathways involving Met and its methylated compounds (e.g. choline [42]), which some data

[25]. The present study confirmed the overfeeding-induced depression of DMI postpartum and hepatic TAG accumulation [25]. Furthermore, despite previous studies reporting that overfed cows were always able to maintain similar levels of milk production as the control-fed counterparts [31], these changes led to worse milk performance including lower milk and energy corrected milk yield [25].

As hypothesized, supplementation of rumen-protected Met to a moderate energy diet was able to overcome the detrimental effects of energy overfeeding. In fact, OVE +

indicate are important for supporting milk fat synthesis in cows [43].

As previously mentioned, overfeeding energy prepartum led to hepatic TAG accumulation [25], a condition that, if excessive, could become a potential burden for proper liver function [2]. OVE cows, in fact, had signs of impaired liver function and inflammatory condition postpartum including lower concentrations of albumin and greater bilirubin, ceruloplasmin, GGT, GOT, and SAA (Table 2, Fig. 3). As hypothesized, supplemental Met was able to correct these effects of the OVE diet. Thus, as a primary outcome, OVE + SM cows had less liver TAG accumulation [25] despite similar NEFA concentration between OVE and OVE + SM [25]. This was at least in part due to greater *PPARA* expression with Met supplementation.

Among the most important metabolic functions coordinated by PPARα are LCFA uptake, intracellular activation, oxidation, and ketogenesis [44]. Thus its greater expression in OVE + SM cows could have improved NEFA handling, i.e. through greater oxidation. Furthermore, PCK1 is also involved in glyceroneogenesis, as it can catalyze the production of glycerol-3-phospate for use during fatty acid esterification [45]. Thus the

Table 2 Effect of feeding a control lower-energy diet (CON, 1.24 Mcal/kg DM; high-straw) during the whole dry period (~50 d), a higher-energy (1.54 Mcal/kg DM) diet without (OVE) rumen-protected methionine during the last 21 d before calving, or OVE plus rumen-protected methionine (Smartamine M; OVE + SM; Adisseo NA) from -21 d before calving through the first 30 d postpartum on biomarker concentrations of metabolism, liver health, and oxidative status in Holstein cows

Items	Diet[1]			SE[2]	P-values[3]		
	CON	OVE	OVE + SM		D	T	D*T
Metabolism							
Cholesterol, mmol/L	3.24	3.16	3.26	0.11	0.76	<.0001	0.71
Creatinine, μmol/L	97.60	98.88	97.68	1.53	0.77	<.0001	0.17
GH, ng/mL	5.75	4.79	6.95	1.08	0.23	<.0001	0.82
IGF-1, ng/mL	56.65	60.03	59.98	6.64	0.91	<.0001	0.69
Leptin, ng/mL	4.44	5.42	4.40	1.62	0.84	<.0001	0.29
Urea, mmol/L	5.20	5.05	5.05	0.18	0.77	<.0001	0.30
Liver health							
Albumin, g/L	35.41	35.54	36.32	0.41	0.24	0.0002	0.05
Bilirubin , μmol/L	2.29[a]	3.38[b]	2.57[ab]	0.41	0.10	<.0001	0.57
Ceruloplasmin, μmol/L	2.77[ab]	2.91[b]	2.61[a]	0.09	0.006	<.0001	0.51
GGT, U/L	22.96[a]	25.21[ab]	26.85[b]	1.17	0.07	<.0001	0.008
GOT, U/L	84.76	90.30	81.71	5.61	0.48	<.0001	0.04
Haptoglobin, g/L	0.42	0.46	0.41	0.06	0.76	0.003	0.86
IL-6, pg/mL	530.63	586.37	412.76	98.56	0.37	0.001	0.67
SAA, μg/mL	35.55[a]	54.00[b]	34.77[a]	7.79	0.10	0.0005	0.58
Oxidative status							
β-carotene, mg/100 mL	0.20	0.19	0.23	0.02	0.14	0.04	0.06
Liver GSH, mmol/L	953[a]	1281[b]	1693[c]	120	0.0002	0.05	0.14
NO₂, μmol/L	6.03	6.66	6.80	0.45	0.44	0.01	0.09
NO₃, μmol/L	18.65[a]	16.90[b]	16.77[b]	0.40	0.002	<.0001	0.08
NOₓ, μmol/L	24.61	23.54	23.67	0.56	0.31	<.0001	0.18
ORAC, TE mol/L	12,731	12,359	12,739	198	0.25	<.0001	0.66
Paraoxonase, U/mL	77.96[a]	68.41[b]	66.74[b]	2.68	0.01	<.0001	0.02
Retinol, μg/100 mL	46.39	41.79	43.42	3.10	0.44	0.0009	0.08
ROM, mg of H₂O₂/100 mL	14.01	12.99	13.44	0.49	0.31	<.0001	0.79
Tocopherol, μg/mL	3.67	3.68	3.16	0.44	0.46	<.0001	0.31

[1]Prepartum dietary treatment: CON = control energy, OVE = moderate energy, OVE + SM = OVE supplemented with rumen-protected methionine (Smartamine M, Adisseo Inc.)

[2]SE = greatest standard error of the mean

[3]D = diet, T = time, D*T = diet by time interaction

a, b, cSignificant difference among dietary groups (P ≤ 0.05). Differen reported for biomarkers with a tendency (P ≤ 0.10) or a significan (P ≤ 0.05) Diet effect

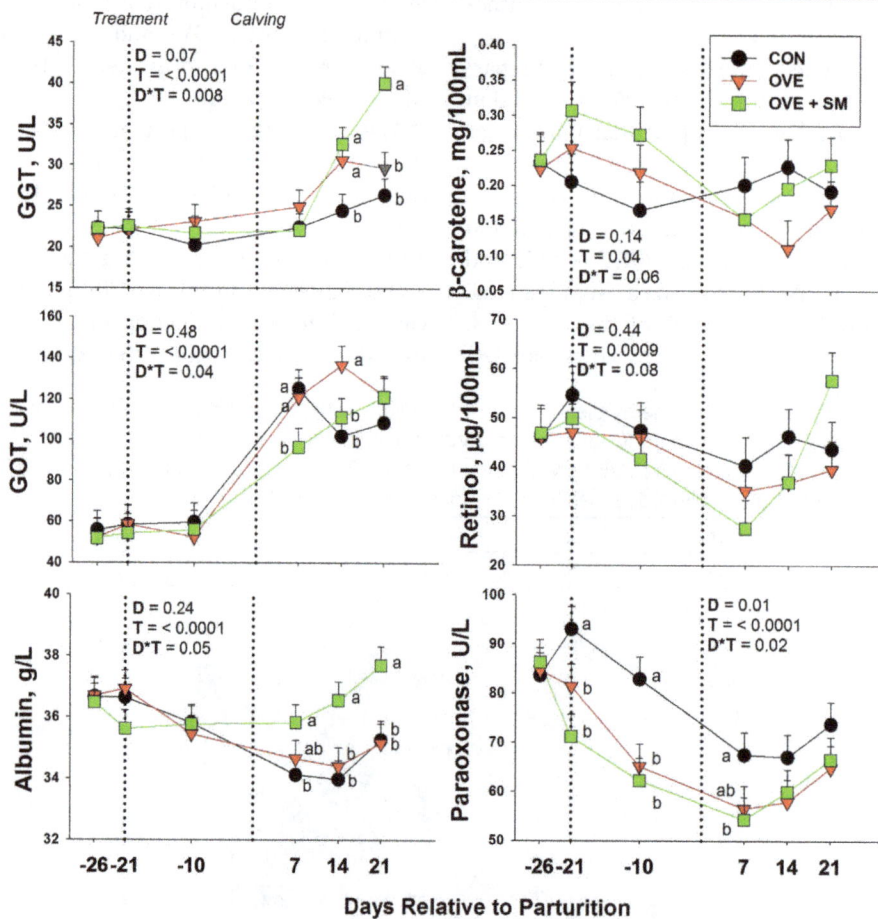

Fig. 3 Effect of feeding a control lower-energy diet (CON, 1.24 Mcal/kg DM; high-straw) during the whole dry period (~50 d), a higher-energy (1.54 Mcal/kg DM) diet without (OVE) rumen-protected methionine during the last 21 d before calving, or OVE plus rumen-protected methionine (Smartamine M; OVE + SM; Adisseo NA) from -21 d before calving through the first 30 d postpartum on blood biomarkers of liver function and antioxidant status in Holstein cows

increase of its expression could have further improved NEFA handling by the liver. The lower expression of *MTTP* in the OVE + SM cows was lower indicated a potentially lower capacity of these cows to synthesize and export VLDL. However, the data from Bernabucci et al. [46] indicated that apolipoprotein mRNA transcription rather than *MTTP* might be the limiting step in the repackaging of TAG into lipoproteins, hence, explaining the increase in concentration of plasma VLDL in OVE + SM cows [25]. As a subsequent outcome, the improved fatty acid metabolism in liver with Met supplementation reduces the risk of liver dysfunction, an idea supported by the biomarkers of liver function (e.g. greater albumin and VLDL, and lower bilirubin) in OVE + SM cows [47].

Metabolic dysfunction and inflammatory events are often linked through oxidative stress, a common outcome to both scenarios [48–50]. The present study partly confirmed the possible molecular mechanisms through which prepartum overfeeding could cause an increased concentration of oxidants proposed by Loor et

al. [51]. OVE did not cause changes in total ROM and NO_x, however, these cows had an impairment of the antioxidant system. Despite similar blood antioxidant capacity, paraoxonase concentration was in fact lower in OVE cows, a condition that not only indicates liver dysfunction, but one that has been proven to lead to an increase in the inflammatory status (confirmed by higher ceruloplasmin and SAA), which notoriously causes an increase in oxidative stress, and a reduction of antioxidative protection during the early postpartum period [27, 52]. As for paraoxonase, postpartum (d 14) concentration of β-carotene, a precursor of vitamin A, which exerts antioxidant effects [53], also was reduced in OVE compared with CON.

Supplementation of rumen-protected Met has been proven to benefit the oxidative status of periparturient cows [19, 20], in large part because it is a precursor for the biosynthesis of glutathione and taurine, two of the most important cellular antioxidants [54, 55]. In the present study, Met supplementation to cows fed a higher energy diet prepartum was able to improve their compromised antioxidant status. In fact, despite the lack of changes in ROM or paraoxonase compared with OVE, OVE + SM cows had greater glutathione concentrations, even compared with CON, together with higher retinol concentrations up to the level of control-fed cows. Concerning retinol, its concentration is also regulated by the hepatic synthesis of its carrier, retinol binding protein [56]. Thus, a greater plasma retinol concentration, besides suggesting a better antioxidant status, could also be a response to the better liver functionality detected in OVE + SM cows. Furthermore, *GSR* expression was decreased in OVE + SM cows to a similar level than OVE. *GSR* encodes the protein glutathione reductase, a central enzyme of cellular antioxidant defense, which reduces oxidized glutathione disulfide to the sulfhydryl form [57]. This further suggests a lesser oxidative status in cows fed methionine, which despite having a greater glutathione concentration seemed to have less of a need to restore the pool of its active form.

Other health benefits of methionine supplementation could also be noticed in the lower somatic cell count in milk. For instance, OVE + SM cows compared with both CON and OVE had lower milk SCC [25], a result that further highlights the immunometabolic effects of methionine and its metabolites [16, 21–23].

At a molecular level, the greater expression *SAHH* prepartum in OVE + SM cows underscores that the increased Met supply to the liver through supplementation was directed through the methionine cycle, leading to the higher glutathione concentrations. However, overfeeding energy prepartum (e.g. OVE and OVE + SM) seemed to reduce the overall expression of *MTR*, as if regenerating Met was not a hepatic priority. This becomes relevant in early lactation, because after calving the decrease in expression of both *MTR* and *SAHH* in all groups indicated that cows might redirect the circulating Met to the mammary gland for milk production. To further complicate this scenario, the greater *DNMT3A* expression in both OVE and OVE + SM cows indicated a role of overfeeding in its regulation. Its greater expression could indicate a higher need of methyl groups from methionine by the liver, hence, in light of the lower hepatic regeneration (e.g. lower *MTR*) but greater utilization (e.g. higher *DNMT3A*) Met supplementation (e.g. OVE + SM) favored the mammary demand. The fact that milk production was restored to the level of CON cows in the OVE + SM cows supports this scenario.

The mechanisms by which prepartal overfeeding causes a greater *DNMT3A* expression, increasing DNA methylation and leading to greater consumption of methyl groups from Met, are not clear. Insulin sensitivity was previously associated with increased global methylation [58], but overfeeding cows prepartum never led to its impairment in our previous experiments [9, 10]. On the other hand, levels of hepatic methylation were associated with fatty liver disease in humans [59, 60]. Because OVE cows had a greater hepatic TAG content [25], *DNMT3A* expression regulation could be explained by the alterations in lipid metabolism.

Conclusions

Current results confirm the detrimental outcome (e.g., reduced DMI, compromised liver function, and higher inflammatory status) of a higher-energy diet during the close up period in dairy cows, thus, supporting the need for energy restriction in the close-up period. However, if the practice persists, dairy producers should improve the diet methionine supply. In fact, supplemental rumen-protected methionine was effective in reducing the aforementioned effects, by (i) stimulating DMI and milk production, (ii) improving hepatic fatty acid metabolism and reducing TAG accumulation, (iii) improving general biomarkers of liver function, and (iv) limiting the post-partal negative effect of inflammation on the cow antioxidant system. Further investigation is needed to assess the effect of methionine supplementation to a prepartal energy restricted diet during the close-up.

Abbreviations

ACOX1: Acyl-CoA oxidase 1; ANGPTL4: Angiopoietin like 4; APOB: Apolipoprotein B; BBOX1: γ-butyrobetaine hydroxylase 1; BHMT: BHMT2, Betaine–homocysteine S-methyltransferase 1, and 2; CBS: Cystathionine-beta-synthase; CP: Ceruloplasmin; CPT1A: Carnitine palmitoyltransferase 1A; CSAD: Cysteine sulfinic acid decarboxylase; CTH: Cystathionine gamma-lyase; DMI: Dry matter intake; DNMT1: DNMT3A, DNA (cytosine-5-)-methyltransferase 1, and 3 α; FGF21: Fibroblast growth factor 21; GCLC: Glutamate-cysteine ligase catalytic subunit; GGT: Gamma-glutamyl-transpeptidase; GH: Growth hormone; GLUT1: GLUT4, Glucose transporter 1, and 4; GOT: Glutamic oxaloacetic transaminase; GPX1: Glutathione peroxidase 1; GSR: Glutathione reductase; GSS: Glutathione synthetase; HMGCS2: 3-hydroxy-3-methylglutaryl-CoA synthase 2; HP: Haptoglobin; IGF1: Insulin like growth factor-1; IL6: Interleukin 6; MAT1A: Methionine adenosyltransferase 1A; Met: Methionine; MTR: 5-methyltetrahydrofolate-homocysteine methyltransferase; MTTP: Microsomal triglyceride transfer protein; NEFA: Non-esterified fatty acids; NEL: Net energy for lactation; NFKB1: Nuclear factor κB subunit 1; NOx: NO2, NO3, Nitric oxides; ORAC: Oxygen radical absorbance capacity; PC: Pyruvate carboxylase; PCK1: Phosphoenolpyruvate carboxykinase 1; PDK4: Pyruvate dehydrogenase kinase 4; PEMT: Phosphatidylethanolamine N-methyltransferase; PPARA: Peroxisome proliferator activated receptor α; ROM: Reactive oxygen metabolites; RXRA: Retinoid X receptor α; SAA: SAA2, Serum amyloid A, and A2; SAHH: S-adenosylhomocysteine hydrolase; SLC22A5: Solute carrier family 22 member 5; SOCS2: Suppressor of cytokine signaling 2; SOD1: SOD2, Superoxide dismutase 1 (soluble), and 2 (mitochondrial); STAT3: STAT5B, Signal transducer and activator of transcription 3, and 5B; TAG: Triacylglycerol; TMLHE: Trimethyllysine hydroxylase, ε; VLDL: Very low density lipoprotein

Acknowledgements

The authors thank Travis Michels and Mike Katterhenry of the University of Illinois Dairy Research Unit (Urbana) staff for help with animal management.

Funding

Financial support for the research was provided in part by Adisseo (Commentry, France) and Hatch funds under project ILLU-538–914, National Institute of Food and Agriculture, Washington, DC, USA.

Authors' contributions

MVR performed the qPCR statistical analysis and wrote the main draft of the manuscript, with inputs from DL, ET and JJL. JLL and DL designed the study. JSO performed the animal study and qPCR. ET and JSO performed the blood biomarker analysis. All authors read and approved the final manuscript.

Competing interests

The authors declare that they have no competing interests.

Author details

[1]Mammalian NutriPhysioGenomics, Department of Animal Sciences and Division of Nutritional Sciences, University of Illinois, Urbana, IL 61801, USA. [2]Istituto di Zootecnica Facoltà di Scienze Agrarie, Alimentari e Ambientali, Università Cattolica del Sacro Cuore, 29122 Piacenza, Italy. [3]Adisseo NA, Alpharetta, GA 30022, USA. [4]Dairy and Food Science Department, South Dakota State University, 1111 College Ave, 113H Alfred DairyScience Hall, Brookings SD 57007, USA.

References

1. Bell AW. Regulation of organic nutrient metabolism during transition from late pregnancy to early lactation. J Anim Sci. 1995;73(9):2804–19.
2. Drackley JK. ADSA Foundation Scholar Award. Biology of dairy cows during the transition period: the final frontier? J Dairy Sci. 1999;82(11):2259–73.
3. van Knegsel ATM, Hammon HM, Bernabucci U, Bertoni G, Bruckmaier RM, Goselink RMA, et al. Metabolic adaptation during early lactation: key to cow health, longevity and a sustainable dairy production chain. CAB Rev. 2014;9(002):1–15.
4. Curtis CR, Erb HN, Sniffen CJ, Smith RD, Kronfeld DS. Path analysis of dry period nutrition, postpartum metabolic and reproductive disorders, and mastitis in Holstein cows. J Dairy Sci. 1985;68(9):2347–60.
5. Boutflour RB. Limiting factors in the feeding and management of milk cows. In: Report of proceedings of 8th world's dairy congress. London; 1928. p. 15-20.
6. Holtenius K, Agenas S, Delavaud C, Chilliard Y. Effects of feeding intensity during the dry period. 2. Metabolic and hormonal responses. J Dairy Sci. 2003;86(3):883–91.
7. Janovick NA, Boisclair YR, Drackley JK. Prepartum dietary energy intake affects metabolism and health during the periparturient period in primiparous and multiparous Holstein cows. J Dairy Sci. 2011;94(3):1385–400.
8. Ji P, Drackley JK, Khan MJ, Loor JJ. Overfeeding energy upregulates peroxisome proliferator-activated receptor (PPAR)gamma-controlled adipogenic and lipolytic gene networks but does not affect proinflammatory markers in visceral and subcutaneous adipose depots of Holstein cows. J Dairy Sci. 2014;97(6):3431–40.
9. Ji P, Osorio JS, Drackley JK, Loor JJ. Overfeeding a moderate energy diet prepartum does not impair bovine subcutaneous adipose tissue insulin signal transduction and induces marked changes in peripartal gene network expression. J Dairy Sci. 2012;95(8):4333–51.
10. Khan MJ, Jacometo CB, Graugnard DE, Correa MN, Schmitt E, Cardoso F, et al. Overfeeding Dairy Cattle During Late-Pregnancy Alters Hepatic PPARalpha-Regulated Pathways Including Hepatokines: Impact on Metabolism and Peripheral Insulin Sensitivity. Gene Regul Syst Bio. 2014;8:97–111.
11. Rukkwamsuk T, Wensing T, Geelen MJ. Effect of overfeeding during the dry period on the rate of esterification in adipose tissue of dairy cows during the periparturient period. J Dairy Sci. 1999;82(6):1164–9.
12. Soliman M, Kimura K, Ahmed M, Yamaji D, Matsushita Y, Okamatsu-Ogura Y, et al. Inverse regulation of leptin mRNA expression by short- and long-chain fatty acids in cultured bovine adipocytes. Domest Anim Endocrinol. 2007;33(4):400–9.
13. Dann HM, Litherland NB, Underwood JP, Bionaz M, D'Angelo A, McFadden JW, et al. Diets during far-off and close-up dry periods affect periparturient metabolism and lactation in multiparous cows. J Dairy Sci. 2006;89(9):3563–77.
14. Graugnard DE, Moyes KM, Trevisi E, Khan MJ, Keisler D, Drackley JK, et al. Liver lipid content and inflammometabolic indices in peripartal dairy cows are altered in response to prepartal energy intake and postpartal intramammary inflammatory challenge. J Dairy Sci. 2013;96(2):918–35.
15. Shahzad K, Bionaz M, Trevisi E, Bertoni G, Rodriguez-Zas SL, Loor JJ. Integrative analyses of hepatic differentially expressed genes and blood biomarkers during the peripartal period between dairy cows overfed or restricted-fed energy prepartum. PLoS One. 2014;9(6):e99757.
16. Osorio JS, Ji P, Drackley JK, Luchini D, Loor JJ. Supplemental Smartamine M or MetaSmart during the transition period benefits postpartal cow performance and blood neutrophil function. J Dairy Sci. 2013;96(10):6248–63.
17. Zhou Z, Vailati-Riboni M, Trevisi E, Drackley JK, Luchini DN, Loor JJ. Better postpartal performance in dairy cows supplemented with rumen-protected methionine compared with choline during the peripartal period. J Dairy Sci. 2016;99(11):8716–32.
18. Zhou Z, Bulgari O, Vailati-Riboni M, Trevisi E, Ballou MA, Cardoso FC, et al. Rumen-protected methionine compared with rumen-protected choline improves immunometabolic status in dairy cows during the peripartal period. J Dairy Sci. 2016;99(11):8956–69.
19. Osorio JS, Ji P, Drackley JK, Luchini D, Loor JJ. Smartamine M and MetaSmart supplementation during the peripartal period alter hepatic expression of gene networks in 1-carbon metabolism, inflammation, oxidative stress, and the growth hormone-insulin-like growth factor 1 axis pathways. J Dairy Sci. 2014;97(12):7451–64.
20. Osorio JS, Trevisi E, Ji P, Drackley JK, Luchini D, Bertoni G, et al. Biomarkers of inflammation, metabolism, and oxidative stress in blood, liver, and milk reveal a better immunometabolic status in peripartal cows supplemented with Smartamine M or MetaSmart. J Dairy Sci. 2014;97(12):7437–50.
21. Grimble RF. The effects of sulfur amino acid intake on immune function in humans. J Nutr. 2006;136(6 Suppl):1660S–5.
22. Grimble RF, Grimble GK. Immunonutrition: role of sulfur amino acids, related amino acids, and polyamines. Nutrition. 1998;14(7-8):605–10.
23. Li P, Yin YL, Li D, Kim SW, Wu G. Amino acids and immune function. Br J Nutr. 2007;98(2):237–52.
24. Redmond HP, Stapleton PP, Neary P, Bouchier-Hayes D. Immunonutrition: the role of taurine. Nutrition. 1998;14(7-8):599–604.
25. Li C, Batistel F, Osorio JS, Drackley JK, Luchini D, Loor JJ. Peripartal rumen-protected methionine supplementation to higher energy diets elicits positive effects on blood neutrophil gene networks, performance and liver lipid content in dairy cows. J Anim Sci Biotechnol. 2016;7:18.
26. Jacometo CB, Osorio JS, Socha M, Correa MN, Piccioli-Cappelli F, Trevisi E, et al. Maternal consumption of organic trace minerals alters calf systemic and neutrophil mRNA and microRNA indicators of inflammation and oxidative stress. J Dairy Sci. 2015;98(11):7717–29.
27. Bionaz M, Trevisi E, Calamari L, Librandi F, Ferrari A, Bertoni G. Plasma paraoxonase, health, inflammatory conditions, and liver function in transition dairy cows. J Dairy Sci. 2007;90(4):1740–50.
28. Trevisi E, Amadori M, Cogrossi S, Razzuoli E, Bertoni G. Metabolic stress and inflammatory response in high-yielding, periparturient dairy cows. Res Vet Sci. 2012;93(2):695–704.
29. Lucini L, Kane D, Pellizzoni M, Ferrari A, Trevisi E, Ruzickova G, et al. Phenolic profile and in vitro antioxidant power of different milk thistle [Silybum marianum (L.) Gaertn.] cultivars. Ind Crop Prod. 2016;83:11–6.
30. Vailati Riboni M, Kanwal M, Bulgari O, Meier S, Priest NV, Burke CR, et al. Body condition score and plane of nutrition prepartum affect adipose tissue transcriptome regulators of metabolism and inflammation in grazing dairy cows during the transition period. J Dairy Sci. 2016;99(1):758–70.
31. Graugnard DE, Bionaz M, Trevisi E, Moyes KM, Salak-Johnson JL, Wallace RL, et al. Blood immunometabolic indices and polymorphonuclear neutrophil function in peripartum dairy cows are altered by level of dietary energy prepartum. J Dairy Sci. 2012;95(4):1749–58.
32. Aschenbach JR, Kristensen NB, Donkin SS, Hammon HM, Penner GB. Gluconeogenesis in dairy cows: the secret of making sweet milk from sour dough. IUBMB Life. 2010;62(12):869–77.
33. Zhao FQ, Keating AF. Expression and regulation of glucose transporters in the bovine mammary gland. J Dairy Sci. 2007;90 Suppl 1:E76–86.
34. Shao Y, Wall EH, McFadden TB, Misra Y, Qian X, Blauwiekel R, et al. Lactogenic hormones stimulate expression of lipogenic genes but not

glucose transporters in bovine mammary gland. Domest Anim Endocrinol. 2013;44(2):57–69.

35. Mackle TR, Dwyer DA, Ingvartsen KL, Chouinard PY, Lynch JM, Barbano DM, et al. Effects of insulin and amino acids on milk protein concentration and yield from dairy cows. J Dairy Sci. 1999;82(7):1512–24.

36. Griinari JM, McGuire MA, Dwyer DA, Bauman DE, Barbano DM, House WA. The role of insulin in the regulation of milk protein synthesis in dairy cows. J Dairy Sci. 1997;80(10):2361–71.

37. Zhou Z, Loor JJ, Piccioli-Cappelli F, Librandi F, Lobley GE, Trevisi E. Circulating amino acids during the peripartal period in cows with different liver functionality index. J Dairy Sci. 2016;99(3):2257-67.

38. Wang C, Liu HY, Wang YM, Yang ZQ, Liu JX, Wu YM, et al. Effects of dietary supplementation of methionine and lysine on milk production and nitrogen utilization in dairy cows. J Dairy Sci. 2010;93(8):3661–70.

39. Broderick GA, Muck RE. Effect of alfalfa silage storage structure and rumen-protected methionine on production in lactating dairy cows. J Dairy Sci. 2009;92(3):1281–9.

40. Broderick GA, Stevenson MJ, Patton RA. Effect of dietary protein concentration and degradability on response to rumen-protected methionine in lactating dairy cows. J Dairy Sci. 2009;92(6):2719–28.

41. Lundquist RG, Otterby DE, Linn JG. Influence of three concentrations of DL-methionine or methionine hydroxy analog on milk yield and milk composition. J Dairy Sci. 1985;68(12):3350–4.

42. Pinotti L, Baldi A, Dell'Orto V. Comparative mammalian choline metabolism with emphasis on the high-yielding dairy cow. Nutr Res Rev. 2002;15(2):315–32.

43. Guretzky NA, Carlson DB, Garrett JE, Drackley JK. Lipid metabolite profiles and milk production for Holstein and Jersey cows fed rumen-protected choline during the periparturient period. J Dairy Sci. 2006;89(1):188–200.

44. Mandard S, Muller M, Kersten S. Peroxisome proliferator-activated receptor alpha target genes. Cell Mol Life Sci. 2004;61(4):393–416.

45. Beale EG, Harvey BJ, Forest C. PCK1 and PCK2 as candidate diabetes and obesity genes. Cell Biochem Biophys. 2007;48(2-3):89–95.

46. Bernabucci U, Ronchi B, Basirico L, Pirazzi D, Rueca F, Lacetera N, et al. Abundance of mRNA of apolipoprotein b100, apolipoprotein e, and microsomal triglyceride transfer protein in liver from periparturient dairy cows. J Dairy Sci. 2004;87(9):2881–8.

47. Bertoni G, Trevisi E, Calamari L, Bionaz M. The inflammation could have a role in the liver lipidosis occurence in dairy cows. In: Joshi N, Herdt TH (Eds.) Production Diseases in Farm Animals. 12th International Conference. Wageningen: Wageningen Academic Publ; 2006. p. 157–58.

48. Sordillo LM, Mavangira V. The nexus between nutrient metabolism, oxidative stress and inflammation in transition cows. Anim Prod Sci. 2014;54(9):1204–14.

49. Sordillo LM, Raphael W. Significance of metabolic stress, lipid mobilization, and inflammation on transition cow disorders. Vet Clin North Am Food Anim Pract. 2013;29(2):267–78.

50. Bradford BJ, Yuan K, Farney JK, Mamedova LK, Carpenter AJ. Invited review: Inflammation during the transition to lactation: New adventures with an old flame. J Dairy Sci. 2015;98(10):6631-50.

51. Loor JJ, Dann HM, Guretzky NA, Everts RE, Oliveira R, Green CA, et al. Plane of nutrition prepartum alters hepatic gene expression and function in dairy cows as assessed by longitudinal transcript and metabolic profiling. Physiol Genomics. 2006;27(1):29–41.

52. Turk R, Juretic D, Geres D, Turk N, Rekic B, Simeon-Rudolf V, et al. Serum paraoxonase activity in dairy cows during pregnancy. Res Vet Sci. 2005;79(1):15–8.

53. Kartha VN, Krishnamurthy S. Antioxidant function of vitamin A. Int J Vitam Nutr Res. 1977;47(4):394–401.

54. Pompella A, Visvikis A, Paolicchi A, De Tata V, Casini AF. The changing faces of glutathione, a cellular protagonist. Biochem Pharmacol. 2003; 66(8):1499–503.

55. Shimada K, Jong CJ, Takahashi K, Schaffer SW. Role of ROS Production and Turnover in the Antioxidant Activity of Taurine. Adv Exp Med Biol. 2015;803:581–96.

56. Goodman DS. Plasma retinol-binding protein. Ann N Y Acad Sci. 1980; 348:378–90.

57. Meister A. On the cycles of glutathione metabolism and transport. Curr Top Cell Regul. 1981;18:21–58.

58. Zhao JY, Goldberg J, Bremner JD, Vaccarino V. Global DNA Methylation Is Associated With Insulin Resistance A Monozygotic Twin Study. Diabetes. 2012;61(2):542–6.

59. Gallego-Duran R, Romero-Gomez M. Epigenetic mechanisms in non-alcoholic fatty liver disease: An emerging field. World J Hepatol. 2015;7(24): 2497–502.

60. Pirola CJ, Gianotti TF, Burgueno AL, Rey-Funes M, Loidl CF, Mallardi P, et al. Epigenetic modification of liver mitochondrial DNA is associated with histological severity of nonalcoholic fatty liver disease. Gut. 2013;62(9):1356–63.

3

Expression of fatty acid sensing G-protein coupled receptors in peripartal Holstein cows

Alea Agrawal, Abdulrahman Alharthi, Mario Vailati-Riboni, Zheng Zhou and Juan J. Loor[*]

Abstract

Background: G-protein coupled receptors (GPCR), also referred as Free Fatty Acid Receptors (FFAR), are widely studied within human medicine as drug targets for metabolic disorders. To combat metabolic disorders prevalent in dairy cows during the transition period, which co-occur with negative energy balance and changes to lipid and glucose metabolism, it may be helpful to identify locations and roles of FFAR and other members of the GPCR family in bovine tissues.

Results: Quantitative RT-PCR (qPCR) of subcutaneous adipose, liver, and PMNL samples during the transition period (-10, +7, and +20 or +30 d) were used for expression profiling of medium- (MCFA) and long-chain fatty acid (LCFA) receptors *GPR120* and *GPR40*, MCFA receptor *GPR84*, and niacin receptor *HCAR2/3*. Adipose samples were obtained from cows with either high (HI; BCS ≥ 3.75) or low (LO; BCS ≤ 3.25) body condition score (BCS) to examine whether FFAR expression is correlated with this indicator of health and body reserves. Supplementation of rumen-protected methionine (MET), which may improve immune function and production postpartum, was also compared with unsupplemented control (CON) cows for liver and blood polymorphonuclear leukocytes (PMNL) samples. In adipose tissue, *GPR84* and *GPR120* were differentially expressed over time, while *GPR40* was not expressed; in PMNL, *GPR40* was differentially expressed over time and between MET vs. CON, *GPR84* expression differed only between dietary groups, and *GPR120* was not expressed; in liver, GPCR were either not expressed or barely detectable.

Conclusions: The data indicate that there is likely not a direct role in liver for the selected GPCR during the transition period, but they do play variable roles in adipose and PMN. In future, these receptors may prove useful targets and/or markers for peripartal metabolism and immunity.

Keywords: Inflammation, Methionine, Neutrophils, Transition cow

Background

The G-protein coupled receptor (GPCR) superfamily is one of the largest families of receptor proteins, comprising 1% or more of the human genome, and as much as 5% in simpler organisms like the nematode [1, 2]. GPCR are also termed as seven transmembrane receptors, from their identifying structure of seven α-helices spanning the membrane. They can receive a variety of ligand classes from the extracellular environment, and stimulate intracellular signaling cascades that may begin with action of associated G-proteins [3]. Although GPCR in general are extensively-researched drug targets due to their abundance and activities, those which have metabolic roles may be targets for treating or preventing metabolic and inflammatory diseases in dairy cows, as well. This could especially be of value in the transition period, where prevalence rates and outcomes of diseases are often at their worst [4].

GPCR with fatty acids (FA), and especially longer-chain FA, ligands are among some of the most interesting targets, due to the ability of saturated versus unsaturated fats to evoke different signals within cells [5, 6], and for signaling potency to vary based on chain length and degree of saturation [7–9]. There is also potential to link nutrition, FA metabolism, and immunity among such signaling pathways [10]. For instance, *GPR40* and *GPR120* may be

* Correspondence: jloor@illinois.edu
Mammalian NutriPhysioGenomics, Department of Animal Sciences and Division of Nutritional Sciences, University of Illinois, 1207 West Gregory Drive, Urbana, IL 61801, USA

implicated in anti-inflammation in macrophages, but are currently of greater interest for their metabolic effects. These receptors have been connected with obesity, insulin responses, and inflammation subsequent to these conditions [10]. *GPR40* may have a unique contribution to immune function, as it has been shown to stimulate calcium mobilization in bovine neutrophils, a necessary signal for neutrophil activation and function [11]. *GPR84* has been identified in cells of both the innate and adaptive immune system, including PMNL, and plays a role in pro-inflammatory responses, e.g., cytokine production [10]. *GPR109A* also links metabolism and immunity; it has been detected in adipose and macrophages and primarily exerts anti-lipolytic effects [12]. For a summary of these genes and their functions, see Table 1 [10, 11].

The present study also involved an examination of GPCR in adipose tissue, and more specifically cows with different degrees of adiposity as evaluated through the body condition score (BCS) at 21 d prior to parturition. BCS represents energy storage status [13], which is vitally important to the peripartal dairy cow, as early lactation proceeds at the expense of stored energy [14]. What is considered as an optimal score can vary, and may instead be expressed as a range, but is currently thought to be around 3.25 near calving [15]. Generally, however, relatively higher prepartum or calving BCS is correlated with greater BCS [16] and/or body weight [17] loss postcalving. Thus, although all cows experience negative energy balance (NEB) after calving [4], higher BCS cows are at higher risk for deeper NEB and excess lipid mobilization – adverse conditions for an optimal transition. Lower BCS may be considered relatively more optimal for health, although very thin cows have been found to produce less milk [18]. Thus, comparison of FA-sensing receptor

expression in both an optimal/thin and a suboptimal/fat group of cows (here, divided as LO = BCS ≤ 3.25, and HI = BCS ≥ 3.75, respectively) could be of value.

The effect of supplementing rumen-protected methionine (MET) was also considered for liver and blood polymorphonuclear leukocytes (PMNL) analyses of GPCR. Previous work by our group demonstrated that MET can improve liver function, immune and antioxidant status, milk yield and protein levels, and may also have beneficial effects on dry matter intake (DMI) around calving [19–21]. Since the GPCR studied here are closely linked to lipid and glucose metabolism and inflammation, it seemed worthwhile to investigate whether gene expression in liver or PMNL is changed with supplementation, and how that relates to observed cow-level effects.

Because the FFAR are not yet well-studied in ruminants [22], the aim of the present study was to assess patterns and levels of receptor expression in periparturient dairy cows. Furthermore, because BCS and supplemental MET can affect production, immune status, and metabolism, we hypothesized that there would be some differential expression between groups.

Methods
Animals and treatments
Cows in the present study were a subset from the experiment of Zhou et al. [21]. Cows were blocked according to lactation and calving measures, and fed the same close-up (1.52 Mcal/kg DM; -21 d until calving) and lactation (1.71 Mcal/kg DM; calving until +30 d) diets as a total mixed ration (TMR) once daily (0630 h). Cows were housed in an enclosed, ventilated barn during the dry period and fed using an individual gate system (American Calan Inc., Northwood, NH), and moved to a tie-stall barn with

Table 1 Genes of interest and their functions

Gene Name	Gene Function	References
GPR40	M/LCFA receptor expressed in pancreatic α- and β-cells, enteroendocrine cells, immune cells, taste buds, and the central nervous system. Stimulation invokes intracellular calcium response and ERK signaling, or increases in cAMP. Mediates insulin release from β-cells, glucagon release from α-cells, and incretins in the gastrointestinal tract in response to free fatty acid (FFA) ligands, playing a central role in glucose homeostasis. Promotes activation and superoxide production in bovine neutrophils. May regulate secretion of brain-derived neurotrophic factor in neuroblastoma cells.	[8, 11, 38, 48, 66]
GPR120	M/LCFA receptor expressed in adipocytes and adipose tissue, macrophages, enteroendocrine cells, pancreatic α-cells, taste buds, and lungs. Stimulation invokes intracellular calcium response and ERK signaling. Stimulation with ω-3 FA leads to β-arrestin2-mediated inhibition of TAK1 (i.e.: anti-inflammatory signaling). Promotes glucose uptake via GLUT4 and G-protein-related insulin stimulatory effects in adipocytes. Promotes incretin (e.g., GLP-1) release in gastrointestinal tract, and glucagon release in α-cells. Reduces inflammatory gene expression in macrophages and adipose, as well as macrophage invasion into adipose tissue.	[38, 44, 49, 67, 68]
GPR84	MCFA receptor expressed in adipocytes and immune cells, including leukocytes. Expression is upregulated in macrophages by LPS. Promotes pro-inflammatory cytokine and chemokine signals.	[38, 69, 70]
HCAR2/3	Nicotinic acid and butyrate/β-hydroxybutyrate receptor expressed in adipose tissue, immune cells, spleen, colon, pancreatic β-cells, and mammary epithelium. Expression is upregulated in adipose and macrophages by LPS and other pro-inflammatory factors. Activation by niacin or BHBA decreases cAMP levels in adipose tissue, thereby reducing lipolysis, plasma FFA, and availability for triglyceride synthesis. Decreases in cAMP also occur in β-cells, inhibiting insulin release. Promotes release of prostaglandins from immune cells, including macrophages. Invokes apoptotic pathways in neutrophils and some cancer cells (e.g., breast and colon cancer).	[12, 46, 71–73]

individual feed bunks after calving. Throughout these periods, diets were top-dressed with either no supplement (CON), or Smartamine M (Adisseo NA) rumen-protected methionine (MET). Complete details of supplementation may be found elsewhere [21]. Lactating cows were milked three times daily (0600, 1400, and 2200 h).

Ten and eleven cows from each group (CON and MET) were used in the present study to compare PMNL and liver gene expression, respectively. Additionally, twenty cows that had received adipose biopsies were retroactively grouped by their BCS (see below) at -3 wk from calving (i.e.: upon entry into the close-up dry period), such that 10 cows with BCS ≥ 3.75 (HI; avg. = 3.83 ± 0.12) were compared to 10 cows with BCS ≤ 3.25 (LO; avg. = 3.11 ± 0.16) for adipose gene expression. The effect of supplementation was not considered in this tissue.

Sample collection

The BCS were assigned weekly during the experiment, along with body weight measurements. BCS was determined on a scale of 1–5 with quarter-point increments, where 1 = thin and 5 = obese. Two scores were given independently each week, so that the average was taken and used for statistical analyses and retroactive grouping. Individual and average BCS loss of cows in each group were also calculated between Prepartum (-21 d) and Postpartum (+21 d) time points. At -10, +7, and +20 d from calving, energy balance (Table 2) was calculated as described previously [23] in the subset of cows which received adipose biopsies.

Blood was only collected from those cows receiving liver biopsies or scheduled for PMNL retrieval. Serum and plasma were collected for analyses into vacutainers (BD Vacutainer; BD and Co., Franklin Lakes, NJ) containing clot activator or lithium heparin, respectively, at -10, +8, and +30 d relative to parturition. Blood for serum samples was kept at 21 °C, and blood for plasma kept on ice, until

Table 2 Least squares means of energy balance (EB) and energy balance as a percentage of requirements (EB % Req) in transition cows with high (HI; BCS ≥ 3.75) or low (LO; BCS ≤ 3.25) body condition score prepartum and postpartum. This data only represents the subset of cows used for adipose gene expression

	BCS[1]		SEM[2]	P-value		
	HI	LO		BCS	Time	BCS × Time
Prepartum						
EB	1.89	2.16	1.08	0.86	-	-
EB % Req	112.35	115.62	7.13	0.75	-	-
Postpartum						
EB	−11.87[a]	−5.52[b]	2.00	0.03	0.03	0.14
EB % Req	70.85[a]	86.20[b]	4.74	0.03	<0.01	0.30

[a,b]Statistical difference (P < 0.05) among time points within the same group
[1]Body condition score
[2]Largest standard error of the mean

centrifugation. Methods for analyses of IL-1β, reactive oxygen metabolites (ROM), and myeloperoxidase (MPO) as indicators of systemic inflammation and oxidative stress was reported previously [20].

Full protocols for relevant tissue biopsies have been previously described [24, 25]. Briefly, cows were given local anesthesia prior to biopsy. The same cows were not used for liver biopsies as for adipose biopsies. Liver was sampled via puncture biopsy at -10, +7, and +30 d from parturition via puncture biopsy, while adipose biopsies from alternate sides of the tail-head region were taken at -10, +7, and +20 d using a blunt dissection method. Both liver and adipose samples were snap-frozen in liquid nitrogen and transferred to a freezer at -80 °C until RNA extraction and further analyses.

PMN Isolation

Blood samples for collection of PMNL were drawn into vacutainers containing acid citrate dextrose (ACD Solution A; Fisher Scientific) from the coccygeal vein at -10, +7, and +30 d relative to parturition. Samples were placed on ice until PMNL isolation by sample centrifugation, cell lysing, and several rounds of centrifugation with PBS washing. Both purity and viability of PMNL were greater than 90%. Complete details of this process can be found in Zhou et al. [26]. Briefly, a 50 μL aliquot of PMNL was incubated for 15 min on ice with 100 μL of primary anti-bovine granulocyte monoclonal antibody (Cat. No. BOV2067, Washington State University, Pullman, USA) solution (15 μg/mL in 1 × PBS). The aliquot was then washed twice with 2 mL 1 × PBS and incubated for another 15 min on ice, protected from light, with 50 μL of secondary phycoerythrin-labeled secondary antibody (Cat. No. 1020–09S, Southern Biotech, Birmingham, AL) (4 μg/mL in 1 × PBS), and 50 μL of Propidium iodide (50 μg/mL) prior to flow cytometry. Isolated PMNL were homogenized at full speed in a solution of 2 mL TRIzol reagent (Invitrogen, Carlsbad, CA) with 1 μL linear acrylamide (Ambion, Inc., Austin, TX). Homogenate was stored at -80 °C in RNA-free microcentrifuge tubes (Fisher Scientific, Pittsburgh, PA).

RNA Isolation

To proceed with RNA extraction, 40 mg liver and 200 mg adipose were thawed and homogenized in QIAzol reagent (Qiagen, Hilden, Germany). Extraction of RNA was performed with the miRNeasy kit (Qiagen) following the manufacturer's protocols. Samples were treated on-column with DNaseI (Qiagen). Prior to storage, RNA purity was confirmed using a NanoDrop ND-1000 (NanoDrop Technologies, Rockland, DE) OD_{260nm}/OD_{280nm} ratio, and RNA quality was recorded using a 2100 Bioanalyzer (Agilent Technologies, Inc., Santa Clara, CA) RNA integrity number (RIN). All liver samples had RIN scores above 8.0. Average RIN scores for the other two tissues were as follows: 6.44 ± 0.21 for adipose and 6.68 ± 0.02 for PMNL.

Real-time Quantitative PCR

Previous publications by our group [27] outlined the full protocols. Briefly, 100 ng of RNA, plus reagents including: 1 µg of dT18 (Operon Biotechnologies, Huntsville, AL), 1 µL of 10 mmol/L dNTP mix (Invitrogen Corp., Carlsbad, CA), 1 µL of random primers (3 mg/µL; Invitrogen Corp.), and 10 µL of DNase-/RNase-free water, were incubated at 65 °C for 5 min, then placed on ice for 3 min. Six µL of master mix, including: 5.5 µL of 5× reaction buffer, 0.25 µL (50 U) of RevertAid reverse transcriptase (Fermentas Inc., Glen Burnie, MD), and 0.25 µL of RNase inhibitor (10 U, Promega, Madison, WI), was then added to complete cDNA synthesis.

Primer design protocols have also been published previously [28]. Except for bovine *GPR40* [29], primer sequences were obtained using Primer Express 3.0. Primer information and obtained products are listed in Additional file 1: Table S1 and S2, respectively. Quantitative PCR was performed using 4 µL of diluted cDNA plus a mixture of 5 µL of 1× SYBR Green master mix (Applied Biosystems, CA), 0.4 µL each of the 10 µmol/L forward and reverse primers, and 0.2 µL of DNase-/RNase-free water in a MicroAmp Optical 384-Well Reaction Plate (Applied Biosystems, Foster City, CA). Each sample was run in duplicate, while a negative control and serially diluted, pooled cDNA were run in triplicate to create a 6-point relative standard curve (User Bulletin #2, Applied Biosystems). PCR reactions were performed in an ABI Prism 7900 HT SDS instrument (Applied Biosystems) under the following conditions: 2 min at 50 °C, 10 min at 95 °C, 40 cycles of 15 s at 95 °C (denaturation), and 1 min at 60 °C (annealing plus extension). Gene expression was normalized using the geometrical mean of three appropriate internal control genes: *GAPDH* and *RPS9* for all tissues, along with *ACTB* for adipose and *UXT* for liver and PMNL [25, 27, 30]. Genes were considered not expressed when the standard curve had slope $-3.50 > y > -3.00$ and $Ct > 30$. At least for the Ct threshold, the criterion is consistent with our established protocols aimed in part at reducing the unreliability of data that often occurs at >30 Ct [25, 27, 28, 30]. In that context, it is important to highlight that in recent studies dealing with the study of FFAR the mean Ct for quantification of GPR40 (which was undetectable using our thresholds) ranged from 31.4 for adipose and liver [31] to 35.7 for adipose [32]. Although it could be possible that the size of the amplicon for the GPR40 primer [29] was too long and led to poor amplification efficiency, Yonezawa et al. [29] verified the identity of bovine GPR40 by sequencing (as did we; see Additional Table S2)

and used it successfully with bovine mammary tissue. Unfortunately, there was no information on amplicon size for the GPR40 primer used in the work of Friedrichs et al. [31, 32]. The qPCR performance is reported for adipose, liver, and PMNL in Additional file 1: Table S3.

Statistical analysis

Prior to analysis, expression data were $\log_2$ normalized. All data sets (i.e.: blood parameters, energy balance, and qPCR) were then subject to ANOVA using repeated measures ANOVA with PROC MIXED in SAS (v 9.2; SAS Institute Inc., Cary, NC). The statistical model for adipose included time (-10, +7, +20 d), BCS (HI, LO), and their interaction as fixed effects. Energy balance was analyzed pre- and postpartum. The model for liver and PMNL gene expression, and blood parameters, included time (-10, +7, +30 d), methionine supplementation (MET, CON), and their interaction as fixed effects. The random effect was cow, nested within treatment. The Kenward-Roger statement was used for computing the denominator degrees of freedom, with sp(pow) as the covariance structure. Energy balance used Compound Symmetry as the covariance structure postpartum. Previous 305 d lactation and parity were used as covariates for blood analysis, and parity was used as a covariate for energy balance. When not significant, covariates were removed from the model. Data were considered significant at $P \leq 0.05$ using the PDIFF statement in SAS. Expression data in Tables 3 and 5 are reported as the $\log_2$ back-transformed least squares means.

Results
Body condition score experiment
Energy balance
Although there was no difference in energy balance by BCS prepartum ($P = 0.86$) (Table 2), cows in the HI BCS group were in more NEB ($P = 0.03$) during the postpartum period.

Adipose

GPR40 was not expressed in adipose tissue. *GPR84* expression was significantly lower ($P < 0.05$) prepartum than postpartum (Table 3). There was also an interaction effect ($P = 0.03$) of BCS × time, such that expression was higher in LO than in HI cows at +7 d. There was also a significant time effect ($P < 0.01$) on expression of *GPR120*, where expression was highest prepartum, and decreased at both of the postcalving time points. Expression of both *GPR120* and *GPR109A* tended ($P = 0.08$ and 0.06) to be greater in LO than in HI. There was also a tendency ($P = 0.08$) due to the time effect on *GPR109A*, mainly for the difference between +7 and +20 d.

Table 3 Log$_2$-backtransformed least squares means of adipose gene expression data in transition cows with high (HI; BCS ≥ 3.75) or low (LO; BCS ≤ 3.25) body condition score at -10, +7, and +20 d from calving

Gene	BCS[1]		Time			BCS × Time HI			LO			SEM[2]	P-value		
	HI	LO	−10	+7	+20	−10	+7	+20	−10	+7	+20		BCS	Time	BCS × Time
GPR40	Not expressed														
GPR84	0.37	0.41	0.19^a	0.69^b	0.46^b	0.21^a	0.39^{a,b}*	0.61^b	0.17^a	1.21^b*	0.34^a	0.40	0.62	<0.01	0.03
GPR120	0.46	0.72	1.69^a	0.49^b	0.22^c	1.36	0.37	0.19	2.09	0.67	0.26	0.64	0.08	<0.01	0.89
GPR109A	1.05	1.30	1.12	1.40	1.02	1.13	1.16	0.89	1.10	1.68	1.17	0.22	0.06	0.08	0.33

[a,b,c]Statistical difference ($P < 0.05$) among time points within the same group
*Statistical difference ($P < 0.05$) between groups within time points
[1]Body condition score
[2]Largest standard error of the mean

Rumen-protected methionine experiment
Inflammation and oxidative stress biomarkers
Overall, cows in MET had lower ($P < 0.05$) concentrations of IL-1β and ROM but had greater ($P < 0.05$) concentration of MPO (Table 4).

Liver
Neither GPR40 nor GPR120 were expressed in liver (Table 5). For GPR84 and GPR109A, there were no significant differences ($P > 0.15$) or tendencies for effects of MET, time, or the interaction.

PMNL
GPR120 was not expressed in PMNL (Table 5). Among the other three genes, there were significant differences ($P < 0.05$) between MET and CON for GPR40 and GPR84, such that GPR40 was lower in MET, and GPR84 was higher in MET. GPR40 expression also differed across time ($P = 0.04$), such that cows had significantly greater expression at -10 d than at +30 d, decreasing (albeit not significantly) in between. GPR109A tended ($P = 0.09$) to be higher in MET PMNL than in CON.

Table 4 Least squares means of immune biomarker concentrations in blood in transition cows supplemented with rumen-protected methionine (MET) or unsupplemented (CON) at -10, +7, and +30 d. This data only represents the subset of cows used for polymorphonuclear leukocyte (PMNL) gene expression

Parameter	Diet		SEM[1]	P-value		
	MET	CON		Diet	Time	Diet × Time
IL-1β[2]	3.63^a	5.74^b	0.69	0.05	0.17	0.70
ROM	12.29^a	14.20^b	0.23	<0.01	<0.01	0.05
MPO	466.87^a	405.18^b	15.89	0.02	0.14	0.44

[a,b]Statistical difference (P ≤ 0.05) among time points within the same group
[1]Largest standard error of the mean
[2]Significance ($P < 0.05$) for parity in the model

Discussion
Body condition score
Adipose tissue
Postpartum, transition cows can experience stress-induced, pathogen-independent inflammation, as marked by pro-inflammatory cytokines and acute-phase proteins (APP) in the blood [33]. Furthermore, expression of chemoattractants and cytokines within adipose tissue indicates some degree of local inflammation, which has been postulated as a homeorhetic mechanism to aid lactation [34, 35]. This theory is supported by the knowledge that cytokines present during inflammation encourage lipolysis [34], the "metabolic hallmark" of transition [36]. Here, both the function and the temporal expression of GPR84 match physiologic adaptations in the cow, indicating that effects of inflammation in adipose are partly mediated through this receptor.

An interaction of BCS × time at +7 d, where LO cows had much greater expression of GPR84 than HI cows also provides evidence of the pro-inflammatory role for GPR84 in cattle. Biomarkers like haptoglobin, bilirubin, and paraoxonase have been used to indicate peripartal inflammation through the first 2-3 wk of lactation [37]. Here, although blood biomarker data were unavailable, the surge in GPR84 expression may also indicate greater inflammation in LO cows, at least within the adipose tissue depot [38]. This may not necessarily be mirrored at a systemic level since cows with higher BCS are normally associated with greater overall inflammation and lower health status [13]. Despite this, thin cows can still be health-compromised during the transition period [13, 39]. Barring clinical or subclinical disease, inflammatory signals resolve toward the end of the transition period [37], which parallels the return of GPR84 expression to prepartum levels.

If the above holds true, GPR84-mediated inflammation could also help explain production in these cows. Inflammation in early lactation has been connected with poorer performance (i.e. milk yield; [33]), and in fact,

Table 5 Log$_2$-backtransformed least squares means of liver and polymorphonuclear leukocyte (PMNL) gene expression data in transition cows supplemented with rumen-protected methionine (MET) or unsupplemented (CON) at -10, +7, and +30 d from calving

Gene	Diet		Time			Diet × Time						SEM[1]	P-value		
	MET	CON	-10	+7	+30	MET			CON				Diet	Time	Diet × Time
						-10	+7	+30	-10	+7	+30				
Liver															
GPR40	Not expressed														
GPR84	0.77	0.84	0.88	0.74	0.80	0.85	0.73	0.75	0.91	0.75	0.86	0.14	0.64	0.25	0.89
GPR120	Not expressed														
GPR109A	0.51	0.53	0.64	0.48	0.46	0.57	0.52	0.44	0.72	0.44	0.47	0.15	0.77	0.22	0.61
PMNL															
GPR40	0.85*	1.13*	1.15[a]	0.95[a,b]	0.85[b]	1.00	0.80	0.77	1.33	1.14	0.94	0.17	0.04	0.04	0.78
GPR84	1.49*	0.65*	0.96	1.05	0.96	1.22	1.82	1.49	0.75	0.60	0.61	0.64	0.04	0.93	0.51
GPR120	Not expressed														
GPR109A	1.34	0.87	1.26	0.91	1.10	1.42	1.31	1.28	1.11	0.63	0.94	0.44	0.09	0.60	0.71

[a,b]Statistical difference (P < 0.05) among time points within the same group
*Statistical difference (P < 0.05) between groups within time points
[1]Largest standard error of the mean

through the first month, milk yield from LO cows was numerically, albeit not statistically, lower than HI cows (39 kg/d vs. 42 kg/d; data not shown). In a recent study [17], Pires et al. obtained similar results: numerically, low BCS cows had lower production than medium or high BCS cows. Further work including more cows, as well as immune biomarkers concurrent with gene expression, will be necessary to elucidate the potential involvement of GPR84 and validity of such relationships.

Contrary to GPR84, GPR120 is primarily an anti-inflammatory receptor [8]. Thus, lower expression over time potentially reinforces evidence of postpartal inflammation, although without a lessening effect by 3 week postpartum. Its metabolic role may explain this pattern: GPR120 stimulates adipogenesis and differentiation, rather than lipolysis [40]. As necessity for lipolysis increases through early lactation NEB [14], expression of GPR120 should decrease, as observed here. Receptor activation in adipose also improves insulin sensitivity [41]. Because insulin resistance in peripheral tissues is important for pushing available glucose to the mammary gland post-calving [42], lower expression of GPR120 in adipose should be expected, and may be a necessary part of the transcriptome adaptation to milk synthesis.

To a lesser extent, limited expression could also be a regulatory mechanism. As lipolysis occurs, releasing primarily long-chain FA (LCFA) (i.e. GPR120 ligands) to the blood [43], signaling may provide negative feedback to prevent hyperactivation. In some cases, GPR120 can activate β-arrestin2, which promotes receptor internalization and prevents continuous ligand-sensing [44]; perhaps gene expression is used as an additional level of control, or acts as a primary regulatory mechanism when alternate pathways are activated.

Both of the above ideas may explain why HI cows tended to have lower overall GPR120 expression. As previously mentioned, milk yield of HI cows was ~3 kg/d greater, numerically. Greater milk production translates to greater glucose requirements [36], and gene expression should reflect a heightened need for peripheral insulin resistance. To produce milk during NEB, HI cows may also mobilize more of their body fat reserves which we also detected in this study, leading to higher FA mobilization and circulating FA [16, 39]. The marked NEB postpartum in HI cows supports this idea. As GPR120 interacts with free LCFA, a negative response on gene expression may prevent further signaling at a time when that could be counterproductive. Certainly, going forward, this receptor could be an interesting metabolic target in the transition period.

The finding that GPR109A tends to have greater expression in LO than in HI cows is not surprising. Studies of GPR109A indicated that it is a primary anti-lipolytic receptor in adipose tissue [45]. The lower BCS in LO cows was indicative of potentially lower amount of stored fat, and greater expression of this receptor could be a mechanism to maintain as much of their already-reduced body condition as possible. Smaller BCS losses between the beginning of the close-up period (-21 d) and both postpartum time points (+21 d) in LO cows compared with HI cows may provide evidence in support of this hypothesis (BCS loss of 0.80 versus 0.40). To address the species-specific effects of GPR109A on lipolysis, that is, whether in ruminants it has temporary and/or rebound effects as in humans, or longer-term FA-lowering effects as in rodents [46], plasma FA from multiple time points should be considered. Although no blood data were available for the subset of cows with adipose biopsies, amount of circulating FA can be correlated with energy balance (EB) [36]. Thus, with better EB (i.e.: more shallow NEB) than HI cows at both time points postpartum, it can be assumed that LO cows had lower circulating FA, meaning that *GPR109A* may be effective over some period of time in cattle. Blood analyses will be useful to confirm this idea.

For this reason, a tendency for greater *GPR109A* expression +7 d after calving, regardless of BCS, was unexpected. The nadir in energy balance, when low DMI coincides with high lactation requirements, typically occurs within three weeks following calving. As previously mentioned, this tends to correspond with lipomobilization; in fact, peak levels of basal and norepinephrine-stimulated lipolysis occur around +10 d [14]. Thus, it appears that high rates of lipolysis can occur in spite of the anti-lipolytic influence of *GPR109A*.

This could be due to a greater net signal for fat mobilization over storage brought on by calving and initiation of lactation. Because *GPR109A* signaling involves a decrease in cAMP levels that inhibits hormone-sensitive lipase and prevents release of FA [45], it is possible that parturition- and lactation-induced flux in hormone levels themselves (e.g., increased catecholamines and decreased insulin) are more influential on hormone-sensitive lipase than GPCR signaling [47]. More information could become available with trials on *GPR109A* in periparturient animals.

Rumen-protected methionine supplementation
Liver
It is not altogether surprising that the genes of interest were not expressed, or were expressed at low levels in the liver. Although these GCPR can be widely expressed throughout the body [8], few studies identified the liver as a major site of expression for these particular genes. In fact, *GPR40* [46, 47] and *GPR109A* [43, 48] have been reported as not detected in liver, and *GPR120* has been reported as not detected, except in Kupffer cells (i.e.: macrophages), in the liver [49, 50]. In the present study, the fact that *GPR40* and *GPR120* were not detected agrees with the literature.

The barely detectable expression of *GPR109A* and *GPR84*, and the lack of differences between groups or across time, indicates that these genes may typically be expressed at very low levels, if at all, in bovine liver. Greater numbers of cows, and/or protein detection methods, should be used in future studies to confirm the presence or absence of these receptors within the liver.

Nonetheless, the genes of interest do play an important role in function and diseases in the liver. It is therefore plausible that effects of these GPCR on the liver are mostly indirect, as a result of signaling that originates in other peripheral tissues or immune cells. Notably, the acute, insulin-promoting effects of GPR120 [41], and

GPR109A signaling to reduce hormone-sensitive lipase [45] in adipose tissue, could protect against excess lipid mobilization preceding fatty liver. Conversely, activation of GPR40 in pancreatic β-cells could signal hyperinsulinemia and increased risk of lipid accumulation in the liver [51]. Therefore, for the present genes of interest, systemic metabolic networks – rather than localized pathways – may provide better insight as to hepatic responses in bovines. Alternatively, other families of receptors, e.g., peroxisome proliferator-activated receptor (PPAR), could contribute more to direct outcomes of relevant ligands (i.e.: LCFA) in the liver [52].

PMNL
Neutrophils are the only tissue in which *GPR40* was detected. Since GPR40 has already been implicated in calcium-dependent degranulation and superoxide production in bovine neutrophils [53], its presence was certainly expected. Its downregulation over time suggests lower PMNL activity postpartum. Indeed, it is well-known that hormones and NEB contribute to immunosuppression post-calving [54–56]. However, it is curious that MET cows had lower expression than CON despite having greater phagocytosis and oxidative burst capacity when challenged in vitro with a bacterial pathogen [20] and higher plasma levels of myeloperoxidase (Table 4) [20]. Instead, it seems that MET could indirectly affect *GPR40* expression through substrate (i.e. fatty acid) availability, because inflammation can promote lipolysis [57], but MET supplementation appears to benefit the inflammatory status (i.e. lower IL-1 β and ROM) [34]. However, MET supplementation did not affect circulating FA levels [21]. Thus, the role of MET supplementation in regulating *GPR40* expression needs further investigation.

In contrast, *GPR84* upregulation in MET agrees with previous data describing enhanced immune response (i.e.: phagocytosis and respiratory burst) in these cows [20]. *GPR84* expression can be stimulated by lipopolysaccharide (LPS), and subsequent signaling produces pro-inflammatory signals, demonstrating that receptor function is tied to immune responses [10]. In agreement, in vivo data revealed that supplementation with MCFA (i.e.: GPR84 ligands) in the transition period lessened neutrophil apoptosis [58], thereby improving antimicrobial capacity of the cells [59]. S Piepers and S De Vliegher [58] hypothesized that GPCR signaling could be at least partly responsible for the improvement in cell viability, and the present study provides evidence that GPR84 could specifically play a role.

Interestingly, *GPR109A* also tended to be upregulated in MET PMNL vs. CON. Like *GPR84*, *GPR109A* expression can be induced by LPS [12], yet with opposite outcome: *GPR109A* activates apoptosis in neutrophils [60]. As noted above, apoptosis should correspond with lower PMNL function, but this did not occur. On the other hand, because apoptosis promotes resolution of inflammation [61], *GPR109A* expression concurs with the largely anti-inflammatory environmental conditions. Possibly, with the influence of MET to lower inflammation (as demonstrated by Zhou et al. [21]), neutrophils maintain their function over a short lifespan [62]. To aid in rapid clearance of aged neutrophils and maintenance of a stable environment, MET cells would induce apoptosis versus necrosis [61]. In this way, *GPR109A* may help protect transition cows from chronic inflammation. More in-depth studies would be needed, but if true, this could provide a new context in which to study niacin as a transition feed supplement [63–65].

Conclusions

Conditions which are present in ruminants and cause metabolic disorders and clinical disease near calving: negative energy balance, lipid mobilization, insulin resistance, and immunosuppression, closely resemble dysregulated metabolic systems in human diseases. Thus, molecular targets in human medicine may translate as targets for the transition cow. The present GPCR show promise for such work. Although none were expressed well in the liver, their contributions to inflammation, insulin resistance, and lipolysis in adipose indicate that they may indirectly affect liver accumulation of fat. The GPCR-modulated environment may also contribute to level of milk production and severity of systemic inflammation in early lactation. Additionally, the differences between BCS groups highlight the role of the transcriptome in coordinating lipid metabolism and energy status, and differences between MET and CON reinforce previous findings and demonstrate potential networks for immune-enhancing action of supplemental methionine. Thus, the present GPCR and related receptors (e.g., *FFAR2* and *FFAR3*) are suggested as continued areas of research in bovine to improve transition health.

Abbreviations
APP: Acute phase protein; BCS: Body condition score; CON: Control-fed; DMI: Dry matter intake; EB: Energy balance; FA: Fatty acid; FFAR: Free fatty acid receptor; GPCR/GPR: G-protein coupled receptor; HCAR: Hydroxycarboxylic acid receptor; HI: High BCS; IL-1B: Interleukin 1β; LCFA: Long-chain fatty acids; LO: Low BCS; LPS: Lipopolysaccharide; MCFA: Medium-chain fatty acids; MET: Methionine-supplemented; NEB: Negative energy balance; PMNL: Polymorphonuclear leukocytes; PPAR: Peroxisome proliferator-activated receptor; qPCR: quantitative polymerase chain reaction; RIN: RNA integrity number; ROM: Reactive oxygen species; TMR: Total mixed ration.

Acknowledgements
Alea Agrawal is a recipient of a Jonathan Baldwin Turner MS fellowship from the University of Illinois (Urbana-Champaign). Z. Zhou is recipient of a fellowship from China Scholarship Council (CSC) to perform his PhD studies at the University of Illinois (Urbana-Champaign). The funders had no role in the design of the study and collection, analysis, and interpretation of data and in writing the manuscript.

Funding
Not applicable.

Authors' contributions
A Agrawal, A Alharthi, MVR, and ZZ performed analyses and analyzed data. JJL conceived the animal experiments. A Agrawal wrote the manuscript. All authors approved the final version of the manuscript.

Authors' information
A. Agrawal, MS, University of Illinois, Urbana, Illinois, 61801, USA. A. Alharthi, MS, University of Illinois, Urbana, Illinois, 61801, USA. M. Vailati-Riboni is PhD candidate at the University of Illinois (Urbana-Champaign). Z. Zhou is PhD degree candidate at University of Illinois, Urbana, Illinois, 61801, USA. J. J. Loor is Associate Professor in the Department of Animal Sciences, University of Illinois, Urbana, Illinois, 61801, USA.

Competing interests
The authors declare that they have no competing interests.

References
1. Perez DM. The evolutionarily triumphant G-protein-coupled receptor. Mol Pharmacol. 2003;63(6):1202–5.
2. Hanlon CD, Andrew DJ. Outside-in signaling–a brief review of GPCR signaling with a focus on the Drosophila GPCR family. J Cell Sci. 2015;128(19):3533–42.
3. Fredriksson R, Lagerstrom MC, Lundin LG, Schioth HB. The G-protein-coupled receptors in the human genome form five main families. Phylogenetic analysis, paralogon groups, and fingerprints. Mol Pharmacol. 2003;63(6):1256–72.
4. Drackley JK. ADSA Foundation Scholar Award. Biology of dairy cows during the transition period: the final frontier? J Dairy Sci. 1999;82(11):2259–73.
5. Kwon B, Lee HK, Querfurth HW. Oleate prevents palmitate-induced mitochondrial dysfunction, insulin resistance and inflammatory signaling in neuronal cells. Biochim Biophys Acta. 2014;1843(7):1402–13.
6. Lee JY, Plakidas A, Lee WH, Heikkinen A, Chanmugam P, Bray G, et al. Differential modulation of Toll-like receptors by fatty acids: preferential inhibition by n-3 polyunsaturated fatty acids. J Lipid Res. 2003;44(3):479–86.
7. Briscoe CP, Tadayyon M, Andrews JL, Benson WG, Chambers JK, Eilert MM, et al. The orphan G protein-coupled receptor GPR40 is activated by medium and long chain fatty acids. J Biol Chem. 2003;278(13):11303–11.
8. Miyamoto J, Hasegawa S, Kasubuchi M, Ichimura A, Nakajima A, Kimura I. Nutritional Signaling via Free Fatty Acid Receptors. Int J Mol Sci. 2016;17(4):450.
9. Stein DT, Stevenson BE, Chester MW, Basit M, Daniels MB, Turley SD, et al. The insulinotropic potency of fatty acids is influenced profoundly by their chain length and degree of saturation. J Clin Invest. 1997;100(2):398–403.
10. Alvarez-Curto E, Milligan G. Metabolism meets immunity: The role of free fatty acid receptors in the immune system. Biochem Pharmacol. 2016;114:3–13.
11. Hidalgo MA, Nahuelpan C, Manosalva C, Jara E, Carretta MD, Conejeros I, et al. Oleic acid induces intracellular calcium mobilization, MAPK phosphorylation, superoxide production and granule release in bovine neutrophils. Biochem Biophys Res Commun. 2011;409(2):280–6.
12. Feingold KR, Moser A, Shigenaga JK, Grunfeld C. Inflammation stimulates niacin receptor (GPR109A/HCA2) expression in adipose tissue and macrophages. J Lipid Res. 2014;55(12):2501–8.
13. Roche JR, Friggens NC, Kay JK, Fisher MW, Stafford KJ, Berry DP. Invited review: Body condition score and its association with dairy cow productivity, health, and welfare. J Dairy Sci. 2009;92(12):5769–801.
14. Bauman DE, Currie WB. Partitioning of nutrients during pregnancy and lactation: a review of mechanisms involving homeostasis and homeorhesis. J Dairy Sci. 1980;63(9):1514–29.
15. Bewley JM, Schutz MM. Review: An interdisciplinary review of body condition scoring for dairy cattle. Prof Anim Sci. 2008;24:507–29.
16. Roche JR, Macdonald KA, Burke CR, Lee JM, Berry DP. Associations among body condition score, body weight, and reproductive performance in seasonal-calving dairy cattle. J Dairy Sci. 2007;90(1):376–91.
17. Pires JA, Delavaud C, Faulconnier Y, Pomies D, Chilliard Y. Effects of body condition score at calving on indicators of fat and protein mobilization of

periparturient Holstein-Friesian cows. J Dairy Sci. 2013;96(10):6423–39.

18. Rukkwamsuk T, Kruip TA, Wensing T. Relationship between overfeeding and overconditioning in the dry period and the problems of high producing dairy cows during the postparturient period. Vet Q. 1999;21(3):71–7.

19. Li C, Batistel F, Osorio JS, Drackley JK, Luchini D, Loor JJ. Peripartal rumen-protected methionine supplementation to higher energy diets elicits positive effects on blood neutrophil gene networks, performance and liver lipid content in dairy cows. J Anim Sci Biotechnol. 2016;7:18.

20. Zhou Z, Bulgari O, Vailati-Riboni M, Trevisi E, Ballou MA, Cardoso FC, et al. Rumen-protected methionine compared with rumen-protected choline improves immunometabolic status in dairy cows during the peripartal period. J Dairy Sci. 2016;99(11):8956–69.

21. Zhou Z, Vailati-Riboni M, Trevisi E, Drackley JK, Luchini DN, Loor JJ. Better postpartal performance in dairy cows supplemented with rumen-protected methionine than choline during the peripartal period. J Dairy Sci. 2016;99(11):8716–32.

22. Bionaz M, Osorio J, Loor JJ. TRIENNIAL LACTATION SYMPOSIUM: Nutrigenomics in dairy cows: Nutrients, transcription factors, and techniques. J Anim Sci. 2015;93(12):5531–53.

23. Osorio JS, Ji P, Drackley JK, Luchini D, Loor JJ. Supplemental Smartamine M or MetaSmart during the transition period benefits postpartal cow performance and blood neutrophil function. J Dairy Sci. 2013;96(10):6248–63.

24. Dann HM, Litherland NB, Underwood JP, Bionaz M, D'Angelo A, McFadden JW, et al. Diets during far-off and close-up dry periods affect periparturient metabolism and lactation in multiparous cows. J Dairy Sci. 2006;89(9):3563–77.

25. Ji P, Osorio JS, Drackley JK, Loor JJ. Overfeeding a moderate energy diet prepartum does not impair bovine subcutaneous adipose tissue insulin signal transduction and induces marked changes in peripartal gene network expression. J Dairy Sci. 2012;95(8):4333–51.

26. Zhou Z, Bu DP, Vailati Riboni M, Khan MJ, Graugnard DE, Luo J, et al. Prepartal dietary energy level affects peripartal bovine blood neutrophil metabolic, antioxidant, and inflammatory gene expression. J Dairy Sci. 2015;98(8):5492–505.

27. Moyes KM, Graugnard DE, Khan MJ, Mukesh M, Loor JJ. Postpartal immunometabolic gene network expression and function in blood neutrophils are altered in response to prepartal energy intake and postpartal intramammary inflammatory challenge. J Dairy Sci. 2014;97(4):2165–77.

28. Bionaz M, Loor JJ. Gene networks driving bovine milk fat synthesis during the lactation cycle. BMC Genomics. 2008;9:366.

29. Yonezawa T, Haga S, Kobayashi Y, Katoh K, Obara Y. Unsaturated fatty acids promote proliferation via ERK1/2 and Akt pathway in bovine mammary epithelial cells. Biochem Biophys Res Commun. 2008;367(4):729–35.

30. Jacometo CB, Zhou Z, Luchini D, Trevisi E, Correa MN, Loor JJ. Maternal rumen-protected methionine supplementation and its effect on blood and liver biomarkers of energy metabolism, inflammation, and oxidative stress in neonatal Holstein calves. J Dairy Sci. 2016;99(8):6753–63.

31. Friedrichs P, Saremi B, Winand S, Rehage J, Danicke S, Sauerwein H, et al. Energy and metabolic sensing G protein–coupled receptors during lactation-induced changes in energy balance. Domest Anim Endocrinol. 2014;48:33–41.

32. Friedrichs P, Sauerwein H, Huber K, Locher L, Rehage J, Meyer U, et al. Expression of metabolic sensing receptors in adipose tissues of periparturient dairy cows with differing extent of negative energy balance. Animal. 2016;10(4):623–32.

33. Bertoni G, Trevisi E, Han X, Bionaz M. Effects of inflammatory conditions on liver activity in puerperium period and consequences for performance in dairy cows. J Dairy Sci. 2008;91(9):3300–10.

34. Farney JK, Mamedova LK, Coetzee JF, KuKanich B, Sordillo LM, Stoakes SK, et al. Anti-inflammatory salicylate treatment alters the metabolic adaptations to lactation in dairy cattle. Am J Physiol Regul Integr Comp Physiol. 2013;305(2):R110–7.

35. Vailati-Riboni M, Kanwal M, Bulgari O, Meier S, Priest NV, Burke CR, et al. Body condition score and plane of nutrition prepartum affect adipose tissue transcriptome regulators of metabolism and inflammation in grazing dairy cows during the transition period. J Dairy Sci. 2016;99(1):758–70.

36. Bell AW. Regulation of organic nutrient metabolism during transition from late pregnancy to early lactation. J Anim Sci. 1995;73(9):2804–19.

37. Bionaz M, Trevisi E, Calamari L, Librandi F, Ferrari A, Bertoni G. Plasma paraoxonase, health, inflammatory conditions, and liver function in transition dairy cows. J Dairy Sci. 2007;90(4):1740–50.

38. Hara T, Kashihara D, Ichimura A, Kimura I, Tsujimoto G, Hirasawa A. Role of free fatty acid receptors in the regulation of energy metabolism. Biochim Biophys Acta. 2014;1841(9):1292–300.

39. Akbar H, Grala TM, Vailati Riboni M, Cardoso FC, Verkerk G, McGowan J, et al. Body condition score at calving affects systemic and hepatic transcriptome indicators of inflammation and nutrient metabolism in grazing dairy cows. J Dairy Sci. 2015;98(2):1019–32.

40. Gotoh C, Hong YH, Iga T, Hishikawa D, Suzuki Y, Song SH, et al. The regulation of adipogenesis through GPR120. Biochem Biophys Res Commun. 2007;354(2):591–7.

41. Ichimura A, Hirasawa A, Poulain-Godefroy O, Bonnefond A, Hara T, Yengo L, et al. Dysfunction of lipid sensor GPR120 leads to obesity in both mouse and human. Nature. 2012;483(7389):350–4.

42. Bell AW, Bauman DE. Adaptations of glucose metabolism during pregnancy and lactation. J Mammary Gland Biol Neoplasia. 1997;2(3):265–78.

43. Contreras GA, Sordillo LM. Lipid mobilization and inflammatory responses during the transition period of dairy cows. Comp Immunol Microbiol Infect Dis. 2011;34(3):281–9.

44. Oh DY, Talukdar S, Bae EJ, Imamura T, Morinaga H, Fan W, et al. GPR120 is an omega-3 fatty acid receptor mediating potent anti-inflammatory and insulin-sensitizing effects. Cell. 2010;142(5):687–98.

45. Tunaru S, Kero J, Schaub A, Wufka C, Blaukat A, Pfeffer K, et al. PUMA-G and HM74 are receptors for nicotinic acid and mediate its anti-lipolytic effect. Nat Med. 2003;9(3):352–5.

46. Kamanna VS, Kashyap ML. Nicotinic acid (niacin) receptor agonists: will they be useful therapeutic agents? Am J Cardiol. 2007;100(11 A):S53–61.

47. Holm C. Molecular mechanisms regulating hormone-sensitive lipase and lipolysis. Biochem Soc Trans. 2003;31(Pt 6):1120–4.

48. Edfalk S, Steneberg P, Edlund H. Gpr40 is expressed in enteroendocrine cells and mediates free fatty acid stimulation of incretin secretion. Diabetes. 2008;57(9):2280–7.

49. Hirasawa A, Tsumaya K, Awaji T, Katsuma S, Adachi T, Yamada M, et al. Free fatty acids regulate gut incretin glucagon-like peptide-1 secretion through GPR120. Nat Med. 2005;11(1):90–4.

50. Raptis DA, Limani P, Jang JH, Ungethum U, Tschuor C, Graf R, et al. GPR120 on Kupffer cells mediates hepatoprotective effects of omega3-fatty acids. J Hepatol. 2014;60(3):625–32.

51. Steneberg P, Rubins N, Bartoov-Shifman R, Walker MD, Edlund H. The FFA receptor GPR40 links hyperinsulinemia, hepatic steatosis, and impaired glucose homeostasis in mouse. Cell Metab. 2005;1(4):245–58.

52. Nakamura MT, Yudell BE, Loor JJ. Regulation of energy metabolism by long-chain fatty acids. Prog Lipid Res. 2014;53:124–44.

53. Manosalva C, Mena J, Velasquez Z, Colenso CK, Brauchi S, Burgos RA, et al. Cloning, identification and functional characterization of bovine free fatty acid receptor-1 (FFAR1/GPR40) in neutrophils. PLoS One. 2015;10(3):e0119715.

54. Goff JP, Horst RL. Physiological changes at parturition and their relationship to metabolic disorders. J Dairy Sci. 1997;80(7):1260–8.

55. Hoeben D, Monfardini E, Opsomer G, Burvenich C, Dosogne H, De Kruif A, et al. Chemiluminescence of bovine polymorphonuclear leucocytes during the periparturient period and relation with metabolic markers and bovine pregnancy-associated glycoprotein. J Dairy Res. 2000;67(2):249–59.

56. Kehrli Jr ME, Nonnecke BJ, Roth JA. Alterations in bovine neutrophil function during the periparturient period. Am J Vet Res. 1989;50(2):207–14.

57. Grimble RF. Nutrition and cytokine action. Nutr Res Rev. 1990;3(1):193–210.

58. Piepers S, De Vliegher S. Oral supplementation of medium-chain fatty acids during the dry period supports the neutrophil viability of peripartum dairy cows. J Dairy Res. 2013;80(3):309–18.

59. Van Oostveldt K, Paape MJ, Dosogne H, Burvenich C. Effect of apoptosis on phagocytosis, respiratory burst and CD18 adhesion receptor expression of bovine neutrophils. Domest Anim Endocrinol. 2002;22(1):37–50.

60. Kostylina G, Simon D, Fey MF, Yousefi S, Simon HU. Neutrophil apoptosis mediated by nicotinic acid receptors (GPR109A). Cell Death Differ. 2008;15(1):134–42.

61. Savill J, Haslett C. Granulocyte clearance by apoptosis in the resolution of inflammation. Semin Cell Biol. 1995;6(6):385–93.

62. Tak T, Tesselaar K, Pillay J, Borghans JA, Koenderman L. What's your age again? Determination of human neutrophil half-lives revisited. J Leukoc Biol. 2013;94(4):595–601.

63. Minor DJ, Trower SL, Strang BD, Shaver RD, Grummer RR. Effects of nonfiber carbohydrate and niacin on periparturient metabolic status and lactation of dairy cows. J Dairy Sci. 1998;81(1):189–200.

64. Morey SD, Mamedova LK, Anderson DE, Armendariz CK, Titgemeyer EC, Bradford BJ. Effects of encapsulated niacin on metabolism and production of periparturient dairy cows. J Dairy Sci. 2011;94(10):5090–104.

65. Yuan K, Shaver RD, Bertics SJ, Espineira M, Grummer RR. Effect of rumen-

protected niacin on lipid metabolism, oxidative stress, and performance of transition dairy cows. J Dairy Sci. 2012;95(5):2673–9.

66. Liou AP, Lu X, Sei Y, Zhao X, Pechhold S, Carrero RJ, et al. The G-protein-coupled receptor GPR40 directly mediates long-chain fatty acid-induced secretion of cholecystokinin. Gastroenterology. 2011;140(3):903–12.

67. Li X, Yu Y, Funk CD. Cyclooxygenase-2 induction in macrophages is modulated by docosahexaenoic acid via interactions with free fatty acid receptor 4 (FFA4). FASEB J. 2013;27(12):4987–97.

68. Suckow AT, Polidori D, Yan W, Chon S, Ma JY, Leonard J, et al. Alteration of the glucagon axis in GPR120 (FFAR4) knockout mice: a role for GPR120 in glucagon secretion. J Biol Chem. 2014;289(22):15751–63.

69. Suzuki M, Takaishi S, Nagasaki M, Onozawa Y, Iino I, Maeda H, et al. Medium-chain fatty acid-sensing receptor, GPR84, is a proinflammatory receptor. J Biol Chem. 2013;288(15):10684–91.

70. Wang J, Wu X, Simonavicius N, Tian H, Ling L. Medium-chain fatty acids as ligands for orphan G protein-coupled receptor GPR84. J Biol Chem. 2006;281(45):34457–64.

71. Chen L, So WY, Li SY, Cheng Q, Boucher BJ, Leung PS. Niacin-induced hyperglycemia is partially mediated via niacin receptor GPR109a in pancreatic islets. Mol Cell Endocrinol. 2015;404:56–66.

72. Elangovan S, Pathania R, Ramachandran S, Ananth S, Padia RN, Lan L, et al. The niacin/butyrate receptor GPR109A suppresses mammary tumorigenesis by inhibiting cell survival. Cancer Res. 2014;74(4):1166–78.

73. Li HM, Zhang M, Xu ST, Li DZ, Zhu LY, Peng SW, et al. Nicotinic acid inhibits glucose-stimulated insulin secretion via the G protein-coupled receptor PUMA-G in murine islet beta cells. Pancreas. 2011;40(4):615–21.

4

Preparation, characterization, antimicrobial and cytotoxicity studies of copper/zinc-loaded montmorillonite

Lefei Jiao[1], Fanghui Lin[1], Shuting Cao[1], Chunchun Wang[1], Huan Wu[1], Miaoan Shu[1*] and Caihong Hu[1,2*]

Abstract

Background: A series of modified montmorillonites (Mt) including zinc-loaded Mt (Zn-Mt), copper-loaded Mt (Cu-Mt), copper/zinc-loaded Mt with different Cu/Zn ratio (Cu/Zn-Mt-1, Cu/Zn-Mt-2, Cu/Zn-Mt-3) were prepared by an ion-exchange reaction, and characterized using X-ray diffraction (XRD), fourier transformed infrared spectroscopy (FTIR) and transmission electron microscopy (TEM). The specific surface areas, antimicrobial activity and cytotoxicity of the modified Mt were investigated.

Results: In the modified Mt, hydrated Cu ions and Zn ions were exchanged in the interlayer space of Mt and the particles were irregular shapes. The results showed that Cu/Zn-Mt enhanced antibacterial and antifungal activity compared with Zn-Mt and Cu-Mt possibly due to the synergistic effect between Cu and Zn. Among the Cu/Zn-Mt with different Cu/Zn raitos, Cu/Zn-Mt with a Cu/Zn ratio of 0.98 or 0.51 showed higher antimicrobial activity against gram-negative bacteria (*Escherichia coli*), gram-positive bacteria (*Staphylococcus aureus*), fungi (*Candida albicans*). Moreover, the antimicrobial activity of Cu/Zn-Mt was correlated with its specific surface area. Cytotoxicity studies on IPEC-J2 cell showed a slight cytotoxicity of Cu/Zn-Mt.

Conclusions: The current data provide clear evidence that in terms of its antimicrobial activity and relatively low toxicity, the Cu/Zn-Mt holds great promise for applications in animal husbandry.

Keywords: Antimicrobial reagent, Modified montmorillonites, Synergistic antimicrobial effect

Background

Recently, various inorganic antimicrobial materials have been developed and have attracted considerable interest in animal husbandry [1–3]. Among various antimicrobial metals,Copper (Cu) and Zinc (Zn) are normally used in animal feed in concentrations in excess of the nutritional requirements of the animals and for prevention of diarrhea disease, and also as an alternative to in-feed antibiotics for growth promotion [4, 5]. However, the strategy has been criticized because high level of Zn and Cu give rise to microbial drug resistance [6–8]. Enteral bacteria, both commensal and pathogenic, in farmed animals have been shown to develop resistance to trace elements (Cu and Zn) and concomitant cross-resistance to antimicrobial agents. Such bacteria may be transferred to other animals and human [7]. Moreover, large quantities of Cu/Zn were excreted, which would pose an environmental problem [4]. Therefore, it is essential to find an alternative to reduce Cu/Zn supplementation for sustaining animal production.

Special attention has been paid to metal ions loading inorganic carrier, which are superior in terms of safety, and long-term antibacterial effectiveness when compared with conventional metal [9–11]. As for inorganic carriers, the use of clay minerals as supports for synthesis of various inorganic antimicrobial materials has attracted considerable interest, owing to their nontoxic, environmentally friendly characteristic, and easy preparation [12–15]. It has been demonstrated that heavy metal (silver, Cu, Zn, and so on) exchanged clay minerals can serve as an antimicrobial reagent in vitro [8, 10, 16]. Until recently, Cu or Zn exchanged clay minerals has been added to the animal feed as an antibiotic

* Correspondence: shuma@zju.edu.cn; chhu@zju.edu.cn
[1]Animal Science College, Zhejiang University, Key Laboratory of Animal Feed and Nutrition of Zhejiang Province, No.866, Yuhangtang Road, Hangzhou 310058, People's Republic of China
Full list of author information is available at the end of the article

alternative [4, 8, 15], with the additive amount of Cu and Zn being quite lower than that in conventional animal diet. Moreover, a few researches have reported that loading two metal ions onto montmorillonite (Mt) displayed obvious synergistic antimicrobial effect in vitro [17–19]. However, there are no data loading Zn^{2+} and Cu^{2+} onto Mt so far. It has been reported that Cu-Mt and Zn-Mt displayed different antibacterial and antifungal activity [10]. Mt with different metal loading capacity had different physical and chemical properties [20–22]. So whether the Cu/Zn- loaded Mt (Cu/Zn-Mt) with different Cu/Zn ratios will affect the antibacterial and antifungal effect in vitro or not need to been explored.

Moreover, since the ultimate location of Cu/Zn-Mt is the intestinal tract of animals, it is necessary to test its cytotoxicity using an in vitro model for safety consideration. Recently, a cell line from jejunum epithelium isolated from a neonatal unsuckled piglet, small intestinal porcine epithelial cell line (IPEC-J2) was characterized and used as an in vitro model system for study [23]. This cell lines exhibited strong similarities to primary intestinal epithelial cells, and can be used as an appropriate model through the advantage of direct comparison with the experimental animals [24]. Therefore, IPEC-J2 was selected for the cytotoxicity studies of Cu/Zn-Mt.

In the present work, a series of modified Mt including Zn-Mt, Cu-Mt, Cu/Zn-Mt with different Cu/Zn ratio (Cu/Zn-Mt-1, Cu/Zn-Mt-2, Cu/Zn-Mt-3) were synthesized. Their synergistic antibacterial and antifungal effect in vitro were compared. Considering the application of Cu/Zn-Mt in animal husbandry, cytotoxicity assay was measured, too.

Methods
Materials
Mt was obtained from the Inner Mongolia Autonomous Region, China. The content of the purified Mt was 99.0%. The cation exchange capacity (CEC) was 1.30 mmol/kg Mt. $ZnSO_4 \cdot 7H_2O$ and $CuSO_4 \cdot 5H_2O$ were purchased from Sinopharm Chemical Reagent Co., Ltd., China. Gram-negative *Escherichia coli* (*E. coli*, ATCC 25922), gram-positive *Staphylococcus aureus* (*S. aureus*, ATCC 29213) and *Candida albicans* (*C. albicans*, ATCC 10231, fungi) were purchased from China Center of Industrial Culture Collection.

Preparation of modified Mt
The Mt (Cu-Mt, Zn-Mt, Cu/Zn-Mt-1, Cu/Zn-Mt-2, Cu/Zn-Mt-3) were prepared by an ion-exchange reaction. Ten grams of the Mt was mixed with 0.1 L of 0.2 mol/L NaCl solution. The dispersion was agitated for 5 h on a magnetic stirrer (700 rpm). The Na-Mt was then separated by centrifugation (15 min, 8000 × g) and washed with deionized water for three times. The washed

Na-Mt was then added to 0.1 L of 0.19 mol/L $ZnSO_4$ solutions, 0.2 mol/L $CuSO_4$ solutions, 0.2 mol/L $CuSO_4$ and $ZnSO_4$ mixed solutions (the ratio of Cu and Zn is 1:1, 1:2, 1:4), respectively. The dispersion was agitated at 60 °C for 6 h on a magnetic stirrer to accelerate the cation exchange. After centrifugation, the sediment was washed with deionized water for three times, dried at 80 °C over night, and ground to a size less than 300 mesh. Zinc or/and copper concentration in the modified Mt were measured by atomic absorption spectroscopy (ICE 3300, Thermo Fisher Scientific, Waltham, USA). The Cu concentration of Cu-Mt was 5.70%. The Zn concentration of Zn-Mt was 5.62%. The Cu and Zn concentration of Cu/Zn-Mt-1 were 2.78%, 2.85%, respectively. The Cu and Zn concentration of Cu/Zn-Mt-2 were 1.89%, 3.72%, respectively. The Cu and Zn concentration of Cu/Zn-Mt-3 were 1.15%, 4.45%, respectively. .

Characterization of the specimens
A PANalytical X'pert PRO powder diffractometer equipped with a Cu Kα radiation source was employed to determine the phase compositions and structures of the samples at 40 kV and 30 mA. Fourier transform infrared spectrometer (FTIR) was performed with a Nicolet Avatar 37- DTGS FT-IR spectrophotometer to study the structure of the materials by analyzing the vibrational frequencies of chemical bonds. Transmission electron microscopy (TEM) and energy-dispersive x-ray spectroscopy (EDX) (Tecnai G2 F20 S-TWIN; FEI Company, Hillsboro, OR, USA) were carried to characterize the microstructure of the samples.

Antimicrobial activity of the specimens
For antimicrobial experiments, the minimum inhibitory concentrations (MIC) of the specimens were estimated by a two-fold diluting method. The typical microorganisms of *E. coli*, *S. aureus* and *C. albicans* were selected as indicators. Luria Bertani (LB) broth was used as a growing medium for *E. coli* and *S. aureus*. *C. albicans* was cultivated in liquid sabouraud medium. Bacterial strains were grown overnight and diluted with fresh medium to achieve an approximate density of 10^7 CFU/mL. Specimens of each material (Mt, Cu-Mt, Zn-Mt and Cu/Zn-Mt) were put into tubes containing 5 mL LB broth or liquid sabouraud medium, and then two-fold diluted into different concentrations. Bacterial inoculums were added to tubes with a final concentration of 10^5 cells/mL. Each specimen was determined in triplicate. The bacteria–mineral mixtures were incubated at 37 °C for 24 h, with continuous shaking at 200 rpm. The MIC of the specimens was determined by the lowest concentration of the specimens that inhibited completely the bacteria or fungi visible growth when judging by eye [25].

Particle size, specific surface areas and antimicrobial assays of Cu/Zn-Mt

Cu/Zn-Mt was ground to a size less than 100, 200, 300, 400 mesh, respectively. The N2 adsorption isotherms and specific Brunauer-Emmitt-Teller (BET) surface areas of Cu/Zn-Mt were measured at 77 K, using a Tristar 3000 specific surface area and porosimetry analyzer (Micromeritics Instrument Corp., USA). All the samples were degassed at 623 K for 1 h under vacuum before analysis.

For antimicrobial assays of Cu/Zn-Mt with different article size and specific surface areas, Cu/Zn-Mt was put into tubes containing nutrient broth to achieve a concentration of 400 mg/L, and then the bacterial inoculums (*E. coli* or *S. aureus* or *C. albicans*) were added with an approximate density of 10^5cells/mL. The tubes without Cu/Zn-Mt served as the control group. After incubation at 37 °C for 24 h, the mixtures were subjected to successive 10-fold serial dilutions in the corresponding medium, mixed with a vortex shaker to ensure dispersion and quantitatively cultured in duplicate onto agar plates to determine the number of viable bacteria. The viable cell counts were expressed as fold changes, calculated relative to the control group.

Cytotoxicity assay of Cu/Zn-Mt

Cytotoxicity of Cu/Zn-Mt was performed on IPEC-J2 cell by tetrazolium dye (MTT) based assay. Briefly, cells were seeded into 96-well culture plate in triplicate (1.2 × 10^5 number of cells in 100 μL DMEM). Mono-layers of cells were treated with Cu/Zn-Mt of increasing concentration (0–0.5 mg/mL). At the end of the culture period, 20 μL of 5 mg/mL MTT stock solution was added in each well. After additional 4 h of incubation at 37 °C, the resultant intracellular formazan crystals were solubilized with acidic isopropanol and the absorbance of the solution was measured at 570 nm using an ELISA reader (Emax, Molecular device, CA, USA).

Results and Discussions

Structure and morphology analysis of the specimens

XRD patterns were obtained to identify the intercalation of Cu^{2+} or and Zn^{2+} into Mt [20, 26]. The measured interlayer spacings d_{001} are shown in Fig. 1. Na-Mt displays a reflection at the 2θ value of 5.76°, which is assigned to d_{001} basal spacing of 1.534 nm. As for modified Mt, the reflection was emerged at lower 2θ values of 5.34°, 5.40°, 5.36°, 5.38°, 5.37° (Cu-Mt, Zn-Mt, Cu/Zn-Mt-1, Cu/Zn-Mt-2, Cu/Zn-Mt-3), corresponding to the increased d_{001} basal spacing of 1.654 nm, 1.638 nm, 1.649 nm, 1.641 nm, 1.645 nm. It can be presumed that these increases after Cu^{2+} or/and Zn^{2+} ion exchange are caused by a difference in the size of hydrated form between Cu^{2+}/Zn^{2+} and Na^+ ion, although

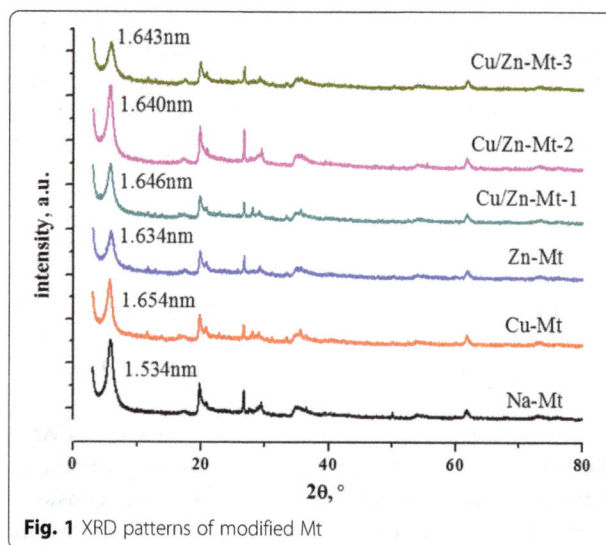

Fig. 1 XRD patterns of modified Mt

ion radius of Cu and Zn (0.072 nm,0.074 nm, respectively) is smaller than that of sodium (0.095 nm). The sodium ions in Na-Mt have been exchanged with $[Zn(H_2O)_6]^{2+}$ or/and $[Cu(H_2O)_6]^{2+}$[27]. Moreover, no difference of d_{001} basal spacing was observed in modified Mt, which might be associated with similar ion radius between Cu^{2+} and Zn^{2+}.

The infrared spectra of the specimens were obtained by the KBr method using an FTIR spectrometer and demonstrated in Fig. 2. Characteristic bands for Na-Mt are present around 3,405 cm^{-1} and 1,633 cm^{-1} (O-H stretching), 1,035 cm^{-1} (Si-O stretching), 529 cm^{-1} and 464 cm^{-1} (Si-O bending vibration) [28, 29]. As for modified Mt (Cu-Mt, Zn-Mt, Cu/Zn-Mt), the O-H stretching vibration band increased slightly, which can be attributed to the hydrated Cu^{2+}/Zn^{2+} ions [26]. In addition, the positions of the Si-O bending vibrations remained basically unchanged at 518 cm^{-1} (Si-O-Al) and 466 cm^{-1}

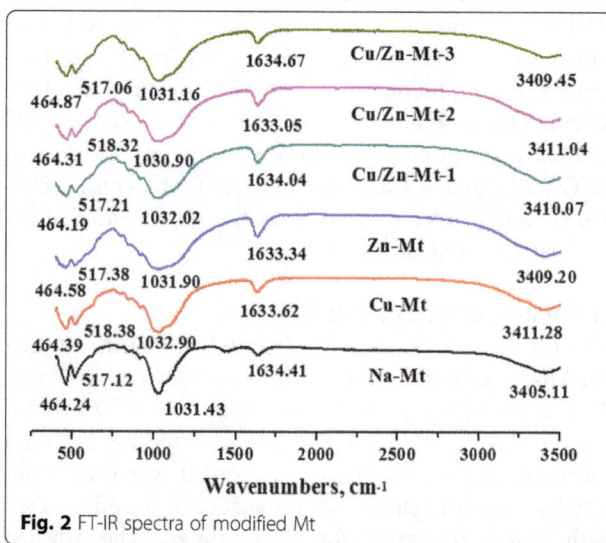

Fig. 2 FT-IR spectra of modified Mt

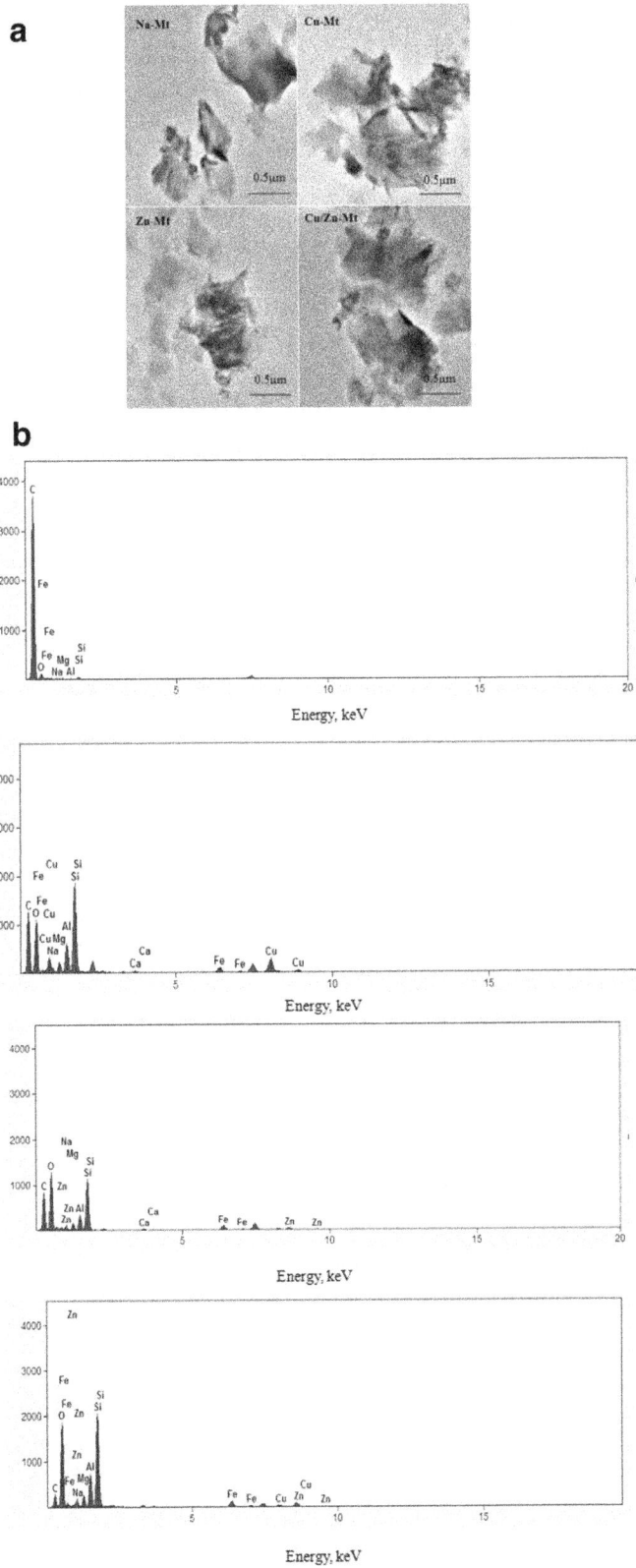

Fig. 3 TEM and EDX spectrum of modified Mt

(Si-O-Si) for the Cu/Zn-Mt, which indicated that the presence of Zn and Cu in the hexagonal cavities didn't affects this vibration in the experiment.

The TEM images reveal the internal structures of the Mt and modified Mt as presented in Fig. 3a. It was observed that the layered crystallites of Mt and modified Mt aggregated in large sized particles. Consistent with XRD, FTIR results, Cu or Zn loaded on the interlayers of Mt did not change the Mt's layered structure. Moreover, Fig. 3b shows the EDX spectrum of the specimens, confirming the presence of Cu or Zn in modified Mt. Apart from elements of Na-Mt (Al, Si, O, C, Fe, Mg and Na), peak of Cu or/and Zn was visible in Cu–Mt, Zn–Mt and Cu/Zn-Mt. Peak of sodium is detected which indicates incomplete exchange with Cu or/and Zn.

Antimicrobial assay of the specimens

The antimicrobial activity of the specimens was investigated in Table 1. In the case of Na-Mt, no antimicrobial activity was detected against the microorganisms (the MIC against the three kinds of microorganisms were all more than 10,000 mg/L). Similar results have also been reported that Mt with large specific surface area could adhere to bacteria by electrostatic forces, but showed no bacteriostatic activity [16, 30]. Moreover, results showed E. coli was more sensitive to modified Mt compared with S. aureus and C. albicans. One of the possible explanations of that difference in sensitivity is the different characteristics of the cell surfaces. It has been reported that the negative charge on the cell surface of gram-negative bacteria was higher than on gram-positive bacteria [12, 31]. Due to a higher negative charge on cell surface, the interaction between E. coli and Mt (positively charged in the interlayer) was definitely stronger than that of S. aureus and C. albicans, which could facilitate contact of Cu^{2+} and Zn^{2+} with bacterial cells wall and thus enable their damaging effect to the bacteria.

In addition, the antimicrobial effect was detected in Zn-Mt, Cu-Mt and Cu/Zn-Mt treatments. Compared with Zn-Mt, Cu-Mt displayed higher antibacterial and antifungal activity, which was corresponded to the previous results [10, 12]. The antibacterial and antifungal

properties of Zn-Mt or Cu-Mt could been attributed to the attraction, by electrostatic forces, of the negatively charged membrane of the bacteria to the surface of the Mt, where the positive charged Cu or Zn ions kills the bacteria or renders them unable to replicate [20, 32]. In the case of Cu/Zn-Mt, the antibacterial and antifungal activity has been improved compared with Cu-Mt and Zn-Mt. Cu/Zn-Mt showed obvious synergistic antimicrobial effect. Similar to our findings, it was reported that compared with Chitosan/Ag and CS/ZnO, Chitosan/Ag/ZnO composite displayed excellent antimicrobial activities against B. subtilis, E. coli, S. aureus, Penicillium, Aspergillus, Rhizopus and yeast [33]. Moreover, it was found that Zn^{2+}-Ce^{3+} loaded Mt presented much higher antibacterial and antifungal efficiency than Zn^{2+} loaded Mt and Ce^{3+} loaded Mt [17]. A recent review indicated that different metals caused discrete and distinct types of injuries to microbial cells as a result of oxidative stress, protein dysfunction or membrane damage [34]. It has been reported that toxicity associated with Cu might be due to impaired membrane function and reactive oxygen species (ROS) mediated cellular damage [35]. Zn could initiate bacteriostasis though oxidation of cellular thiols or damaging Fe–S-containing dehydratases in vitro independently of ROS and inhibit these enzymes activity [36]. Therefore, we speculated that different antimicrobial mechanism might result in more effective damage to bacteria, which might require further study. What is more, the antibacterial and antifungal properties of Cu/Zn-Mts varied with different Cu/Zn atomic ratios. Compared with Cu/Zn-Mt-3, Cu/Zn-Mt-1 and Cu/Zn-Mt-2 showed higher antibacterial and antifungal activity. So the study was especially focused on the structure and properties of Cu/Zn-Mt-2 in the following research.

Particle size, surface properties and antimicrobial activity of the Cu/Zn-Mt-2

The effect of particle size on the antimicrobial activity of Cu/Zn-Mt (Cu/Zn-Mt-2) was seen in Fig. 4. The antimicrobial activity increased with the enhancement of the specific surface area when the particle size decreased, which suggested that the overall antimicrobial effect is not only related to the presence and quantity of metal ions but also is affected by the surface characteristics of the modified Mt [16]. It was reasonable to state that the binding of Cu/Zn-Mt particles to the bacteria depended on the surface area available for interaction. Smaller particles having the larger surface area available for interaction would give higher antimicrobial effect than the larger particles [35].

Cytotoxicity assay of the Cu/Zn-Mt-2

In order to utilize the antimicrobial effect of Cu/Zn-Mt for therapeutic purposes, it is absolutely essential to

Table 1 The antimicrobial activity of modified Mt

Sample	MIC, mg/L		
	E. coli	S. aureus	C. albicans
Na-Mt	>10,000	>10,000	>10,000
Cu-Mt	411.63	657.78	1,315.56
Zn-Mt	823.26	1,315.56	2631.12
Cu/Zn-Mt-1	328.89	411.63	823.26
Cu/Zn-Mt-2	328.89	411.63	823.26
Cu/Zn-Mt-3	411.63	657.78	1,315.56

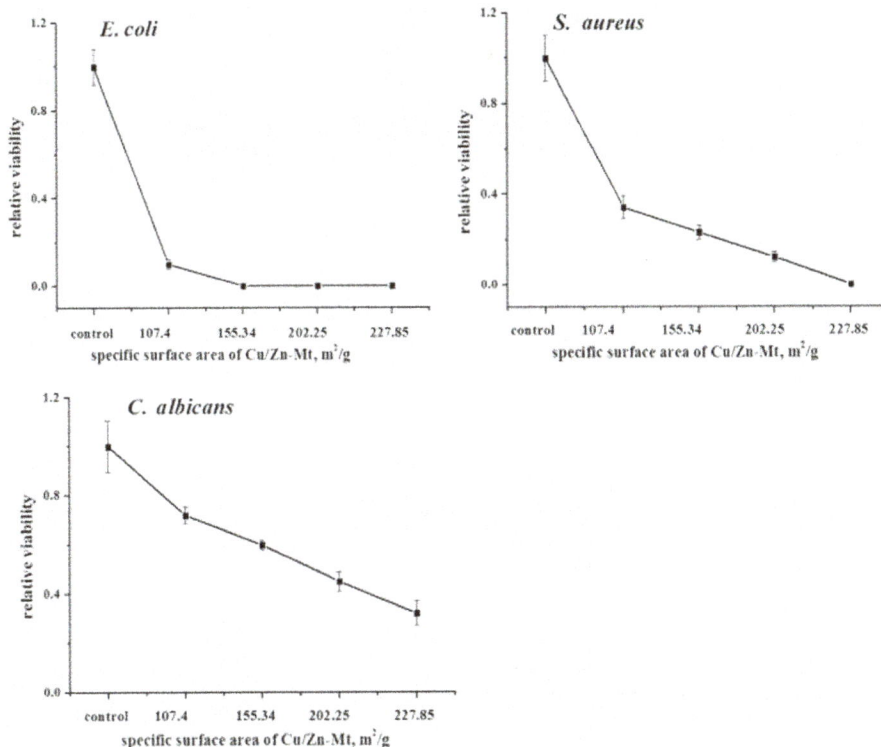

Fig. 4 The antimicrobial activity of Cu/Zn-Mt-2 with different specific surface area

perform a cytotoxicity study (Fig. 5). Cu/Zn-Mt-2 (0.1 mg/mL) exhibited a slight cytotoxicity (~10%) to IPEC-J2 cell within 24 h incubation by MTT assays. Cu/Zn-Mt-2 of higher concentration (0.3 mg/mL) led to increased cytotoxicity (~20%) within 24 h. Overall, the concentrations of the cell population showed stabilization at concentrations beyond the MIC value. This may be explained as follows, lower concentration of Cu or Zn had minimal adverse effect on cells in vitro [32, 37]. Moreover, Mt has been regarded as a dermatological and gastrointestinal protector by attaching to the cells and absorbing any toxic molecules due to their high absorptive capacity, which played an important role in reducing toxicity of Cu/Zn-Mt-2 [38]. Therefore, Cu/Zn-Mt was good inorganic antimicrobial materials with slight cytotoxicity.

Conclusion
Cu/Zn-Mt with different Cu/Zn ratios could be prepared by an ion-exchange reaction and showed synergistic antimicrobial effect and relatively low cell toxicity. Moreover, the antimicrobial activity of Cu/Zn-Mt was correlated with its specific surface area and Cu/Zn ratios. Therefore, Cu/Zn-Mt holds great promise for applications in animal husbandry.

Abbreviations
Cu/Zn-Mt: Copper/zinc loaded montmorillonite; Cu-Mt: Copper loaded montmorillonite; Mt: Montmorillonite; Zn-Mt: Zinc loaded montmorillonite

Acknowledgements
Not applicable.

Funding
This work was jointly supported by the Special Fund for Agro-scientific Research in the Public Interest (No. 201403047), Science Technology Department of Zhejiang Province (2015C02022). The funders had no role in study design, data collection and analysis, decision to publish, or preparation of the manuscript.

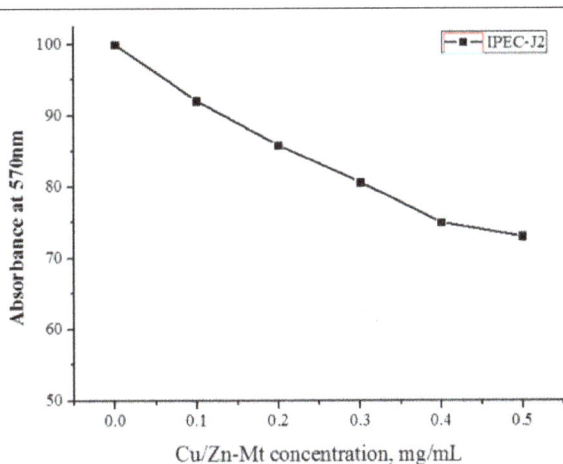

Fig. 5 The cytotoxicity of Cu/Zn-Mt-2

Authors' contributions

LFJ conceived and designed the experimental plan. FHL and STC collected the experiments data. CCW and HW analyzed the data. LFJ wrote the paper. CHH and MAS had primary responsibility for the final content. All authors read and approved the final manuscript.

Competing interests

The authors declare that they have no competing interests.

Author details

[1]Animal Science College, Zhejiang University, Key Laboratory of Animal Feed and Nutrition of Zhejiang Province, No.866, Yuhangtang Road, Hangzhou 310058, People's Republic of China. [2]Key Laboratory of Animal Nutrition and Feed in East China, Ministry of Agriculture, No.866, Yuhangtang Road, Hangzhou 310058, People's Republic of China.

References

1. Davis ME, Parrott T, Brown DC, de Rodas BZ, Johnson ZB, Maxwell CV, et al. Effect of a *Bacillus*-based direct-fed microbial feed supplement on growth performance and pen cleaning characteristics of growing-finishing pigs. J Anim Sci. 2008;86:1–7.
2. Hu CH, Xia MS. Adsorption and antibacterial effect of copper-exchanged montmorillonite on *Escherichia coli K-88*. Appl Clay Sci. 2006;b31:180–4.
3. Saavedra MJ, Dias CSP, Martinez-Murcia A, Bennett RN, Aires A, Rosa EAS. Antibacterial effects of glucosinolate-derived hydrolysis products against *Enterobacteriaceae* and *Enterococci* isolated from pig ileum segments. Foodborne Pathog Dis. 2012;9:338–45.
4. Song J, Li YL, Hu CH. Effects of copper-exchanged montmorillonite, as alternative to antibiotic, on diarrhea, intestinal permeability and proinflammatory cytokine of weanling pigs. Appl Clay Sci. 2013;77–78:52–5.
5. Huang YL, Ashwell MS, Fry RS, Lloyd KE, Flowers WL, Spears JW. Effect of dietary copper amount and source on copper metabolism and oxidative stress of weanling pigs in short-term feeding. J Anim Sci. 2015;93:2948–55.
6. Jondreville C, Revy PS, Dourmad JY. Dietary means to better control the environmental impact of copper and zinc by pigs from weaning to slaughter. Livest Prod Sci. 2003;84:147–56.
7. Yazdankhah S, Rudi K, Bernhoft A. Zinc and copper in animal feed - development of resistance and co-resistance to antimicrobial agents in bacteria of animal origin. Microb Ecol Health Dis. 2014;25:1–7.
8. Jiao LF, Ke YL, Xiao K, Song ZH, Lu JJ, Hu CH. Effects of zinc-exchanged montmorillonite with different zinc loading capacities on growth performance, intestinal microbiota, morphology and permeability in weaned piglets. Appl Clay Sci. 2015;112:40–3.
9. Zhao DF, Zhou J, Liu N. Preparation and characterization of Mingguang palygorskite supported with silver and copper for antibacterial behavior. Appl Clay Sci. 2006;33:161–70.
10. Malachova K, Praus P, Rybkova Z, Kozak O. Antibacterial and antifungal activities of silver, copper and zinc montmorillonites. Appl Clay Sci. 2011;53:642–5.
11. Krishnani KK, Zhang Y, Xiong L, Yan YS, Boopathy R, Mulchandani A. Bactericidal and ammonia removal activity of silver ion-exchanged zeolite. Bioresource Techn. 2012;117:86–91.
12. Stanic V, Dimitrijevic S, Antic-Stankovic J, Mitric M, Jokic B, Plecas IB, et al. Synthesis, characterization and antimicrobial activity of copper and zinc-doped hydroxyapatite nanopowders. Appl Surf Sci. 2010;256:6083–9.
13. Cai X, Zhang J, Ouyang Y, Ma D, Tan SZ, Peng YL. Bacteria-adsorbed palygorskite stabilizes the quaternary phosphonium salt with specific-targeting capability, long-term antibacterial activity, and lower cytotoxicity. Langmuir. 2013;29:5279–85.
14. Gaskell EE, Hamilton AR. Antimicrobial clay-based materials for wound care. Future Med Chem. 2014;6:641–55.
15. Wang LC, Zhang TT, Wen C, Jiang ZY, Wang T, Wang T, et al. Protective effects of zinc-bearing clinoptilolite on broilers challenged with Salmonella pullorum. Poult Sci. 2012;91:1838–45.
16. Magana SM, Quintana P, Aguilar DH, Toledo JA, Angeles-Chavez C, Cortes MA, et al. Antibacterial activity of montmorillonites modified with silver. J Mol Catal a-Chem. 2008;281:192–9.
17. Tan SZ, Zhang KH, Zhang LL, Xie YS, Liu YL. Preparation and characterization of the antibacterial Zn(2+) or/and Ce(3+) loaded montmorillonites. Chinese J Chem. 2008;26:865–9.
18. Ramesh A, Hasegawa H, Maki T, Ueda K. Adsorption of inorganic and organic arsenic from aqueous solutions by polymeric Al/Fe modified montmorillonite. Sep Purif Technol. 2007;56:90–100.
19. Cai X, Dai GJ, Tan SZ, Ouyang Y, Ouyang YS, Shi QS. Synergistic antibacterial zinc ions and cerium ions loaded alpha-zirconium phosphate. Mater Lett. 2012;67:199–201.
20. Shi QS, Tan SZ, Yang QH, Jiao ZP, Ouyang YS, Chen YB. Preparation and characterization of antibacterial Zn^{2+}-exchanged montmorillonites. J Wuhan Univ Technol. 2012;25:725–9.
21. Dahn R, Baeyens B, Bradbury MH. Investigation of the different binding edge sites for Zn on montmorillonite using P-EXAFS - The strong/weak site concept in the 2SPNE SC/CE sorption model. Geochim Cosmochim Ac. 2011;75:5154–68.
22. Churakov SV, Dahn R. Zinc adsorption on clays inferred from atomistic simulations and EXAFS spectroscopy. Environ Sci Technol. 2012;46:5713–9.
23. Geens MM, Niewold TA. Optimizing culture conditions of a porcine epithelial cell line IPEC-J2 through a histological and physiological characterization. Cytotechnology. 2011;63:415–23.
24. Schierack P, Nordhoff M, Pollmann M, Weyrauch KD, Amasheh S, Lodemann U, et al. Characterization of a porcine intestinal epithelial cell line for in vitro studies of microbial pathogenesis in swine. Histochem Cell Biol. 2006;125:293–305.
25. Xu GN, Qiao XL, Qiu XL, Chen JG. Preparation and characterization of nano-silver loaded montmorillonite with strong antibacterial activity and slow release property. J Mater Sci Technol. 2011;27:685–90.
26. Tanaka M, Itadani A, Abe T, Taguchi H, Nagao M. Observation of characteristic IR band assignable to dimerized copper ions in montmorillonite. J Colloid Interface Sci. 2007;308:285–8.
27. Kozák O, Praus P, Machovič V, Klika Z. Adsorption of zinc and copper ions on natural and ethylenediamine modified montmorillonite. Ceram-Silikaty. 2010;54:78–84.
28. Darder M, Colilla M, Ruiz-Hitzky E. Biopolymer-clay nanocomposites based on chitosan intercalated in montmorillonite. Chem Mater. 2003;15:3774–80.
29. Tyagi B, Chudasama CD, Jasra RV. Determination of structural modification in acid activated montmorillonite clay by FT-IR spectroscopy. Spectrochim Acta A. 2006;64:273–8.
30. Malachova K, Praus P, Pavlickova Z, Turicova M. Activity of antibacterial compounds immobilised on montmorillonite. Appl Clay Sci. 2009;43:364–8.
31. Du WL, Niu SS, Xu YL, Xu ZR, Fan CL. Antibacterial activity of chitosan tripolyphosphate nanoparticles loaded with various metal ions. Carbohydr Polym. 2009;75:385–9.
32. Bagchi B, Kar S, Dey SK, Bhandary S, Roy D, Mukhopadhyayc TK, et al. In situ synthesis and antibacterial activity of copper nanoparticle loaded natural montmorillonite clay based on contact inhibition and ion release. Colloid Surface B. 2013;108:358–65.
33. Li LH, Deng JC, Deng HR, Liu ZL, Li XL. Preparation, characterization and antimicrobial activities of chitosan/Ag/ZnO blend films. Chem Eng J. 2010;160:378–82.
34. Lemire JA, Harrison JJ, Turner RJ. Antimicrobial activity of metals: mechanisms, molecular targets and applications. Nat Rev Microbiol. 2013;11:371–84.
35. Hong R, Kang TY, Michels CA, Gadura N. Membrane lipid peroxidation in copper alloy-mediated contact killing of Escherichia coli. Appl Environ Microbiol. 2012;78:1776–84.
36. Xu FF, Imlay JA. Silver(I), mercury(II), cadmium(II), and zinc(II) target exposed enzymic iron-sulfur clusters when they toxify. Escherichia coli. 2012;10:3614–21.
37. Lodemann U, Einspanier R, Scharfen F, Martens H, Bondzio A. Effects of zinc on epithelial barrier properties and viability in a human and a porcine intestinal cell culture model. Toxicol In Vitro. 2013;27:834–43.
38. Williams LB, Haydel SE. Evaluation of the medicinal use of clay minerals as antibacterial agents. Int Geol Rev. 2010;52:745–70.

5

Effects of *Saccharomyces cerevisiae* fermentation products on performance and rumen fermentation and microbiota in dairy cows fed a diet containing low quality forage

Wen Zhu[1], Zihai Wei[1], Ningning Xu[1], Fan Yang[1], Ilkyu Yoon[2], Yihua Chung[2], Jianxin Liu[1*] and Jiakun Wang[1]

Abstract

Background: A possible option to meet the increased demand of forage for dairy industry is to use the agricultural by-products, such as corn stover. However, nutritional value of crop residues is low and we have been seeking technologies to improve the value. A feeding trial was performed to evaluate the effects of four levels of *Saccharomyces cerevisiae* fermentation product (SCFP; Original XP; Diamond V) on lactation performance and rumen fermentation in mid-lactation Holstein dairy cows fed a diet containing low-quality forage. Eighty dairy cows were randomly assigned into one of four treatments: basal diet supplemented with 0, 60, 120, or 180 g/d of SCFP per head mixed with 180, 120, 60, or 0 g of corn meal, respectively. The experiment lasted for 10 wks, with the first 2 weeks for adaptation.

Results: Dry matter intake was found to be similar ($P > 0.05$) among the treatments. There was an increasing trend in milk production (linear, $P \leq 0.10$) with the increasing level of SCFP supplementation, with no effects on contents of milk components ($P > 0.05$). Supplementation of SCFP linearly increased ($P < 0.05$) the N conversion, without affecting rumen pH and ammonia-N ($P > 0.05$). Increasing level of SCFP linearly increased ($P < 0.05$) concentrations of ruminal total volatile fatty acids, acetate, propionate, and butyrate, with no difference in molar proportion of individual acids ($P > 0.05$). The population of fungi and certain cellulolytic bacteria (*Ruminococcus albus*, *R. flavefaciens* and *Fibrobacter succinogenes*) increased linearly ($P < 0.05$) but those of lactate-utilizing (*Selenomonas ruminantium* and *Megasphaera elsdenii*) and lactate-producing bacteria (*Streptococcus bovis*) decreased linearly ($P \leq 0.01$) with increasing level of SCFP. The urinary purine derivatives increased linearly ($P < 0.05$) in response to SCFP supplementation, indicating that SCFP supplementation may benefit for microbial protein synthesis in the rumen.

(Continued on next page)

* Correspondence: liujx@zju.edu.cn
[1]Institute of Dairy Science, College of Animal Sciences, Zhejiang University, 866 Yuhangtang Road, Hangzhou 310058, People's Republic of China
Full list of author information is available at the end of the article

(Continued from previous page)

Conclusions: The SCFP supplementation was effective in maintaining milk persistency of mid-lactation cows receiving diets containing low-quality forage. The beneficial effect of SCFP could be attributed to improved rumen function; 1) microbial population shift toward greater rumen fermentation efficiency indicated by higher rumen fungi and cellulolytic bacteria and lower lactate producing bacteria, and 2) rumen microbial fermentation toward greater supply of energy and protein indicated by greater ruminal VFA concentration and increased N conversion. Effects of SCFP were dose-depended and greater effects being observed with higher levels of supplementation and the effect was more noticeable during the high THI environment.

Keywords: Corn stover, Lactating cow, Rumen fermentation, Rumen microbiota, *Saccharomyces cerevisiae* fermentation product

Background

The increase in human consumption of milk has increased the demand for high quality forages for lactating dairy cows. However, China does not have enough high quality forages for dairy cows. A possible option to meet the increased demand of forage is to use the agricultural by-products that are generated in large amounts globally, with estimated 265 million tons of corn stover produced annually in China [1]. These by-products are rich in carbohydrates representing a large potential dietary energy source for ruminants. However, nutritional value of crop residues is low due to their high content of fiber, low digestibility, and low contents of crude protein (CP), metabolizable energy (ME), minerals, and vitamins [2]. Therefore, it is essential to develop nutritional strategies to maintain high milk production when crop residues are used as the main forage source. Feeding ruminal fermentation modifiers has been shown as a cost-effective and safe way to maximize feed utilization of low-quality forage, and thereby improve milk production [3].

Saccharomyces cerevisiae fermentation product (SCFP; Original XP; Diamond V, Cedar Rapids, IA, USA) is one of the most widely used rumen fermentation modifiers. A recent meta-analysis of dairy cow studies indicated that SCFP increased milk yield and feed utilization of lactating dairy cows [4]. These effects may be the results of SCFP favorably altered ruminal microbial fermentation, by stimulating the growth and activity of fiber-digesting bacteria, increasing fiber digestion, increasing microbial protein (MCP) synthesis, stimulating growth of lactate-utilizing bacteria; and decreasing accumulation of lactate [5–8]. It is reported that SCFP improved the rumen fermentation of both low quality forages and their mixed diets by stimulating the number of fiber-digesting rumen microbes, especially fungi populations in vitro [9]. It was hypothesized that SCFP may be beneficial when cows are fed low quality forages, by promoting a better forages utilization and therefore by improving nutrient availability. Thus, the objective of the current study was to investigate the effects of several levels of SCFP on dry matter intake (DMI), milk production, rumen fermentation and microbial communities in dairy cows fed a diet containing low quality forage.

Methods

Animals

The Animal Care Committee of Zhejiang University approved the use of animals for this experiment (Hangzhou, China). The experiment was conducted as a randomized complete block design with repeated measurements. Eighty Holstein cows (initial mean ± SD: parity 3.23 ± 0.81, 655 ± 65 of kg BW, 180 ± 45 d in milk (DIM), 26.6 ± 0.79 kg/d milk) were divided into 20 blocks based on milk yield and DIM and then randomly assigned within block to one of four treatments. The cows were fed individually and supplemented with 0 (control), 60, 120, or 180 g/d of SCFP per head mixed with 180, 120, 60, or 0 g of corn meal, respectively. The SCFP (Original XP), a fully fermented yeast culture containing residual yeast cells, fermentation metabolites and growth media, was from Diamond V (Cedar Rapids, Iowa, USA). Daily, SCFP was individually top-dressed with 1/3 of the supplement provided at each of the 3 feedings. Each cow was observed for 20 min after feeding to ensure complete consumption of the supplements.

Diets

Diets were designed according to nutrient requirements for mid-lactation Holstein cows weighing 600 kg and producing 30 kg/d of milk [10]. The ingredients and chemical composition of the experimental diet are presented in Table 1. The forage was comprised (DM basis) of 150 g/kg corn stover (pelletized), 70 g/kg Chinese ryegrass, and 173 g/kg corn silage. These forage sources were considered to be of low quality based on their chemical composition as described in the footnote of Table 1. Samples of forage and concentrate were collected weekly and analyzed to adjust diets to account for DM fluctuation. Feed was offered ad libitum to allow for 10% orts.

Table 1 Ingredient and chemical composition of basal diets used in the experiment

Items	Contents
Ingredient, g/kg DM	
Corn silage[a]	173
Chinese ryegrass[a]	70
Corn stover (pelletized)[a]	150
Ground corn	148
Steam-flaked corn	74
Barley	49
Soybean meal	123
Cottonseed meal	49
Beet pulp	85
Brewer's grains	28
Calcium Carbonate	2
Premix[b]	49
Chemical composition, g/kg DM	
OM	925
CP	149
NDF	415
ADF	246
Non-fibrous carbohydrates (NFC)[c]	346
Ca	6.8
P	4.6
NE_L^d, Mcal/kg DM	1.57

[a]Chemical compositions (% of DM) of forages are as follows ($n = 5$): corn silage: OM 94.7, CP 8.1, NDF 69.5, ADF 34.0, and NFC 14.6; Chinese ryegrass: OM 92.3, CP 7.72, NDF 67.5, ADF 42.6, and NFC 16.5; corn stover (pelletized): OM 90.0, CP 6.0, NDF 56.8, ADF 26.7, and NFC 28.8
[b]Formulated to provide (DM basis): 1% CP, 15% ether extracts, 6% crude fiber, 7% Ca, 1.3% P, 10% salt, 3% Mg, 1.5% K, 1% Met, 260 mg/kg Cu, 260 mg/kg Fe, 1,375 mg/kg Zn, 500 mg/kg Mn, 112,000 IU/kg vitamin A, 29,500 IU/kg vitamin D_3, and 700 IU/kg vitamin E
[c]NFC = 100 - %NDF - %CP - %EE - %ash
[d]Net energy for lactation, calculated based on Ministry of Agriculture of P.R. China recommendations [10]

Experimental design

The feeding trial was conducted for 8 wk, following a 2-week adaptation when cows were fed a common basal total mixed ration (TMR) (Table 1) without SCFP. Cows were housed in a tie-stall barn and fed 3 times daily at 0630, 1330 and 2000 h with free access to drinking water. Cows were milked 3 times daily at 0700, 1400 and 2030 h. The basal TMR was mixed on site at every feeding, with the grain mix prepared every 2 wk.

Sampling, measurement, and analyses

In order to calculate the temperature-humidity index (THI), temperature and relatively humidity (RH) inside the barn were measured as described by Zhu et al. [11]. The THI was calculated as: THI = TD − (0.55-0.55 RH/100) (TD-58), where TD was the dry bulb temperature in °F (°F = 32 + 1.8°C) and RH was expressed as a percentage [12]. The average daily THI were calculated.

Feed offered and refused were weighed daily for each cow. Representative samples (1 kg) of all dietary ingredients, TMR and orts were collected on d 14 of the adaptation period and d 3 of each week during the experimental period. Sample preparation and analysis were performed according to Zhu et al. [13]. All samples were analyzed for contents of DM, CP, crude ash and acid detergent fiber (ADF) according to AOAC methods as described by procedures #934.01, #988.05, #927.02, and #973.18, respectively, and neutral detergent fiber (NDF) was assayed without a heat stable amylase and inclusive of residual ash [14, 15].

Milk samples were collected at each milking on d 11 to 14 of the adaptation period and on d 4 and 5 of each week during the experimental period. At each milking 50 mL milk samples were collected and pooled in a proportion of 4:3:3 considering the ratio of milk yield for the 3 times milking. Potassium dichromate (milk preservative, D&F Control Systems, San Ramon, CA, USA) was added at 0.6 g/kg to milk samples prior to storage. Milk samples were sent to Shanghai Dairy Herd Improvement testing center (Shanghai, China) for analysis of milk fat, protein, lactose, somatic cell count, and milk urea nitrogen by infrared analysis using a spectrophotometer (Foss-4000, Foss, Hillerød, Denmark) [16].

Rumen fluid (approximately 100 mL) was sampled using a stomach tube from 40 cows randomly (10 per treatment) at approximately 3 h after the morning feeding on d 7 of wk 4 and 8 during the experimental period as described by Shen et al. [17]. Fifty milliliters of collections were squeezed through 4 layers of cheesecloth, and rumen pH was measured immediately using a portable pH meter (Starter 300, Ohaus Instruments Co. Ltd., Shanghai, China). Two 5.0 mL subsamples of each rumen filtrate were collected and stored at -20°C for further analysis of ammonia nitrogen (N) and volatile fatty acids (VFA). One subsample was acidified with 1.0 mL of 250 g/kg orthophosphoric acid for VFA determination. Concentrations of VFA were determined by gas chromatography (GC-8A, Shimadzu Corp., Kyoto, Japan) according to the method described by Chaney and Marbach [18]. Another subsample was used to determine the ammonia-N by colorimetric method, as described by Hu et al. [19]. Another 50 mL of unfiltered rumen fluid was stored at -80°C for DNA extraction to determine the relative quantity of ruminal fibrolytic bacteria, fungi, protozoa, and lactate-utilizing and lactate-producing bacteria to 16S rDNA of total bacteria (as described below).

Cow body weight (BW) was estimated on two consecutive days at the beginning and end of the experiment based on the measurement of heart girth and body length using the following equation: Estimated BW (kg) = heart girth^2 (m) × body length (m) × 90 [20]. A scaled score (5-point scale, where 1 = thin and 5 = fat) was used to determine body condition score (BCS) on d 7 of wk 2, 4, 6, and 8 by 2 experienced investigators blinded to treatments according to Edmonson et al. [21]. Values from investigators were averaged for each scoring date and the mean BCS were used for statistical analysis.

Estimation of microbial protein synthesis in the rumen

Microbial protein (MCP) synthesis in the rumen was estimated by urinary purine derivatives (PD) as reported by Chen and Gomes [22]. Spot urine samples (10 mL each time) were collected twice daily at approximately 3 h and 6 h after the morning feeding on d 6 of wk 4 and 8 during the experimental period. The daily urine samples were pooled at equal portion by cow and 15 mL of subsamples were acidified immediately with 60 mL 0.036 mol/L H_2SO_4 (1:4) and stored at -20°C for later analysis. For analysis of PD, allantoin and uric acid were analyzed by the procedure of Chen and Gomes [22]. Creatinine was detected using a picric acid assay [23]. Creatinine was used as the marker to estimate urine volume, and was assumed to be excreted at a rate of 29 mg/kg of BW for calculating the urine volume excretion rate [24, 25].

Determination of rumen microbial population

A total of 2 mL of unfiltered rumen fluid was used for the genomic DNA extraction with the RBB + C method as described by Yu and Morrison [26]. The RNA and protein were removed by sequential digestion with RNase A and proteinase K. Then DNA was purified using columns from the QIAamp DNA Stool Mini Kit (QIAGEN, Valencia, CA, USA). The purity and concentration of total DNA were determined by spectroscopy (NanoDrop 2000, Thermo Scientific Inc., Wilmington, DE, USA), and the extracted DNA was diluted to 10 ng/μL. The integrity of the total DNA was assessed by 10 g/kg agarose gel electrophoresis.

Quantitative real-time PCR was used to determine the relative abundance of protozoa, fungi and 6 bacterial populations using previously validated primers listed in Table 2 [27–32]. The real-time qPCR assays were performed on an ABI 7500 Real-Time PCR System (Applied Biosystems, Foster City, CA, USA) using a kit (SYBR Premix Ex Taq kit; TaKaRa Biosystems Co. Ltd, Dalian, China). The template DNA used possessed an A_{260}/A_{280} ratio within the range of 1.7 to 1.9. The qPCR assays was performed in a total volume of 20 μL solution contained 1 μL of genomic DNA (10 ng/μL), 0.2 μmol/L of each primer, 10 μL of SYBR Premix Ex Taq (2×), 0.4 μL of ROX II (50×), and double-distilled water. All qPCR assays were performed in triplicate for each sample. For all primers, amplification started with a denaturalization at 94°C for 10 s followed by 40 cycles of 95°C for 5 s and 60°C for 34 s.

Table 2 Primers used in this study

Target		Primer sequences (5′→3′)	Product size, bp	Reference
Total bacteria	F	CGGCAACGAGCGCAACCC	130	[27]
	R	CCATTGTAGCACGTGTGTAGCC		[27]
Protozoa	F	GCTTTCGWTGGTAGTGTATT	223	[28]
	R	CTTGCCCTCYAATCGTWCT		[28]
Fungi	F	GAGGAAGTAAAAGTCGTAACAAGGTTTC	120	[27]
	R	CAAATTCACAAAGGGTAGGATGATT		[27]
Ruminococcus albus	F	CCCTAAAAGCAGTCTTAGTTCG	175	[29]
	R	CCTCCTTGCGGTTAGAACA		[29]
R. flavefaciens	F	CGAACGGAGATAATTTGAGTTTACTTAGG	132	[27]
	R	CGGTCTCTGTATGTTATGAGGTATTACC		[27]
Fibrobacter succinogenes	F	GTTCGGAATTACTGGGCGTAAA	121	[27]
	R	CGCCTGCCCCTGAACTATC		[27]
Selenomonas ruminantium	F	TGCTAATACCGAATGTTG	515	[30]
	R	TCCTGCACTCAAGAAAGA		[30]
Megasphaera elsdenii	F	GACCGAAACTGCGATGCTAGA	128	[31]
	R	CGCCTCAGCGTCAGTTGTC		[31]
Streptococcus bovis	F	ATGTTAGATGCTTGAAAGGAGCAA	90	[32]
	R	CGCCTTGGTGAGCCGTTA		[32]

Table 3 Effects of *Saccharomyces cerevisiae* fermentation product on lactation performance of dairy cows fed corn stover-containing diet

Item[a]	SCFP[b], g/d				SEM	P-value[c]			
	0	60	120	180		L	Q	W	T × W
DMI, kg/d	20.1	20.3	19.9	19.9	0.22	0.26	0.54	<0.01	0.92
Milk yield, kg/d	22.1	21.8	22.6	22.3	0.28	0.08	0.89	<0.01	<0.01
ECM, kg/d	25.4	25.8	26.0	25.7	0.35	0.52	0.35	<0.01	<0.01
Milk composition, %									
Fat	4.35	4.41	4.35	4.36	0.075	0.95	0.81	<0.01	<0.01
Protein	3.41	3.44	3.41	3.44	0.039	0.74	0.87	<0.01	<0.01
Lactose	4.78	4.79	4.73	4.75	0.036	0.33	0.87	<0.01	0.65
Total solids	13.7	13.7	13.4	13.6	0.19	0.55	0.59	<0.01	0.98
SCC, ×10^4	57.9	35.2	46.9	47.6	11.4	0.63	0.78	0.95	0.75
MUN, mg/dL	13.9	13.8	13.9	13.8	0.37	0.84	0.86	<0.01	0.75
Initial BW, kg	654	643	670	651	13.7	0.78	0.47		
BW gain, kg/d	0.17	0.22	0.15	0.05	0.14	0.62	0.27		
BCS	2.98	2.77	2.99	2.80	0.10	0.47	0.94	0.04	0.20
Feed efficiency	1.28	1.28	1.32	1.28	0.025	0.17	0.50	<0.01	0.67
N conversion	24.7	24.3	25.2	25.1	0.21	0.04	0.82	<0.01	0.17

[a]DMI = dry matter intake; ECM (energy-corrected milk, kg) = 0.3246 × milk yield (kg) + 13.86 × fat yield (kg) + 7.04 × protein yield (kg) [44]; SCC = somatic cell counts; MUN = milk urea nitrogen; Feed efficiency = kg of ECM/kg of DMI; BCS = body condition score; N conversion = kg of milk protein yield/ kg of dietary crude protein intake × 100
[b]*Saccharomyces cerevisiae* fermentation product (Diamond V Original XP, Cedar Rapids, Iowa, USA)
[c]T = treatment effect; W = week effect; T × W = interaction between treatment and week; L = linear effect of treatment; Q = quadratic effect of treatment

Calculations and statistical analysis

The quantification of protozoa, fungi, *Ruminococcus albus*, *R. flavefaciens*, *Fibrobacter succinogenes*, *Selenomonas ruminantium*, *Megasphaera elsdenii*, and *Streptococcus bovis* was expressed as a ratio to 16S rDNA of total bacteria. The $2^{-\Delta\Delta CT}$ method was used to analyze the relative changes in each gene expression, where Ct represented threshold cycle [33].

Data analyses were carried out using SAS software (version 9.0, SAS Institute Inc., Cary, NC, USA) [34]. All data except for BW gain were analyzed through the PROC MIXED program of SAS with the covariance type AR (1) for repeated measures. To determine differences in lactation performance [DMI, milk yield, ECM, and feed efficiency] among treatments, calculated mean values of DMI and milk yield on d 11 to 14 of the 14-d adaptation period were used as the covariate for treatment response. A randomized block design with repeated measurements was used. The model included week, treatment, interaction of treatment × week and block as fixed effects. Cow was included as a random effect to provide the error term to test the significance of the differences. Means were separated using the PDIFF option in the LSMEANS statement.

Data on BW gain were analyzed using the PROC GLM of SAS. The statistical model was the same as indicated above except that week and treatment × week were omitted. Linear and quadratic effects of treatment were tested for all data using orthogonal polynomial contrasts.

Results are reported as least squares means. Probability values of $P \leq 0.05$ were defined as statistically significant and the values $0.05 < P \leq 0.10$ were defined as trend.

Results

Feed intake and lactation performance

Body weight gain and BCS were not affected ($P > 0.05$) by SCFP supplementation. Dry matter intake was similar ($P > 0.05$) among the treatments. Milk yield had an increasing trend (linear, $P = 0.08$) with the increasing amount of SCFP supplementation (Table 3). Energy-corrected milk (ECM) was not affected by SCFP supplementation ($P > 0.05$). Feed efficiency (ECM/DMI) was

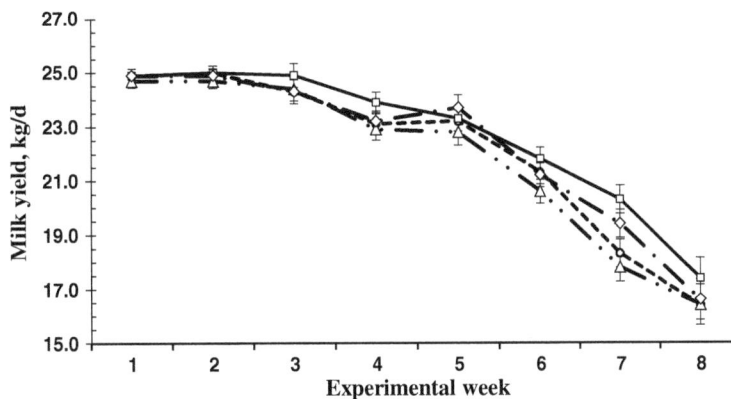

Fig. 1 Change in milk yield of lactating cows fed a diet containing low-quality forage with supplementation of a *Saccharomyces cerevisiae* fermentation products at 0 (o), 60 (Δ), 120 (□), or 180 (◊) g/d. Bars indicate standard error of mean

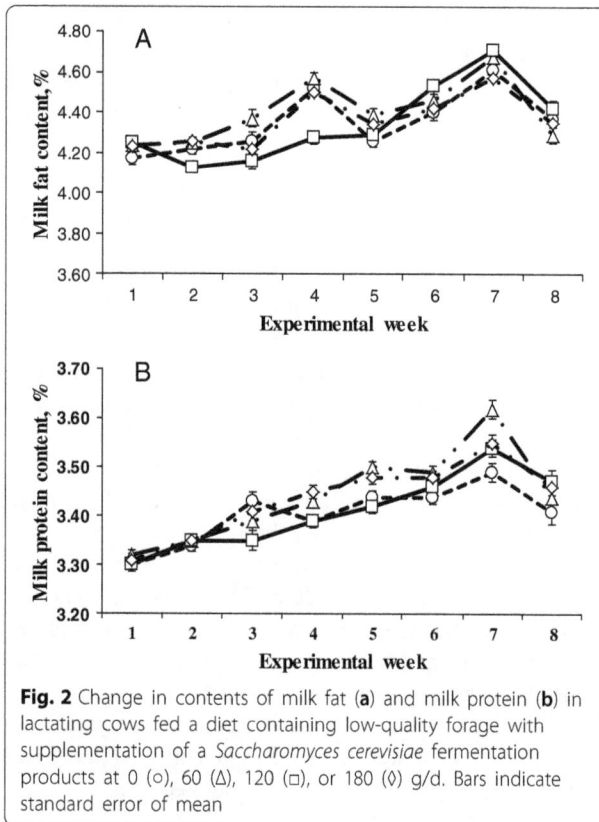

Fig. 2 Change in contents of milk fat (**a**) and milk protein (**b**) in lactating cows fed a diet containing low-quality forage with supplementation of a *Saccharomyces cerevisiae* fermentation products at 0 (o), 60 (Δ), 120 (□), or 180 (◊) g/d. Bars indicate standard error of mean

were supplemented to SCFP for an extended period of time.

Rumen fermentation parameters

Rumen pH and ammonia-N concentration were similar ($P > 0.05$) among the treatments (Table 4). Greater feeding rate of SCFP increased (linear, $P < 0.01$) concentrations of total VFA, acetate, propionate, and butyrate, with no difference ($P > 0.05$) in molar proportion of individual acids and acetate to propionate ratio (Table 4).

Rumen microbial population

The population of fungi increased (quadratic, $P < 0.05$) while protozoa decreased (linear, $P < 0.01$; quadratic, $P < 0.05$) with increasing amount of SCFP (Table 5). In response to SCFP supplementation, cellulolytic bacteria, including *R. flavefaciens* (linear, $P < 0.01$; quadratic, $P = 0.01$) and *F. succinogenes* (linear, $P < 0.01$) increased, but *R. albus* (linear, $P < 0.05$) decreased. *M. elsdenii* (linear, $P < 0.01$; quadratic, $P < 0.01$), *S. ruminantium* (linear, $P < 0.01$; quadratic, $P < 0.01$) and *S. bovis* (linear, $P < 0.01$; quadratic, $P < 0.01$) decreased with increasing amount of SCFP.

Estimated MCP synthesis in the rumen

Estimated urine volume was not influenced by the SCFP supplementation ($P > 0.05$, Table 6). Concentrations of uric acid and endogenous PD were similar ($P > 0.05$) among the treatments. Allantoin (linear, $P = 0.01$) concentration and the sum of urinary PD (linear, $P = 0.05$) increased linearly with the increasing amount of SCFP. Thus, the ruminal MCP estimated from urinary PD increased (linear, $P < 0.01$) with the increasing amount of

not affected ($P > 0.05$), whereas N conversion (milk protein yield/dietary CP intake) increased linearly ($P < 0.05$) with the increasing amount of SCFP. Contents of milk components were similar ($P > 0.05$) among the treatments. Treatment × week interactions for milk yield (Fig. 1) and contents of milk fat and milk protein (Fig. 2) with an enhanced SCFP effect being found when cows

Table 4 Effects of *Saccharomyces cerevisiae* fermentation product on rumen fermentation characteristics of dairy cows fed corn stover-containing diet

Item[a]	SCFP[b], g/d				SEM	P-value	
	0	60	120	180		Linear	Quadratic
pH	6.31	6.32	6.41	6.30	0.06	0.82	0.32
Ammonia-N, mg/dL	17.2	17.0	16.9	17.1	0.24	0.52	0.39
Total VFA[a], mmol/L	93.6	109.3	110.4	114.9	4.56	<0.01	0.23
Acetate, mmol/L	63.3	73.6	73.9	81.1	2.80	<0.01	0.59
Propionate, mmol/L	17.4	20.6	20.7	22.5	0.71	<0.01	0.35
Butyrate, mmol/L	12.8	15.7	15.7	16.5	0.52	<0.01	0.05
Molar proportion, mmol/L/100 mmol/L							
Acetate (Ac)	67.8	67.5	66.9	67.1	0.53	0.30	0.60
Propionate (Pr)	18.6	18.8	18.9	19.0	0.33	0.36	0.84
Butyrate	13.7	13.7	14.2	13.9	0.24	0.32	0.42
Ac:Pr	3.66	3.64	3.55	3.53	0.070	0.22	0.86

[a]Total VFA = Total volatile fatty acids, which is the sum of acetate, propionate and butyrate
[b]*Saccharomyces cerevisiae* fermentation product (Diamond V Original XP, Cedar Rapids, Iowa, USA)

Table 5 Effects of *Saccharomyces cerevisiae* fermentation product on rumen microbial populations of dairy cows fed corn stover-containing diet

Item[a]	SCFP[b], g/d				SEM	P-value	
	0	60	120	180		Linear	Quadratic
Fungi, 10^{-3}	1.79	1.89	2.04	1.88	0.063	0.12	0.05
Protozoa, 10^{-1}	1.59	1.24	1.30	1.14	0.044	<0.01	0.03
Ruminococcus albus, 10^{-3}	5.81	5.37	4.86	4.99	0.199	<0.01	0.16
Ruminococus flavefaciens, 10^{-2}	1.33	1.44	2.69	2.36	0.75	<0.01	0.01
Fibrobacter succinogenes, 10^{-2}	1.47	1.73	3.04	3.24	0.078	<0.01	0.74
Selenomonas ruminantium, 10^{-7}	4.23	3.32	3.06	3.03	0.139	<0.01	0.003
Megasphaera elsdenii, 10^{-5}	17.6	12.7	7.27	12.0	0.52	<0.01	<0.01
Streptococcus bovis, 10^{-6}	3.84	3.34	2.07	2.92	0.98	<0.01	<0.01

[a] Rumen microbial population was expressed as the ratio to total bacterial 16S rDNA
[b]*Saccharomyces cerevisiae* fermentation product (Diamond V Original XP, Cedar Rapids, Iowa, USA)

SCFP, with 12.8% or 166 g/d higher MCP in the cows fed 120 g/d SCFP than that in the control.

Discussion

There was a linear trend of increasing milk yield in response to SCFP supplementation in the present study, in agreement with the results of the meta-analysis by Poppy et al. [4], where an overall positive effect of SCFP on milk yield was reported. One noticeable observation is that the positive effect of SCFP on milk production was more apparent during the second half (wk 5 to 8) of the study (Fig. 1). It may suggest that cows need to adapt to the supplement for several weeks before production response is detectable. Another possible explanation is the changes in weather pattern during the trial period. Weather became hot from the sixth week of the study and the average daily mean thermal-humidity index [35] started moving upward (Fig. 3) while milk production

Table 6 Effects of *Saccharomyces cerevisiae* fermentation product on rumen microbial protein (MCP) supply of dairy cows fed corn stover-containing diet

Item[a]	SCFP[b], g/d				SEM	P-value	
	0	60	120	180		Linear	Quadratic
Urine volume[a], L/d	23.0	22.9	22.1	22.3	0.96	0.54	0.90
Urinary PD[b], mmol/d							
Allantoin	304	280	351	330	12.6	0.01	0.90
Uric acid	27.0	28.0	25.3	27.3	1.15	0.74	0.68
Endogenous PD	50.6	45.2	47.7	44.7	3.97	0.63	0.84
Sum	330	307	373	346	12.7	0.05	0.89
MCP, g/d	1,293	1,120	1,459	1,418	47.6	<0.01	0.59

[a]Urine volume (L/d) = body weight (kg) × 29 (mg/d)/creatinine (mg/L) [25]; Endogenous purine derivatives (PD) = 0.385 × $BW^{0.75}$; Sum = allantoin + uric acid – endogenous PD; Microbial protein (MCP), indirectly calculated based on PD [22]
[b]*Saccharomyces cerevisiae* fermentation product (Diamond V Original XP, Cedar Rapids, Iowa, USA)

decreased sharply during that time (Fig. 1). The heat wave may cause additional stress to dairy cows particularly when they were consuming diets containing low quality forages, which contributed to the decreased milk production. The lactation curve for the last four weeks showed that milk production in cows fed 120 and 180 g/d SCFP declined slower than that of the control. This suggests that increasing levels of SCFP might be more effective in maintaining milk persistency of mid-lactation cows under hot environment. Improved feed efficiency has been reported consistently when mid-lactation dairy cows were supplemented with SCFP during summer months [11, 36]. In our study, feed efficiency of SCFP supplemented cows was more apparent as the length of the supplementation was extended.

Improving rumen microbial community, especially those that are involved in fiber digestion would be the key components for maintaining productivity of lactating dairy cows when fed low quality forages. The increased total VFA concentrations with SCFP supplementation indicated stimulated rumen microbial fermentation activities, and help support the milk production and persistency as was observed in the present study. Increased VFA production could be attributed to an increased rumen fungi and fiber-digesting bacteria population. In vivo and in vitro studies have documented positive effects of SCFP on rumen fermentation [9, 37]. Increases in rumen propionate have also been observed in previous studies when SCFP was fed [5, 38]. Higher propionic acid concentrations observed with SCFP in the current study would lead to an increased glucogenic potential of the diet and milk production.

It is reported that variable response in VFA production and pattern with yeast culture supplementation is a consequence of yeast culture's effect on the growth of different species of rumen microbes [39]. Rumen microbial populations detected in the current

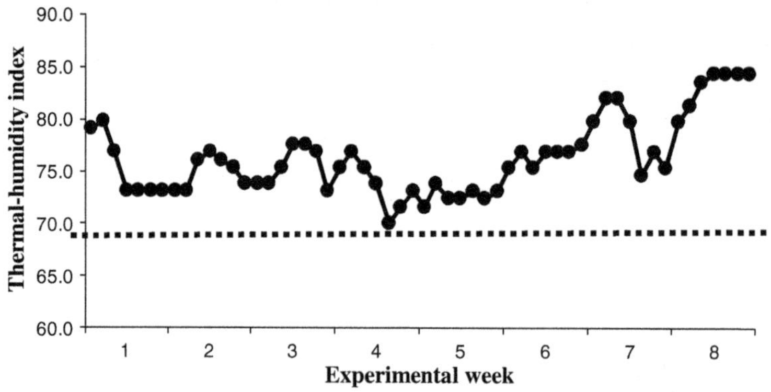

Fig. 3 Daily mean thermal-humidity index (THI) during the trial period. Dashed line represents THI = 68, when cows are expected to suffer from heat stress [35]

study were altered in response to SCFP supplementation. Supplementation with SCFP could provide various growth factors, pro-vitamins, and/or micro-nutrients that help stimulate the growth of ruminal bacteria [40]. Vitamins such as biotin and thiamine that could be provided by SCFP supplementation are reported to be required for fungal growth and activity [41]. In the current study, supplementation of SCFP increased the rumen fungi population, and similar results were reported in a previous in vitro study [9]. In the present study, SCFP supplementation stimulated the growth of cellulolytic bacterial population (*R. flavefaciens* and *F. succinogenes*). Such effects have also been reported in earlier studies [8, 37]. The reduction of lactate-producing bacteria (*S. bovis*) in the present study could contribute to the stability of rumen fermentation. The stabilized rumen condition allows the increased growth and activity of fiber-digesting bacteria, resulting in increased ruminal VFA concentrations [5]. The decrease in number of protozoa in response to SCFP supplementation may decrease the bacterial preying and allows more microbial protein to reach the small intestine. It is reported that steers supplemented with SCFP decreased the proportion of Entodinium from 87.7 to 69.6% [42]. Entodinium is a rumen protozoa responsible for engulfing bacteria and reducing microbial protein supply to the small intestine [43]. Microbial protein is a high quality protein source for dairy cows, and was increased in response to SCFP supplementation in our study. Such effects have been reported in some other studies [7, 9].

Overall, SCFP supplementation manipulated rumen microbial population that resulted in improved energy supply (enhanced VFA production) and improved protein nutrition (greater microbial protein synthesis and more efficient conversion of dietary N to milk N) of lactating cows consuming diets containing low quality forages.

Conclusion

Supplementation of SCFP shifted rumen microbial population to a greater energetic and nitrogen efficiency of dairy cows consuming diets containing low quality forages. These changes with SCFP supplementation support maintaining better milk persistency of mid-lactation cows particularly when cows were under hot environment. The effects of SCFP were dose-dependent and greater effects being observed with higher levels.

Abbreviations
ADF: Acid detergent fiber; BW: Body weight; CP: Crude protein; DIM: Days in milk; DM: Dry matter; MCP: Microbial crude protein; ME: Metabolizable energy; N: Nitrogen; NDF: Neutral detergent fiber; PD: Purine derivatives; SCFP: *Saccharomyces cerevisiae* fermentation product; TMR: Total mixed ration; VFA: Volatile fatty acids

Acknowledgments
The authors gratefully thank all of the staff of the Hangjiang Dairy Farm (Hangzhou, China) for their assistance in milking and care of the animals. We also acknowledge the members of the Institute of Dairy Science, Zhejiang University (Hangzhou, China) for their assistance in the sampling and analysis of the feeds and urine.

Funding
This study was supported by funds from Diamond V (Cedar Rapids, IA) and from the China Agriculture (Dairy Cow) Research System (CARS-37).

Authors' contributions
WZ carried out the study design, data interpretation and manuscript writing and editing; JKW and JXL were involved in the study design, data interpretation and manuscript editing; ZHW, FY, and NNX were involved in the animal experiment; IY and YHC were involved in the study design and manuscript editing. All authors read and approved the final manuscript.

Author details
[1]Institute of Dairy Science, College of Animal Sciences, Zhejiang University, 866 Yuhangtang Road, Hangzhou 310058, People's Republic of China.
[2]Diamond V, Cedar Rapids, IA 52405, USA.

References

1. Ministry of Agriculture (MOA), P.R.C. Report on investigation and evaluation of crop straws and stovers resources in China. Agric Engineering Technol. New Energy Industry. 2011;2:2–5.

2. Kebede G. Effect of urea-treatment and leucaena (Leucaena leucocephala) addition on the utilization of wheat straw as feed for sheep. Haramaya Ethiopia: MSc.Thesis. Haramaya Univ; 2006.

3. Eastridge ML. Major advances in dairy cattle nutrition. J Dairy Sci. 2006;89:1311–23.

4. Poppy GD, Rabiee AR, Lean IJ, Sanchez WK, Dorton KL, Morley PS. A meta-analysis of the effects of feeding yeast culture produced by anaerobic fermentation of Saccharomyces cerevisiae on milk production of lactating dairy cows. J Dairy Sci. 2012;95:6027–41.

5. Harrison GA, Hemken RW, Dawson KA, Harmon RJ, Barker KB. Influence of addition of yeast culture supplement to diets of lactating cows on ruminal fermentation and microbial populations. J Dairy Sci. 1988;71:2967–75.

6. Yoon I, Garrett JE. Yeast culture and processing effects on 24-h in situ ruminal degradation of corn silage. Proceeding of World Animal Production Conference. 1998; 1. p. 322-3.

7. Hristov AN, Varga G, Cassidy T, Long M, Heyler K, Kaenati SKR, et al. Effect of Saccharomyces cerevisiae fermentation product on ruminal fermentation and nutrient utilization in dairy cows. J Dairy Sci. 2010;93:682–92.

8. Callaway ES, Martin SA. Effect of Saccharomyces cerevisiae culture on ruminal bacteria that utilize lactate and digest cellulose. J Dairy Sci. 1997;80: 2035–44.

9. Mao HL, Mao HL, Wang JK, Liu JX, Yoon I. Effects of Saccharomyces cerevisiae fermentation product on in vitro fermentation and microbial communities of low-quality forages and mixed diets. J Anim Sci. 2013;91:3291–8.

10. Ministry of Agriculture (MOA), P.R.C. Feeding Standard of Dairy Cattle (NY/T 34 - 2004). Beijing; 2004.

11. Zhu W, Zhang BX, Yao KY, Yoon I, Chung YH, Wang JK, et al. Effects of supplemental levels of Saccharomyces cerevisiae fermentation product on lactation performance in dairy cows under heat stress. Asian-Aust J Animal Sci. 2015;29:801–6.

12. National Oceanic and Atmospheric Administration (NOAA). Livestock Hot Weather Stress. Operations Manual Letter C-31–76. Kansas: Department of Commerce, NOAA, National Weather Service Central Region; 1976.

13. Zhu W, Fu Y, Wang B, Wang C, Ye JA, Wu YM, et al. Effects of dietary forage sources on rumen microbial protein synthesis and milk performance in early-lactating dairy cows. J Dairy Sci. 2013;96:1727–34.

14. AOAC. Official Methods of Analysis. 17th ed. Arlington: Association of Official Analytical Chemists; 2012.

15. Van Soest PJ, Robertson JB, Lewis BA. Methods of dietary fiber, neutral detergent fiber, and nonstarch polysaccharides in relation to animal nutrition. J Dairy Sci. 1991;74:3583–97.

16. Laporte MF, Paquin P. Near-infrared analysis of fat, protein, and casein in cow's milk. J Agric Food Chem. 1999;47:2600–5.

17. Shen JS, Chai Z, Song LJ, Liu JX, Wu YM. Insertion depth of oral stomach tubes may affect the fermentation parameters of ruminal fluid collected in dairy cows. J Dairy Sci. 2012;95:5978–84.

18. Chaney AL, Marbach EP. Modified reagents for determination of urea and ammonia. Clin Chem. 1962;8:130–2.

19. Hu WL, Liu JX, Ye JA, Wu YM, Guo YQ. Effect of tea saponin on rumen fermentation in vitro. Anim Feed Sci Technol. 2005;120:333–9.

20. Modern WJQ, Science DP. China Agricultural Press. China: Beijing; 2006 (In Chinese).

21. Edmonson AJ, Lean IJ, Weaver LD, Farver T, Webster G. A body condition scoring chart for Holstein dairy cows. J Dairy Sci. 1989;72:68–78.

22. Chen XB, Gomes MJ. Estimation of microbial protein supply to sheep and cattle based on urinary excretion of purine derivatives: An overview of technical details. Aberdeen, UK: Int'l. Feed Res. Unit, Occasional Publ. Rowett Research Institute; 1992.

23. Oser BL. Hawk's Physiological Chemistry. 14th ed. New York: McGraw-Hill; 1965.

24. Leonardi C, Stevenson M, Armentano LE. Effect of two levels of crude protein and methionine addition on performance of dairy cows. J Dairy Sci. 2003;86:4033–42.

25. Valadares RFD, Broderick GA, Valadares Filho SC, Clayton MK. Effect of replacing alfalfa silage with high moisture corn on ruminal protein synthesis estimated from excretion of total purine derivatives. J Dairy Sci. 1999;82:2686–96.

26. Yu Z, Morrison M. Improved extraction of PCR-quality community DNA from digesta and fecal samples. Biotechnique. 2004;36:808–12.

27. Denman SE, McSweeney CS. Development of a real-time PCR assay for monitoring anaerobic fungal and cellulolytic bacterial populations within the rumen. FEMS Microbiol Ecol. 2006;58:572–82.

28. Sylvester JT, Karnati SKR, YU Z, Morrison M, Firkins LJ. Development of an assay to quantify rumen ciliate protozoal biomass in cows using real-time PCR. J Nutr. 2004;134:3378–84.

29. Koike S, Kobayashi Y. Development and use of competitive PCR assays for the rumen cellulolytic bacteria: Fibrobacter succinogenes, Ruminococcus albus and Ruminococcus flavefaciens. FEMS Microbiol Lett. 2001;204:361–6.

30. Tajima K, Aminov RI, Nagamine T, Matsui H, Nakamura M, Benno Y. Diet-dependent shifts in the bacterial population of the rumen revealed with real-time PCR. Appl Environ Microbiol. 2001;67:2766–74.

31. Ouwerkerk D, Klieve AV, Forster RJ. Enumeration of Megasphaera elsdenii in rumen contents by real-time Taq nuclease assay. J Appl Microbiol. 2002;92:753–8.

32. Klieve AV, Hennessy D, Ouwerkerk D, Forster RJ, Mackie RI, Attwood GT. Establishing populations of Megasphaera elsdenii YE 34 and Butyrivibrio fibrisolvens YE 44 in the rumen of cattle fed high grain diets. J Appl Microbiol. 2003;95:621–30.

33. Livak KJ, Schmittgen TD. Analysis of relative gene expression data using real-time quantitative PCR and the 2-ΔΔCT method. Methods. 2001;25:402–8.

34. SAS Institute. SAS User's Guide: Statistics, version 8.01. Cary: SAS Inst. Inc.; 2000.

35. Burgos-Zimbelman R, Collier RJ. Feeding strategies for high-producing dairy cows during periods of elevated heat and humidity. Tri-State Dairy Nutrition Conference. Indiana, USA; 2011. p. 111–126.

36. Schingoethe DJ, Linke KN, Kalscheur KF, Hippen AR, Rennich DR, Yoon I. Feed efficiency of mid-lactation dairy cows fed yeast culture during summer. J Dairy Sci. 2004;87:4178–81.

37. Yoon I, Stern MD. Effects of Saccharomyces cerevisiae and Aspergillus oryzae cultures on ruminal fermentation in dairy cows. J Dairy Sci. 1996;79:411–7.

38. Yoon IK, Garrett JE, Cox DJ. Effect of yeast culture supplementation to alfalfa-grass hay diet on microbial fermentation in continuous culture of rumen contents. J Anim Sci. 1997;75 Suppl 1:98. Abstr.

39. Lascano GJ, Heinrichs AJ. Rumen fermentation pattern of dairy heifers fed restricted amounts of low, medium, and high concentrate diets without and with yeast culture. Livest Sci. 2009;124:48–57.

40. Newbold CJ, Wallace RJ, Chen XB, McIntosh FM. Different strains of Saccharomyces cerevisiae differ in their effects on ruminal bacterial numbers in vitro and in sheep. J Anim Sci. 1995;73:1811–9.

41. Akin DE, Borneman WS. Role of rumen fungi in fiber degradation. J Dairy Sci. 1990;73:3023–32.

42. Arakaki LC, Stahringer RC, Garrett JE, Dehority BA. The effects of feeding monensin and yeast culture, alone or in combination, on the concentration and generic composition of rumen protozoa in steers fed on low-quality pasture supplemented with increasing levels of concentrate. Anim Feed Sci Technol. 2000;84:121–7.

43. Ivan M, Neill L, Foster R, Alimon R, Rode LM, Entz T. Effects of Isotricha, Dasytricha, Entodinium, and total fauna on rumen fermentation and duodenal flow in weathers fed different diets. J Dairy Sci. 2000;83:776–87.

44. Orth R. Sample day and lactation report, DHIA 200. Fact Sheet A-2. Ames: Mid-States Dairy Records Processing Center (DRPC); 1992.

6

Fluoroacetate in plants - a review of its distribution, toxicity to livestock and microbial detoxification

Lex Ee Xiang Leong[1†], Shahjalal Khan[2†], Carl K. Davis[1], Stuart E. Denman[3] and Chris S. McSweeney[3*]

Abstract

Fluoroacetate producing plants grow worldwide and it is believed they produce this toxic compound as a defence mechanism against grazing by herbivores. Ingestion by livestock often results in fatal poisonings, which causes significant economic problems to commercial farmers in many countries such as Australia, Brazil and South Africa. Several approaches have been adopted to protect livestock from the toxicity with limited success including fencing, toxic plant eradication and agents that bind the toxin. Genetically modified bacteria capable of degrading fluoroacetate have been able to protect ruminants from fluoroacetate toxicity under experimental conditions but concerns over the release of these microbes into the environment have prevented the application of this technology. Recently, a native bacterium from an Australian bovine rumen was isolated which can degrade fluoroacetate. This bacterium, strain MFA1, which belongs to the Synergistetes phylum degrades fluoroacetate to fluoride ions and acetate. The discovery and isolation of this bacterium provides a new opportunity to detoxify fluoroacetate in the rumen. This review focuses on fluoroacetate toxicity in ruminant livestock, the mechanism of fluoroacetate toxicity, tolerance of some animals to fluoroaceate, previous attempts to mitigate toxicity, aerobic and anaerobic microbial degradation of fluoroacetate, and future directions to overcome fluoroacetate toxicity.

Keywords: Aerobic, Anaerobic, Degradation, Dehalogenase, Fluoroacetate, 1080, Synergistetes, TCA, Toxicity

Background

Sodium monofluoroacetate (referred to as fluoroacetate hereafter), has the chemical formula $FCH_2COO^-Na^+$, and is a highly toxic compound primarily used as a pesticide known commercially as Compound 1080. Despite having a strong carbon-fluorine bond (one of the strongest bonds in nature), fluoroacetate appears to be rather labile in the environment being readily degraded by different microorganisms [1] or anabolised by higher organisms. This is in contrast to polyfluorinated compounds (such as Teflon) which are very recalcitrant and can persist in the environment for many years [2]. It is well suited as a pesticide because it is virtually tasteless and odourless, which enables it to be easily disguised within bait material targeted towards a specific pest

species [3]. However, due to its non-specific poisoning of other animals and accidental human ingestion, this pesticide is currently used under strict control by governments around the world.

Fluoroacetate was first synthesised in the laboratory in 1896 but it was only first isolated from "gifblaar" (a South African plant) by Marais in 1943 [4]. These plants were believed to naturally produce this toxic compound as a defence mechanism against grazing by herbivores. Ingestion by livestock often results in fatal poisonings, which causes significant economic problems to commercial farmers in many countries such as Australia, Brazil and South Africa [5–8]. In Brazil, 60% of the cattle losses are due to fluoroacetate poisoning from grazing fluoroacetate-producing plants [9]. Fluoroacetate toxicity costs the Australian livestock industry around 45 million dollars (AUD) annually due to the increased death rates and associated productivity impacts [10]. In this paper, we will focus on the natural fluoroacetate found in plants impacting ruminant livestock industries, mechanism of its

* Correspondence: chris.mcsweeney@csiro.au
†Equal contributors
3CSIRO Agriculture and Food, Queensland Bioscience Precinct, St Lucia 4072, QLD, Australia
Full list of author information is available at the end of the article

toxicity, previous attempts to mitigate toxicity, aerobic and anaerobic microbial degradation of fluoroacetate, tolerance of some animals to fluoroaceate, and future directions to overcome fluoroacetate toxicity.

Fluoroacetate in the environment

Fluoroacetate containing plants grow worldwide and cause sudden death in livestock. The southern continents of Africa, Australia and South America are the common locations of these plants. All of the plants containing fluoroacetate belong to the families Fabaceae, Rubiaceae, Bignoniaceae, Malpighiaceae and Dichapetalaceae [11].

Fluoroacetate is found in these tropical and subtropical plants generally at low concentrations although some are able to accumulate fluoroacetate in high concentrations [12]. These plants grow on a variety of soil types, including acidic, heavier soils or sandy loams but rarely in deep sandy soil [7]. In Africa, most fluoroacetate-accumulating plants belong to the genus *Dichapetalum*. The seeds of *D. braunii* can contain levels of fluoroacetate up to 8000 mg/kg, which is the highest ever recorded [13]. Fluoroacetate is also present in plants from South America, particularly *Palicourea marcgravii*, which can contain levels up to 500 mg/kg [14]. Other South American plants that are known to contain fluoroacetate are from the *Amorimia* genus, which has lower concentration of fluoroacetate than *P. marcgravii* [15]. Although plants from South America may not contain high concentration of fluoroacetate, they are still

responsible for many livestock deaths due to the high toxicity of fluoroacetate.

In Australia, about 40 species of plants can generate fluoroacetate and most of them belong to the genus *Gastrolobium* [16]. Later these plants were classified as three genera *Gastrolobium*, *Oxylobium* and *Acacia*. After reclassification, many of the "nontoxic" *Gastrolobium* spp. haven been transferred to the genus *Nemcia* and the "toxic" *Oxylobium* spp. have all been placed in *Gastrolobium* [17, 18]. These fluoroacetate-containing plants are widely distributed in Australia (Fig. 1). Heart-leaf bush, *Gastrolobium grandiforum*, can contain as much as 2600 mg/kg fluoroacetate, while the 50% lethal dose (LD$_{50}$) of fluoroacetate is only 0.4 mg/kg of cattle body weight [12]. Although it contains less fluoroacetate than some other species, they are responsible for most of the livestock deaths in Australia because of their high abundance in cattle-producing regions [19].

In South America, especially in Brazil, around 500,000 cattle die every year by poisonous plants which cause sudden death [20]. *Palicourea marcgravii* and *Amorimia rigida* are the two most common toxic plants in Brazil [21]. Fluroacetate was found to be the principle toxin in these two plants [22]. In South Africa, *Dichapetalum cymosum* is the third most important poisonous plant causing livestock deaths particularly during spring and episodes of drought [23]. The biosynthesis pathway of fluoroacetate by these plants is still largely unknown. This is the result of the inability to produce stable fluoroacetate-

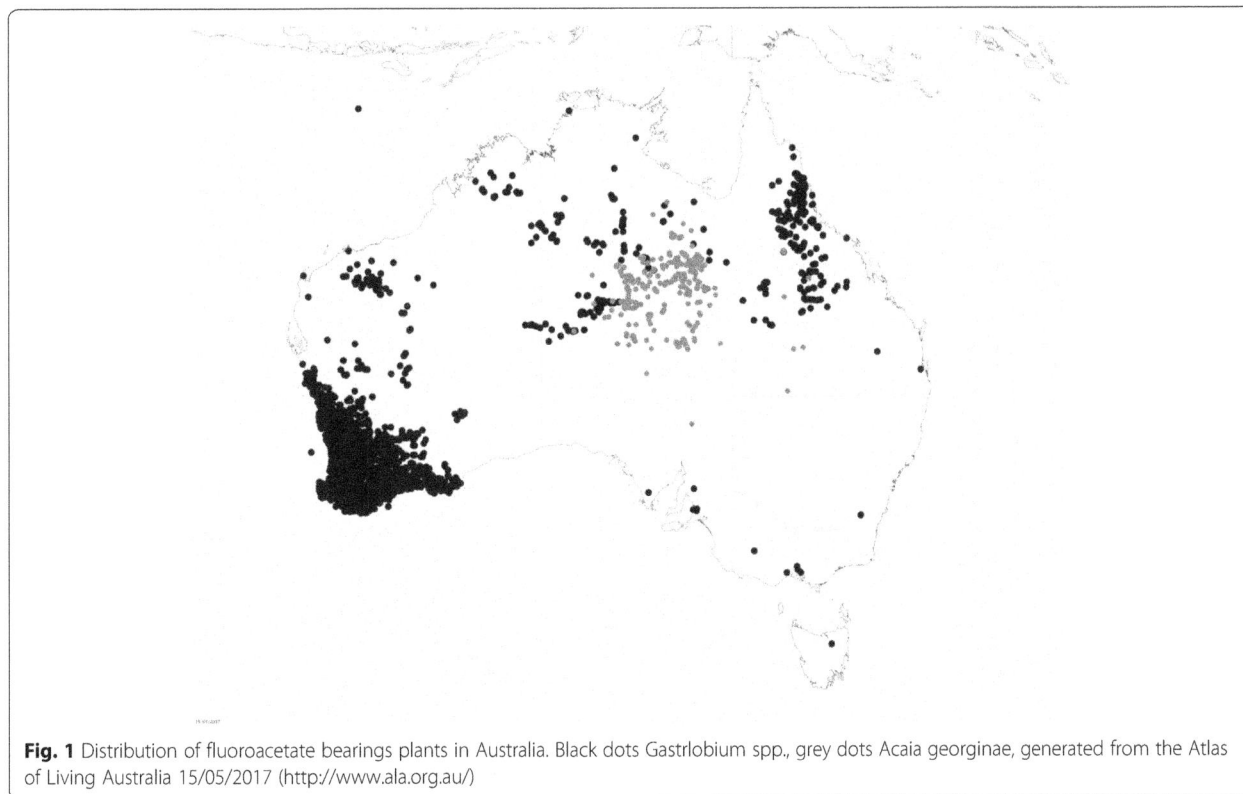

Fig. 1 Distribution of fluoroacetate bearings plants in Australia. Black dots Gastrlobium spp., grey dots Acaia georginae, generated from the Atlas of Living Australia 15/05/2017 (http://www.ala.org.au/)

degrading plant cell lines [24, 25]. Although a cell-free extract of *Dicepatalum cymosum* is able to convert fluoropyruvate to fluoroacetate, researchers could not identify the mechanism and enzymes required [26]. Analysis of soils in which some fluoroacetate-accumulating plants are found show that biosynthesis of fluoroacetate occurs even when total soil inorganic fluoride is very low [14]. Fluoroacetate biosynthesis seems to be relatively widespread, however some plants clearly have evolved to accumulate high concentrations, giving them a selective advantage from predation by animals.

This review will focus mainly on toxicity of fluoroacetate but some plants also contain fluorocitrate, fluoroacetone and fluorofatty acid compounds. Fluorinated natural products, for example, the seeds of *Dichapetalum toxicarium*, an indigenous shrub of West Africa, cause death of animals after ingestion and the symptoms are similar to fluoroacetate poisoning [27]. The seeds of *D. toxicarium* contain up to 1800 µg/g organic fluorine and the main fluorinated component was ω-fluorooleic acid (C18:1 F) [28]. Additional fluorofatty acids including o ~ –fluoro-palmitoleic, -stearic, -linoleic, -arachidic and -eicosenoic acids and 18-fluoro-9,10-epoxystearic acid have since been identified [29].

Some bacteria have been identified that can produce fluoroacetate in the environment. For example the soil bacterium *S. cattleya*, possess fluorinase (fluorination enzyme) which catalyses a nucleophilic substitution reaction between fluoride ion and S-adenosyl-L-methionine to produce 5′-fluorodeoxyadenosine (FDA). FDA is then processed to fluoroacetate and 4 -fluorothreonine (4-FT). By incorporating isotopically labelled glycerol it has been determined that the C5′ fluoromethyl and C4′

carbon of FDA are converted to fluoroacetate and C3 and C4 of 4-FT. It has also been established that both hydrogens of the fluoromethyl group of FDA are reserved in the conversion to the fluoromethyl groups of fluoroacetate and 4-FT [30] (Fig. 2).

Fluoroacetate toxicity mechanism

The tricarboxylic acid (TCA) cycle is central to cellular energy production in the mitochondria of higher organisms and fluoroacetate interrupts the TCA cycle. Fluoroacetate poisoning has been well-documented in animals since its application as a pesticide. Following oral administration and absorption through the gut, fluoroacetate is converted to fluorocitrate by citrate synthase (EC 4.1.3.7) [31] which strongly binds to the aconitase enzyme (EC 4.2.1.3), that converts citrate to succinate in the citric acid cycle [31]. This results in the termination of cellular respiration due to a shortage of aconitase [32, 33], and an increase in concentration of citrate in body tissues including the brain [32]. The build-up of citrate concentration in tissues and blood also causes various metabolic disturbances, such as acidosis which interferes with glucose metabolism through inhibition of phosphofructokinase, and citric acid also binds to serum calcium resulting in hypocalcaemia and heart failure [32, 34–37] (Fig. 3).

Despite a common mechanism of poisoning in all vertebrates, there are differences in the signs and symptoms of fluoroacetate toxicity. In general, carnivores (dogs) show primarily central nervous system (CNS) signs including convulsions and running movements with death due to respiratory failure. Herbivores (rabbit, goat, sheep, cattle, horse) show mostly cardiac effects with ventricular

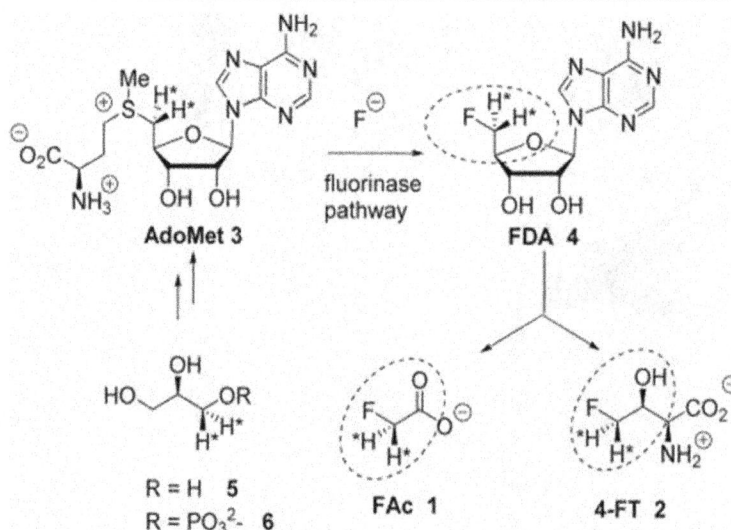

Fig. 2 Production of 5′-fluorodeoxyyadenosine (FDA) from S-adenosyl-L-l-methionine (Adomet) by Fluorinase reaction (3–4). Formation of Fluoroaceate (FAc) and 4-fluorothreonine (4-FT) from (4 to 1–2). Incorporation of Isotope labelled Glycerol (5 and 8 to 3)

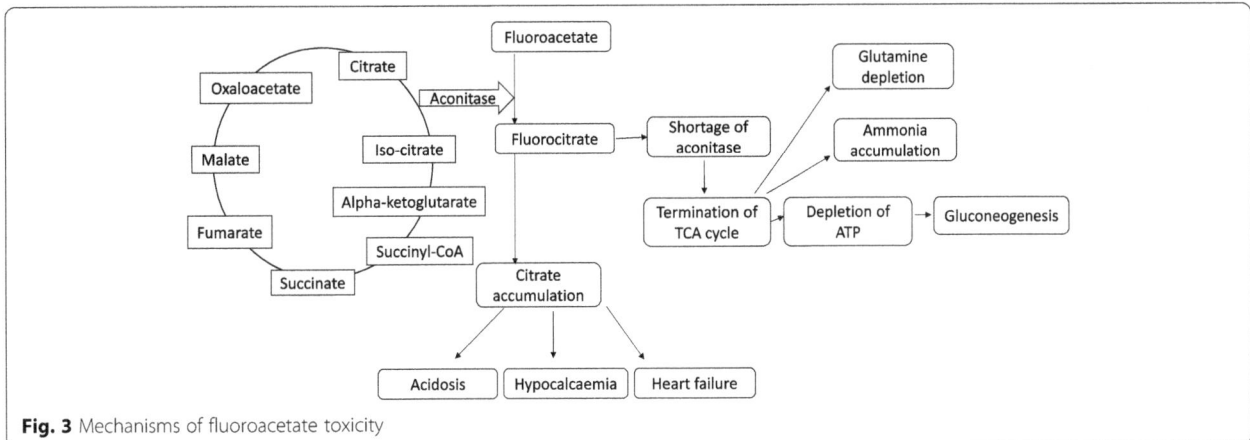

Fig. 3 Mechanisms of fluoroacetate toxicity

fibrillation and little or no CNS signs. The clinical symptoms of omnivores similarly consist of both cardiac and respiratory failure and central nervous system depression [38].

In the pig (omnivores), the clinical symptoms consist of ventricular fibrillation, tremors, violent myotonic convulsions, and respiratory depression [39]. Moreover, the onset of these symptoms can vary between animals of the same species [3]. The symptoms of fluoroacetate poisoning in cattle consist of urinary incontinence, loss of balance, muscle spasms, and in-place running lasting 3 to 20 min or convulsion followed by death of the animal [40]. In Robison's [40] report, symptoms were undetected for up to 29 h following ingestion of fluoroacetate and occurred just before death, hence the term "sudden death" described by some researchers [5]. The clinical symptoms of fluoroacetate poisoning in sheep are relatively similar to cattle, including abnormal posturing, urinary incontinence, muscle spasms and convulsions. They are also known to have severe respiratory distress and extremely rapid heart rate [39, 41].

Diagnosis is generally made on the basis of verified exposure, clinical signs, necropsy findings and chemical analysis. Samples for analysis are, vomitus, liver, stomach or rumen contents and kidney. Increased citric acid levels in kidney and serum is an indicator of fluoroacetate poisoning when correlated with clinical history. Differential diagnosis can be made amongst compounds such as strychnine, chlorinated hydrocarbons, plant alkaloids and lead. A number of other non-specific biochemical changes are suggestive including hyperglycaemia, hypocalcaemia, hypokalaemia and metabolic acidosis [10].

Fluoroacetate tolerance

Many species of animal possess an innate tolerance to fluoroacetate even when there is no evidence of evolutionary exposure. Dogs and other carnivores and rodents and many wildlife species are highly susceptible. Mammalian herbivores have intermediate sensitivity. Reptiles and amphibians are the most tolerant within the animal kingdom. Fish are generally more resistant. This tolerance is likely due to the reduced metabolic rate of these animals. It has been demonstrated that a lower metabolic rate results in less fluoroacetate being converted to fluorocitrate thus allowing more time for excretion and detoxification [42]. The skink (*Tiliqua rugosa*) has a metabolic rate about 10 fold less than a rat of similar size, but has approximately 100 fold greater tolerance to fluoroacetate [43]. Mammals with lower metabolic rate such as the bandicoot also possess a greater tolerance to fluoroacetate [44].

Interestingly, some Australian animals that live in areas where there are fluoroacetate accumulating plants have acquired a remarkable tolerance to fluoroacetate [45, 46]. The degree of tolerance is most apparent in herbivores, especially seed eating birds, which are most likely to have more direct exposure to the toxin compared to carnivorous animals [47]. Other factors which influence the degree of tolerance within a species or population may include the length of time exposed to toxic vegetation, the broadness of both diet and habitat, the size of the resident habitat and the degree of mobility. The emu, which is Australia's oldest seed eating bird, can be up to 150-times more tolerant than the same species of emu outside of areas with fluoroacetate-accumulating plants [48]. This phenomenon has also been observed in other animals such as the possum [42]. Tolerance to fluoroacetate is also demonstrated in insects. Some insects not only utilise the vegetation in their diet, but some actually store the toxin, probably in vacuoles, and use it as defence against predation [49].

The biochemical nature of acquired tolerance to fluoroacetate in animals is not fully understood. It is proposed that there are four obvious biochemical factors that may affect the metabolism of fluoroacetate: (1) the rate of conversion of fluoroacetate to fluorocitrate; (2) the sensitivity of aconitase to fluorocitrate; (3) the citrate transport system in mitochondria, and; (4) the ability to detoxify fluoroacetate [42, 43]. A study compared two distant populations of possums, one having prior exposure to fluoroacetate vegetation and the other having no prior exposure. No differences were found in the defluorination rate of liver extracts between the two populations [42]. Despite a number of other studies attempting to address the biochemical mechanisms for tolerance and fluoroacetate detoxification, there is still inadequate information available.

The soil bacterium *Streptomyces cattleya* is able to produce both fluoroacetate and fluorothreonine but has pathways that possibly confer resistance to these compounds [50]. A fluoroacetyl-CoA-specific thioesterase (FlK) in *S. cattleya* selectively hydrolyzes fluoroacetyl-CoA over acetyl-CoA and exhibits a 10^6-fold higher catalytic efficiency for fluoroacetyl-CoA compared to acetyl-CoA [51]. The FlK gene is located in the same cluster as the C-F bond-forming fluorinase (flA), raising the probability that FlK-catalyzed hydrolysis of fluoroacetyl-CoA plays a role in fluoroacetate resistance in *S. cattleya* by inhibiting the entrance of fluoroacetyl-CoA into the TCA cycle [52].

Degradation of fluoroacetate

Studies to isolate, purify and characterise fluoroacetate-detoxifying enzymes from animals have generally been unsuccessful and contradictory in their findings. Nonetheless, it is generally appreciated from early studies that the vast majority of fluoroacetate is defluorinated within the liver by an enzyme termed fluoroacetate specific defluorinase [53, 54]. This enzyme has been purified from mouse liver cytosol but it is distinct from multiple cationic and anionic glutathione S-transferase isozymes [55]. However, there has been no definitive classification of the enzyme [56]. The enzyme appears to act via a glutathione-dependent mechanism [57]. The focus of the most recent studies has been to determine the relationship between fluoroacetate specific defluorinase and glutathione S-transferase family enzymes to gain a better understanding of the mechanism of fluoroacetate detoxification.

Mead and co-workers [58] characterized a glutathione-dependent dehalogenation pathway in the liver of possum utilizing fluoroacetate as substrate. In the urine of fluoroacetate-treated animals, they found S-carboxymethyl-cysteine which indicates defluorination was catalyzed by an enzyme of the glutathione S-transferase group.

Microbial aerobic degradation

Contrary to the animal studies on fluoroacetate detoxification, microbial degradation of fluoroacetate has been extensively studied. Moreover, the mechanism for aerobic fluoroacetate degradation is well characterised and documented [59–64]. Microorganisms from the soil have been identified with ability to aerobically degrade fluoroacetate. The bacterial communities involved in fluoroacetate degradation vary significantly depending on the areas studied. In Western Australia, species of *Bacillus*, *Pseudomonas*, *Aspergillus*, *Penicillium* and *Streptomyces* were isolated from soil in a of temperate climate [64], while *Burkholderiaceae*, *Ancylobacter* sp., *Paenibacillus* sp., *Staphylococcus* sp. and *Stenotrophomonas* sp. were isolated from the soil of Brazilian areas where the fluoroacetate-containing plants *Mascagnia rigida* and *Palicourea aenofusca* are found [65].

Microorganisms have also been isolated from bait containing the 1080 poison (fluoroacetate) that is used for vertebrate pest control [66]. Bacteria, particularly *Pseudomonas fluorescens*, were isolated from the 1080 bait when mixed with ground kangaroo meat, while both bacteria and soil fungi such as *Fusorium oxysporum* have been isolated from the bait mixed with oats [66, 67]. The bacteria and soil fungi degraded fluoroacetate in the presence and absence of another carbon source. However in the presence of peptone, degradation was higher.

In Western Australia, several microorganisms were isolated from soil with and without previous exposure to fluoroacetate. These included (*Aspergillus fumigatus*, *Fusarium oxysporum*, *Pseudomonas acidovorans*, *Pseudomonas fluorescens 1*, an unidentified *Pseudomonas* sp., *Penicillium purpurescens* and *Penicillium restriction*. These microbes can degrade fluoroacetate, presumably utilising it as a carbon source when grown in solution (2 to 89%) [67]. Recently, two other fluoroacetate degrading-bacteria were isolated from the Brazilian caprine rumen which had the ability to degrade fluoroacetate under aerobic conditions [68]. The bacteria were closely related to *Pigmentiphaga kullae* and *Ancylobacter polymorphus*. Fluoroacetate was degraded to fluoride ions, but the end products containing the carbon atoms from fluoroacetate were not discussed. Moreover, these bacteria might potentially be facultative anaerobes, and it was speculated that degradation occurred through the aerobic process.

Walker and Lien [59] were first to identify two fluoroacetate-degrading enzymes (initially termed haloacetate halidohydrolase) from *Pseudomonas* species and a fungus *Fusarium solani*. At the same time, a fluoroacetate dehalogenase was isolated from a fluoroacetate-dehalogenating bacterium in industrial wastewater, and tentatively named *Moraxella* sp. strain B [62]. It has now been reclassified as *Delftia acidovorans* strain B. Other soil bacteria which play a role in defluorination of fluoroacetate are *Burkholderia* sp. strain FA1, *P. fluorescens*, *Rhodopseudomonas palustris* CGA009 and different strains of *Pseudomonas* species [61, 66, 69, 70]. The fluoroacetate dehalogenase enzymes identified in some of these bacteria appear to degrade fluoroacetate via a similar mechanism, where an ester is produced as an intermediate which is hydrolyzed by a water molecule to form glycolate (Fig. 4).

[69]. Although all the enzymes have a similar degradation mechanism, the size of these enzymes varies significantly. *Pseudomonas* sp. strain A and *P. fluorescens* enzymes are presumed to be monomers, and have an estimated molecular weight of 42 and 32.5 kDa, respectively. Conversely *Burkholderia* sp. FA1 and *D. acidovorans* strain B are dimers of two identical subunits with an estimated molecular mass of 79 and 67 kDa, respectively [61, 72].

All these enzymes release inorganic fluoride from fluoroacetate, but some also cleave chlorinated and brominated analogues, albeit at slower rates [59, 61, 73]. To date, *D. acidovorans* strain B is the only fluoroacetate-dehalogenating bacterium which harbours two haloacetate dehalogenase enzymes; Fluoroacetate dehalogenase H-1 (*dehH1*) and fluoroacetate dehalogenase H-2 (*dehH2*) which are encoded by two different genes on its 65 kb plasmid pUO1. Fluoroacetate dehalogenase H-1 acts predominately on fluoroacetate, while fluoroacetate dehalogenase H-2 has a broader range of substrate specificity for haloacetate, but not fluoroacetate [73].

Two other fluoroacetate dehalogenase enzymes which were purified and tested for their substrate specificities are fluoroacetate dehalogenases from *Burkholderia* sp. FA1 (Fac-dex) and *R. palustris* CGA009 (RPA1163) [61, 70]. When compared to DelH1 of *D. acidovorans* strain B, the two fluoroacetate dehalogenases were more specific to fluoroacetate than to other halogenated analogues [61, 70].

To date, the mechanism of fluoroacetate degradation by fluoroacetate dehalogenase has been extensively studied in *Burkholderia* sp. strain FA1 and *D. acidovorans* strain B [63, 70, 72, 74–76]. Several catalytic regions were identified by comparing the amino acid sequence with that of a haloalkane dehalogenase from *Xanthobacter autotrophicus* [60], and the specific amino acids have been identified by mutagenic studies [63]. It has been found that the active site of the H-1 enzyme contains a conserved Asp105 and His272.

In the initial steps of the pathway for fluoroacetate degradation to glycolate, the carboxylate group of Asp105 acts as a nucleophile to form an ester

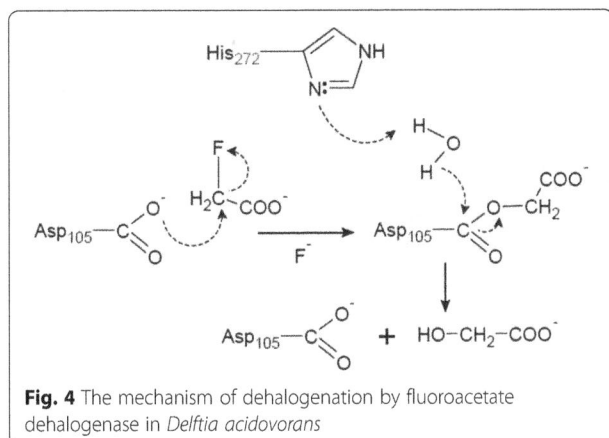

Fig. 4 The mechanism of dehalogenation by fluoroacetate dehalogenase in *Delftia acidovorans*

In spite of their novel mechanisms, limited work has been conducted on these enzymes. The biochemical studies show (Table 1) relatively similar properties between these dehalogenases. All the bacterial enzymes have optimal activities at a slightly alkaline pH around pH 8.0 to 9.0 [59, 61, 69]. However, defluorinating activities in fungi have a wider optimal pH range, with pH 7-8 for *F. solani* compared to pH 5-8 for *F. oxysporum* [59, 67].

The thermal stability of these enzymes differs significantly depending on the species of the microorganisms. Fluoroacetate dehalogenase in *Pseudomonas sp.* from the New Zealand soil was shown to have higher thermal stability, approximately 55 °C, than the fluoroacetate dehalogenase in *F. solani* [59]. However, this notion of high thermal stability was not observed in some *Psuedomonas* species, *P. fluorescens* DSM 8341 was shown to have thermal stabilities to 30 °C [69].

The dehalogenases were shown to use water as the sole co-substrate, and no evidence indicates the involvement of metal ions in their catalytic activity [59, 71]. However, an increase in fluoroacetate degradation activity with addition of low concentration metals ion such as Mg^{2+}, Fe^{2+} and Mn^{2+} has been demonstrated but higher concentration of these metals were inhibitory

Table 1 Physical and biochemical properties of fluoroacetate dehalogenase isolated from different aerobic microorganisms

Microbial source	Number of genes [a]	Gene location	Native enzyme sizes,kDa	Subunit Composition	Optimal pH	Optimal temperature	Reference
Delftia acidovorans strain B	2,*deH1, deH2*	Plasmid	67	Dimer	9.5	50	[60]
Pseudomonas fluorescens DSM 8341	N.D.	N.D.	32.5	monomer	8	30	[69]
Burkholderia sp. FA1	1,*fac-dex*	Chromosome	79	Dimer	9.5	N.D.	[61]
Rhodopseudomonas palustris CGA009	1,RPA1163 [b]	Chromosome	N.D	Dimer	N.D.	N.D.	[70]
Pseudomonas sp. strain A	N.D.	N.D.	42	Monomer	9	50	[62]
Pseudomonas sp.	N.D.	N.D.	62	N.D.	8	N.D.	[59]
Fusarium solani	N.D	N.D.	62	N.D.	7-8	N.D.	[59]

[a] gene names were described in parentheses
[b] gene name identified in the form of locus tag

intermediate around the beta carbon atom of fluoroace-tate to displace the fluorine atom [63, 75]. Then the acetate intermediate is hydrolysed by a deprotonated water molecule formed by a conserved His272. The net result of the reaction is a displacement of a fluoride ion producing glycolate and regeneration of the carboxylate group belonging to Asp105 (Fig. 4).

The catalytic sites of *D. acidovorans* strain B are also conserved as Asp105 and His271 in *Burkholderia* sp. strain FA1 [72]. Moreover, release of fluoride was found to be stabilised by the hydrogen bonds to His149, Trp150 and Tyr212 of *Burkholderia* sp. strain FA1 [75]. This stabilisation effect reduces the activation barrier, where the energy required to cleave the C-F bond was calculated to be only 2.7 kcal/mol, despite the strong C-F bond. A similar structure was also noted in the fluoroacetate dehalogenase from *R. palustris* CGA009 [70].

Due to the fact that the fluoroacetate dehalogenase of *Burkholderia* sp. strain FA1 has a preference for fluoroacetate compared to chloroacetate, the substrate specificity was tested using this enzyme [76]. Using docking stimulations and quantum mechanics/molecular mechanics (QM/MM), Nakayama and colleagues [76] managed to show that fluoroacetate and chloroacetate were incorporated into the active site of fluoroacetate dehalogenase in different conformations. Moreover, the hydrogen bonds of the chloroacetate-enzyme complex do not sufficiently reduce the activation barrier for chloroacetate, which is in a good agreement with the observed high specificity of this enzyme towards fluoroacetate.

Li et al. [77] worked on the structural requirements for defluorination by fluoroacetate degalogenase or FAcD (from bacterium *Rhodopseudomonas palustris* CGA009, PDB code 3R3V) in enabling defluorination rather than dechlorination. They have shown that conformational variations relating to neutrally charged histidine are Hsd155 and Hse155 may cause differences in enzymatic preference. They found that the structure FAcDHse155 is more energetically feasible than the structure FAcDHsd155 for enzyme FAcD, whereas FAcDHse155 prefers defluorination rather than the dechlorination process. Besides the residues Arg111, Arg114, His155, Trp156, and Tyr219, the important role of residues His109, Asp134, Lys181, and His280 during the defluorination process were also emphasized in their experiment. In addition, they found that conformational variations may cause different enzymatic preferences toward competitive pathways.

Microbial anaerobic degradation

Compared with aerobic degradation of fluoroacetate, there is a lack of studies on the isolation of anaerobic microorganisms that have the ability to degrade fluoroacetate.

However recently, a native bacterium from the Australian bovine rumen was isolated using anaerobic agar plates containing fluoroacetate as a carbon source [1]. This bacterium, strain MFA1, which belongs to the Synergistetes phylum has the ability to degrade fluoroacetate, producing fluoride and acetate, as opposed to glycolate from aerobic fluoroacetate-degrading bacteria. Similar observations were noted from other studies on anaerobic degradation of trifluoroacetic acid in anoxic sediments, where acetate was produced from the degradation of this compound [78, 79]. Moreover, similar mechanisms were also noted with anaerobic dechlorinating bacteria. An anaerobic microbial enrichment culture containing *Dehalococcoides ethenogenes* 195 was capable of completely dechlorinating tetrachloroethene to chlorides and ethene [80].

Acetate is not used by strain MFA1 for growth, unlike aerobic fluoroacetate dehalogenating bacteria which utilise the end product, glycolate, as an energy source. Strain MFA1 appears to degrade fluoroacetate via the reductive dehalogenation pathway utilising it as terminal electron acceptor rather than a carbon source. Reductive dehalogenation occurs in anaerobic bacteria, where a halogen substituent is released from a molecule with concurrent addition of electrons to that molecule [81].

There appeared to be a consumption of hydrogen and formate during the growth of strain MFA1 in fluoroacetate [1]. This observation was also noted from reductive dehalogenation of other halogenated compounds in anoxic environment. A net loss of hydrogen was measured from anoxic sediment microcosms dosed with various halogenated compounds [82], and hydrogen was consumed by a *Dehalococcoides ethenogenes* strain 195 with degradation of tetrachloroethene and vinyl chlorides to ethene [83]. However, there is not yet any enzyme identified in strain MFA1 responsible for the degradation of fluoroacetate.

Biotechnological-derived methods for fluoroacetate detoxification in cattle

There have been several attempts to reduce the toxic effects of fluoroacetate in ruminant livestock production. A biotechnological approach to the problem did provide some evidence that detoxifying fluoroacetate by microbial metabolism was possible in the rumen [84]. Gregg and colleagues [84] transformed the rumen bacterium *Butyrivibrio fibrisolvens* with the fluoroacetate dehalogenase gene (DelH1) from *Delftia acidovorans* strain B, and the recombinant bacteria demonstrated active dehalogenation of fluoroacetate in vitro.

The fluoroacetate dehalogenase H1 gene from *D. acidovorans* strain B was incorporated into the plasmid pBHf for transfection into *Butyrivibrio fibrisolvens* [84]. The transfection was relatively stable, with the pBHf plasmid remaining detectable after 500 generations under non-selective conditions. Gregg and colleagues [84] also performed an in vitro study, where a growing population of the recombinant bacterium was able to release fluorine from fluoroacetate at the rate of 9.9 nmol/min/mg [84]. However, dehalogenase activity was not detected outside the bacterial cells, and so it was predicted that fluoroacetate in the media diffused readily into the cells [84]. The genetically modified *B. fibrisolvens* strain expressed dehalogenase enough to detoxify fluoroacetate from the surrounding medium at a rate of 10 nmol/(min·mg) bacterial protein in in vitro testing. The plasmid that carries the dehalogenase gene appears to be very stable and was retained by 100% of the transformed bacteria after 500 generations of growth in non-selective media [84].

In an in vivo study conducted by Gregg and colleagues [85], one group of sheep were inoculated with the recombinant bacteria before being fed fluoroacetate-injected snow-peas, while a control group was not inoculated with the recombinant bacteria. This study showed a significant difference between groups, where the inoculated sheep appeared to be relatively normal despite a 0.4 mg dose of fluoroacetate per kg of animal, while the control sheep died of the fluoroacetate poisoning [85]. The modified bacteria were able to colonise the rumens of two sheep and were shown to persist for an experimental period of 5 months.

In another in vivo study conducted using 20 Angus steers, animals orally inoculated with seven different strains of *Butyrivibrio fibrisolvens* (*B. fibrisolvens* 0/10, 10/1, 85, 149/83, 156, 291, 52/10 strains respectively) containing the plasmid (pBHf)-bearing the fluoroacetate dehalogenase gene DelH1 did not develop the acute symptoms of fluoroacetate toxicity compared to the controls [86]. PCR analysis of rumen fluid collected at 7, 12 and 15 days post-inoculation confirmed the presence of the recombinant bacteria in the rumen at 10^4 to 10^7 cells/ mL. Post-mortem PCR analysis of the rumen fluid from all test animals showed approximately 10^6 colony forming units (CFU) per mL of recombinant *B. fibrisolvens* for several of the strains, 20 days after inoculation [86]. The dose of recombinant bacteria used was able to significantly diminish the effects of fluoroacetate poisoning. Therefore, these in vivo tests showed significant protection of livestock from fluoroacetate using the recombinant bacteria approach. However, in Australia, this technology has not been adopted because approval has not been granted due to strict government regulations regarding release of genetically modified organisms.

In order to prevent animals from unintentional fluoroacetate poisoning, one of the therapies involves the adsorption of fluoroacetate with activated charcoal or other resins. These agents were investigated for their abilities to absorb fluoroacetate from gastrointestinal fluid, thus potentially preventing the conversion of fluoroacetate to fluorocitrate [87]. Moreover, the doses of 2 g/kg of such resins are impractical for preventing fluoroacetate poisoning in livestock. Acetate donor therapy has also been investigated as a treatment for poisoning. Early studies on the effect of fluoroacetate poisoning revealed that fluoroacetate inhibits acetate metabolism in poisoned animals [88]. This led to other studies to investigate whether acetate in the animal at high concentration would provide protection to the animals from fluoroacetate poisoning [89]. This treatment was only effective when provided immediately after the ingestion of the toxin and therefore not practical for treating grazing livestock due to limited surveillance of animals in a rangeland production system. In some cases, animals have died after consumption of fluoroacetate due to the severity of symptoms caused by the depletion of tissue citrate. Therefore, by relieving the symptoms of fluoroacetate poisoning using citrate therapy, researchers have been able to enhance the survival rate of poisoned animals [90]. However, these symptom-reversing therapies would need to be administrated immediately to the poisoned animals to show any effect. Furthermore, some of the poisoned animals in these studies died of other complications even though the major symptoms were suppressed [90].

Rumen microbial transfer

Amorimia pubiflora is one of the main causes of fluoroacetate poisoning in Brazil. In a recent study researchers were able to induce resistance to toxicity by feeding non-toxic doses of this plant to sheep. In addition transferring rumen contents from the resistant animals to naïve sheep was able to confer protection from toxicity [91].

Conclusions

To date, attempts to prevent fluoroacetate toxicity have been unsuccessful except for physically preventing access to toxic plants in the grazing environment. Animal house studies have demonstrated in principle that rumen bacteria engineered to hydrolyse the toxin could prevent toxicity but approvals for the release of these organisms into the environment are unlikely due to current government regulatory restrictions. However the recent discovery of a naturally occurring rumen bacterium (Synergistetes strain MFA1) capable of degrading fluoroacetate may provide a biotechnological solution to the problem of toxicity in rangeland animals. Even though Synergistetes strain MFA1 appears to be ubiquitous throughout the digestive systems of animals such as emus, kangaroos and other cattle, they are present in low numbers which may limit their ability to protect the animal from a lethal dose of the toxin [1]. However it is possible that there are other rumen bacteria able to degrade fluoroacetate which are at higher abundance or could act in concert with other rumen microorganisms to ameliorate the full impact of the toxin. Therefore, further surveys for the presence of other fluoroacetate degrading rumen bacteria and studies on increasing the numbers of these bacteria and expression of the genes responsible for degrading the toxin seems a logical approach for developing a practical strategy to protect livestock from fluoroacetate poisoning. Recent studies demonstrating tolerance to toxicity by adapting the rumen microbiota to non-toxic doses of fluoroacetate further supports a 'rumen detoxification' approach.

Abbreviations
AUD: Australian Dollar; CNS: Central nervous system; MM: Molecular mechanics; QM: Quantum mechanics; TCA: Tricarboxylic acid

Acknowledgements
This work was partly supported by Meat & Livestock Australia (MLA).

Funding
Not applicable.

Authors' contributions
First three authors wrote the review and the remaining authors supervised these students, provided advice and edited the manuscript. All authors read and approved the final manuscript.

Competing interests
Not applicable.

Author details
[1]School of Chemistry and Molecular Bioscience, University of Queensland, St Lucia 4072, QLD, Australia. [2]School of Agriculture and Food Sciences, University of Queensland, St Lucia 4072, QLD, Australia. [3]CSIRO Agriculture and Food, Queensland Bioscience Precinct, St Lucia 4072, QLD, Australia.

References
1. Davis CK, Webb RI, Sly LI, Denman SE, Mcsweeney CS. Isolation and survey of novel fluoroacetate-degrading bacteria belonging to the phylum Synergistetes. FEMS Microbiol Ecol. 2012;80:671–84.
2. Adams DEC, Halden RU. Fluorinated Chemicals and the Impacts of Anthropogenic Use. Contaminants of Emerging Concern in the Environment: Ecological and Human Health Considerations. J Am Chem Soc. 2010;1048:539–60.
3. Minnaar PP, McCrindle RI, Naude TW, Botha CJ. Investigation of biological samples from monofluoroacetate and Dichapetalum cymosum poisoning in southern Africa. Onderstepoort J Vet Res. 2000;67:27–30.
4. Marais JSC. Isolation of the toxic principle "K cymonate" from "Gifblaar" Dichapetalum cymosum. Onderstepoort J Vet Sci Anim Ind. 1943;18:203–6.
5. Medeiros RMT, Geraldo Neto SA. Sudden bovine death from Mascagnia rigida in Northeastern Brazil. Vet Hum Toxicol. 2002;44:286–8.
6. Bell AT, Newton LG, Everist SL, Legg J. Acacia georginae poisoning of cattle and sheep. Aust Vet J. 1955;31:249–57.
7. Aplin TEH. Poison plants of Western Australia. The toxic species of the genera Gastrolobium and Oxylobium. Western Australia: Department of Agriculture; 1969. p. 1–66.
8. Nwude N, Parson LE, Adaudi AO. Acute toxicity of the leaves and extracts of Dichapetalum barteri in mice, rabbits and goats. Toxicology. 1977;7:23–9.
9. Tokarnia CH, Peixoto PV, Dobereiner J. Poisonous plants affecting heart function of cattle in Brazil. Pesqui Vet Bras. 1990;10:1–10.
10. Ian Perkins AP. Thomas Perkins and Nigel Perkins. Impact of fluoroacetate toxicity in grazing cattle. North Sydney: Meat and Livestock Australia Limited; 2015.
11. Lee ST, Cook D, Pfister JA, Allen JG, Colegate SM, Riet-Correa F, et al. Monofluoroacetate-Containing Plants That Are Potentially Toxic to Livestock. J Agric Food Chem. 2014;62:7345–54.
12. Twigg LE, King DR. The impact of fluoroacetate-bearing vegetation on native Australian fauna: a review. Oikos. 1991;61:412–30.
13. O'Hagan D, Perry R, Lock JM, Meyer JJM, Dasardhi L, Hamilton JTG. High levels of monofluoroacetate in Dichapetalum braunii. Phytochemistry. 1993; 33:1043–5.
14. Hall RJ. The distribution of organic fluorine in some toxic tropical plants. New Phytol. 1972;71:855–71.
15. Lee ST, Cook D, Riet-Correa F, Pfister JA, Anderson WR, Lima FG, et al. Detection of monofluoroacetate in Palicourea and Amorimia species. Toxicon. 2012;60:791–6.
16. Mead AJ, King DR, Hubach PH. The co-evolutionary role of fluoroacetate in plant-animal interactions in Australia. Oikos. 1985;44,55–60.
17. Crisp MD, Weston PH. Cladistics and legume systematics, with an analysis of the Bossiaeeae, Brongniartieae and Mirbelieae. Advances in Legumes Systematics, Part 3. (Ed. C.H.Stirton). 1987;65-130.
18. Chandler GT, Bayer RJ, Gilmore SR. Oxylobium/Gastrolobium (Fabaceae: Mirbelieae) condundrum: further studies using molecular data and a reappraisal of morphological characters. Plant Species Biol. 2003;18:91–101.
19. Oelrichs PB, McEwan T. Isolation of the toxic principle in Acacia georginae. Nature. 1961;190:808–9.
20. Camboim EKA, Almeida AP, Tadra-Sfeir MZ, Junior FG, Andrade PP, Mcsweeney CS, et al. Isolation and Identification of Sodium Fluoroacetate Degrading Bacteria from Caprine Rumen in Brazil. Sci World J. 2012;2012:1–6.
21. Fernando H, Edson M, Ricardo AA, et al. Poisonous Plants Affecting Cattle in Central-Western Brazil. Int J Poisonous Plant Res. 2012;2:1-13.
22. Krebs HC, Kemmerling W, Habermehl G. Qualitative and quantitative determination of fluoroacetic acid in Arrabidea bilabiata and Palicourea marcgravii by 19 F-NMR spectroscopy. Toxicon. 1994;32:909–13.
23. Kellerman TS. Poisonous plants. Onderstepoort J Vet Res. 2009;74:1.
24. Grobbelaar N, Meyer JJM. Fluoroacetate Production by Dichapetalum cymosum. J Plant Physiol. 1990;135:550–3.
25. Bennett LW, Miller GW, Yu MH, Lynn RI. Production of fluoroacetate by callus tissue from leaves of Acacia georginae. Fluoride. 1983;16:111–7.
26. Meyer JJM, O'Hagan D. Conversion of 3-fluoropyruvate to fluoroacetate by cell-free extracts of Dichapetalum cymosum. Phytochemistry. 1992;31:2699–701.
27. O'Hagan D, Harper DB. Fluorine-containing natural products. J Fluor Chem. 1999;100(1–2):127–33.
28. Peters RA, Hall PJ, Ward PFV, Sheppard N. The chemical nature of the toxic compounds containing fluorine in the seeds of Dichapetalum toxicarium. Biochem J. 1960;77(1):17–22.
29. Hamilton JTG, Harper DB. Fluoro fatty acids in seed oil of Dichapetalum toxicarium. Phytochemistry. 1997;44(6):1129–32.
30. Bartholomé A, Janso JE, Reilly U, O'Hagan D. Fluorometabolite biosynthesis: isotopicall labelled glycerol incorporations into the antibiotic nucleocidin in Streptomyces calvus. Org Biomol Chem. 2017;15:61–4.
31. Morrison JF, Peters RA. Biochemistry of fluoroacetate poisoning: the effect of fluorocitrate on purified aconitase. Biochem J. 1954;58:473–9.
32. Sherley M. The traditional categories of fluoroacetate poisoning signs and symptoms belie substantial underlying similarities. Toxicol Lett. 2004;151:399–406.

33. Lauble H, Kennedy MC, Emptage MH, Beinert H, Stout CD. The reaction of fluorocitrate with aconitase and structure of the enzyme-inhibition complex. Proc Natl Acad Sci U S A. 1996;93:13699–703.

34. Goh CSS, Hodgson DR, Fearnside SM, Heller J, Malikides N. Sodium monofluoroacetate (Compound 1080) poisoning in dogs. Aust Vet J. 2005;83:474–9.

35. Omara F, Sisodia CS. Evaluation of potential antidotes for sodium fluoroacetate in mice. Vet Hum Toxicol. 1990;32:427–31.

36. Shapira AR, Taitelman U, Bursztein S. Evaluation of the role of ionized calcium in sodium fluoroacetate ("1080") poisoning. Toxicol Appl Pharmacol. 1980;56:216–20.

37. Proudfoot AT, Bradberry SM, Vale JA. Sodium fluoroacetate poisoning. Toxicol Rev. 2006;25:213–9.

38. Calver MC, King DR. Controlling vertebrate pests with fluoroacetate: lessons in wildlife managemen, bio-ethics, and co-evolution. J Biol Educ. 1986;20:257–62.

39. Mcilroy JC. The sensitivity of Australian animals to 1080 poison 1. Intraspecific variation and factors affecting acute toxicity. Aust Wildly Res. 1981;8:369–83.

40. Robison WH. Acute toxicity of sodium monofluoroacetate to cattle. J Wildl Manag. 1970;34:647–8.

41. Schultz RA, Coetzer JA, Kellerman TS, Naude TW. Observations on the clinical, cardiac and histopathological effects of fluoroacetate in sheep. Onderstepoort J Vet Res. 1982;49:237–45.

42. King DR, Oliver AJ, Mead RJ. The adaptation of some Western Australian mammals to food plants containing fluoroacetate. Aust J Zool. 1978;26:699-712.

43. Twigg LE, Mead RJ, King DR. Metabolism of fluoroacetate in the skink (Tiliqua-rugosa) and the rat (Rattus-norvegicus). Aust J Biol Sci. 1986;39:1–15.

44. Macmille RE, Nelson JE. Bioenergetics and body size in dasyurid marsupials. Am J Physiol. 1969;217:1246–51.

45. Oliver AJ, King DR, Mead RJ. Fluoroacetate Tolerance, a Genetic Marker in some Australian Mammals. J Zool. 1979;27:363–72.

46. Twigg LE, Martin GR, Eastman AF, King DR, Kirkpatrick WE. Sensitivity of some Australian animals to sodium fluoroacetate (1080): additional species and populations, and some ecological considerations. Aust J Zool. 2003;51:515–31.

47. Twigg LE, King DR. Tolerance to sodium fluoroacetate in some Australian birds. Aust Wildl Res. 1989;16:49–62.

48. Twigg LE, King DR, Davis HM, Saunders DA, Mead RJ. Tolerance to, and metabolism of, fluoroacetate in the emu. Aust Wildl Res. 1988;15:239–47.

49. Meyer JJM, O'Hagan D. Rare fluorinated natural products. Chem Br. 1992;28:785–8.

50. Murphy CD, Moss SJ, O'Hagan D. Isolation of an Aldehyde Dehydrogenase Involved in the Oxidation of Fluoroacetaldehyde to Fluoroacetate in Streptomyces cattleya. Appl Environ Microbiol. 2001;67:4919–21.

51. Huang F, Haydock SF, Spiteller D, Mironenko T, Li T-L, O'Hagan D, Leadlay PF, Spencer JB. The Gene Cluster for Fluorometabolite Biosynthesis in Streptomyces cattleya: a Thioesterase Confers Resistance to Fluoroacetyl-Coenzyme A. Chem Biol. 2006;13:475–84.

52. Dong C, Huang F, Deng H, Schaffrath C, Spencer JB, O'Hagan D, Naismith JH. Crystal structure and mechanism of a bacterial fluorinating enzyme. Nature. 2004;427:561–5.

53. Kostyniak PJ, Bosmann HB, Smith FA. Defluorination of fluoroacetate in vitro by rat-liver subcellular fractions. Toxicol Appl Pharmacol. 1978;44:89–97.

54. Mead RJ, Moulden DL, Twigg LE. Significance of sulfhydryl compounds in the manifestation of fluoroacetate toxicity to the rat, brush-tailed possum, woylie and western grey-kangaroo. Aust J Bio Sci. 1985;38:139–49.

55. Kostyniak Q, Soiefer AI. Purification of a Fluoroacetate-specific Defluorinase from Mouse Liver Cytosol. J Biol Chem. 1984;259:10787–92.

56. Tu LQ, Wright PFA, Rix CJ, Ahokas JT. Is fluoroacetate-specific defluorinase a glutathione S-transferase? Comp Biochem Phys C. 2006;143:59–66.

57. Tu LQ, Wright PFA, Rix CJ, Ahokas JT. Comparatve study of detoxication enzymes in catalysing defluorination. APS. 2006;27:441.

58. Mead RJ, Oliver AJ, King DR. Metabolism and defluorination of fluoroacetate in the brushtail possum (Trichosurus vulpecula). Aust J Bio Sci. 1979;32:15–26.

59. Walker JRL, Lien BC. Metabolism of fluoroacetate by a soil Pseudomonas sp. and Fusarium solani. SBS. 1981;13:231–5.

60. Liu J, Kurihara T, Ichiyama S, Miyagi M, Tsunasawa S, Kawasaki H, Soda K, Esaki N. Reaction mechanism of fluoroacetate dehalogenase from Moraxella sp. B. J Biol Chem. 1998;273:30897–902.

61. Kurihara T, Yamauchi T, Ichiyama S, Takahata H, Esaki N. Purification, characterization, and gene cloning of a novel fluoroacetate dehalogenase from Burkholderia sp. FA1. J Mol Catal B Enzym. 2003;23:347–55.

62. Kawasaki H, Tone N, Tonomura K. Plasmid-determined dehalogenation of haloacetates in Moraxella species. Agric Biol Chem. 1981;45:29–34.

63. Ichiyama S, Kurihara T, Kogure Y, Tsunasawa S, Kawasaki H, Esaki N. Reactivity of asparagine residue at the active site of the D105N mutant of fluoroacetate dehalogenase from Moraxella sp B. Biochim Biophys Acta. 2004;1698:27–36.

64. Twigg LE, Socha LV. Defluorination of sodium monofluoroacetate by soil microorganisms from central Australia. Soil Biol Biochem. 2001;33:227–34.

65. Camboim EK, Tadra-Sfeir MZ, de Souza EM, Pedrosa Fde O, Andrade PP, Mcsweeney CS, et al. Defluorination of sodium fluoroacetate by bacteria from soil and plants in Brazil. Sci World J. 2012;2012:149893.

66. Wong DH, Kirkpatrick WE, Kinnear JE, King DR. Defluorination of sodium monofluoroacetate (1080) by microorganisms found in bait materials. Wild Life Res. 1991;18:539–45.

67. Wong DH, Kirkpatrick WE, King DR, Kinnear JE. Environmental factors and microbial inoculum size, and their effect on biodefluorination of sodium monofluoroacetate (1080). Soil Biol Biochem. 1992;24:839–43.

68. Camboim EK, Almeida AP, Tadra-Sfeir MZ, Junior FG, Andrade PP, Mcsweeney CS, et al. Isolation and identification of sodium fluoroacetate degrading bacteria from caprine rumen in Brazil. Sci World J. 2012;2012:178254.

69. Donnelly C. SanadaCD. Purification and properties of fluoroacetate dehalogenase from Pseudomonas fluorescens DSM 8341. Biotechnol Lett. 2009;31:245–50.

70. Chan PWY, Yakunin AF, Edwards EA, Pai EF. Mapping the reaction coordinates of enzymatic defluorination. J Am Chem Soc. 2011;133:7461–8.

71. Janssen DB, Pries F, van der Ploeg JR. Genetics and biochemistry of dehalogenating enzymes. Annu Rev Microbiol. 1994;48:163–91.

72. Jitsumori K, Omi R, Kurihara T, Kurata A, Mihara H, Miyahara I, et al. X-ray crystallographic and mutational studies of fluoroacetate dehalogenase from Burkholderia sp. strain FA1. J Bacteriol. 2009;191:2630–7.

73. Kawasaki H, Tsuda K, Matsushita I, Tonomura K. Lack of homology between two haloaceate dehalogenase genes encoded on a plasmid from Moraxella sp. strain B. J Gen Microbiol. 1992;138:1317–23.

74. Zhang Y, Li ZS, Wu JY, Sun M, Zheng QC, Sun CC. Homology modeling and SN2 displacement reaction of fluoroacetate dehalogenase from Burkholderia sp. FA1. Biochem Biophys Res Commun. 2004;325:414–20.

75. Kamachi T, Nakayama T, Shitamichi O, Jitsumori K, Kurihara T, Esaki N, et al. The catalytic mechanism of fluoroacetate dehalogenase: a computational exploration of biological dehalogenation. Chemistry. 2009;15:7394–403.

76. Nakayama T, Kamachi T, Jitsumori K, Omi R, Hirotsu K, Esaki N, et al. Substrate specificity of fluoroacetate dehalogenase: an insight from crystallographic analysis, fluorescence spectroscopy and theoretical computations. Chem Eur J. 2012;18:8392–402.

77. Li Y, Zhang R, Du L, Zhang Q, Wang W. Catalytic mechanism of C-F bond cleavage: insights from QM/MM analysis of fluoroacetate dehalogenase. Cat Sci Tech. 2016;6:73–80.

78. Kim BR, Suidan MT, Wallington TJ, Du X. Biodegradability of trifluoroacetic acid. Environ Eng Sci. 2000;17:337–42.

79. Visscher PT, Culbertson CW, Oremland RS. Degradation of trifluoroacetate in oxic and anoxic sediments. Nature. 1994;369:729–31.

80. Magnuson JK, Stern RV, Gossett JM, Zinder SH, Burris DR. Reductive dehclorination of tetrachloroethene to ethene by a two-component enzyme pathway. Appl Environ Microbiol. 1998;64:1270–5.

81. Mohn WW, Tiedje JM. Microbial reductive dehalogenation. Microbiol Rev. 1992;56:482–507.

82. Mazur CS, Jones WJ, Tebes-Stevens C. H2 consumption during the microbial reductive dehalogenation of chlorinated phenols and tetrachloroethene. Biodegradation. 2003;14:285–95.

83. Maymo-Gatell X, Chien Y, Gossett JM, Zinder SH. Isolation of a bacterium that reductively dechlorinates tetrachloroethene to ethene. Science. 1997;276:1568–71.

84. Gregg K, Cooper CL, Schafer DJ, Sharpe H, Beard CE, Allen B, et al. Detoxification of the plant toxin fluoroacetate by a genetically modified rumen bacterium. Biotechnology. 1994;12:1361–5.

85. Gregg K, Hanmdof B, Henderson K, Kopecny J, Wong C. Genetically modified ruminal bacteria protect sheep from fluoroacetate poisoning. Appl Environ Microbiol. 1998;64:3496–8.

86. Padmanabha J, Gregg K, Ford M, Prideaux C, Mcsweeney CS. Protection of

cattle from fluoroacetate poisoning by genetically modified ruminal bacteria. Proc Aust Soc Anim Prod. 2004;25:293.

87. Norris WR, Temple WA, Eason CT, Wright GR, Ataria J, Wickstrom ML. Sorption of fluoroacetate (compound 1080) by Colestipol, activated charcoal and anion-exchange in resins in vitro and gastrointestinal decontamination in rats. Vet Hum Toxicol. 2000;42:269–75.

88. Bartlett GR, Barron ESG. The effect of fluoroacetate on enzymes and on tissue metabolism. Its use for the study of the oxidative pathway of pyruvate metabolism. J Biol Chem. 1947;170:67–82.

89. Chenoweth MB. Monofluoroacetic acid and related compounds. Pharmacol Rev. 1949;1:383–424.

90. Tourtellotte WW, Coon JM. Treatment of fluoroacetate poisoning in mice and dogs. J Pharmacol Exp Ther. 1951;101:82–91.

91. Beckerl M, Carneiro FM, de Oliveira LP, da Silva MIV, Riet-Correa F, Lee T, et al. Induction and transfer of resistance to poisoning by Amorimia pubiflora in sheep whith non-toxic dosis of the plant and ruminal content. Cienc Rural. 2016;46:674–80.

7

Effects of feeding untreated, pasteurized and acidified waste milk and bunk tank milk on the performance, serum metabolic profiles, immunity, and intestinal development in Holstein calves

Yang Zou[1,2], Yajing Wang[1], Youfei Deng[1], Zhijun Cao[1], Shengli Li[1*] and Jiufeng Wang[3*]

Abstract

Background: The present experiment was performed to assess the effects of different sources of milk on the growth performance, serum metabolism, immunity, and intestinal development of calves. Eighty-four Holstein male neonatal calves were assigned to one of the following four treatment groups: those that received bunk tank milk (BTM), untreated waste milk (UWM), pasteurized waste milk (PWM), and acidified waste milk (AWM) for 21 d.

Results: Calves in the BTM and AWM groups consumed more starter ($P < 0.05$) than those in the UWM group. Average daily gain in the UWM group was the highest ($P < 0.05$). Calves exhibited the highest ($P < 0.05$) serum total protein, albumin, total cholesterol, high density lipoprotein, triglycerides, growth hormone, immunoglobulin (Ig) A and IgM concentrations in the UWM group, highest malondialdehyde and tumor necrosis factor-α in the PWM group ($P < 0.05$), and highest glutathione peroxidase and IgG in the BTM group ($P < 0.05$). The jejunum and ileum of the calves in all treatments presented a slight inflammatory response. The jejunal inflammation scores were higher ($P < 0.05$) in the UWM and AWM groups than the BTM group; the ileal inflammation scores increased more ($P < 0.05$) in the AWM group than the BTM group. Jejunal immunohistochemical scores (IHS) were higher ($P < 0.05$) in the PWM and AWM groups than the BTM group. Compared to the other three groups, calves feeding on BTM had lower ($P < 0.05$) ileal IHS. Jejunal *interleukin(IL)-1β, IL-8*, and *IL-10* mRNA expression in the UWM group was the highest ($P < 0.05$). Calves fed AWM increased ($P < 0.05$) mRNA expression of *IL-8* and *toll like receptor 4 (TLR-4)* in the jejunum and *IL-8, IL-1β, IL-6, IL-8*, and *IL-10* in the mesenteric lymph nodes.

Conclusions: Overall, bunk tank milk is the best choice for calf raising compared to waste milk. The efficiency of feeding pasteurized and acidified waste milk are comparable, and the acidification of waste milk is an acceptable labor-saving and diarrhea-preventing feed for young calves.

Keywords: Acidified waste milk, Calf, Intestinal development, Pasteurized waste milk, Serum metabolism, Waste milk

* Correspondence: lisheng0677@163.com; jiufeng_wang@hotmail.com
[1]State Key Laboratory of Animal Nutrition, Beijing Engineering Technology Research Center of Raw Milk Quality and Safety Control, College of Animal Science and Technology, China Agricultural University, Beijing 100193, China
[3]College of Veterinary Medicine, China Agricultural University, Beijing 100193, China
Full list of author information is available at the end of the article

Background

All non-saleable milk on farms, including colostrum, transitional milk, high somatic cell content milk, and milk from cows treated with veterinary drugs due to diseases, is classified as waste milk (WM) [1]. The use of WM in calf feeding is considered economical but controversial [2, 3]. The major concern in using WM is the excessive amounts of bacteria and antibiotics [1]. The pasteurization of WM has been suggested to minimize the occurrence of pathogens such as *Salmonella* [4–6]. Additionally, different sterilization method of WM may result in varition of growth performance and development of calves. Pasteurized milk feeding could improve weight gain and reduce sickness in calves [2, 3, 7]. For pasteurized milk, a longer holding time, instead of a higher temperature, was more effective in the inactivation of pathogenic organisms [8]. The acidification of milk has also been approved to be a labor-saving, simple and cost-effective method for calf feeding [6] and has been shown to prevent the rapid growth of pathogenic organisms in the digestive tract and reduce the incidence of infectious scours in calves 3 weeks of age and less [9, 10].

To our knowledge, considerable literature has been published concerning the difference between feeding pasteurized milk or acidified milk replacer [11–13], but there is a lack of published data on the difference between feeding pasteurized waste milk and acidified waste milk to calves. Furthermore, whole milk is generally considered to be the best feed for young calves. Therefore, this investigation was performed to elucidate the effects of feeding bunk tank milk (BTM), untreated waste milk (UWM), pasteurized waste milk (PWM) or acidified waste milk (AWM) on the growth performance, serum metabolism, immunity, and intestinal development in calves.

Methods

Animals, treatments and management

Eighty-four Holstein neonatal male calves with similar birth weight (43.6 ± 5.1 kg) were selected from the Modern Farming Feidong Farm (Hefei, Anhui, China). Each calf was fed 4.0 L colostrum immediately after birth and then assigned to one of four treatment groups with three calves/pen fed individually with seven replicates/treatment for each kind of milk. For each treatment, calves received the following different sources of milk from d 2 after birth for 21 d: bunk tank milk (BTM), untreated waste milk (UWM), pasteurized waste milk (PWM), and acidified waste milk (AWM). Bunk tank milk (BTM) was taken from the milking line twice daily. Untreated waste milk (UWM), which was composed of surplus colostrum, transitional milk, and milk from cows treated with veterinary drugs due to

mastitis or other diseases, was collected twice daily into a specific tank. Pasteurized waste milk (PWM) was prepared by pasteurization of untreated waste milk at 72 °C for 15 s. Acidified waste milk (AWM) was acidified by the addition of 30 mL formic acid into 1 L of untreated waste milk, formic acid solution was diluted by 85% formic acid (Fucheng Chemical Reagent Factory, Tianjin, China) with water.

Calves received the milk at equal volumes twice daily via nipple buckets at 0700 h and 1700 h from d 2 in the amount of 3.0 L/meal, from d 8 in the amount of 3.5 L/meal, and from d 15 in the amount of 4.0 L/meal. The same pelleted calf starter was offered from d 4 until d 21. The nipple buckets were cleaned twice daily after milk feeding with a brush using hot tap water and a commercial detergent followed by rinsing with clear water. Calves were housed in pens with three calves as a replicate group on straw bedding and had free access to water during the experimental period. The experimental protocol was conducted in accordance with the practices outlined in the Guide for the Care and Use of Agriculture Animals in Agriculture Research and Teaching [14].

Experimental sampling, measurements and chemical analysis

Milk composition

Milk samples were taken daily before feeding, preserved with potassium dichromate and stored at 4 °C until analyzed for milk fat, protein and solid non-fat (SNF) percentages using a near-infrared reflectance spectroscopy analyzer (Milk Scan 605, Foss Electric, Hillerød, Denmark) at the Beijing Dairy Cattle Center. The pH values were measured using a digital pH meter (PHB-4, Shanghai Hongyi instrument Limited, China). Milk acidity was determined by titration with NaOH, and phenolphthalein was used as the indicator. Total viable count (TVC) was calculated using the colony count on plates in the laboratory of Modern Farming Feidong Farm (Hefei, Anhui, China).

Growth performance measurement and diarrhea incidence

The feed intake of each pen was recorded every morning. The fact BW of all calves in the BTM, UWM, PWM and AWM groups were taken initially and on d 22 before feeding, and average daily feed intake (ADFI) and average daily gain (ADG) were calculated.

Fecal consistency scores for calves were determined daily based on a 1 to 5 system described as follows: 1 = normal, thick in consistency; 2 = normal, soft, less thick; 3 = abnormally thin but not watery; 4 = watery; and 5 = watery with abnormal coloring [15]. A fecal score of 2 and above was considered diarrhea, and gentamycin sulfate injections were administered to calves when fecal scores were 4 or higher.

Serum collection and analysis

Serum samples of all calves were taken on 48 h after born to detect TP concentration and assess passive immunity. On d 22 before morning feeding, 6 calves per treatment (BTM, UWM, PWM, AWM) were randomly chosen, and blood samples were collected via jugular venipuncture into Vacutainer tubes (Becton Dickinson, Franklin Lakes, NJ) before feeding. Samples were centrifuged at $3,000 \times g$ for 15 min, and the supernatants (serum) were collected and frozen at −20 °C until analysis for the determination of total protein (TP), albumin (ALB), high density lipoprotein (HDL), low density lipoprotein (LDL), urea nitrogen (UN), creatinine (Cr), triglycerides (TG), total cholesterol (TC), bilirubin (BIL), glucagon (GC), growth hormone (GH), superoxide dismutase (SOD), malondialdehyde (MDA), glutathione peroxidase (GSH-px), immunoglobulin G (IgG), immunoglobulin A (IgA), immunoglobulin M (IgM), complement 3 (C3), complement 4 (C4), interleukin-1β (IL-1β), interleukin-6 (IL-6), interleukin-7 (IL-7), interleukin-8 (IL-8), interleukin-10 (IL-10), interleukin-22 (IL-22), and tumor necrosis factor-α (TNF-α).

The concentrations of TP, ALB, HDL, LDL, UN, Cr, TG, TC, BIL, SOD, MDA, GSH-px, IgA, IgM, C3, and C4 were determined using corresponding assay kits from Shandong BioBase Biotechnologies Inc. (Shandong, China) and Nanjing Jiancheng Bioengineering Institute (Jiangsu, China) in an automated biochemistry analyzer (Model GF-D200, Rainbow, Shandong, China). Aliquots of plasma serum were used to measure GC, GH and IgG using radioimmunoassay kits from HTA Co. Ltd. (Beijing, China) in an absorbance microplate reader (EXL 800, Bio-tek, Vermont, USA). The serum immune factors (IL-1β, IL-6, IL-7, IL-8, IL-10, IL-22, TNF-α) were analyzed using commercial ELISA kits from Shanghai HoraBio Inc. (Shanghai, China).

Small intestinal tissue sampling

On d 22 after morning feeding, the six calves selected for blood collection were harvested by exsanguination. The intestinal tracts of these calves were excised and divided into the following three segments: the duodenum, jejunum and ileum. The jejunal and ileal pH values were immediately measured in the mid-segment using a digital pH meter (PHB-4, Shanghai Hongyi instrument Limited, China). For gut histopathological and immuno-histochemical analysis, approximately 2-cm lengths of the mid-jejunum and ileum were removed and flushed with ice-cold buffered PBS at pH 7.4 and immediately placed in a 4% formalin solution [16]. For mRNA analysis, jejunal mucosa and mesenteric lymph nodes were rinsed in saline and transferred to plastic vials, snap-frozen in liquid N, and stored at −80 °C until analysis [17].

mRNA abundance

Samples were selected for analysis of *IL-1β, IL-6, IL-8, IL-10, TNF-α*, and toll like receptor 4 (*TLR-4*). Total RNA was isolated using the method of Al-Trad et al. [18]. Total RNA was extracted from 100 mg of each sample using TRIzol reagent (Invitrogen, Carlsbad, CA) according to the manufacturer's instructions. The quantity and quality of the isolated RNA were determined by absorbance at 260 and 280 nm. Total RNA (1 μg) was reverse transcribed using 1 μL Oligo-dT primer (Promega, Madison, WI, USA) and 0.625 μL RNasin (Promega, Madison, WI, USA) in a 25-μL reaction for 5 min at 72 °C and 60 min at 42 °C. The resulting first-strand cDNA was stored at −20 °C until used for real-time PCR. Semi-quantitative real-time PCR was carried out on an ABI7500 Real-time PCR (Applied Biosystems, Inc., Carlsbad, US) using GAPDH (forward: 5′-GTCTTCACTACCATGGAGAAGG-3′, reverse: 5′-TCATGGATGACCTTGGCCAG-3′) as an internal reference gene. Primers for quantitative rt-PCR were designed as follows: IL-1β: 5′-CCTCGGTTCCA TGGGAGATG-3′ (forward) and 5′-AGGCACTGTT CCTCAGCTTC-3′ (reverse); IL-6: 5′-GCTGAATC TTCCAAAAATGGAGG-3′ (forward) and 5′-GCTTCAG GATCTGGATCAGTG-3′ (reverse); IL-8: 5′-ACACATT CCACACCTTTCCAC-3′ (forward) and 5′-ACCTTCTG CACCCACTTTTC-3′ (reverse); IL-10: 5′-CCTTGTCGG AAATGATCCAGTTTT-3′ (forward) and 5′-TCAGGCC CGTGGTTCTCA-3′ (reverse); TNF-α: 5′-TCCAGAAGT TGCTTGTGCCT-3′ (forward) and 5′-CAGAGGGCTG TTGATGGAGG- 3′ (reverse); and TLR-4: 5′-TATGAAC-CACTCCACTCGCTC-3′ (forward) and 5′-CATCATTT GCTCAGCTCCCAC-3′ (reverse). For each sample, the target gene and the control gene were run under duplex reaction conditions in duplicate. All reactions used the following protocol according to the instructions: 2 μL of sample cDNA, 10 μL 2× premix ExTaqTm II (Takara Bio, Inc., Kusatsu, Shiga, Japan), 0.8 μL of 10 μmol/L forward primers, 0.8 μL of 10 μmol/L reverse primers, 0.4 μL ROX correction dye, and the total volume was adjusted with nuclease-free H_2O to 20 μL. For amplification, the following cycling conditions were used: 95 °C for 5 s, 60 °C for 30 s for 40 cycles after an initial denaturation step of 30 s at 95 °C, and then an elongation step at 70 °C for 15 s. A dissociation curve was achieved by melting DNA at 95 °C for 15 s, incubating at 60 °C for 1 min, ramping up to 95 °C for 30 s, and cooling down to 60 °C for 15 s. The relative gene expression was determined by quantitative RT-PCR and expressed using $2^{-\Delta\Delta CT}$ methods as described by Livak and Schmittgen [19]; BTM was established as the control group. The relative expression of the target gene mRNA in each group was calculated as follows: $\Delta CT = CT$ (target gene) - CT (GAPDH), and $\Delta\Delta CT = \Delta CT$ (treated group) - ΔCT (BTM group).

Histopathological scoring

The intestinal segments fixed with formalin solution were prepared using paraffin embedding, and 4-µm thickness sections were cut and then stained with hematoxylin-eosin (HE) for microscopic examination under 10×, 20×, and 60× magnifications using an Olympus BX41 microscope (Olympus, Tokyo, Japan) equipped with a Canon EOS 550D camera head (Canon, Tokyo, Japan) [20]. The bunk tank milk group was established as the control group.

Inflammation was scored using four aspects, including epithelial necrosis (0 = none; 1 = mild, <10%; 2 = moderate, 11-20%; 3 = severe, ulceration), leukocyte infiltration (0 = normal, <10%; 1 = mild, 11-15%; 2 = moderate, 16-20%; 3 = severe, >20%), central lacteal expansion (0 = none; 1 = mild; 2 = moderate; 3 = severe), and submucosal edema (0 = none; 1 = mild; 2 = moderate; 3 = severe). The sum of the four parameter scores represents the overall inflammation status of each intestinal segment, which were evaluated as follows: 0 = normal, 1–5 = mild, 6–10 = moderate, and 11–12 = severe.

Immunohistochemical scoring

After deparaffinization and hydration using xylenes, slides were subjected to a sodium citrate buffer solution for antigen retrieval. Endogenous peroxidase activity was quenched with 3% H_2O_2 in methanol for 20 min, after which slides were incubated with goat serum (Zhongshan Golden Bridge Biotechnology Co., Beijing, China) of the species for 15 min.

Next, sections were incubated for 1 h at 37 °C with a mouse monoclonal TLR-4 antibody (76B357.1, Abcam, England, United Kingdom). Afterwards, slides were incubated with a biotinylated secondary antibody for 30 min at 37 °C. Bound antibody conjugates were visualized using 3,3′-DAB (Zhongshan Golden Bridge Biotechnology Co., Beijing, China) as a chromogen. The slides were counterstained with hematoxylin for microscopic examination under a 40× magnification using an Olympus BX41 microscope (Olympus, Tokyo, Japan) equipped with a Canon EOS 550D camera head (Canon, Tokyo, Japan).

Immunohistochemical scores were evaluated using two aspects, including the percentage of immunoreactive cells (0 = 0–1%; 1 = 1–10%; 2 = 10–50%; 3 = 50–80%; 4 = 80–100%) and staining intensity (0 = negative; 1 = weak; 2 = moderate; 3 = strong), according to Soslow et al. [21]. The multiplied score of the two parameters represents the overall immunoreactivity of each intestinal segment,

which was evaluated as follows: 0 = negative, 1–4 = weak, 5–8 = moderate, and 9–12 = strong.

Statistical analysis

Data were analyzed using the GLM procedure in SAS 9.0. Values are expressed as the means ± SE. The model for treatment differences was:

$$Y_{ij} = \mu + T_i + C_j + e_{ij}$$

where

μ = overall mean;
T_i = the effect of treatment (i = 1 to 4);
C_j = the effect of calf (j = 1 to 21 of body weight and fecal score; j = 1 to 7 of feed intake; j = 1 to 6 of serum and intestinal index);
e_{ij} = random error.

Significant differences in mean values were evaluated using Duncan's multiple range test. A significance level of $P < 0.05$ was used.

Results
Milk composition

Milk composition of BTM, UWM, PWM and AWM is shown in Table 1. There were no significant differences ($P > 0.05$) in milk protein among the four kinds of milk. Milk fat and SNF percentage in UWM were significantly higher than BTM, PWM, and AWM. The pH value in AWM decreased and acidity increased with the addition of formic acid. TVC in AWM was higher than PWM.

Growth performance and health status

All calves in three groups had similar BW at birth (Table 2). Body weight for calves in the UWM group was higher than the BTM group, and weight gain in the UWM group was higher than the BTM, PWM, and AWM groups. ADG in the BTM group was lower than the UWM, PWM and AWM groups. Calves in the BTM and AWM groups consumed more starter than the

Table 1 Milk composition of different sources of milk

Items	Treatments[c]			
	BTM	UWM	PWM	AWM
Milk protein, %	3.26 ± 0.04	3.54 ± 0.08	3.62 ± 0.05	3.35 ± 0.16
Milk fat, %	4.22 ± 0.24[b]	6.30 ± 0.44[a]	4.50 ± 0.07[b]	4.33 ± 0.10[b]
SNF[d]	13.0 ± 0.04[b]	15.1 ± 0.08[a]	13.3 ± 0.05[b]	13.4 ± 0.20[b]
pH	6.71 ± 0.02[a]	6.63 ± 0.05[a]	6.58 ± 0.02[a]	4.65 ± 0.06[b]
Acidity	12.7 ± 0.18[b]	14.3 ± 0.65[b]	15.4 ± 0.33[b]	72.9 ± 1.96[a]
TVC[d], × 10³	91.0 ± 51.4[ab]	87.8 ± 49.6[ab]	0.18 ± 0.15[b]	520 ± 300[a]

[a,b]Means within a row not sharing a common superscript letter are significantly different (P < 0.05)
[c]BTM bunk tank milk group, UWM untreated waste milk group, PWM pasteurized waste milk group, AWM acidified waste milk group
[d]SNF solid non-fat, TVC total viable count

Table 2 Growth performance and health status of calves feeding on different sources of milk

Items[d]	Treatments[c]			
	BTM	UWM	PWM	AWM
Initial body weight, kg	43.7 ± 1.13	43.4 ± 1.28	43.1 ± 0.87	44.4 ± 0.91
Body weight, kg	49.3 ± 1.26[b]	54.4 ± 1.85[a]	52.5 ± 1.36[ab]	52.5 ± 1.37[ab]
Weight gain, kg	5.43 ± 0.85[b]	11.0 ± 1.09[a]	9.53 ± 0.92[b]	8.56 ± 0.95[b]
ADG, g/d	258 ± 40.4[b]	525 ± 51.9[a]	454 ± 43.9[a]	408 ± 45.2[a]
ADFI, g/d	34.4 ± 6.64[a]	21.3 ± 3.89[b]	29.4 ± 4.72[ab]	36.2 ± 4.33[a]
Fecal score	2.99 ± 0.07[a]	2.94 ± 0.06[a]	2.94 ± 0.10[a]	2.72 ± 0.05[b]
Diarrhea, no. of animals	12	14	12	11

[a,b]Means within a row not sharing a common superscript letter are significantly different ($P < 0.05$)
[c]BTM bunk tank milk group, UWM untreated waste milk group, PWM pasteurized waste milk group, AWM acidified waste milk group
[d]ADG average daily gain, ADFI average daily feed intake

UWM group. The Fecal score in the AWM group was significantly lower than the BTM, UWM, and PWM groups. Twelve calves experienced diarrhea in the BTM and PWM groups, while 14 or 11 calves had diarrhea in the UWM or AWM groups, respectively.

Serum metabolites and growth index

Serum TP content were similar ($P \geq 0.05$) in the four experimental groups (Table 3) at 48 h after feeding. There were no significant differences ($P > 0.05$) among the four treatments for LDL, UN, Cr, and BIL. Serum TP, ALB, TC, and GC concentrations were significantly higher in the UWM group than the BTM, PWM, and AWM groups. Untreated waste milk feeding calves expressed

higher HDL, TG and GH concentrations, the BTM group expressed lower HDL and GH, and the AWM group expressed lower TG.

Immune and antioxidant performance

There were no significant differences ($P > 0.05$) in SOD, C3, C4, IL-6, IL-7, IL-8, IL-10, and IL-22 concentrations among the four treatments (Table 4) before feeding on d 22. The MDA concentration was the highest in the PWM group and higher in the AWM group compared to the BTM and UWM groups. Compared with the other three groups, calves expressed higher GSH-px and IgG concentrations in the BTM group and IgA in the UWM group. Serum IgM concentration was higher in the UWM group than the AWM group, and IL-1β and TNF-α were higher in the UWM and PWM groups compared to the BTM and AWM groups (Table 5).

Table 3 Serum metabolites and growth index of calves feeding on different sources of milk

Items[d]	Treatments[c]			
	BTM	UWM	PWM	AWM
TP on 48 h, g/L	6.02 ± 0.14	5.98 ± 0.13	5.55 ± 0.14	5.61 ± 0.13
TP, g/L	56.2 ± 5.16[b]	71.3 ± 2.46[a]	48.5 ± 2.91[b]	45. 7 ± 4.70[b]
ALB, g/L	16.3 ± 2.01[b]	22.5 ± 1.91[a]	13.8 ± 0.60[b]	13.0 ± 1.39[b]
HDL, mmol/L	1.03 ± 0.16[b]	1.62 ± 0.28[a]	1.12 ± 0.13[ab]	1.11 ± 0.08[ab]
LDL, mmol/L	0.32 ± 0.18	0.47 ± 0.11	0.29 ± 0.08	0.19 ± 0.04
UN, mmol/L	2.86 ± 0.23	3.17 ± 0.16	2.83 ± 0.29	2.49 ± 0.28
Cr, μmol/L	79.0 ± 8.07	67.3 ± 3.86	63.8 ± 3.28	71.8 ± 8.62
TG, mmol/L	0.30 ± 0.05[ab]	0.36 ± 0.05[a]	0.26 ± 0.03[ab]	0.18 ± 0.02[b]
TC, mmol/L	1.48 ± 0.25[b]	2.25 ± 0.33[a]	1.52 ± 0.18[b]	1.38 ± 0.09[b]
BIL, μmol/L	6.90 ± 0.77	6.93 ± 0.71	4.78 ± 0.43	6.12 ± 1.32
GC, pg/mL	559 ± 84.6[b]	811 ± 21.4[a]	320 ± 76.2[b]	203 ± 27.0[b]
GH, ng/mL	1.26 ± 0.27[b]	2.00 ± 0.18[a]	1.58 ± 0.09[ab]	1.82 ± 1.15[ab]

[a,b]Means within a row not sharing a common superscript letter are significantly different ($P < 0.05$)
[c]BTM bunk tank milk group, UWM untreated waste milk group, PWM pasteurized waste milk group, AWM acidified waste milk group
[d]TP total protein, ALB albumin, HDL high density lipoprotein, LDL low density lipoprotein, UN urea nitrogen, Cr creatinine, TG triglyceride, TC total cholesterol, BIL bilirubin, GC glucagon, GH growth hormone

Table 4 Antioxidant and immunestatus of calves feeding on different sources of milk

Items[d]	Treatments[c]			
	BTM	UWM	PWM	AWM
SOD, U/mL	56.5 ± 0.51	54.1 ± 2.03	57.2 ± 0.88	57.9 ± 0.84
MDA, nmol/L	1.28 ± 0.22[c]	1.37 ± 0.29[c]	1.91 ± 0.09[a]	1.58 ± 0.22[b]
GSH-px, μmol/mL	79.1 ± 33.2[a]	44.5 ± 23.8[b]	51.7 ± 27.0[b]	43.1 ± 22.6[b]
IgG, pg/mL	0.87 ± 0.28[a]	0.58 ± 0.08[b]	0.57 ± 0.10[b]	0.43 ± 0.13[b]
IgA, g/L	0.66 ± 0.03[b]	0.82 ± 0.04[a]	0.71 ± 0.05[b]	0.69 ± 0.03[b]
IgM, g/L	2.49 ± 0.17[ab]	2.89 ± 0.20[a]	2.53 ± 0.18[ab]	2.17 ± 0.05[b]
C3, mg/dL	0.20 ± 0.04	0.31 ± 0.05	0.25 ± 0.04	0.22 ± 0.05
C4, mg/dL	2.23 ± 0.31	2.11 ± 0.23	2.26 ± 0.26	2.54 ± 0.23

[a,b]Means within a row not sharing a common superscript letter are significantly different ($P < 0.05$)
[c] BTM bunk tank milk group, UWM untreated waste milk group, PWM pasteurized waste milk group, AWM acidified waste milk group
[d]SOD superoxide dismutase, MDA malondialdehyde, GSH-px glutathione peroxidase, IgG immunoglobulin G, IgA immunoglobulin A, IgM immunoglobulin M, C3 complement 3, C4 complement 4

Table 5 Serum immune factors of calves feeding on different sources of milk

Items[d]	Treatments[c]			
	BTM	UWM	PWM	AWM
IL-1β, ng/L	60.0 ± 5.49[b]	70.3 ± 7.98[a]	72.3 ± 3.04[a]	61.7 ± 4.14[b]
IL-6, ng/L	610 ± 12.8	583 ± 26.6	599 ± 49.4	602 ± 36.0
IL-7, ng/L	24.9 ± 1.27	26.6 ± 1.22	26.3 ± 1.18	24.8 ± 1.28
IL-8, ng/L	86.9 ± 2.80	83.8 ± 3.44	91.6 ± 4.26	88.9 ± 2.38
IL-10, ng/L	264 ± 21.5	269 ± 26.8	277 ± 20.1	283 ± 13.9
IL-22, ng/L	0.49 ± 0.02	0.49 ± 0.05	0.51 ± 0.01	0.54 ± 0.02
TNF-α, ng/L	258 ± 20.7[b]	299 ± 27.2[a]	301 ± 13.8[a]	273 ± 6.07[b]

[a,b]Means within a row not sharing a common superscript letter are significantly different (P < 0.05)
[c]BTM bunk tank milk group, UWM untreated waste milk group, PWM pasteurized waste milk group, AWM acidified waste milk group
[d]IL-1β interleukin-1β, IL-6 interleukin-6, IL-7 interleukin-7, IL-8 interleukin-8, IL-10 interleukin-10, IL-22 interleukin-22, TNF-α tumor necrosis factor-α

mRNA expression

There were no differences (P > 0.05) in mRNA expression of jejunal IL-6 and TNF-α, and the TNF-α and TLR-4 in the mesenteric lymph nodes (Table 6). Compared with the other three groups, calves feeding on UWM increased jejunal IL-10 expression, and feeding on AWM increased IL-8 mRNA expression in the mesenteric lymph nodes.

For jejunum, IL-1β mRNA expression was higher in the UWM group than the BTM group, IL-8 mRNA

Table 6 MRNA expression of immune factors on jejunal mucosa and mesenteric lymph nodes of calves

Items[d]	Treatments[c]			
	BTM	UWM	PWM	AWM
Jejunal mucosa				
IL-1β	1.00 ± 0.22[b]	6.32 ± 0.92[a]	1.67 ± 0.67[ab]	3.00 ± 0.90[ab]
IL-6	1.00 ± 0.27	4.38 ± 0.15	1.86 ± 0.46	3.76 ± 1.30
IL-8	1.00 ± 0.63[b]	6.11 ± 0.87[a]	1.80 ± 0.47[b]	4.41 ± 1.18[a]
IL-10	1.00 ± 0.64[b]	10.49 ± 0.40[a]	1.41 ± 0.72[b]	3.06 ± 0.83[b]
TNF-α	1.00 ± 0.08	4.88 ± 0.45	0.98 ± 0.05	1.45 ± 0.26
TLR-4	1.00 ± 0.16[b]	0.88 ± 0.06[b]	7.31 ± 0.67[ab]	10.09 ± 1.82[a]
Mesenteric lymph nodes				
IL-1β	1.00 ± 0.23[b]	1.73 ± 0.24[ab]	1.73 ± 0.26[ab]	3.30 ± 0.40[a]
IL-6	1.00 ± 0.13[b]	1.13 ± 0.37[ab]	1.03 ± 0.30[ab]	1.95 ± 0.28[a]
IL-8	1.00 ± 0.70[b]	1.33 ± 0.58[b]	0.72 ± 0.21[b]	6.30 ± 0.83[a]
IL-10	1.00 ± 0.33[b]	1.55 ± 0.16[ab]	1.35 ± 0.18[b]	2.22 ± 0.08[a]
TNF-α	1.00 ± 0.12	0.98 ± 0.30	0.94 ± 0.1	0.56 ± 0.24
TLR-4	1.00 ± 0.09	1.36 ± 0.26	0.82 ± 0.54	1.60 ± 0.47

[a,b]Means within a row not sharing a common superscript letter are significantly different (P < 0.05)
[c]BTM, bunk tank milk group, UWM untreated waste milk group, PWM pasteurized waste milk group, AWM acidified waste milk group
[d] IL-1β interleukin-1β, IL-6 interleukin-6, IL-8 interleukin-8, IL-10 interleukin-10, TNF-α tumor necrosis factor-α, TLR-4 toll-like receptor-4

expression was improved in the UWM and AWM groups than the BTM group, and TLR-4 mRNA expression was higher in the AWM group than the UWM and BTM groups. In the mesenteric lymph nodes, mRNA expression of IL-1β and IL-6 was higher in the AWM group than the BTM group, and IL-10 was higher in the AWM group than the BTM and PWM groups.

Intestinal development

Jejunal and ileal pH values were similar (P > 0.05) among the four experimental groups. The jejunum of the calves in the four treatments were well developed with intact structure, with long, intensive and uniform villi and deep crypts, with a few tissue abscesses in the villus apex (Fig. 1b). The ileal basal structure of the calves in the four treatments was clear and intact, but villi of the ileum from calves in the BTM group were less uniform, where villi of the ileum in the UWM, PWM and AWM groups were more uniform with some apical abscesses (Fig. 1d). The jejunum and ileum in all four experimental treatments expressed mild submucosa edema and no central lacteal expansion (Fig. 1a, c). Slight epithelial lesions were observed in the jejunum and ileum of the BTM and UWM calves, and leukocyte infiltration was found in the UWM and AWM calves. Numerically, the inflammation scores of the jejunum increased in the UWM and AWM groups compared to the BTM group and increased in the ileum in the AWM group compared to the BTM group (Table 7).

Immunohistochemical analysis was used to localize TLR-4 cells in the jejunum and ileum of calves subjected to different kinds of milk. The analyses of jejunum and ileum sections from all calves revealed that TLR-4 cells exhibited brown particles in the epithelium, which were clearly detected in the BTM group, a few existed in the UWM group, and more were observed in the PWM and AWM groups (Fig. 2b, d). The jejunal IHS was higher in the PWM and AWM groups than the BTM group. Compared with the other three groups, calves feeding on BTM had lower ileal IHS (Fig. 2a, c).

Discussion

Growth and health

On d 22, calves fed untreated waste milk had gained more body weight than those fed bunk tank milk, but more calves experienced diarrhea. We speculated the weight gain due to the high amount of milk fat and SNF percentages in the UWM, which contained excessive colostrum and transitional milk from the farm [22–24]. An experiment conducted on calves from d 3 to d 56 concluded that weight gain and health parameters were not influenced by feeding on untreated waste milk, pasteurized waste milk, or bulk milk [1]. Similar results have also been reported in the literature [7, 25–27],

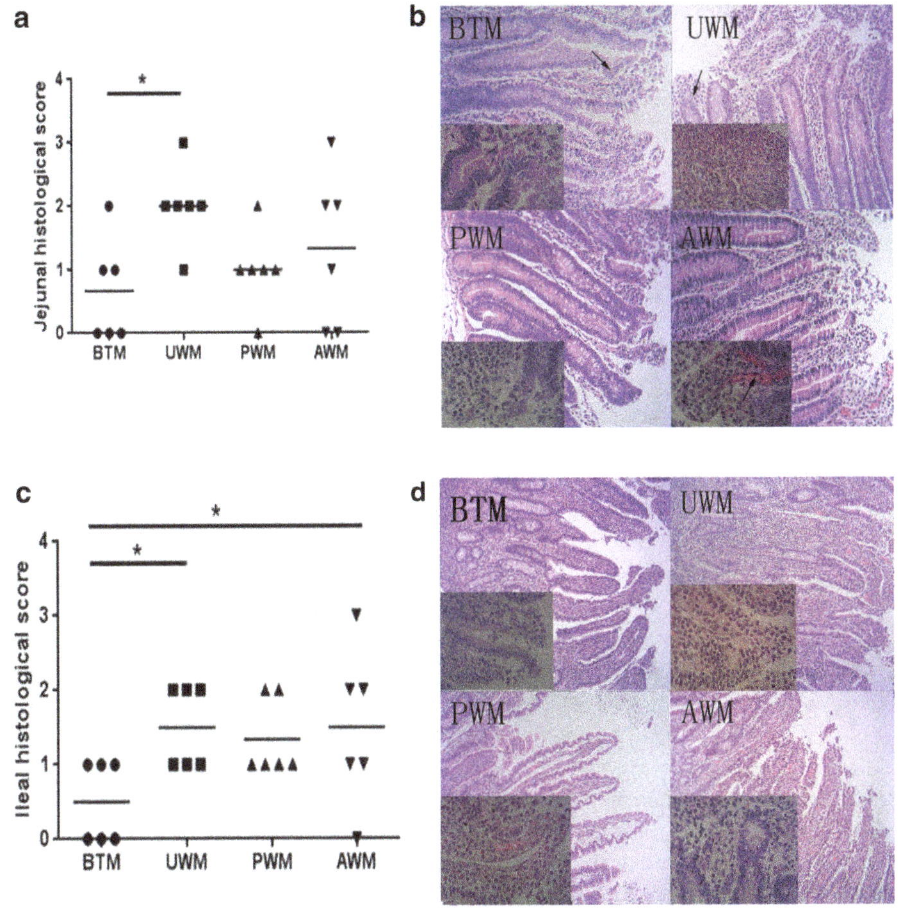

Fig. 1 Histological scores and light micrographs of hematoxylin and eosin stained jejunal and ileal sections of calves feeding on different sources of milk. **a** Jejunal histological scores. **b** Representative photomicrographs of hematoxylin and eosin stained jejunal sections in the BTM, UWM, PWM, and AWM groups (H & E 20× and 60× original magnification). **c** Ileal histological scores. **d** Representative photomicrographs of hematoxylin and eosin stained ileal sections in the BTM, UWM, PWM, and AWM groups (H & E 10× and 60× original magnification). Intestine pictures from calves receiving bunk tank milk, untreated waste milk, pasteurized waste milk, or acidified waste milk after feeding on colostrum, respectively. The jejunal and ileal basal structure of the calves in the four treatments was clear and intact. Slight epithelial lesions (arrows) were observed in the jejunum and ileum of the BTM and UWM fed calves, and leukocyte infiltration (arrows) was found in the UWM and AWM fed calves

which did not observe detrimental effects of waste milk on growth performance and calf health.

The advantages of feeding pasteurized waste milk have been observed in previous studies with 300 or 438 calves [2, 3], which demonstrated higher mean weight gain and lower diarrhea incidence. Similarly, growth performance and calf health were comparable between the BTM and PWM groups in this study, except for a higher ADG in the PWM group.

The acidification of milk replacer for use in rearing calves has been studied by several investigators [28–30], and the digestibility of dietary nutrients and growth performance could be improved when the pH was decreased to a proper range [11]. The most apparent difference was the decrease in pH in waste milk and the fecal score of calves fed with AWM compared to the increase in ADFI and ADG, which may be because the

addition of formic acid enhanced dietary flavor thus promoting animal appetite [31].

Serum metabolites and growth index

When the serum TP level is lower than 5.2 g/dL, 24–48 h after birth then it is referred to as failure of passive transfer [32]. All calves in the present study achieved passive transfer. The changes in serum TP, ALB, and BUN reflects the utilization efficiency of protein [33] and changes in TG and TC reflects lipid metabolism [34, 35]. Serum total protein can be inhibited when dietary nutrients are imbalanced or when the feed intake is decreased [34, 36]. The untreated waste milk fed calves had higher serum TP, ALB, HDL, TG and TC than the BTM fed calves. This mirrored milk protein content and ADFI demonstrates that a highly nutritional milk could enhance protein and lipid synthesis ability [37]. The

Table 7 Intestinal development of calves feeding on different sources of milk

Items[d]	Treatments[c]			
	BTM	UWM	PWM	AWM
pH				
Jejunum	6.05 ± 0.07	5.94 ± 0.12	5.95 ± 0.17	6.13 ± 0.10
Ileum	6.68 ± 0.10	6.53 ± 0.32	6.44 ± 0.47	6.84 ± 0.14
Inflammation scores				
Jejunum	0.67 ± 0.67[b]	2.33 ± 0.33[a]	1.33 ± 0.33[ab]	2.33 ± 0.33[a]
Ileum	0.67 ± 0.33[b]	1.33 ± 0.33[ab]	1.00 ± 0.00[ab]	2.00 ± 0.58[a]
IHS				
Jejunum	1.33 ± 0.33[b]	2.67 ± 0.67[ab]	4.00 ± 1.15[a]	4.00 ± 0.00[a]
Ileum	1.67 ± 0.33[b]	4.00 ± 0.00[a]	4.67 ± 0.67[a]	5.33 ± 0.67[a]

[a,b]Means within a row not sharing a common superscript letter are significantly different ($P < 0.05$)
[c]BTM bunk tank milk group, UWM untreated waste milk group, PWM pasteurized waste milk group, AWM acidified waste milk group
[d]IHS immunohistochemical scores

untreated waste milk had more of a beneficial effect than the BTM in GH and GC, which was consistent with weight gain and average daily gain.

Serum antioxidant and immune performance

The body antioxidant system, which prevents the toxic effects of free oxygen and its metabolites, is normally under a dynamic equilibrium between the generation and removal of free oxidative radicals and is highly related to animal health. Calves could acquire an antioxidative defense ability from milk after birth. Superoxide dismutase, which is the main parameter to assess oxidative status [38], catalyzes the dismutation of the superoxide radical anion [39]. Bovine milk contains Cu-Zn superoxide dismutase (Cu/Zn-SOD) [40, 41]. In the present study, serum SOD concentration in calves was not influenced by different sources of milk. The serum MDA level is used to monitor the extent of lipid peroxidation by reactive oxygen species [42] and reflects the degree of damage directly caused by free radicals. Calves feeding on PWM and AWM exhibited a high degree of destruction by free radicals based on high serum MDA level, especially in the pasteurized waste milk feeding group. Superoxide can be catalyzed and converted into water by GSH-px [43]. Glutathione peroxidase concentration in calves fed BTM was much higher compared to those fed UWM, PWM and AWM. We postulate that when the antioxidant mechanisms of SOD and GSH-px were activated by bunk tank milk, the oxidation by MDA production was lowered. This finding indicates that a high quality nutrient content and microbial activity in bunk tank milk is helpful in establishing antioxidative defense mechanisms in calves.

Complement 3 and C4 are the intrinsic composition of the complement system. The immune system will function properly if C3 and C4 content is maintained within a suitable range, otherwise, immunity defense mechanisms can be reduced. Complement 3 and C4 concentrations of the four groups did not vary in this study, reflecting a stable immunity for all of the calves.

The regulation mechanism of humoral immunity can be directly reflected by serum immune-globulins levels. Researchers have documented that the serum IgG content of calves with diarrhea was clearly lower than healthy calves and that there is a positive relationship between serum IgG concentration and diarrhea incidence of calves [44]. Additionally, Tu [11] noted serum IgA and IgM was highest when milk pH replacer reached 5.0. However, higher IgG was observed in calves receiving BTM, and higher IgA and IgM was observed in calves receiving UWM, and no positive correlation was observed between IgG content and calf diarrhea.

Serum immune factors and mRNA expression

Interleukins are a class of immune factors that play a role in regulating inflammation and immune response initiated by infection and injury [45]. Therefore, interleukins, such as IL-6, IL-8 and IL-10, are indicators of inflammation [46, 47]. The tumor necrosis factor participates in cell mediated immune response [48] and plays an important role in resisting and defending against intracellular viruses and mycoplasma [49, 50]. Toll-like receptor-4 recognizes microbial and inflammatory responses.

Studies that observed the effects of strains on the serum and mucosal immune factors varied, where *E. coli* caused diarrhea in piglets and increased the serum TNF-α, lipopolysaccharide stimulated IL-1, IL-6, IL-8 and IL-12 [51], but *Lactobacillus* improved IL-6 and IL-10 [52]. In the present study, calves feeding on UWM, PWM or AWM exhibited some up-regulation of immune factors in the serum, jejunal mucosa and mesenteric lymph nodes, which may be the results of different species and quantities of microbial activity, such as serum IL-1β and TNF-α of UWM and PWM feeding groups, four immune-factor of jejunal mucosa in UWM or AWM groups, and IL-1β, IL-6, IL-8, IL-10 in mesenteric lymph nodes of AWM feeding group.

Intestinal development

Woodford [28] indicated that an acidified diet may decrease the pH only in the abomasum and that it would be neutralized by the pancreatic juice when the digesta reached the small intestine. The different sources of milk did not change the jejunal and ileal pH in calves, which agrees with the findings of a previous study on acidified milk replacer [11].

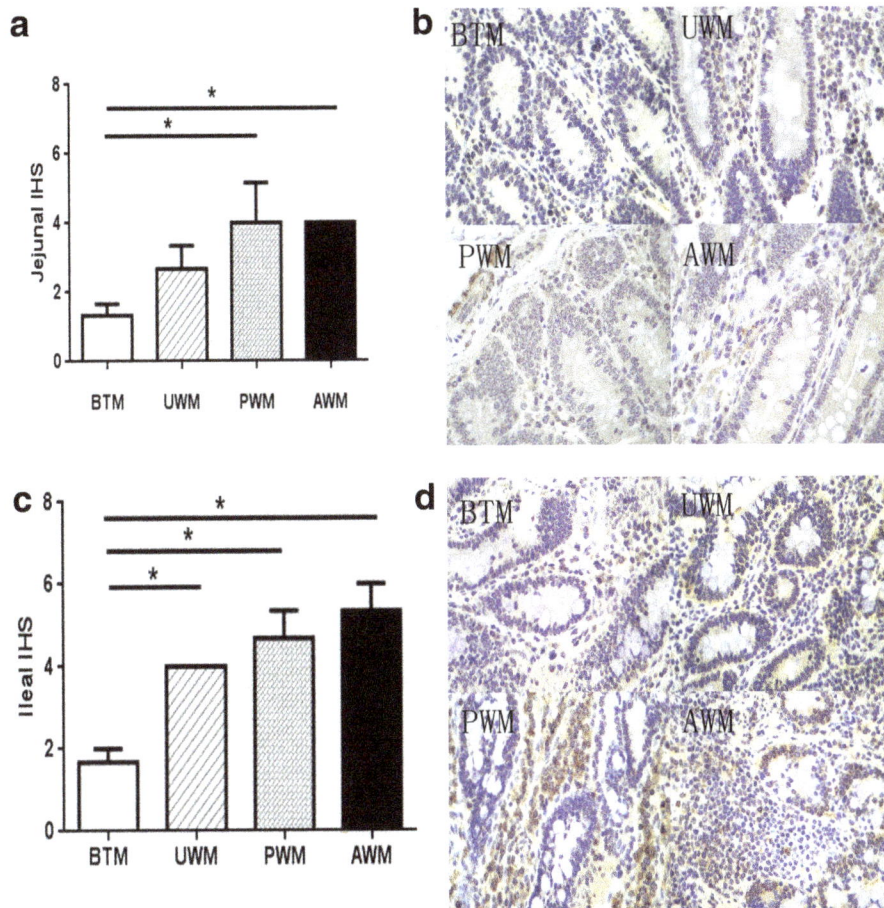

Fig. 2 Immunohistochemical scores and light micrographs of DAB stained jejunal and ileal sections of calves. **a** Jejunal immunohistochemical scores. **b** Representative photomicrographs of DAB stained jejunal sections in the BTM, UWM, PWM, and AWM groups (H & E 40 × original magnification). **c** Ileal immunohistochemical scores. **d** Representative photomicrographs of DAB stained ileal sections in the BTM, UWM, PWM, and AWM groups (H & E 40 × original magnification). Intestine pictures from the calves receiving bunk tank milk, untreated waste milk, pasteurized waste milk, or acidified waste milk after feeding colostrum, respectively. Analyses of the jejunum and ileum sections from all the calves revealed that TLR-4 cells exhibited brown particles in the epithelium, which was clearly detected in the BTM group, a few detected in the UWM group, and more were observed in the PWM and AWM groups

Normally, there is a dynamic balance between the intestinal microbial activity and the host, and calves can experience a series of diseases when this balance was destroyed [53]. Injury to the mucosa could damage the intestinal epithelial barrier function, which then can induce enteritis and diarrhea [54]. In the present study, slight apical abscissions, epithelial lesions, leukocyte infiltration and submucosal edemas were observed in all four milk feeding groups. In an experiment conducted on acidified milk replacer, Tu [11] noted the intestinal architecture can be well preserved when liquid pH equals 5.5 or 5.0. However, it appears from the histological scores that the acidified waste milk and untreated waste milk caused inflammation of the jejunum and ileum, and the intestine of the calves in the bunk tank milk feeding group were relatively healthier.

Immunohistochemical analyses of the jejunal and ileal tissues revealed that TLR-4 cells were primarily localized in the lamina propria and scattered in the epithelium [55]. Correspondingly, waste milk feeding increased the percentage of TLR-4 cells in the jejunum and ileum, which indicated an inflammatory response [55].

Conclusions

Growth performance was not improved by bunk tank milk feeding as we expected, and untreated waste milk demonstrated the highest weight gain and average daily gain in the calves. Additionally, the acidified waste milk promoted daily feed intake. However, calves in the bunk tank milk feeding group had better antioxidant capacity than that in the pasteurized waste milk group. In addition, the small intestine of calves receiving bunk tank milk was healthier. From a nutritional and health point of view, bunk tank milk is the best choice for calf raising. All calves feeding on waste milk experienced varying degrees of enteritis. The efficiency of feeding pasteurized and acidified waste milk are comparable, and the acidification of waste milk is an acceptable labor-saving and diarrhea preventing feed for young calves.

Abbreviations

ADFI: Average daily feed intake; ADG: Average daily gain; ALB: Albumin; AWM: Acidified waste milk; BIL: Bilirubin; BTM: Bunk tank milk; C3: Complement 3; C4: Complement 4; Cr: Creatinine; Cu/Zn-SOD: Cu-Zn superoxide dismutase; GC: Glucagon; GH: Growth hormone; GSH-px: Glutathione peroxidase; HDL: High density lipoprotein; HE: Hematoxylin-eosin; IgA: Immunoglobulin A; IgG: Immunoglobulin G; IgM: Immunoglobulin M; IHS: Immunohistochemical scores; IL-10: Interleukin-10; IL-1β: Interleukin-1β; IL-22: Interleukin-22; IL-6: Interleukin-6; IL-7: Interleukin-7; IL-8: Interleukin-8; LDL: Low density lipoprotein; MDA: Malondialdehyde; PWM: Pasteurized waste milk; SNF: Solid non-fat; SOD: Superoxide dismutase; TC: Total cholesterol; TG: Triglycerides; TLR-4: Toll like receptor-4; TNF-α: Tumor necrosis factor-α; TP: Total protein; TVC: Total viable count; UN: Urea nitrogen; UWM: Untreated waste milk

Acknowledgements

We thank Modern Farming Feidong Farm and College of Veterinary Medicine of China Agricultural University help me for calf rearing and intestinal samples analysis.

Funding

Not applicable.

Authors' contributions

YW, ZC, SL and JW participated in the design of the study. YZ and YD collected the experiments data. YZ analyzed the data and wrote the first draft of the manuscript. All authors read and approved the final manuscript.

Competing interests

The authors declare that they have no competing interests.

Author details

[1]State Key Laboratory of Animal Nutrition, Beijing Engineering Technology Research Center of Raw Milk Quality and Safety Control, College of Animal Science and Technology, China Agricultural University, Beijing 100193, China. [2]Beijing Dairy Cattle Center, Beijing 100192, China. [3]College of Veterinary Medicine, China Agricultural University, Beijing 100193, China.

References

1. Aust V, Knappstein K, Kunz HJ, Kaspar H, Wallmann J, Kaske M. Feeding untreated and pasteurized waste milk and bulk milk to calves: effects on calf performance, health status and antibiotic resistance of faecal bacteria. J Anim Physiol An N. 2013;97(6):1091–103.
2. Jamaluddin AA, Carpenter TE, Hird DW, Thurmond MC. Economics of feeding pasteurized colostrum and pasteurized waste milk to dairy calves. J Am Vet Med Assoc. 1996;209:751–6.
3. Godden SM, Fetrow JP, Feirtag JM, Green LR, Wells SJ. Economic analysis of feeding pasteurized nonsaleable milk versus conventional milk replacer to dairy calves. J Am Vet Med Assoc. 2005;226:1547–54.
4. Selim SA, Cullor JS. Number of viable bacteria and presumptive antibiotic residues in milk fed to calves on commercial dairies. J Am Vet Med Assoc. 1997;211:1029–35.
5. Butler JA, Sickles SA, Johanns CJ, Rosenbusch RF. Pasteurization of discard mycoplasma mastitic milk used to feed calves: Thermal effects on various mycoplasma. J Dairy Sci. 2000;83:2285–8.
6. Yanar M, Güler O, Bayram B, Metin J. Effects of feeding acidified milk replacer on the growth, health and behavioural characteristics of holstein friesian calves. Turk J Vet Anim Sci. 2006;30(2):235–41.
7. Chardavoyne JR, Ibeawuchi JA, Kesler EM. Waste milk from antibiotic treated cows as feed for young calves. J Dairy Sci. 1979;62:1285–9.
8. Grant IR, Ball HJ, Neill SD, Rowe MT. Effect of higher pasteurization temperatures, and longer holding times at 72 °C, on the inactivation of mycobacterium paratuberculosis in milk. Lett Appl Microbiol. 1999;28:46–5.
9. Muller C. An acidified milk replacer for calves compared clinically with a traditional replacer with special reference to the effect on intestinal flora. Giessen: German Federal Republic, Justus Liebig University; 1986. p. 154.
10. Jaster EH, McCoy GC, Tomkins T, Davis CL. Feeding acidified or sweet milk replacer to dairy calves. J Dairy Sci. 1990;73:3563–6.
11. Tu Y. Effect of milk replacer acidity and acidity adjusting on growth performance, blood gas parameters and gastrointestinal tract development in pre-ruminant calves. Beijing: Chinese academy of agricultural sciences; 2011.
12. Moore DA, Taylor J, Hartman ML, Sischo WM. Quality assessments of waste milk at a calf ranch. J Dairy Sci. 2009;92(7):3503–9.
13. Medina M, Johnson LW, Knight AP, Olson JD, Lewis LD. Evaluation of milk replacers for dairy calves. Compend Contin Educ Pract Vet. 1983;5:S148–52.
14. FASS. Guide for the care and use of agricultural animals in research and teaching. 3rd ed. Texas: The federation of animal science societies; 2010.
15. Hill TM, Aldrich JM, Schlotterbeck RL, Bateman HG. Amino acids, fatty acids, and fat sources for calf milk replacers. Prof Anim Sci. 2007;23(4):401–8.
16. Yang Y, Iji PA, Kocher A, Mikkelsen LL, Choct M. Effects of mannanoligosaccharide on growth performance, the development of gut microflora, and gut function of broiler chickens raised on new litter. J Appl Poult Res. 2007;16:280–8.
17. Greenfield RB, Cecava MJ, Donkin SS. Changes in mRNA expression for gluconeogenic enzymes in liver of dairy cattle during the transition to lactation. J Dairy Sci. 2000;83:1228–36.
18. Al-Trad B, Wittek T, Penner GB, Reisberg K, Gäbel G, Fürll M, et al. Expression and activity of key hepatic gluconeogenesis enzymes in response to increasing intravenous infusions of glucose in dairy cows. J Anim Sci. 2010; 88:2998–3008.
19. Livak KJ, Schmittgen TD. Analysis of relative gene expression data using real-time quantitative PCR and the 2 − ΔΔCT methods. Methods. 2001;25:402–8.
20. Zhou D. Effects of *Bacillus* on intestinal T cells immune responses in F4R-weaned piglets challenged F4+ *Escherichia coli*. Beijing: China Agricultural University; 2015.
21. Soslow RA, Dannenberg AJ, Rush D, Woerner BM, Khan KN, Masferrer J, et al. COX-2 is expressed in human pulmonary, colonic, and mammary tumors. Cancer. 2000;89(12):2637–45.
22. Davis CL, Drackley JK. The development, nutrition, and management of the young calf. 1st ed. Ames: Iowa State University Press; 1998.
23. Foley JA, Otterby DE. Availability, storage, treatment, composition, and feeding value of surplus colostrum: a review. J Dairy Sci. 1978;61(8):1033–60.
24. Kehoe SI, Jayarao BM, Heinrichs AJ. A survey of bovine colostrum composition and colostrum management practices on pennsylvania dairy farms. J Dairy Sci. 2007;90(9):4108–16.
25. Kesler EM. Feeding mastitic milk to calves: review. J Dairy Sci. 1981;64(5):719–23.
26. Keys JE, Pearson RE, Weinland BT. Performance of calves fed fermented mastitic milk, colostrum, and fresh whole milk. J Dairy Sci. 1980;63(7):1123–7.
27. Keith EA, Windle LM, Keith NK, Gough RH. Feeding value of fermented waste milk with or without sodium bicarbonate for dairy calves. J Dairy Sci. 1983;66(4):833–9.
28. Woodford ST, Whetstone HD, Murphy MR, Davis CL. Abomasal pH, nutrient digestibility, and growth of Holstein bull calves fed acidified milk replacer. J Dairy Sci. 1987;70(4):888–91.
29. Vajda V, Magic D. An acidified milk feeding programme and its effects on the health, metabolism and growth intensity of suckling calves. Slovensky Veterinarsky Casopis. 1993;18:45–9.
30. Chernuho CV. The efficiency of an acidified milk replacer for calves. Zootech Sci Belarus. 2001;36:261–5.
31. Tu Y, Meng SY, Diao QY, Qi D, Zhou Y, Yun Q. Effects of a complex acidifier on growth performance and blood parameters in pre-ruminant calves. Feed Industry. 2010;A02:42–6.
32. Weaver DM, Tyler JW, VanMeter DC, Hostetler DE, Barrington GM. Passive transfer of colostral immunoglobulins in calves. J Vet Intern Med. 2000;14:569–77.

33. Zhou Y, Diao QY, Tu Y, Yun Q. Effects of yeast β-glucan on gastrointestinal development in early-weaning calves. Chin J Anim Nutri. 2009;21(6):846–52.

34. Ye J. A study on the effect of different processed soybean flours in milk replacer on calves. Yangzhou: Yangzhou University. Yangzhou; 2006.

35. Zhang X. Effects of supplementation of lactoferrin in milk replacer on the growth and development of calves. Yangzhou: Yangzhou University; 2007.

36. Yuangklang C, Wensing T, Van DBL, Jittakhot S, Beynen AC. Fat digestion in veal calves fed milk replacers low or high in calcium and containing either casein or soy protein isolate. J Dairy Sci. 2004;87(4):1051–6.

37. Egli CP, Blum JW. Clinical, haematological, metabolic and endocrine traits during the first three months of life of suckling Simmentaler calves held in a cow-calf operation. J Vet Med A. 1998;45:99–118.

38. Wang YZ, Xu CL, An ZH, Liu JX, Feng J. Effects of dietary bovine lactoferrin on performance and antioxidant status of piglets. Anim Feed Sci Technol. 2008;140:326–36.

39. Jiang ZY, Jiang SQ, Lin YC, Xi PB, Yu DQ, Wu TX. Effects of soybean isoflavone on growth performance, meat quality, and antioxidation in male broilers. Poult Sci. 2007;86:1356–62.

40. Asada K. Occurrence of superoxide dismutase in bovine milk. Agric Biol Chem. 1976;40:1659–60.

41. Korycka-Dahl M, Richardson T, Hicks CL. Superoxide dismutase activity in bovine milk serum. J Food Prot. 1979;42:867–71.

42. Sumida SK, Tanaka KK, Akadomo FN. Exercise induced lipid peroxidation and leakage of enzyme before and after vitamin E supplementation. Int J Biochem. 1989;21:835–8.

43. Halliwell B, Chirico S. Lipid peroxidation: Its mechanism, measurement, and significance. Am J Clin Nutr. 1993;57:715S–24S.

44. Wang CW, Yang WD, Yang WY, Lv WF, Yang LY. Effect of acanthopanax synbiotics on serum TAOC, IgG and IgA of lactation calves. Chin J Vet Med. 2015;51(2):48–50.

45. Van Snick J. Interleukin-6: An overview. Annu Rev Immunol. 1990;8:253–78.

46. Maizels R, Yazdanbakhsh M. Immune regulation by helminth parasites: cellular and molecular mechanisms. Nat Rev Immunol. 2003;3(9):733–44.

47. Nizar S, Copoer J, Meyer B, Bodman-Smith M, Galustian C, Kumar D, et al. T-regulatory cell modulation: the future of cancer immunotherapy? Brit J Cancer. 2009;100(11):1697–703.

48. Tarradas J, Argilaguet JM, Rosell R, Nofrarias M, Crisci E, Cordoba L, et al. Interferon-gamma induction correlates with protection by DNA vaccine expressing E2 glycoprotein against classical swine fever virus infection in domestic pigs. Vet Microbiol. 2010;142(1–2):51–8.

49. Bensaude E, Turner JL, Wakeley PR, Sweetman DA, Pardieu C, Drew TW, et al. Classical swine fever virus induces proinflammatory cytokines and tissue factor expression and inhibits apoptosis and interferon synthesis during the establishment of long-term infection of porcine vascular endothelial cells. J Gen Virol. 2004;85(4):1029–37.

50. Katze MG, He Y, Jr GM. Viruses and interferon: a fight for supremacy. Nat Rev Immunol. 2002;2(9):675–87.

51. Kopp E, Medzhitov R, Carothers J, Xiao C, Douglas I, Janeway CA, et al. ECSIT is an evolutionarily conserved intermediate in the Toll/IL-1 signal transduction pathway. Genes Dev. 1999;13(16):2059–71.

52. Miettinen M, Vuopiovarkila J, Varkila K. Production of human tumor necrosis factor alpha, interleukin-6, and interleukin-10 is induced by lactic acid bacteria. Infect Immun. 1997;64(12):5403–5.

53. Clemente J, Ursell L, Parfrey LW, Knight R. The impact of the gut microbiota on human health: An integrative view. Cell. 2012;148(6):1258–70.

54. Broz P, Ohlson MB, Moacke DM. Innate immune response to Salmonella typhimurium, a model enteric pathogen. Gut Microbes. 2012;3(2):62–70.

55. Ungaro R, Abreu MT, Fukata M, Fukata M. Practical techniques for detection of Toll-like receptor-4 in the human intestine. Methods Mol Biol. 2009;517:345–61.

8

Variance components and correlations of female fertility traits in Chinese Holstein population

Aoxing Liu[1,2], Mogens Sandø Lund[2], Yachun Wang[1*], Gang Guo[3], Ganghui Dong[3], Per Madsen[2] and Guosheng Su[2*]

Abstract

Background: The objective of the present study was to estimate (co)variance components of female fertility traits in Chinese Holsteins, considering fertility traits in different parities as different traits. Data on 88,647 females with 215,632 records (parities) were collected during 2000 to 2014 from 32 herds in the Sanyuan Lvhe Dairy Cattle Center, Beijing, China. The analyzed female fertility traits included interval from calving to first insemination, interval from first to last insemination, days open, conception rate at first insemination, number of inseminations per conception and non-return rates within 56 days after first insemination.

Results: The descriptive statistics showed that the average fertility of heifers was superior to that of cows. Moreover, the genetic correlations between the performances of a trait in heifers and in cows were all moderate to high but far from one, which suggested that the performances of a trait in heifers and cows should be considered as different but genetically correlated traits in genetic evaluations. On the other hand, genetic correlations between performances of a trait in different parities of cows were greater than 0.87, with only a few exceptions, but variances were not homogeneous across parities for some traits. The estimated heritabilities of female fertility traits were low; all were below 0.049 (except for interval from calving to first insemination). Additionally, the heritabilities of the heifer interval traits were lower than those of the corresponding cow interval traits. Moreover, the heritabilities of the interval traits were higher than those of the threshold traits when measuring similar fertility functions. In general, estimated genetic correlations between traits were highly consistent with the biological categories of the female fertility traits.

Conclusions: Interval from calving to first insemination, interval from first to last insemination and non-return rates within 56 days after first insemination are recommended to be included in the selection index of the Chinese Holstein population. The parameters estimated in the present study will facilitate the development of a genetic evaluation system for female fertility traits to improve the reproduction efficiency of Chinese Holsteins.

Keywords: Chinese Holsteins, Female fertility, Genetic correlation, Heritability

* Correspondence: wangyachun@cau.edu.cn; guosheng.su@mbg.au.dk
[1]Laboratory of Animal Genetics, Breeding and Reproduction, Ministry of Agriculture of China, National Engineering Laboratory of Animal Breeding, College of Animal Science and Technology, China Agricultural University, Beijing 100193, China
[2]Center for Quantitative Genetics and Genomics, Department of Molecular Biology and Genetics, Aarhus University, 8830 Tjele, Denmark
Full list of author information is available at the end of the article

Background

Female fertility traits are the most economically functional traits affecting the efficiency of the dairy industry. Cows with excellent fertility resume cycles soon after calving and become pregnant when properly inseminated [1]. According to previous studies, poor reproduction has become a worldwide problem and the most probable cause for culling [2, 3]. Due to the unfavorable correlations between yield and female fertility traits, intensive selection for milk production over the past five decades has led to a strong decline in cow reproduction, particularly in high-producing populations [4, 5]. Moreover, the heritabilities of most fertility traits are typically below 5% [6–8], thus, selection for female fertility is not as efficient as selection for production traits. Therefore, the reproductive decline in modern dairy cattle cannot be ignored, as the costs of extra inseminations, fertility treatments and involuntary culling are increasing.

Although the heritabilities of reproduction traits are relatively low, their additive genetic variances have remained sufficiently high to perform effective selection. Nordic countries (including Denmark, Finland and Sweden) are pioneers in applying fertility traits in dairy cattle breeding and have implemented fertility traits in their selection index since 1994. The Nordic Cattle Genetic Evaluation has shown that the genetic trend for the female fertility index of Holstein bulls has increased by approximately 1% per year since 2000 (http://www.sweebv.info/ba52nycknav.aspx). Furthermore, the results of routine genetic evaluations in Canada (https://www.cdn.ca/) and the United States (http://www.holsteinusa.com/) have shown that the decline in the genetic trend of fertility traits has flattened and even slightly reversed after increasing the weight of fertility traits in the selection index [9]. In a recent test run of the multiple across-country evaluation (MACE) by Interbull in August 2016, 21 countries participated in the genetic evaluation of female fertility traits [10]. In China, however, an insemination and calving record collection system has not yet been implemented on the national scale, therefore, female fertility traits are still not included in the current selection index for Chinese Holsteins. Preliminary results from a genetic evaluation of five fertility traits (age at first insemination, number of inseminations per conception, days from calving to first insemination, days open and calving interval) in Chinese Holsteins have been recently published [11]. However, that dataset was relatively small and did not include several internationally recommended female fertility traits, such as conception rate and non-return rate at first insemination. Moreover, the main limitation of the previous analysis is that the performances of a trait in heifers and cows were simply treated as the same trait, which differs from various previous genetic evaluation studies [6, 12]. The physiological status of cows changes considerably after first calving [13], thus, the performances of heifers and cows have typically been considered biologically distinct [14, 15]. Nevertheless, the genetic relationships among female fertility traits across multiple parities in dairy cattle remain less well investigated.

With respect to the models and algorithms typically used in the genetic evaluation of female fertility traits, many studies have shown that linear models perform similarly to threshold models but with less computational demand, consequently, linear models have been widely used in routine genetic evaluations [12, 16]. In addition, animal and sire models are currently the most frequently implemented models in routine genetic evaluations. Compared with sire models, animal models are theoretically superior for estimating variance components and breeding values. Several previous studies have reported that animal models provide more accurate predictions than sire models [17–19].

The aim of the present study was to assess the genetic relationships across multiple parities, estimate the variance and covariance components, and calculate the heritabilities and genetic correlations of female fertility traits in Chinese Holsteins. The results of this study will facilitate the development of a routine genetic evaluation system for female fertility traits, resulting in a more balanced dairy cattle breeding program in China.

Methods

Data

Field records of insemination and calving from 2000 to 2014 were collected from 32 herds in the Sanyuan Lvhe Dairy Cattle Center, Beijing, China. The following traits were recorded: interval from calving to first insemination (ICF), days open (DO), interval from first to last insemination (IFL), conception rate at first insemination (CR), number of inseminations per conception (AIS) and non-return rate within 56 days after first insemination (NRR56). For binary traits, CR was coded as 0 (failure) or 1 (success), and NRR56 was coded as 1 when there was no subsequent insemination within 56 days after the first insemination and 0 otherwise. All traits recorded before first calving were considered as heifer traits and were coded as parity 0. Traits measured in first-lactation cows were coded as parity 1, and the same strategy was used for the remaining lactations. Among these traits, ICF and DO were available only for cows. All the remaining traits were measured for both heifers and cows. In this context, h and c were used as suffixes in trait abbreviations to distinguish heifers and cows.

Most fertility traits included censored records, as mistakes in recording and some culling events arose during data collection or fertility events were still in progress. The last insemination with a positive pregnancy test result or with the following calving record was considered a verified conception. In the present data, 5.92% of heifers and 13.92% of cow-lactations showed no verified conception from the last insemination. In such cases, the IFL and AIS were calculated according to the last insemination, and the expectation of the trait was added as a penalty. Based on uncensored data, the expectations of IFLh, IFLc, AISh and AISc were 21 days, 43 days, 1.5, and 2.0, respectively. Similarly, for cows without ascertained conception, the penalty for DO was the same as that for IFL. To further reduce the proportion of censored records, we only used data for females with birth years from 1999 to 2013, insemination years from 2000 to 2014, and calving years from 2000 to 2014. To reduce possible selection bias, cows without heifer records were also removed from the data. Moreover, the data in the analysis contained only females inseminated by AI bulls. Other criteria for data editing included age at first insemination between 270 and 900 days, age at first calving between 500 and 1100 days, gestation length between 240 and 300 days, ICF between 20 and 230 days, IFL less than 365 days and AIS less than eight times per parity. Records with values smaller than the lower limit were deleted as outliers, while those with values larger than the upper limit were set as the upper limit.

Based on the distribution of records across parities for each trait, only data from heifers and the first three lactations were used in the analysis. After editing, the total number of cows in the final dataset was 88,647, and the total number of records (parities) was 215,632. The numbers of observations available for heifers and the first three parities of cows were 88,647, 60,022, 41,858 and 25,105, respectively. Heifer records comprised 41.11% of the entire dataset.

Full pedigree data were provided by the Dairy Association of China (Beijing, China), and each animal was traced back as many generations as possible. Unknown parents were assigned into 16 phantom groups, based on sex and birth year. In total, the pedigree included 179,939 females and 4310 males, with birth years from 1930 to 2013. Considering the number of heifers/cows in the present study, the size of pedigree was small. This was determined by the data structure. In the current study, the phenotypic data covered a long period, and the females came from 32 herds during birth year 1999 to 2013. Thus, the phenotypic data already covered a large number of female ancestors (such like mother, grandmother and so on). Therefore, the number of animals in the pedigree was only about two times as large as the number of animals with fertility records, even though the heifers/cows were traced back as many generations as possible.

Models

Five linear Gaussian animal models were used to estimate the (co)variance components of female fertility traits. First, data across multiple parities were analyzed using a multi-trait model in which the performances of a trait in different parities were treated as different traits. For each of the traits defined both in heifers and cows (IFL, CR, AIS and NRR56), a four-trait model was used.

$$
\begin{bmatrix} y_0 \\ y_1 \\ y_2 \\ y_3 \end{bmatrix} = \begin{bmatrix} X_0 & 0 & 0 & 0 \\ 0 & X_1 & 0 & 0 \\ 0 & 0 & X_2 & 0 \\ 0 & 0 & 0 & X_3 \end{bmatrix} \begin{bmatrix} \beta_0 \\ \beta_1 \\ \beta_2 \\ \beta_3 \end{bmatrix} + \begin{bmatrix} Z_0 & 0 & 0 & 0 \\ 0 & Z_1 & 0 & 0 \\ 0 & 0 & Z_2 & 0 \\ 0 & 0 & 0 & Z_3 \end{bmatrix} \begin{bmatrix} a_0 \\ a_1 \\ a_2 \\ a_3 \end{bmatrix} + \begin{bmatrix} e_0 \\ e_1 \\ e_2 \\ e_3 \end{bmatrix} \tag{1}
$$

For traits only defined in cows (ICF and DO), a three-trait model was used.

$$
\begin{bmatrix} y_1 \\ y_2 \\ y_3 \end{bmatrix} = \begin{bmatrix} X_1 & 0 & 0 \\ 0 & X_2 & 0 \\ 0 & 0 & X_3 \end{bmatrix} \begin{bmatrix} \beta_1 \\ \beta_2 \\ \beta_3 \end{bmatrix} + \begin{bmatrix} Z_1 & 0 & 0 \\ 0 & Z_2 & 0 \\ 0 & 0 & Z_3 \end{bmatrix} \begin{bmatrix} a_1 \\ a_2 \\ a_3 \end{bmatrix} + \begin{bmatrix} e_1 \\ e_2 \\ e_3 \end{bmatrix} \tag{2}
$$

In addition, cow traits (ICF, DO, IFLc, CRc, AISc and NRR56c) were further analyzed by pooling data over parities and using a single-trait model treating records in multiple parities as repeated measurements.

$$ y = X\beta + Za + Wpe + e \tag{3} $$

Correlations between a pair of heifer traits and a pair of cow traits were estimated using a two-trait model. For a pair of heifer traits (IFLh, CRh, AISh and NRR56h), the following two-trait model was used:

$$
\begin{bmatrix} y_1 \\ y_2 \end{bmatrix} = \begin{bmatrix} X_1 & 0 \\ 0 & X_2 \end{bmatrix} \begin{bmatrix} \beta_1 \\ \beta_2 \end{bmatrix} + \begin{bmatrix} Z_1 & 0 \\ 0 & Z_2 \end{bmatrix} \begin{bmatrix} a_1 \\ a_2 \end{bmatrix} + \begin{bmatrix} e_1 \\ e_2 \end{bmatrix} \tag{4}
$$

For a pair of cow traits (ICF, DO, IFLc, CRc, AISc and NRR56c), the following two-trait model was used:

$$
\begin{bmatrix} y_1 \\ y_2 \end{bmatrix} = \begin{bmatrix} X_1 & 0 \\ 0 & X_2 \end{bmatrix} \begin{bmatrix} \beta_0 \\ \beta_1 \end{bmatrix} + \begin{bmatrix} Z_1 & 0 \\ 0 & Z_2 \end{bmatrix} \begin{bmatrix} a_1 \\ a_2 \end{bmatrix} + \begin{bmatrix} W_1 & 0 \\ 0 & W_2 \end{bmatrix} \begin{bmatrix} pe_1 \\ pe_2 \end{bmatrix} + \begin{bmatrix} e_1 \\ e_2 \end{bmatrix} \tag{5}
$$

In the above models, y_i was the vector of observations (i = 0, 1, 2, 3 representing heifers and parities 1, 2 and 3 of cows in models (1) and (2); i = 1, 2 representing fertility traits 1 and 2 in models (4) and (5)); β_i was the vector of fixed effects for the ith trait, including age group (in terms of age at first insemination, divided into 270–439 days, 440–469 days, 470–499 days, 500–529 days and 530–900 days), herd-year of calving (for ICF and DO) or herd-year of first insemination within parity (for IFL, CR, AIS and NRR56), year-month of calving (for ICF and DO) or year-month of first insemination within parity (for IFL, CR, AIS and NRR56), AI technician, sexed semen (coded as 1 for the use of sexed semen and 0 for the use of non-sexed semen), and parity (available only for models (3) and (5); a_i was the vector of random additive genetic effects for the ith trait; pe_i was the vector of permanent environmental effects for the ith trait (available only for models (3) and (5)); and e_i was the vector of random residual effects for the ith trait. X_i, Z_i and W_i were the incidence matrices connecting β_i, a_i and pe_i to y_i. It was

assumed that
$$\begin{bmatrix} a_1 \\ \vdots \\ a_n \end{bmatrix} \sim N \left(0, A \otimes \begin{bmatrix} \sigma^2_{a_1} & \cdots & \sigma_{a_1 a_n} \\ & \ddots & \\ & & \sigma^2_{a_n} \end{bmatrix} \right),$$

$$\begin{bmatrix} pe_1 \\ \vdots \\ pe_n \end{bmatrix} \sim N \left(0, I \otimes \begin{bmatrix} \sigma^2_{pe_1} & \cdots & \sigma_{pe_1 pe_n} \\ & \ddots & \vdots \\ & & \sigma^2_{pe_n} \end{bmatrix} \right), \quad \begin{bmatrix} e_1 \\ \vdots \\ e_n \end{bmatrix} \sim N$$

$$\left(0, I \otimes \begin{bmatrix} \sigma^2_{e_1} & \cdots & \sigma_{e_1 e_n} \\ & \ddots & \vdots \\ & & \sigma^2_{e_n} \end{bmatrix} \right).$$

In the above distributions, A is the matrix of additive genetic relationships between individuals in the pedigree; I is the identity matrix; $\sigma^2_{a_i}$, $\sigma^2_{pe_i}$ and $\sigma^2_{e_i}$ were the additive genetic variance, permanent environmental variance and residual variance of the ith trait, respectively; $\sigma_{a_i a_j}$, $\sigma_{pe_i pe_j}$ and $\sigma_{e_i e_j}$ $(i \neq j)$ were the additive genetic covariance, permanent environmental covariance and residual covariance between the ith trait and the jth trait, respectively.

The (co)variance components in the above models were estimated using average information-restricted maximum likelihood (AI-REML) implemented in the DMU package [20]. The asymptotic standard errors of the variances and covariances were obtained from the average information matrix. Standard errors of heritabilities and correlations were calculated according to an expansion of the Taylor series, as previously described [21].

Results

The descriptive statistics for female fertility traits regarding heifers and cows are presented in Table 1. In general, all fertility capacities of the heifers were superior to those of the cow: IFL were 47.4 days shorter, CR was 27.0% higher, NRR56 was 14.8% higher, and AIS was 1.035 less. For all traits of both heifers and cows, large phenotypic variations were observed.

Table 1 Descriptive statistics of female fertility traits in a Chinese Holstein population

Trait[a]	N	Mean	SD	Minimum	Maximum
IFLh	88,647	32.1	69.2	0	365
CRh	88,647	0.626	0.484	0	1
AISh	88,647	1.742	1.286	1	8
NRR56h	88,647	0.732	0.443	0	1
ICF	126,985	82.1	37.3	20	230
DO	126,985	161.6	104.9	20	595
IFLc	126,985	79.5	100.1	0	365
CRc	126,985	0.356	0.479	0	1
AISc	126,985	2.777	1.994	1	8
NRR56c	126,985	0.584	0.493	0	1

[a]*IFL* interval from first to last insemination, *CR* conception rate at first insemination, *AIS* number of inseminations per conception, *NRR56* non-return rate at 56 days after first insemination, *ICF* interval from calving to first insemination, *DO* days open. For traits expressed in both heifers and cows, a suffixes h (for heifers) or c (for cows) was attached to the trait abbreviations

Estimates of the variance components and heritabilities of fertility traits using a multi-trait model and a single-trait model are shown in Table 2. Estimates of heritability of female fertility traits were generally low, ranging from 0.005 (first parity of cows for NRR56) to 0.102 (first parity of cows for ICF). The approximate standard errors for the heritability estimates ranged from 0.001 to 0.011. The estimated heritabilities all differed significantly from zero. As shown in Table 2, binary traits (CR and NRR56) did not exhibit large differences in terms of variance components between heifers and cows. However, for IFL and AIS, the additive variances of heifers (56.6 for IFLh and 0.018 for AISh) were only approximately one = fifth those of cows (257.8 for IFLc and 0.095 for AISc in an analysis of data pooled over parities). For cow traits, additive variances for ICF, DO and IFLc in the first parity were higher than those in the second and third parities, but CRc, AISc and NRR56c exhibited no clear trends. In addition, the three interval traits of cows (ICF, IFL and DO) had higher heritabilities (ranging from 0.027 to 0.083) than the other three traits (ranging from 0.006 to 0.026). On the other hand, the heritabilities of the four heifer traits did not differ dramatically (ranging from 0.008 to 0.013).

The genetic and residual correlations between parities of the performances of a trait using the multi-trait model are presented in Table 3. Generally, the genetic correlations between the performances in heifers and in cows were moderate but far from one, ranging from 0.25 for AIS between heifers and cows at the third parity to 0.74 for NRR56 between heifers and cows at the second parity. However, the genetic correlations between parities of cows were generally high, except for those between the third parity and the other two parities which were 0.63 and 0.66 and had large standard errors, the other genetic

Table 2 Variance components[a] using a multi-trait model and a single-trait model

Parameter[b]	IFL	CR	AIS	NRR56	ICF	DO
$\sigma^2_{a_0}$	56.6	1.851 E-03	0.018	1.352 E-03		
$\sigma^2_{a_1}$	325.8	2.921 E-03	0.101	1.120 E-03	120.5	539.7
$\sigma^2_{a_2}$	282.3	1.937 E-03	0.100	1.292 E-03	69.1	380.0
$\sigma^2_{a_3}$	238.0	2.821 E-03	0.101	2.697 E-03	59.5	396.6
$\sigma^2_{e_0}$	4326.4	0.178	1.392	0.162		
$\sigma^2_{e_1}$	9398.3	0.204	3.390	0.210	1059.5	10,520.4
$\sigma^2_{e_2}$	9073.3	0.187	3.642	0.217	818.8	9862.0
$\sigma^2_{e_3}$	8944.9	0.189	3.642	0.221	798.6	9660.6
$\hat{h}^2_0$	0.013(0.003)	0.010(0.002)	0.013(0.002)	0.008(0.002)		
$\hat{h}^2_1$	0.034(0.005)	0.014(0.003)	0.029(0.005)	0.005(0.002)	0.102(0.008)	0.049(0.006)
$\hat{h}^2_2$	0.030(0.005)	0.010(0.003)	0.027(0.005)	0.006(0.003)	0.078(0.009)	0.037(0.006)
$\hat{h}^2_3$	0.026(0.006)	0.015(0.005)	0.027(0.006)	0.012(0.005)	0.069(0.011)	0.039(0.007)
$\sigma^2_{a_{123}}$	257.8	2.314 E-03	0.095	1.215 E-03	86.5	433.8
$\sigma^2_{pe_{123}}$	804.5	5.285 E-03	0.317	5.664 E-03	41.3	1173.8
$\sigma^2_{e_{123}}$	8438.8	0.191	3.214	0.210	911.3	9034.1
$\hat{h}^2_{123}$	0.027(0.004)	0.012(0.002)	0.026(0.003)	0.006(0.001)	0.083(0.006)	0.041(0.005)

[a] $\sigma^2_{a_i}$ ($i = 0, 1, 2, 3$) = additive genetic variance of ith parity using multi-trait models (model 1); $\sigma^2_{e_i}$ ($i = 0, 1, 2, 3$) = residual variance of ith parity using multi-trait models (model 1); $\hat{h}^2_i$ = estimated heritability of ith parity using multi-trait models (model 1); $\sigma^2_{a_{123}}$ = additive genetic variance of a cow trait over parities using single-trait models (model 4); σ^2_{pe123} = permanent environmental variance of a cow trait over parities using single-trait models; $\sigma^2_{e_{123}}$ = residual variance of a cow trait over parities using single-trait models (model 4); $\hat{h}^2_{123}$ = estimated heritability of acow trait over parities using single-trait models

[b] IFL interval from first to last insemination, CR conception rate at first insemination, AIS number of inseminations per conception, NRR56 non-return rate at 56 days after first insemination, ICF interval from calving to first insemination, DO days open

correlation ranged from 0.87 to 1. On the other hand, the residual correlations between heifers and cows and between parities of cows were very low, ranging from −0.002 to 0.14, indicating small permanent effect on fertility traits. Since all fertility traits had low heritabilities, the phenotypic correlations (not presented) were just very slightly higher than the residual correlations.

Table 4 presents the genetic and phenotypic correlations between heifer traits. The genetic correlations between IFLh, CRh, AISh and NRR56h were strong. The absolute values of these genetic correlations ranged from 0.63 (negative between IFLh and NRR56h) to 0.94 (negative between CRh and AISh). The phenotypic correlations between these traits were moderate to high. The absolute

Table 3 Correlations[a] between performances of a trait in different parities using a multi-trait model

Trait[b]	IFL	CR	AIS	NRR56	ICF	DO
r_{a_0,a_1}	0.64 (0.09)	0.45(0.15)	0.48(0.11)	0.65(0.18)		
r_{a_0,a_2}	0.49 (0.12)	0.52(0.18)	0.40(0.13)	0.74(0.22)		
r_{a_0,a_3}	0.61 (0.14)	0.25(0.22)	0.55(0.14)	0.37(0.23)		
r_{a_1,a_2}	0.89 (0.06)	1.00(0.11)	0.92(0.06)	0.99(0.24)	0.91(0.03)	0.89(0.05)
r_{a_1,a_3}	0.93 (0.08)	0.66(0.19)	0.92(0.08)	0.92(0.23)	0.92(0.04)	0.96(0.05)
r_{a_2,a_3}	0.94 (0.07)	0.63(0.21)	0.94(0.07)	0.87(0.24)	0.96(0.04)	0.96(0.05)
r_{e_0,e_1}	0.06 (0.01)	0.03(0.004)	0.07(0.005)	0.004(0.004)		
r_{e_0,e_2}	0.04 (0.01)	0.01(0.01)	0.05(0.01)	−0.002(0.005)		
r_{e_0,e_3}	0.04 (0.01)	0.01(0.01)	0.05(0.01)	0.01(0.01)		
r_{e_1,e_2}	0.09 (0.01)	0.02(0.01)	0.09(0.01)	0.02(0.01)	0.05(0.01)	0.12(0.01)
r_{e_1,e_3}	0.07 (0.01)	0.02(0.01)	0.08(0.01)	0.02(0.01)	0.01(0.01)	0.08(0.01)
r_{e_2,e_3}	0.11 (0.01)	0.04(0.01)	0.13(0.01)	0.03(0.01)	0.07(0.01)	0.14(0.01)

[a] r_{a_i,a_j} = genetic correlation between performances of a trait in parities i and j, r_{e_i,e_j} = residual correlation between performances of a trait in parities i and j. Phenotypic correlations were similar to residual correlations (not presented)

[b] IFL interval from first to last insemination, CR conception rate at first insemination, AIS number of inseminations per conception, NRR56 non-return rate at 56 days after first insemination, ICF interval from calving to first insemination, DO days open

Table 4 Genetic (above the diagonal) and phenotypic (below the diagonal) correlations between heifer traits using a two-trait model

Trait[a]	IFLh	CRh	AISh	NRR56h
IFLh		−0.86	0.85	−0.63
		(0.06)	(0.04)	(0.12)
CRh	−0.58		−0.94	0.84
	(2.24 E-03)		(0.03)	(0.06)
AISh	0.80	−0.70		−0.87
	(1.24 E-03)	(1.72 E-03)		(0.06)
NRR56h	−0.26	0.76	−0.56	
	(3.14 E-03)	(1.42 E-03)	(2.32 E-03)	

[a]IFLh interval from first to last insemination in heifers, CRh conception rate at first insemination in heifers, AISh number of inseminations per conception in heifers, NRR56h non-return rate within 56 days after first insemination in heifers

values of these phenotypic correlations ranged from 0.26 (negative between IFLh and NRR56h) to 0.80 (negative between IFLh and AISh).

The genetic and phenotypic correlations between cow traits, estimated using data pooled over three parities, are shown in Table 5. There were strong correlations between DO, IFLc, CRc and AISc. The absolute values of genetic correlations between the four traits ranged from 0.82 (positive between DO and AISc) to 0.98 (negative between DOc and IFLc), and the phenotypic correlations ranged from 0.55 (negative between DO and CRc) to 0.97 (positive between DOc and IFLc). ICF had high genetic correlation and moderate phenotypic correlation with DO, while NRR56c had moderate genetic and phenotypic correlations with CRc and AISc.

Discussion

This study investigated the relationships among parities and variance components in each parity for female fertility traits in Chinese Holsteins. In addition, the present study estimated genetic and phenotypic correlations between heifer traits and between cow traits. The results obtained from the present study will be useful for genetic evaluations in dairy cattle breeding programs, particularly for developing a routine genetic evaluation system for female fertility traits in Chinese Holstein population.

The average performances of fertility traits in the present study were similar to those of other populations in general [6, 12, 22]. It was also observed that all the fertility performances of heifers were better than those of cows, consistent with previous studies [23]. The differences in the reproductive physiology between maiden heifers and cows could be caused by first calving and the negative energy balance resulting from high milk yield [13]. In addition, the larger amount of censored data for cows (13.92%) than for heifers (5.92%) could imply that cows had poorer fertility conditions than heifers.

The heritabilities obtained in the present study were consistent with the previous estimates for fertility traits in various populations [6, 12, 24]. For example, Veerkamp and Beerda [25] reviewed the heritabilities estimated in 17 studies and reported that the average heritability estimates for IFL, CR, AIS, NRR56, ICF and DO were less than 5%. Similarly, in the present study, the estimates of heritabilities of IFL, CR, AIS and NRR56 in various parities ranged from 0.005 to 0.034. Furthermore, the estimates of heritabilities of ICF (0.069-0.102) and DO (0.037-0.049) were higher than those of the other traits, which was consistent with previous studies [12, 24].

Table 5 Genetic (above the diagonal) and phenotypic (below the diagonal) correlations between cow traits using a two-trait model

Trait[a]	ICF	DO	IFLc	CRc	AISc	NRR56c
ICF		0.77	0.48	−0.43	0.35	0.40
		(0.03)	(0.07)	(0.09)	(0.07)	(0.11)
DO	0.33		0.98	−0.87	0.82	−0.01
	(2.67 E-03)		(1.57 E-03)	(0.04)	(0.03)	(0.13)
IFLc	2.06 E-03	0.97		−0.90	0.90	−0.15
	(3.00 E-03)	(2.33E-04)		(0.04)	(0.02)	(0.14)
CRc	0.01	−0.55	−0.58		−0.95	0.31
	(2.93 E-03)	(2.00 E-03)	(1.88 E-03)		(0.03)	(0.14)
AISc	1.61 E-03	0.81	0.85	−0.63		−0.51
	(3.00 E-03)	(1.01 E-03)	(7.92 E-04)	(1.70 E-03)		(0.11)
NRR56c	0.03	−0.18	−0.21	0.59	−0.42	
	(2.91 E-03)	(2.76 E-03)	(2.71 E-03)	(1.83 E-03)	(2.35 E-03)	

[a]ICF interval from calving to first insemination, DO days open, IFLc interval from first to last insemination in cows, CRc conception rate at first insemination in cows, AISc number of inseminations per conception in cows, NRR56c non-return rate at 56 days after first insemination in cows

For traits defined in both heifers and cows, variance components for IFL and AIS differed considerably between heifers and cows, and heritability estimates of the two heifer traits were much lower than those of the corresponding cow traits. These results were consistent with previous observations in Canadian Holsteins [6], Brown Swiss [23] and US Holsteins [26]. Moreover, genetic correlations between the performances of heifers and cows were moderate but far from one, consistent with previous studies [6, 14, 23, 26]. These results further indicate that traits in heifers and cows should be treated as different but genetically correlated traits in genetic evaluations. In a breeding program, both heifer and cow traits should be considered. In addition, information for heifer traits is useful for early selection, as heifer traits can be measured during a relatively early period of life.

Genetic correlations between the performances of a trait across parities of cows were high (over 0.87) with the exception for CRc, for which the genetic correlation between the first and second parities was 0.66 and between the second and third parities was 0.63 but with a relatively large standard error. The results were consistent with previous literature [15, 23, 27]. Despite the high genetic correlation, the additive genetic variances of ICFc and DOc in the first parity were substantially larger than those in later parities. The additive genetic variance of IFLc in the first parity was also clearly larger than those in later parities. The differences in additive genetic variances led to differences in heritabilities between the first parity and the other two parities, but the differences in heritabilities were smaller than the differences in additive genetic variances. Differences in additive genetic variances and heritabilities between parities of cows for fertility traits have also been reported in previous studies [15, 23]. The results from the present study suggest that even if there is a high genetic correlation between parities of cows, the performance of a trait in the first parity should be treated different from those in later parities in genetic evaluations, especially for ICF, DO and IFL. To increase reliability and reduce bias in routine genetic evaluations for female fertility traits, Nordic countries have treated parities from 1 to 3 as separate traits instead of repeated observations since May 2015 (http://www.nordicebv.info).

Although the genetic correlations among traits varied substantially, moderate or strong correlations were obtained between traits in groups with similar definitions of fertility functions. Therefore, four groups of traits defined based on reproductive biology highly consistent with the patterns of genetic correlations. The first group of traits measured the abilities of the maiden heifers to conceive and maintain pregnancy, including IFLh, CRh, AISh and NRR56h, which had genetic correlations ranging from 0.63 to 0.94 in absolute values. The second trait group measured the abilities of lactating cows to recycle after calving, indicated by ICF in the present study. The third group of traits described the capacities of cows to conceive and maintain pregnancy, including IFLc, CRc, AISc and NRR56c. Most genetic correlations between these traits were moderate to high, ranging from 0.51 to 0.95 in absolute values, except for the correlations of NRR56 with IFLc (−0.15) and CRc (0.31). The fourth group of traits was DO, which measured the combined abilities to recycle, conceive and maintain pregnancy. This grouping of traits is consistent with the current Interbull concepts of grouping fertility traits.

The traits IFL, CR, AIS and NRR56 all reflect the ability to conceive and maintain pregnancy. According to the estimated genetic parameters, the expected responses to selection given the same selection intensity can be calculated. Among these traits, selection for IFL would achieve the largest genetic progress in the ability to conceive and maintain pregnancy in both heifers and cows. Since IFL had a very high genetic correlation with CR and AIS, it is not necessary to include CR and AIS in a fertility index that contains IFL. NRR56, which measures the capacity for conception and early embryo loss, had a low genetic correlation with IFL for cows. It would be reasonable to include NRR56 in the fertility index. There is no need to include DO in the fertility index given its component traits, ICF and IFL, are already in the index. Thus, IFL, NRR56 and ICF are recommended to be included in the selection index of the Chinese Holstein population.

The present study also estimated genetic and phenotypic correlations between a pair of traits in each parity, using a two-trait model (results not shown). Compared with the results from the analysis using pooled data over parities and treating them as repeated records, the patterns of estimated correlations in each parity were somewhat different from those estimated using the repeatability model (model 5). Moreover, the estimated correlations were not stable across different parities. One possible reason could be bias caused by selection on previous parities, especially the selection for heifer fertility. To avoid this bias, a multi-trait model, which used data from all parities and treated fertility traits from different parities as different traits, was implemented to estimate correlations between a pair of traits for each parity. However, this multi-trait model had difficulty reaching convergence. Therefore, only correlations estimated using data pooled over parities and treating them as repeated records were presented in this study.

For traits with binary phenotypes, a linear model was used in the present study due to low computational requirements. The results showed that the heritability estimates of binary traits were generally lower than those of interval traits measuring conception abilities, consistent with previous studies [6, 8]. Many sophisticated models, such as the threshold model and the generalized linear mixed model, have been proposed for the genetic evaluation of categorical fertility traits. Differences between linear and non-linear models were negligible when phenotypes were polychotomous or the incidences of the binary response were not extreme, e.g. between 25 and 75% [28]. Sun and Su [7] reported that probit and logit models performed slightly better than linear models in terms of the prediction of breeding values for non-return rate and success in first insemination, but the correlations between estimated breeding values from different models were high (up to 0.99). Furthermore, a study assessing NRR56 and ICF using a bivariate threshold-linear and a bivariate-linear model showed that no difference was observed between two models in predictive ability [29]. Nevertheless, due to its intensive computational costs and possible numerical difficulties, the threshold model was not recommended for practical use [29]. Moreover, a simulation study comparing computing aspects of linear and non-linear models illustrated that the computing cost of a threshold model analysis was three to five times larger than that of a linear model [7]. Hence, linear models are widely used in practical routine genetic evaluations of categorical fertility traits in many countries, which is also the reason that a linear model was used in the present study. In fact, four types of generalized linear mixed models, including a probit animal model, a probit sire model, a logit animal model and a logit sire model, were also applied for CR and NRR56 (results not shown). The estimates of heritability from these models were similar with those of the linear animal models. Therefore, only the results of the linear animal models were presented here.

One large challenge in genetic evaluations of fertility interval traits is the presence of censoring, which reflects data truncated through culling or fertility events still in progress. Regarding censored data as uncensored or simply excluding them from the dataset would favor bulls with many daughters having censored records [30], and would fail to account for selection bias [14]. A simple strategy to reduce bias is to add a penalty to censored records. In previous studies of Danish Holsteins, a penalty of 21 days was added to censored ICF, IFL and DO, and a penalty of 1 was added to censored AIS, assuming that cows failing to become pregnant would conceive when provided an extra estrus cycle [19, 24, 31]. In the present study, a penalty of the expected value for a fertility trait was added to a censored record, assuming that the success of an insemination was independent of the result of previous inseminations. The methods for adding a penalty to censored records are simple and offer a potentially sub-optimal strategy to manage censored data. Other methods, such as proportional hazard models [24, 32], may be more efficient for handling data that

include censored records. However, such models also have some limitations, e.g., high computational demand or extra work to obtain animal model solutions. Moreover, choosing the optimum editing criteria using data truncation with different threshold limits could also be an alternative strategy for dealing with censored data [30].

Conclusions
The parameters estimated in the present study will facilitate the development of a genetic evaluation system for fertility traits to improve the female fertility abilities of Chinese Holsteins. The performances of a trait in heifers and cows were significantly different in genetic and should be considered as distinct but correlated traits in genetic evaluations. In addition, even though high genetic correlations between the performances of a trait in different parities of cows, additive genetic variances and heritabilities differed between the first parity and the later parities, which suggests that the performances of a trait in the first and the later parities should be treated as different traits in genetic evaluations. Generally, the estimated heritabilities of all female fertility traits were low, but selection to improve fertility is promising since the genetic variations are large. IFL, NRR56 and ICF are recommended to be included in the selection index of the Chinese Holstein population.

Abbreviations
AI-REML: Average information-restricted maximum likelihood; AIS: Number of inseminations per conception; CR: Conception rate at first insemination; DO: Days open; ICF: Interval from calving to first insemination; IFL: Interval from first to last insemination; MACE: Multiple across-country evaluation; NRR56: Non-return rate within 56 days after first insemination

Acknowledgments
We thank the Dairy Association of China (Beijing, China) for providing pedigrees.

Funding
This work was supported by the earmarked fund for the Modern Agro-industry Technology Research System (CARS-37); the Genomic Selection in Plants and Animals (GenSAP) research project financed by the Danish Council of Strategic Research (Aarhus, Denmark); and the Program for Changjiang Scholar and Innovation Research Team in University (IRT1191). The first author acknowledges the scholarship provided by the China Scholarship Council (CSC).

Authors' contributions
AL conceived the study, collected the data, performed the statistical analysis and wrote the manuscript. GS, YW and ML conceived and supervised the study, made substantial contributions to the interpretation of the results, and revised the manuscript. PM provided statistical analysis tools and revised the manuscript. GG conceived the study and provided the data. GD participated in the data collection. All the authors read and approved the manuscript.

Competing interests
The authors confirm they have read Biomed Central's guidelines on competing interests and declare no competing interests.

Author details

[1]Laboratory of Animal Genetics, Breeding and Reproduction, Ministry of Agriculture of China, National Engineering Laboratory of Animal Breeding, College of Animal Science and Technology, China Agricultural University, Beijing 100193, China. [2]Center for Quantitative Genetics and Genomics, Department of Molecular Biology and Genetics, Aarhus University, 8830 Tjele, Denmark. [3]Beijing Sunlon Livestock Development Co., Ltd, Beijing 100176, China.

References

1. Schneider MD, Strandberg E, Ducrocq V, Roth A. Survival analysis applied to genetic evaluation for female fertility in dairy cattle. J Dairy Sci. 2005;88(6):2253–9.

2. Lucy MC. ADSA Foundation Scholar Award - Reproductive loss in high-producing dairy cattle: Where will it end? J Dairy Sci. 2001;84(6):1277–93.

3. Capitan A, Michot P, Baur A, Saintilan R, Hoze C, Valour D, et al. Genetic tools to improve reproduction traits in dairy cattle. Reprod Fertil Dev. 2015;27(1):14–21.

4. Wall E, Brotherstone S, Woolliams JA, Banos G, Coffey MP. Genetic evaluation of fertility using direct and correlated traits. J Dairy Sci. 2003;86(12):4093–102.

5. Walsh SW, Williams EJ, Evans ACO. A review of the causes of poor fertility in high milk producing dairy cows. Anim Reprod Sci. 2011;123(3–4):127–38.

6. Jamrozik J, Fatehi J, Kistemaker GJ, Schaeffer LR. Estimates of genetic parameters for Canadian Holstein female reproduction traits. J Dairy Sci. 2005;88(6):2199–208.

7. Sun C, Su G. Comparison on models for genetic evaluation of non-return rate and success in first insemination of the Danish Holstein cows. Livest Sci. 2010;127(2–3):205–10.

8. Ghiasi H, Pakdel A, Nejati-Javaremi A, Mehrabani-Yeganeh H, Honarvar M, Gonzalez-Recio O, et al. Genetic variance components for female fertility in Iranian Holstein cows. Livest Sci. 2011;139(3):277–80.

9. Norman HD, Wright JR, Hubbard SM, Miller RH, Hutchison JL. Reproductive status of Holstein and Jersey cows in the United States. J Dairy Sci. 2009;92(7):3517–28.

10. Interbull. 2016. MACE Evaluations Archive. Female fertility. http://www.interbull.org/static/web/fertdoc1608r.pdf. Accessed 10 Oct 2016.

11. Guo G, Guo XY, Wang YC, Zhang X, Zhang SL, Li XZ, et al. Estimation of genetic parameters of fertility traits in Chinese Holstein cattle. Can J Anim Sci. 2014;94(2):281–5.

12. Liu Z, Jaitner J, Reinhardt F, Pasman E, Rensing S, Reents R. Genetic Evaluation of Fertility Traits of Dairy Cattle Using a Multiple-Trait Animal Model. J Dairy Sci. 2008;91(11):4333–43.

13. Wiltbank M, Lopez H, Sartori R, Sangsritavong S, Gumen A. Changes in reproductive physiology of lactating dairy cows due to elevated steroid metabolism. Theriogenology. 2006;65(1):17–29.

14. Raheja KL, Burnside EB, Schaeffer LR. Heifer fertility and its relationship with cow fertility and production traits in Holstein dairy cattle. J Dairy Sci. 1989;72(10):2665–9.

15. Roxström ASE, Berglund B, Emanuelson U, Philipsson J. Genetic and environmental correlations among female fertility traits and milk production in different parities of Swedish Red and White dairy cattle. Acta Agric Scand Sect A Anim Sci. 2001;51(1):7–14.

16. Ranberg IMA, Heringstad B, Klemetsdal G, Svendsen M, Steine T. Heifer fertility in Norwegian dairy cattle: Variance components and genetic change. J Dairy Sci. 2003;86(8):2706–14.

17. Schaeffer LR. Effectiveness of Model for Cow Evaluation Intraherd. J Dairy Sci. 1983;66(4):874–80.

18. Ramirez-Valverde R, Misztal I, Bertrand JK. Comparison of threshold vs linear and animal vs sire models for predicting direct and maternal genetic effects on calving difficulty in beef cattle. J Anim Sci. 2001;79(2):333–8.

19. Sun C, Madsen P, Nielsen US, Zhang Y, Lund MS, Su G. Comparison between a sire model and an animal model for genetic evaluation of fertility traits in Danish Holstein population. J Dairy Sci. 2009;92(8):4063–71.

20. Madsen P, Jensen J. A User's Guide to DMU. DMU, Version 6, release 5.1 2012.

21. Su G, Lund MS, Sorensen D. Selection for litter size at day five to improve litter size at weaning and piglet survival rate. J Anim Sci. 2007;85(6):1385–92.

22. Haile-Mariam M, Morton JM, Goddard ME. Estimates of genetic parameters for fertility traits of Australian Holstein-Friesian cattle. Anim Sci. 2003;76:35–42.

23. Tiezzi F, Maltecca C, Cecchinato A, Penasa M, Bittante G. Genetic parameters for fertility of dairy heifers and cows at different parities and relationships with production traits in first lactation. J Dairy Sci. 2012;95(12):7355–62.

24. Hou Y, Madsen P, Labouriau R, Zhang Y, Lund MS, Su G. Genetic analysis of days from calving to first insemination and days open in Danish Holsteins using different models and censoring scenarios. J Dairy Sci. 2009;92(3):1229–39.

25. Veerkamp RF, Beerda B. Genetics and genomics to improve fertility in high producing dairy cows. Theriogenology. 2007;68:S266–S73.

26. Hansen LB, Freeman AE, Berger PJ. Association of heifer fertility with cow fertility and yield in dairy cattle. J Dairy Sci. 1983;66(2):306–14.

27. Bagnato AO, P. A. Genetic study of fertility and production traits in different parities in Italian Friesian cattle. J Anim Breed Genet. 1993;110:126–34.

28. Meijering A, Gianola D. Linear Versus Nonlinear Methods of Sire Evaluation for Categorical Traits - a Simulation Study. Genet Sel Evol. 1985;17(1):115–31.

29. Andersen-Ranberg IM, Heringstad B, Gianola D, Chang YM, Klemetsdal G. Comparison Between Bivariate Models for 56-Day Nonreturn and Interval from Calving to First Insemination in Norwegian Red. J Dairy Sci. 2005;88(6):2190–8.

30. Oseni S, Tsuruta S, Misztal I, Rekaya R. Genetic parameters for days open and pregnancy rates in US Holsteins using different editing criteria. J Dairy Sci. 2004;87(12):4327–33.

31. Sun C, Madsen P, Lund MS, Zhang Y, Nielsen US, Su G. Improvement in genetic evaluation of female fertility in dairy cattle using multiple-trait models including milk production traits. J Anim Sci. 2010;88(3):871–8.

32. Ducrocq V, Casella G. A Bayesian analysis of mixed survival models. Genet Sel Evol. 1996;28(6):505–29.

9

Genomic regions and pathways associated with gastrointestinal parasites resistance in Santa Inês breed adapted to tropical climate

Mariana Piatto Berton[1*], Rafael Medeiros de Oliveira Silva[1], Elisa Peripolli[1], Nedenia Bonvino Stafuzza[1], Jesús Fernández Martin[2], Maria Saura Álvarez[2], Beatriz Villanueva Gavinã[2], Miguel Angel Toro[3], Georgget Banchero[4], Priscila Silva Oliveira[5], Joanir Pereira Eler[5], Fernando Baldi[1] and José Bento Sterman Ferraz[5]

Abstract

Background: The aim of this study was to estimate variance components and to identify genomic regions and pathways associated with resistance to gastrointestinal parasites, particularly *Haemonchus contortus,* in a breed of sheep adapted to tropical climate. Phenotypes evaluations were performed to verify resistance to gastrointestinal parasites, and were divided into two categories: i) farm phenotypes, assessing body condition score (BCS), degree of anemia assessed by the famacha chart (FAM), fur score (FS) and feces consistency (FC); and ii) lab phenotypes, comprising blood analyses for hematocrit (HCT), white blood cell count (WBC), red blood cell count (RBC), hemoglobin (HGB), platelets (PLT) and transformed (log_{10}) egg per gram of feces (EPG_{log}). A total of 576 animals were genotyped with the Ovine SNP12k BeadChip (Illumina, Inc.), that contains 12,785 bialleleic SNP markers. The variance components were estimated using a single trait model by single step genomic BLUP procedure.

Results: The overall linkage disequilibrium (LD) mean between pairs of markers measured by r^2 was 0.23. The overall LD mean between markers considering windows up to 10 Mb was 0.07. The mean LD between adjacent SNPs across autosomes ranged from 0.02 to 0.10. Heritability estimates were low for EPG_{log} (0.11), moderate for RBC (0.18), PLT (0. 17) HCT (0.20), HGB (0.16) and WBC (0.22), and high for FAM (0.35). A total of 22, 21, 23, 20, 26, 25 and 23 windows for EPG_{log} for FAM, WBC, RBC, PLT, HCT and HGB traits were identified, respectively. Among the associated windows, 10 were shown to be common to HCT and HGB traits on OAR1, OAR2, OAR3, OAR5, OAR8 and OAR15.

Conclusion: The traits indicating gastrointestinal parasites resistance presented an adequate genetic variability to respond to selection in Santa Inês breed, and it is expected a higher genetic gain for FAM trait when compared to the others. The level of LD estimated for markers separated by less than 1 Mb indicated that the Ovine SNP12k BeadChip might be a suitable tool for identifying genomic regions associated with traits related to gastrointestinal parasite resistance. Several candidate genes related to immune system development and activation, inflammatory response, regulation of lymphocytes and leukocytes proliferation were found. These genes may help in the selection of animals with higher resistance to parasites.

Keywords: Gwas, Linkage disequilibrium, Parasites resistance, Santa Inês breed

* Correspondence: mapberton@gmail.com
[1]Departamento de Zootecnia, Faculdade de Ciências Agrárias e Veterinárias, Universidade Estadual Paulista, Via de acesso Prof. Paulo Donato Castellane, s/no, Jaboticabal, SP CEP 14884-900, Brazil
Full list of author information is available at the end of the article

Background

Small ruminants, like sheep (*Ovis aries*) and goats (*Capra aegagrus hircus*) were the first animals domesticated by human for food supply, being the most important group of ruminants raised in temperate and tropical regions [1] . Sheep are multi-purpose animals, raised for meat, milk, wool, hides and skins, having a huge socioeconomic importance worldwide. However, there are several productive drawbacks associated with gastrointestinal parasites infection in small ruminants, since it represents the type of disease with the highest impact on animal health and productivity [2]. Thus, the losses caused by gastrointestinal parasites and the costs due to excessive use of anthelmintic drugs are a struggling problem that restricts the sheep production in many regions of the world.

The main gastrointestinal nematodes of small ruminants belong to the *Trichostrongylidae* family. These parasites occur in the gastrointestinal tract and are seen as the main obstacle in the sheep industry, since they lead to significant economic losses due to high mortality and productivity losses. Within the *Trichostrongylidae* family, the *Haemonchus* genera has great pathogenic action and is the most prevalent parasite affecting small ruminants, mainly in tropical regions, where environmental conditions are characterized by high temperature and humidity, and abundant rainfall during summer [3–6].

To reduce the economic losses caused by nematode infections there are several management alternatives to minimize the damage, such as raise breeds or animals that are more resistant to these infections [5]. In this regard, Santa Inês, an American hair breed, showed higher resistance to gastrointestinal nematode infections when compared to European sheep breeds [5, 7–9]. Several studies have shown that selection for gastrointestinal parasite resistance is possible in sheep, since genetic progress in research and commercial herds [10–16].

The identification of genetic markers associated with gastrointestinal parasites resistance could increase the genetic response for this trait by marker-assisted selection [15]. Furthermore, the identification of genomic regions associated with resistance or susceptibility to gastrointestinal parasites will help to deeper understand the biological and physiological processes underlying this trait [17]. Several studies reported genetic markers associated with gastrointestinal parasites resistance close to the Major Histocompatibility Complex (MHC) [18–20] and *IFN-gama* genes [18–22]. Recently, genome-wide association studies have identified genetic variants for gastrointestinal parasite resistance in some sheep breeds [23–26]. The identification of genomic regions that play a role in gastrointestinal parasite resistance may become an important tool to improve the resistance of Santa Inês or other sheep breeds. Therefore, the aim of this study was to estimate variance components and to identify genomic regions and pathways associated with resistance to gastrointestinal parasites in a Santa Inês population adapted to tropical conditions.

Methods

Data

The phenotypic records were collected from 700 naturally infected animals of Santa Inês breed belonging to four flocks located in the Minas Gerais and São Paulo southeast states and Sergipe northeast state of Brazil. The samples collection and phenotypes evaluations were performed from October to November of 2011. Phenotypes evaluations were achieved to verify resistance to gastrointestinal parasites, and were divided into two categories: i) farm phenotypes, assessing body condition score (BCS), degree of anemia assessed by the famacha card (FAM), fur score (FS) and feces consistency (FC); and ii) lab phenotypes, comprising blood analyses for hematocrit (HCT), white blood cell count (WBC), red blood cell count (RBC), hemoglobin (HGB), platelets (PLT) and the egg counts per gram of feces (EPG_{log}). The EPG values were transformed to $\log_{10}(n + 1) = EPG_{log}$ to meet the basic requirements of normality and homogeneity in an attempt to stabilize the variance prior to analyses, where n is the number of EPG_{log} per sample.

Blood samples were collected by puncture of the jugular vein using vacuum tubes with anticoagulant EDTA and clot activator. Subsequently, samples were submitted to the Veterinary hematology analyzer Sysmex PocH-100iV Diff to perform a complete blood cell count for HCT, WBC, RBC, HGB and PLT. Fecal samples used for EPG evaluation were taken directly from the rectum of the animals and sent to the laboratory for analyses. The EPG count was assessed using the modified McMaster technique [27] and the parasites' genera were identified using the morphometric key by Van Wyk et al. [28]. The feces were classified through visual inspection according to its consistency appearance (FC) developed by Gordon et al. [29] and modified by the authors considering three out of the original five-value scale: 1 for normal stool, 2 for pasty stool, and 3 for diarrheal feces.

The BCS was assessed after evaluating the prominences of the spinous and transverse bones of the spine, fat coverage, and muscle development between the last rib and the ileum wing, as described by Russel et al. [30]. Coloration of the ocular mucosa was measured by trained people by observing the medial part of the lower conjunctiva and comparing it with the famacha chart (FAM; FAMACHA© System) considering a five-value scale: 1 for red robust, 2 for rosy red, 3 for pink, 4 for pale pink, and 5 for white [31]. The descriptive statistics for the analyzed traits are presented in Table 1.

Genotyping of animals

The extraction of the genomic DNA from each animal was performed from blood samples collected with EDTA.

Table 1 Descriptive statistics for eggs per gram of feces (EPG$_{log}$), red blood cells (RBC), famacha (FAM), platelet (PLT), white blood cells (WBC), hematocrit (HCT), and hemoglobin (HGB)

Trait	n	Mean	SD[a]	CV[b], %	Min	Max
EPG$_{log}$	517	2.57	0.53	20.6	1.71	4.04
RBC, 10^6/μL	513	9.97	2.12	21.2	4.39	19.25
FAM	518	3.08	0.97	31.16	1.0	5.0
PLT, 10^3/μL	514	383.8	241.6	62.9	8.0	2101.0
WBC, 10^3/μL	514	10.96	3.73	34.0	3.60	44.40
HCT, %	514	29.05	6.79	23.4	10.90	58.00
HGB, g/dL	514	8.43	1.94	23.0	3.30	17.30

[a]SD: Standard deviation; [b]CV: Coefficient of variation

For this, red blood cells were lysed using 1 mL of whole blood and 500 μL of lysis solution (0.32 mol/L Sucrose, 12 mmol/L Tris-HCl pH 7.5, 5 mmol/L MgCl$_2$, 1% Triton X-100) followed by centrifugation at 13,000 r/min. The supernatant was discarded to reach the leucocytes. The pellets were slowly washed adding 1 mL of ultrapure water and then the microtube was poured out of the water and held for a few minutes on absorbent paper to completely dry the pellet. This washing step was repeated approximately three times until a clean white pellet was obtained. The white pellet was then kept frozen (–20 °C) until sending to the DEOXI laboratory in Araçatuba-SP, where the DNA was completely extracted. The DNA of each sample was quantified and the degree of purity (ratio of optical densities between 260 and 280 nm) (Table 1) was evaluated. After these processes, the samples were stored at –20 °C until genotyping was performed.

A total of 576 animals were genotyped with the Ovine SNP12k BeadChip (Illumina, Inc.), that contained 12,785 biallelic SNP markers. Quality control consisted of excluding markers with unknown genomic position, located on sex chromosomes, monomorphic, with minor allele frequency (MAF) lower than 0.05, call rate lower than 90%, and with excess heterozygosity. After quality control, there were 11,602 SNPs and 574 samples left for analyses.

Quantitative genetic analyses
The variance components were estimated using a single animal trait model by single step genomic BLUP (ssGBLUP) procedure, under Bayesian inference [32]. For all studied traits the fixed effects considered in the model were contemporary groups (farm and management group), month of sample collection, sex, covariable body condition (linear effect), and age at the collection (linear and quadratic effect). All phenotypes were tested for data consistency and contemporary groups with less than three animals and records out of three standard deviations from the contemporary group mean were discarded, remaining 500 animals with phenotypes records. The ssGBLUP is a modified

version of the animal model (BLUP) with additive relationship matrix A^{-1} replaced by H^{-1} [33]:

$$H^{-1} = A^{-1} + \begin{bmatrix} 0 & 0 \\ 0 & G^{-1} - A_{22}^{-1} \end{bmatrix}$$

where A_{22} is a numerator relationship matrix for genotyped animals and G is the genomic relationship matrix created as described by [34]:

$$G = ZDZ'q$$

where Z is the gene matrix containing allele frequency adjustment; D is the matrix that have the SNP weight (initially $D = I$); and, q is a weighting/standardization factor. According to [35], such factors can be obtained by ensuring that the G average diagonal is next to A_{22}. The pedigree file has a total of 1196 animals and the last three generations of animals with records were considered. A linear model was used to analyze HCT, WBC, RBC, HGB, PLT, and EPG$_{log}$. The model can be represented by the following matrix equation:

$$y = X\beta + Za + e$$

where y is the observations vector; β is the vector of fixed effects; a is the additive direct vector; X is the incidence matrix; Z is the incidence genetic random effects additive direct matrix (the β vector associated with the y vector); e is the residual effect vector. The priori distributions of vectors y, a and e were given by:

$$y \sim MVN(X\beta + Za)$$
$$a|G \sim MVN(0, H \otimes G)$$
$$e|R \sim MVN(0, I \otimes R)$$

where H is the relationship coefficients matrix among animals obtained from the single-step analyses (single-step); R is the residual variance matrix; I is the identity matrix; G is genetic additive variance matrix and $\otimes$ is the Kronecker product. An inverted chi-square distribution was used for the prior values of the direct and residual genetic variances. A uniform distribution was used a priori for the fixed effects.

A threshold model was applied to analyze the FAM trait. The scores of FAM for each individual i, were defined by Ui in the underlying scale yi = (1) $t_0 < Ui < t_1$; (2) $t_1 < Ui < t_2$; (3) $t_2 < Ui < t_3$; (4) $t_3 < Ui < t_4$; $i = 1, ... n$, where n is the number of observations; t_1 to t_4 were the threshold values; and U is the unobservable continuous variable, in underlying scale, limited between two unobservable thresholds. After specifying the thresholds t_0 to t_4 it is necessary to adjust one of the thresholds (from t_1 to t_4) into an arbitrary constant. In the present study, it was assumed that $t_1 = 0$, in such a way that the vector of estimable thresholds was defined as $t = t_2, t_3, t_4$. After defining the model parameters, the link between categorical and continuous scales could be established based on the contribution of the probability of an observation that fitted the first category, which is proportional to:

$$P(yv = 0|t, 0) = P(Uv < t|t, 0) = \Phi\left[\left(t - w'vo\right)/\sigma_e\right]$$

where yv is the response variable for the vth observation; t is the threshold value arbitrarily assigned as the true value is unobservable; Uv is the value of the underlying variable for the vth observation; (ϕ) is the cumulative distribution function of a standard normal variable; and $w'v$ is the scale of the incidence matrix that linked θ to the vth observation. As the observations are conditionally independent, given θ, the likelihood function is defined by the product of contributions of each record.

The analyses were performed using the GIBBS2F90 and THRGIBBSF90 programs [33, 36]. The a posteriori variance component estimates were obtained using the POST-GIBBSF90 program [36]. The analysis consisted of a single chain of 500,000 cycles discarding the first 20,000 cycles, taking a sample at every 100 iterations. Thus, 48,000 samples were used to obtain the parameters. The data convergence was verified with the graphical evaluation of sampled values versus interactions according to the criteria proposed by several authors [37–39], using the Bayesian Output Analysis (BOA) of the R 2.9.0 software [40].

Linkage disequilibrium estimation

The linkage disequilibrium (LD) between two SNPs was evaluated using r^2 as follows:

$$r^2 = \frac{(freq.AB^* freq.ab - freq.Ab^* freq.aB)^2}{(freq.A^* freq.a^* freq.B^* freq.b)}$$
$$= \frac{(D)^2}{(freq.A^* freq.a^* freq.B^* freq.b)}$$

where,

$$D = freq.AB - freq.A^* freq.B$$

And

$$D' = \begin{cases} \dfrac{D}{min(freq.A^* freq.b, freq.a^* freq.B)} & if \quad D < 0 \\ \dfrac{D}{min(freq.A^* freq.B, freq.a^* freq.b)} & if \quad D < 0 \end{cases}$$

where $freq.A$, $freq.a$, $freq.B$ and $freq.b$ are the frequencies of alleles A, a, B and b, respectively, and $freq.AB$, $freq.ab$, $freq.aB$ and $freq.Ab$ are the frequencies of haplotypes Ab, ab, aB and Ab in the population, respectively. The expected frequency of haplotype AB ($freq.AB$) is calculated as the product between $freq.A$ and $freq.B$. The r^2 takes values close to zero when the alleles A and B segregate

independently. The $freq.AB$ higher or lower than the expected value indicates that these two loci in particular tend to segregate together and are in LD, with a maximum value for r^2 of one.

Principal component analysis

The principal component analyses (PCAs) were obtained from the genomic relationship matrix through the preGS90 program. The preGS90 is an interface program to process genomic information for the BLUPF90 family [36]. Efficient methods for the creation of the genomic relationship matrix, relationship based on genomic data, and their inverses are described by [41]. The PCAs applied to genotype data can be used to calculate principal components (PCs) that explain differences among individual samples in the genetic data. The top PCs are viewed as continuous axes of variation that reflect genetic variation due to ancestry in the sample. Individuals with similar values for a particular principal component will have a similar ancestry for that axes.

Genome-wide association analysis

The genome-wide association analysis for each trait was performed using the single-step GWAS (ssGWAS) methodology [42]. The same linear animal model for HCT, WBC, RBC, HGB, PLT and EPG_{log}, and the threshold model for FAM used to estimate the variance components were applied. The effects were decomposed in genotyped (a_g) and non-genotyped (a_n) animals as describe by [42], considering the effect of genotyped animals as:

$$a_g = Zu,$$

where Z is a matrix that relates genotypes of each locus and u is a vector of marker effects, and the variance of animal effects was assumed as:

$$var(a_g) = var(Zu) = ZDZ'\sigma_u^2 = G^*\sigma_a^2$$

where D is a diagonal matrix of weights for variances of markers ($D = I$ for GBLUP), σ_u^2 is the genetic additive variance captured by each SNP marker when no weights are present, and G^* is the weighted genomic relationship matrix. The ratio of covariance of genetic effects (a_g) and SNPs (u) is:

$$var\begin{bmatrix} a_t \\ u \end{bmatrix} = \begin{bmatrix} ZDZ' & ZD' \\ DZ' & D \end{bmatrix}\sigma_u^2,$$

Sequentially:

$$G^* = \frac{var(a_g)}{\sigma_a^2} = \frac{var(Zu)}{\sigma_a^2} = ZDZ'\lambda$$

where λ is a variance ratio or a normalizing constant. According to [34],

$$\lambda = \frac{\sigma_a^2}{\sigma_a^2} = \frac{1}{\sum_{i=1}^{M} 2p_i(1-p_i)},$$

where M is the number of SNP and p_i is the allele frequency of the second allele of the ith SNP. According to [43], the markers effect can be described by:

$$\hat{u} = \frac{\sigma_u^2}{\sigma_u^2} \mathbf{DZ'G}*^{-1}\hat{a}_g = \mathbf{DZ'}\left[\mathbf{ZDZ'}\right]^{-1}\hat{a}_{g'}$$

The estimated SNP effects can be used to estimate the variance of each individual SNP effect [44] and apply a different weighting for each marker, such as:

$$\hat{\sigma}_{u,i}^2 = \hat{u}_i^2 2p_i(1-p_i)$$

The following iterative process described by [42] was used considering D to estimate the SNP effects:

1. $D = I$;
2. To calculate the matrix G = ZDZ'q$\mathbf{G} = \mathbf{ZDZ'}$q

3. To calculate GEBVs for all animals in data set using ssGBLUP;
4. To calculate the SNP effect: $\hat{u} = \lambda DZ'G^{*-1}\hat{a}_g$
5. To calculate the variance of each SNP: $d_i = \hat{u}_i^2 2p_i$ $(1-p_i)$, where I is the i-th marker;
6. To normalize the values of SNPs to keep constant the additive genetic variance;
7. Exit, or loop to step 2.

The markers effects were obtained by two iterations from step 2 to 7. The percentage of genetic variance explained by i-th region was calculated as described by [45].

$$\frac{\mathrm{Var}(a_i)}{\sigma_a^2} \times 100 = \frac{\mathrm{Var}\left(\sum_{j=1}^{10} Z_j\hat{u}_j\right)}{\sigma_a^2} \times 100$$

where a_i is genetic value of the i-th region that consists of 10 consecutive SNPs, σ_a^2 is the total genetic variance, $Z_jZ_j$is vector of gene content of the j-th SNP for all individual, and $\hat{u}_j$ $\hat{u}_j$ is marker effect of the j-th within the i-th region. Considering a length of the windows size not exceeding 3 to 4 Mb, the results were presented as the proportion of genetic additive variance explained by windows of 10 consecutive SNPs.

Search for genes

The chromosome regions that explained more than 1.0% of the additive genetic variance were selected to explore and determine possible quantitative trait loci (QTL). The Map Viewer tool of ovine (*Ovis aries*) genome was used for identification of genes, available at "National Center for Biotechnology Information" (NCBI - http://www.ncbi.nlm.nih.gov) database, using the bank references of HuRef assembly,

CHM1 1.0, CRA TCAGchr7v2 and Ensembl Genome Browser (http://www.ensembl.org/index.html). The classification of the genes according to its biological function, identification of metabolic pathways and gene enrichment was performed by DAVID tool v6.8 [46] and GeneCards (http://www.genecards.org/). Gene ontology (GO; $P < 0.01$) and Kyoto Encyclopedia of Genes and Genomes (KEGG; $P < 0.05$) pathways were identified by DAVID tool v6.8 [46]. All annotated genes in the *Ovis aries* genome were used as background.

Results and discussion

Linkage disequilibrium and population structure

The Fig. 1 represents the first two principals (PC1 and PC2) component analysis (PC) based on the genomic relationship matrix. Despite the animals being originated from flocks located in different regions of Brazil, it is important to note that there was no subpopulation structure in this population.

The predicted values of LD vs. linkage distance between genetic markers were presented in Fig. 2. On the basis of this figure, it is possible to state that when considering a distance between markers lower than 1 Mb, the level of LD indicates that the Ovine SNP12k BeadChip may be a suitable tool for identifying genomic regions associated with traits related to gastrointestinal parasites resistance. Most tightly linked SNP pairs have the highest r^2 and average r^2 rapidly decreases as linkage distance increases (Fig. 2). The overall LD mean between markers considering windows up to 10 Mb was 0.07. The LD mean between adjacent SNPs across autosomes ranged from 0.02 to 0.10 (Table 2). Several authors have study the pattern of LD between markers in the genome of various sheep breeds [47–52]. It is difficult to compare the level of LD obtained in different studies since different sample sizes, LD measures, marker types, marker densities and historical population demographics [51] were used. However, the level of LD obtained in the present study was similar to those reported in previous studies for wool sheep breeds using the Ovine SNP50 BeadChip [50–52].

Genetic parameters estimates

The descriptive statistics for FAM indicates an incidence of animals with some degree of anemia (mainly levels 4 and 5) and animals that are not affected (levels 1 and 2) (Table 1). In the present study, animals were free of other sources that lead to anemia, i.e. fluke or minerals deficiencies such as copper, so, it is possible to asseverate that the presence of anemia in these animals was probably due to the prevalence of a *Haemonchus* population and the susceptibility or not of these ewes to infection. Similarly, the descriptive statistics for EPG$_{log}$ indicates an infection by gastrointestinal parasites (Table 1). Most of the gastrointestinal parasites

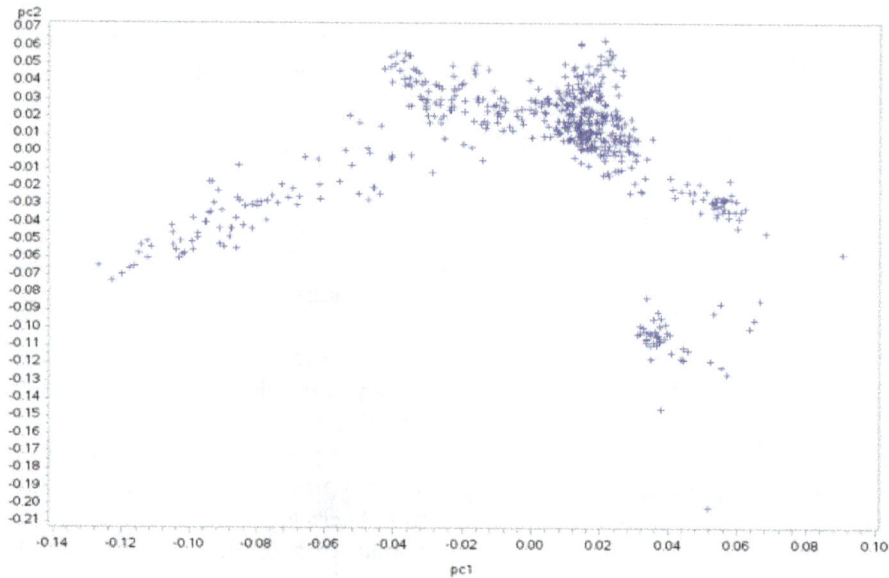

Fig. 1 First two principal components (pc1: 3.16%;pc2:15%) of the genomic relationship matrix of genotyped animals

belonging to the *Trichostrongylidae* family have similar shape and eggs size, and unless the EPG$_{log}$ is accompanied by a larvae cultivar, the eggs can belong to any specie. Moreover, the larva 4 and 5 of *Haemonchus* develops in the wall of the stomach with two particularities: (i) they feed with blood and (ii) do not lay eggs, which make FAM more precise than EPG$_{log}$ to predict infection by *Hemonchus contortus* in all damaging stages.

The variance components and heritability estimates for EPG$_{log}$, RBC, FAM, WBC, PLT, HCT, and HGB were described in Table 3. The parameter estimates

convergence was verified by inspection of trace-plots [37, 38] in which the convergence diagnosis indicates the convergence of the chain. Thus, the burn-in period used was considered enough to achieve the convergence of the estimates of all parameters. The marginal posterior distributions of heritability estimates for the traits showed accurate values, tending to a normal distribution, since the mean and the median were similar (Table 3). Symmetric distributions of central tendency statistics were an indicative that the analysis is reliable.

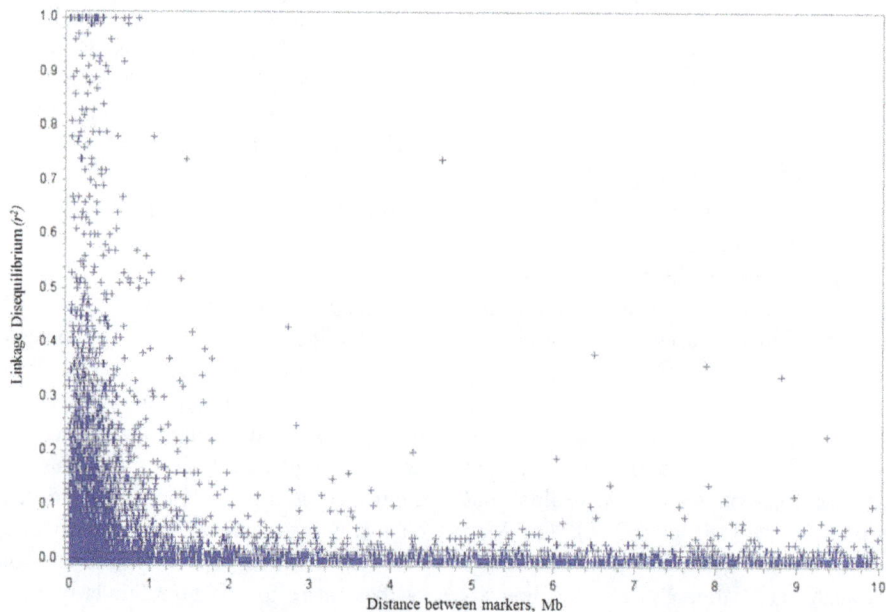

Fig. 2 Linkage disquilibrium (r^2) between markers considering windows up yo 10 Mb

Table 2 Summary of SNP markers analyzed and average linkage disequilibrium (LD; r^2) between synthetic adjacent markers obtained for each autosome (OAR)

OAR	n	Mean LD	SD[a] LD	Mean Distance, Mb	SD[a] Distance, Mb
1	707	0.09	0.17	0.75	1.5
2	683	0.09	0.15	1.01	1.9
3	222	0.04	0.04	2.60	3.0
4	297	0.05	0.09	2.64	3.1
5	369	0.10	0.19	0.72	1.5
6	112	0.03	0.10	2.7	2.5
7	79	0.02	0.04	3.4	2.8
8	288	0.06	0.13	1.9	2.8
9	313	0.07	0.14	1.3	2.3
10	108	0.04	0.10	3.7	2.8
11	171	0.06	0.11	1.5	2.5
12	224	0.06	0.12	1.9	2.7
13	228	0.07	0.14	2.1	2.6
14	131	0.07	0.14	3.0	3.1
15	253	0.08	0.16	1.5	2.5
16	257	0.08	0.16	1.7	2.4
17	152	0.07	0.16	2.6	3.1
18	190	0.07	0.12	1.7	2.5
19	87	0.03	0.09	2.6	2.9
20	137	0.03	0.10	3.4	2.9
21	168	0.06	0.16	2.1	2.8
22	137	0.03	0.06	2.2	2.7
23	201	0.08	0.12	0.9	1.7
24	125	0.09	0.17	1.6	2.1
25	84	0.02	0.11	3.2	2.6
26	175	0.07	0.13	0.93	1.7

[a]SD: Standard deviation

Table 3 Estimates of additive genetic variance (Va), residual variance (Vr), and heritability (h^2) for $\log_{10}$ of eggs per gram of feces (EPG$_{\log}$), red blood cells (RBC), famacha (FAM), platelets (PLT), white blood cells (WBC), hematocrit (HCT), and hemoglobin (HGB)

Trait	Va	Vr	h^2	SD[a]	Median	HPDi	HPDs
EPG$_{\log}$	0.13	1.03	0.11	0.08	0.09	0.001	0.28
RBC	0.45	2.01	0.18	0.10	0.17	0.004	0.37
FAM	0.03	0.06	0.35	0.11	0.35	0.14	0.58
PLT	6.14	28.77	0.17	0.09	0.17	0.02	0.35
WBC	2.71	9.27	0.22	0.10	0.22	0.03	0.41
HCT	4.93	19.04	0.20	0.11	0.19	0.001	0.41
HGB	0.29	1.44	0.16	0.10	0.15	0.001	0.36

[a]SD: heritability standard deviation; HPDi: lower limit of credibility interval (95%) for heritability posterior distribution; HPDs: upper limit of credibility interval (95%) for heritability posterior distribution

Heritability estimates were low for EPG$_{\log}$, moderate for RBC, PLT, HCT, HGB and WBC, and high for FAM. Studies involving Santa Inês breed are scarce in the literature and most of them refers to genetic parameters for traits related to parasites resistance performed in wool sheep breeds. Bisset et al. [53] reported heritability estimate for resistance and resilience of sheep to *Haemonchus contortus* in a South African Merino flock and observed high heritability for EPG, FAM and HCT, with values ranging from 0.47 to 0.55. Contrasting with our results, moderate heritability estimates for EPG among different studies were reported. Bishop et al. [54] studying Texel lambs observed a weighted average heritability of 0.26 and 0.38 for *Strongyle* and *Nematodirus* nematode resistance, respectively. Pickering et al. [25] observed an estimate of 0.25 in New Zealand sheep, and more recently, Benavides et al. [55] reported a heritability estimate of 0.36 in Australian Merino flock. Riley et al. [56] found a low heritability estimate for FAM in a Merino flock in South Africa, with values ranging from 0.06 to 0.24. The heritability estimate obtained for FAM pointed out that genetic progress for this trait is feasible, and this trait should be included in Santa Inês breeding programs. Moreover, the FAM has low cost and it is easily measured. In general, all traits showed genetic variability, therefore, it is important to investigate the presence of genomic regions or genes affecting these traits so as to elucidate and better understand their genetic architecture, especially for *Haemonchus contortus*, since it is one of the most predominant, highly pathogenic and economically important gastrointestinal parasite in sheep 5.

GWAS, genomic regions and enrichment analysis

The SNPs windows regions which accounted for more than 1% of the genetic additive variance were used to search for candidate genes (CG), which are described in Additional file 1. A total of 22, 21, 23, 20, 26, 25 and 23 windows for EPG$_{\log}$, FAM, WBC, RBC, PLT, HCT and HGB traits were identified, respectively. Among the associated windows, 10 are common for HCT and HGB traits, located on OAR1, OAR2, OAR3, OAR5, OAR8 and OAR15 (Additional file 1: Table S9 and S10 – Figs. 3 and 4).

The genomic regions associated with EPG$_{\log}$ are detailed in Additional file 1: Table S4 and illustrated in Fig. 5. Genes such as *CYP11A1* and *CYP1A1* located on chromosome 18, and *CYP19* and *SFXN1* located on chromosome 7 have functions associated with transportation or construction of iron molecules, and its absence in the blood can be an indicator of anemia in animals. The *B2M*, *SFXN1*, *IL25*, *BMP4*, *TSHR*, *CCL28*, *PIK3R1*, *FGF10*, *IL15*, *IL2*, *TP-1*, *BPMG*, *BCL10*, *HSPD1*, and *MALT1* genes are described to have functions related to the body's immune and defense response. Indeed, these

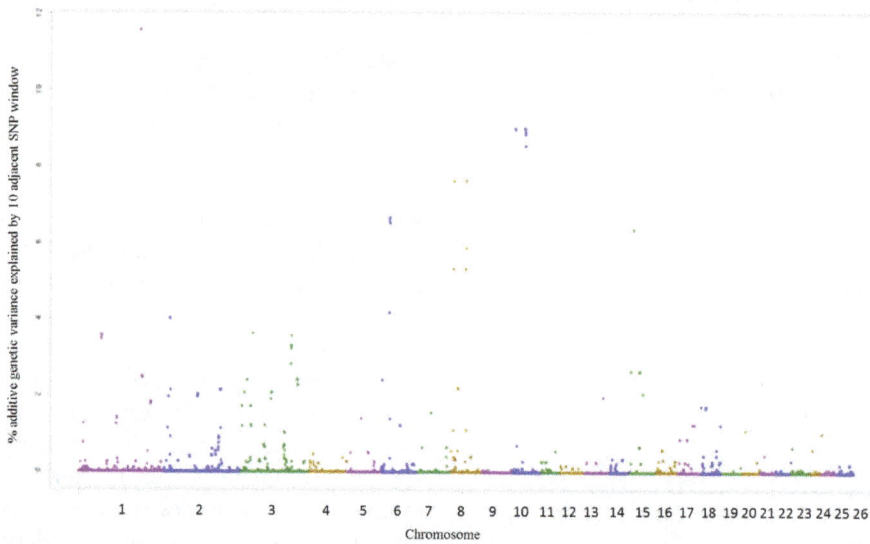

Fig. 3 Manhattan plot of the additive genetic variance explained by windows of 10 adjacent SNPs for hematocrit (HCT) trait

genes participate in metabolic pathways related to the development of the immune system and in the regulation of the immune process effects. Atlija et al. [26] also observed that the *IL25* gene is involved in EPG trait in adult sheep, with functions also linked to immune response. The enrichment analysis revealed the PI3K-Ark signaling pathway, which controls several cellular responses and has important functions in the immune system by regulating many key events in the inflammatory response related to damage and infection [57]. The genomic regions associated with FAM are described in Additional file 1: Table S5 and illustrated in Fig. 6. The

identified CGs have functions related to the development of the immune system and the body's defense response. Genes, such as *ADAM10, IL6ST, TNFRF13B, SIVA1, JUN, PAX1, PIK3R1, SIT1*, and *AKT1* are related to the differentiation of T cells, a group of white blood cells (leukocytes) responsible for defending the body against antigens. The *GUCY1A2* gene, which interacts selectively and noncovalently with iron Fe and the *SIVA1* gene, both are expressed in different subpopulations of T and B cells and provide co-stimulatory signals for the proliferation of T and B cells and immunoglobulin production by B cells [58]. The *GUCY1A2* gene operates in the process of regulating

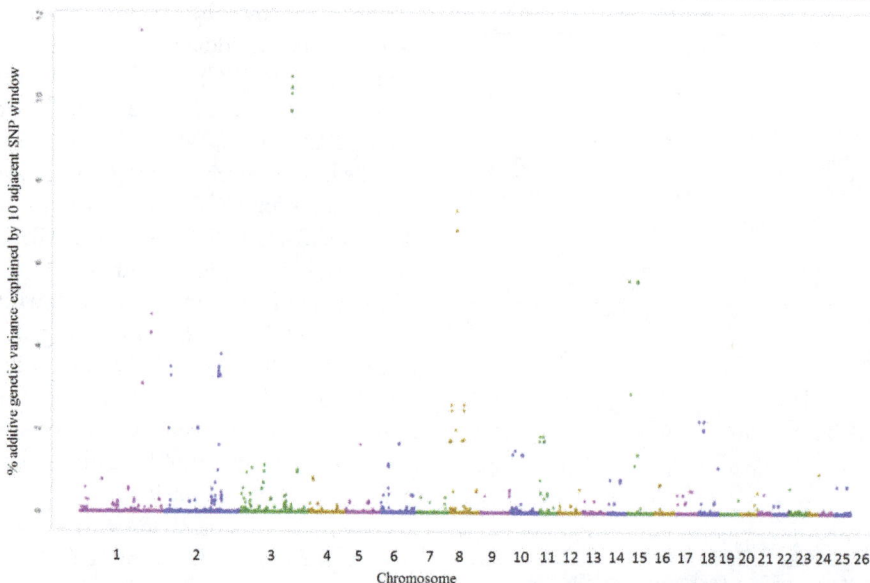

Fig. 4 Manhattan plot of the additive genetic variance explained by windows of 10 adjacent SNPs for hemoglobin (HGB) trait

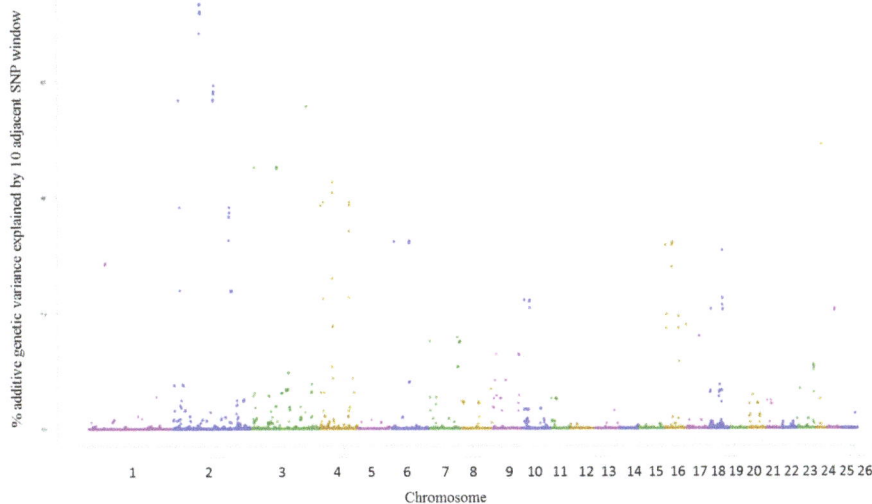

Fig. 5 Manhattan plot of the additive genetic variance explained by windows of 10 adjacent SNPs for egg count per gram of feces (EPG$_{log}$) trait

the proliferation and elimination of T cells, maintaining its number stable in the absence of external stimulus.

The results of the enriched genes and functional grouping analyses showed that genes associated with FAM are involved in functions related to body's immune response. The biological processes (Additional file 2 – Table S11 and S12) showed that these genes are significantly ($P < 0.05$) involved in lymphocytes and leukocytes homeostasis.

The *CCL28* gene located on OAR16 identified both for EPG and FAM acts as a chemotactic for CD4 and CD8 inactive T cells [59]. The metabolic pathways associated with the *CCL28* gene and with others associated with EPG$_{log}$ and FAM traits, whose functions were found

related to immunoglobulin synthesis in the intestine, are presented in Fig. 7.

The genomic regions associated with WBC are described in Additional file 1: Table S6 and illustrated in Fig. 8. The genes found in the SNP windows encode proteins that are involved in many immune system processes, responding to any potential internal or invasive threat. The vertebrate's immune system is composed by the innate immune system (defense responses mediated by germline encoded components that directly recognize components of potential pathogens) and by the adaptive immune system which consists of T- and B-lymphocytes [60]. Many different innate immune mechanisms are deployed for host defense, a unifying

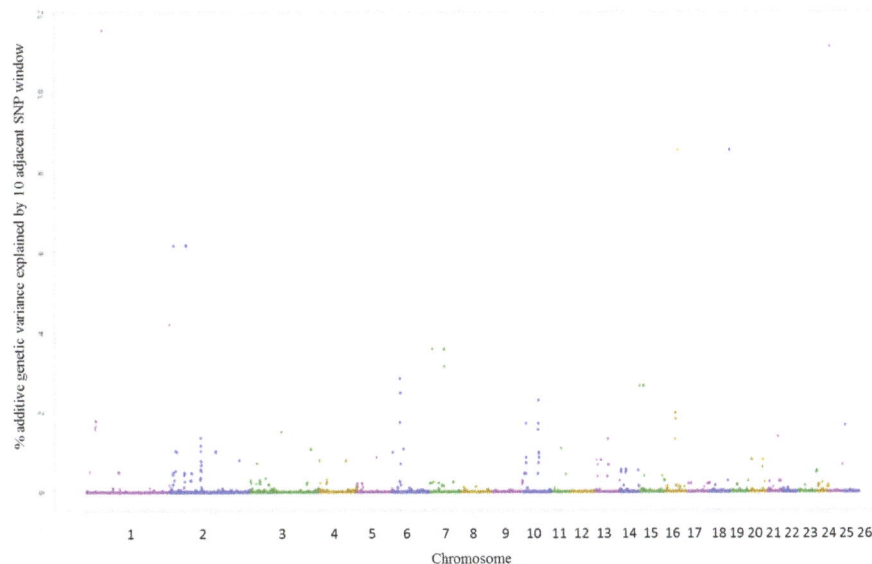

Fig. 6 Manhattan plot of the additive genetic variance explained by windows of 10 adjacent SNPs for famacha (FAM) trait

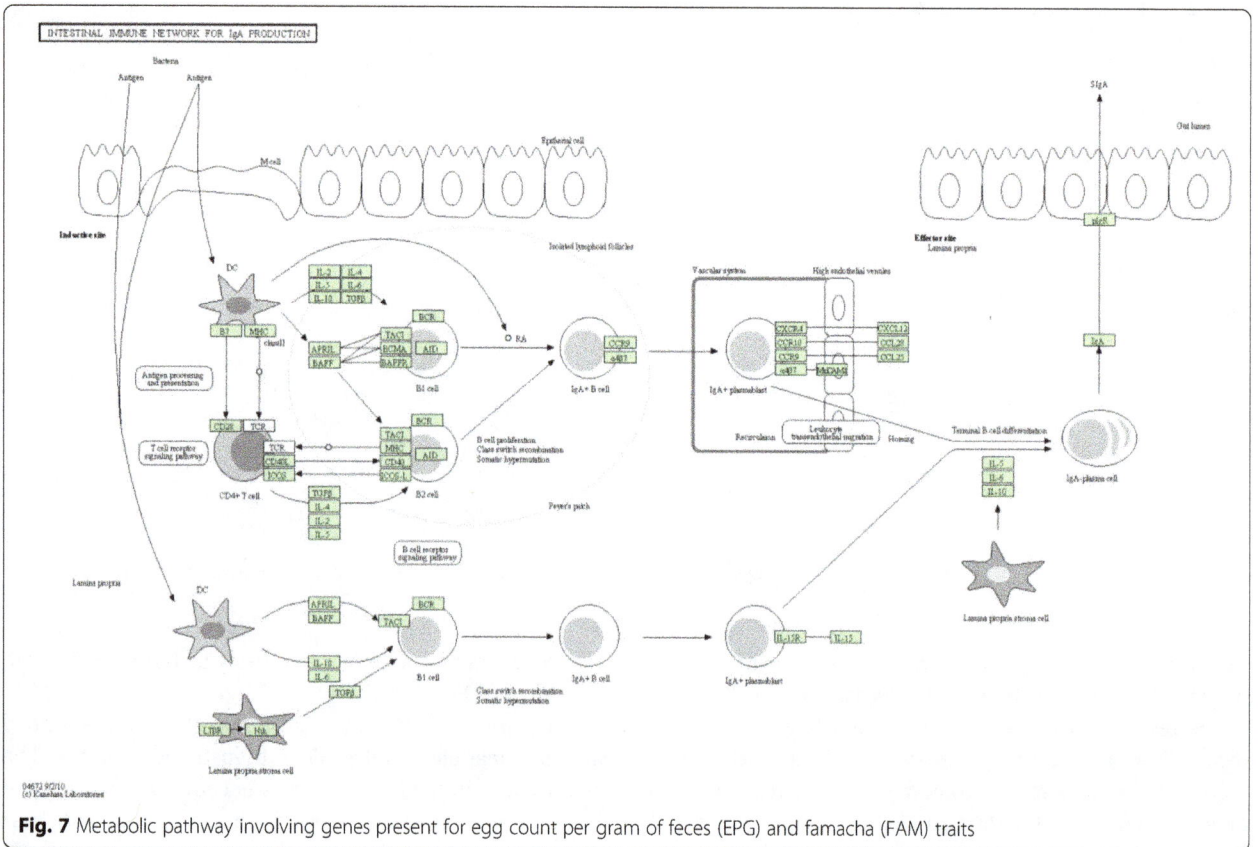

Fig. 7 Metabolic pathway involving genes present for egg count per gram of feces (EPG) and famacha (FAM) traits

theme of innate immunity is the use of germline-encoded pattern recognition receptors for pathogens or damaged self-components, such as the toll-like receptors, nucleotide-binding domain leucine-rich repeat (LRR)-containing receptors, retinoic acid-inducible gene I-like RNA helicases and C-type lectin receptors [61], like the CRCP (CGRP receptor component (CRCP)).

Some of these genes (Additional file 1: Table S6) have already been documented in others species and been associated with immune response. Beside the functions described above, they also present some particularities, i.e. the *EPBH1* (EPH Receptor B1), which is involved in the GPCR pathway and developmental biology. The first process involves the G protein–coupled receptors

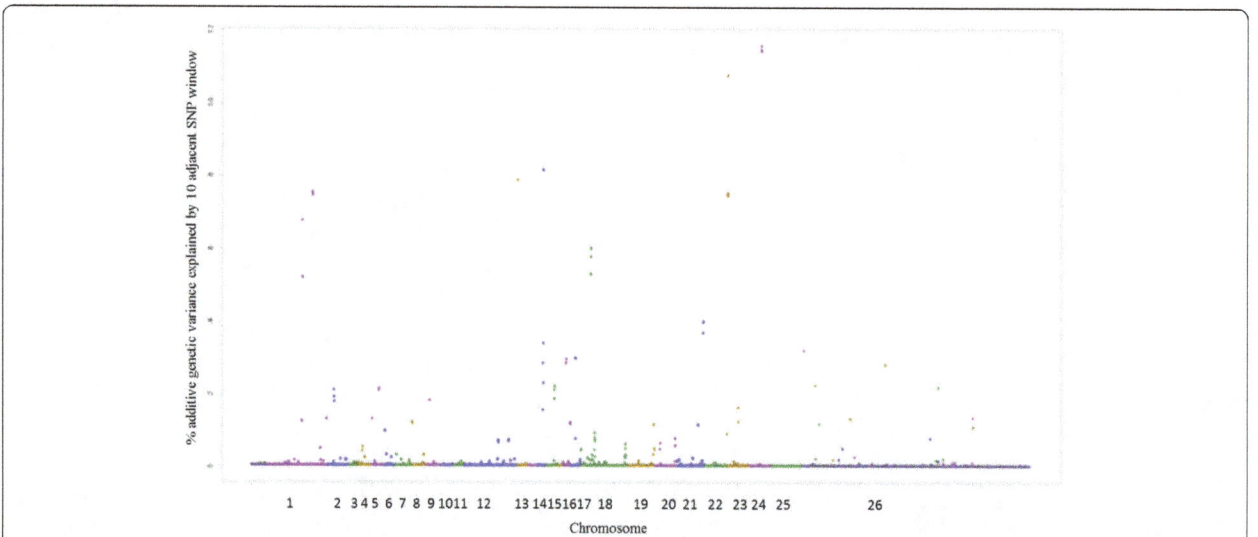

Fig. 8 Manhattan plot of the additive genetic variance explained by windows of 10 adjacent SNPs for white blood cells (WBC) trait

(GPCRs) that constitute a large protein family of receptors that recognize molecules outside the cell and activate inside signal transduction pathways and, finally, cellular responses [62].

The *GATA3* (GATA binding protein 3) gene also acts on the defense response, reacting to the presence of a foreign body or to the occurrence of an injury, restricting the injury/damage extension or preventing/recovering it from the infection caused by the foreign body. The *GATA3* gene also positively regulates T cell differentiation and any process that activates or increases the frequency, rate or extent of T cell differentiation. An additional gene having a peculiar function is the *SAMHD1* (SAM domain and HD domain 1), which is also associated with the regulation of the innate immune responses. The *SLA2* (Src-like-adaptor 2) gene acts in the T cell regulation and negative regulation of B cell activation, and in any process that stops, prevents, or reduces the frequency, rate or extent of B cell activation.

The genes in the enrichment pathways (Additional file 2), such as the *MYPN* (myopalladin), *PDGFRL* (platelet derived growth factor receptor like), *ROR1* (receptor tyrosine kinase-like orphan receptor 1), *CNTN1* (Contactin 1) and *FANK1* (fibronectin type 3, and ankyrin repeat domains 1) are associated with the Immunoglobulin-like domains that are related in both sequence and structure, and can be found in several diverse protein families. Ig-like domains are involved in a variety of functions, including cell-cell recognition, cell-surface receptors, muscle structure and the immune system [63].

Additional file 1: Table S7 and S8 show the genes with functions associated with immune response, innate, and adaptative immune response or those participating in the regulation of innate immune response for the genomic regions associated with RBC and PLT traits, respectively

The genomic regions and genes associated with RBC are presented in Additional file 1: Table S7 and illustrated in Fig. 9. The KEGG pathays analysis revealed PI3K-Ark and toll-like receptor signaling pathways significantly enriched for WBC trait (Additional File 2). The *CD109* gene (CD109 antigen) is a GPI-linked cell surface antigen expressed by CD34+ acute myeloid leukemia cell lines, T-cell lines, activated T lymphoblasts, endothelial cells, and activated platelets [64]. This gene was also reported by [26], indicating that *CD109* gene potentially contributes to resistance to gastrointestinal nematodes in sheep.

The *IL2RB* gene (interleukin 2 receptor), which is involved in T cell-mediated immune responses, is present in three forms with respect to ability to bind interleukin 2. The low affinity form is a monomer of the alpha subunit and it is not involved in signal transduction. The intermediate affinity form consists of an alpha/beta subunit heterodimer, while the high affinity form consists of an alpha/beta/gamma subunit heterotrimer. Both the intermediate and high affinity forms of the receptor are involved in receptor-mediated endocytosis and transduction of mitogenic signals from interleukin 2 [provided by RefSeq, Jul 2008]. Kondo et al. [65] showed that a clonogenic common lymphoid progenitor, a bone marrow-resident cell that gives rise exclusively to lymphocytes (T, B, and natural killer (NK) cells), can be redirected to the myeloid lineage by stimulation through exogenously expressed interleukin (IL)-2 and GM-CSF (granulocyte/macrophage colony-stimulating factor) receptors. As the *IL2RB* is one of the receptors that act in IL-15 expression, it has been proposed to be a critical cytokine for NK cell development. This gene can affect this

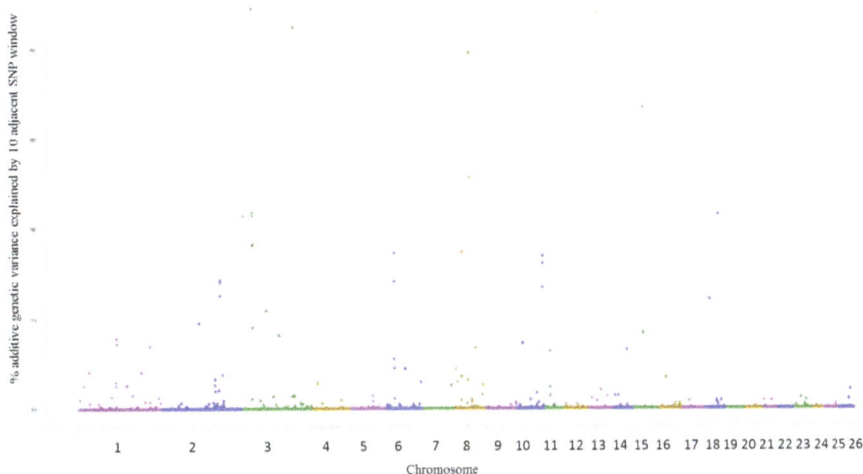

Fig. 9 Manhattan plot of the additive genetic variance explained by windows of 10 adjacent SNPs for red blood cells (RBC) trait

redirections in cytokine signaling and can regulate cell-fate decisions. Additionally, a critical step in lymphoid commitment is downregulation of cytokine receptors that drive myeloid cell development.

Poliovirus receptor-like proteins (PVRLs), such as PVRL4, are adhesion receptors of the immunoglobulin superfamily and function in cell-cell adhesion [66]. The encoded protein contains two immunoglobulin-like (Ig-like) C2-type domains and one Ig-like V-type domain. It is involved in cell adhesion through trans-homophilic and -heterophilic interactions. It is a single-pass type I membrane protein [provided by RefSeq, Jan 2011].

The *CXCR4* gene may influence the immune system under physiologic and pathologic conditions through negative regulation of MHC class II expression, possibly through PKA and SRC kinase [67]. In a study, Contento et al. [68] observed that while human T cell activation by antigen-presenting cells is taking place, the *CCR5* and *CXCR4* chemokine receptors are recruited into the immunological synapse, where they deliver costimulatory signals. This gene also participates in the intestinal immune network for IgA production pathway.

The *ASCC3, CRLF2RB, FNDC3A, FNDC4, IFNAR1, IFNAR2, LEPR, MYLK, NCAM2, PVRL4 SLAMF1, TXNIP, CD109, ACAN, NTRK3, CCDC141, LEPR, MYLK, PVLR4, IL12RB2, IGSF10,* and *JAM2* genes (Additional file 1: Table S7) are domains with an Ig-like fold that can be found in several proteins in addition to immunoglobulin molecules. For example, Ig-like domains occur in several different types of receptors (such as various T-cell antigen receptors), several cell adhesion molecules, MHC class I and II antigens, as well as the hemolymph protein hemolin, and the muscle proteins titin, telokin and twitchin (IPR013783).

The KEGG pathway analysis revealed platelet activation, toll-like receptor and PI3K-Ark signaling pathways for PLT (Additional file 2). The *CXCL1, CXCL8* and *CXCL10* genes identified for PLT (Additional file 1: Table S8; Fig. 10) belong to a subfamily of chemokines that basically are structurally related to molecules that regulate cell trafficking of various types of leukocytes through interactions with a subset of 7-transmembrane and G protein-coupled receptors. They also play a fundamental role in the development, homeostasis, and functionality of the immune system [69].

Different genes were also identified for PLT, in these regard, the Fc-gamma receptors (FCGRs), such as *FCGR1A, FCGR3A,* and *FCGR2B,* that are integral membrane glycoproteins which exhibit complex activation or inhibitory effects on cell functions after aggregation by complexed immunoglobulin G (IgG) [70]. These genes encode proteins that play an important role in the immune response, a receptor for the Fc portion of immunoglobulin G, and participate in the removal of antigen-antibody complexes from the circulation, as well as other antibody-dependent responses. It is also involved in the phagocytosis of immune complexes and in the regulation of antibody production by B-cells, respectively [provided by RefSeq, Jul 2008; RefSeq, Jun 2010].

Atlija et al. [26] observed some significant chromosome-wise QTL detected by linkage analysis and from the combined LD and linkage analysis. These authors reported some CG involved in immune response. Theirs findings support our results, since genes such as *MBN, BTC, CXCL1, CXCL10, EREG, RASSF6, SCARB2, TMPRSS11D, AMBN,* and *AREG* were observed in both studies, indicating that these genes can be used and deply studied as CG for resistance to gastrointestinal nematodes in sheep. Some of the genes cited in Additional file 1: Table S9 and

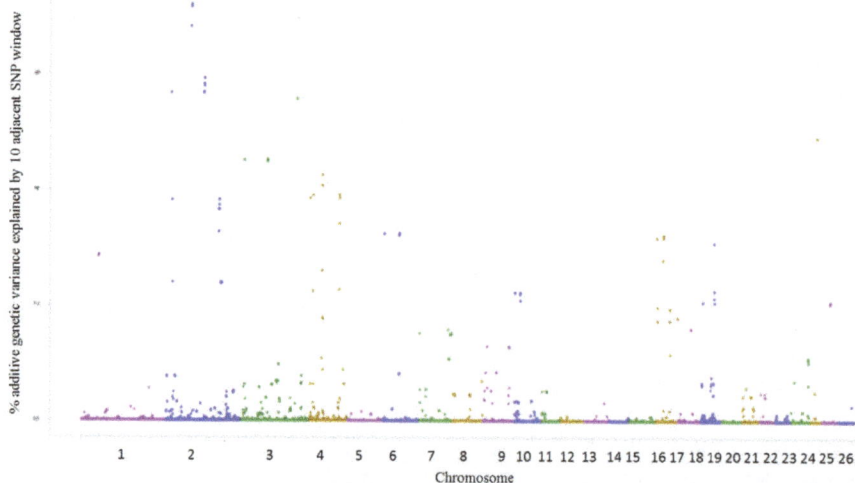

Fig. 10 Manhattan plot of the additive genetic variance explained by windows of 10 adjacent SNPs for platelet (PLT) trait

S10 have functions related to immune response or to immunoglobulin development, and were associated with HCT and HBG traits, respectively.

For HCT trait (Table S9), genes like *IGF1R*, *IFNAR1*, *IFNAR2*, *IL12RB2*, *IL2RB*, *C2F2RB*, *PVRL4*, *LEPR*, *MYLK*, *SLAMF1*, *ADGRF3*, and *TXNIP* were identified. These genes encode domains with an Ig-like fold that can be found in several proteins in addition to immunoglobulin molecules. For example, Ig-like domains occur in several different types of receptors (such as T-cell antigen receptors), many cell adhesion molecules, MHC class I and II antigens, as well as the hemolymph protein hemolin, and the muscle proteins titin, telokin and twitchin.

The *IL1A* and *IL1B* genes have the protein encoded as a member of the interleukin 1 cytokine family. This cytokine is a pleiotropic cytokine involved in several immune responses, inflammatory processes, and hematopoiesis [provided by RefSeq, Jul 2008]. Also the *IL1B* gene, initially discovered as the major endogenous pyrogen, induces prostaglandin synthesis, neutrophil influx and activation, T- and B-cells activation, cytokine production, antibody production, fibroblast proliferation and collagen production. Also it was observed a function related to ligand-binding domain that displays similarity to C2-set immunoglobulin domains (antibody constant domain 2) in the *LEPR* gene. It has a specific effect on T lymphocyte responses, differentially regulating the proliferation of naive and memory T-cells.

For HCT and HGB traits, genes related to immune response were identified, i.e. the *RAC2* gene, that seems to increase the resistance to parasites. In order to understand the function of RAC2 GTPase in regulating the cellular immune response, Williams et al. [71] assayed the effects of hemocytes in parasitized *RAC2* mutant larvae, as well as characterized the effect of over-expressing *RAC2* gene in hemocytes. These authors reported that this gene has an important role in the cellular immune response, being necessary for hemocyte spreading and cell-cell contact formation during immune surveillance against the parasitoid *L. boulardi*. When an invading parasitoid is recognized as a foreign body, circulating hemocytes should recognize them and attach to the egg chorion. After the attachment, the hemocytes should then spread out to form a multilayered capsule surrounding the invader. In *RAC2* mutants this process was disrupted. Besides of that, this gene also acts on the B cell receptor signaling pathway, like the *PPP3CA* gene.

The *HSPD1* gene encodes a mitochondrial protein that plays a role as a signaling molecule in the innate immune system. Zanin-Zhorov et al. [72] reported that this gene, as well as a synthetic peptide derived from it, acted as a costimulatory of human

regulatory CD4-positive/CD25 (*IL2RA*) positive T cells (Tregs), which inhibit lympho proliferation and IFNG and TNF secretion by CD4-positive and CD8-positive T cells. The authors concluded that the self-molecule *HSPD1* can downregulate adaptive immune responses by upregulating Tregs through *TLR2* signaling.

The *CD80* and *CD86* genes identified in almost all traits (RBC, PLT and HCT) probably participate effectively in the activation of T cells, which requires engagement of two separate T-cell receptors. The antigen-specific T-cell receptor (TCR) binds foreign peptide antigen-MHC complexes, and the CD28 receptor binds to the B7 (*CD80*/*CD86*) costimulatory molecules expressed on the surface of antigen-presenting cells (APC). The simultaneous triggering of these T-cell surface receptors with their specific ligands results in the activation of this cell. Many in vitro and in vivo studies demonstrated that both CD80 and CD86 ligands have an identical role in the activation of T cells. Recently, functions of B7 costimulatory molecules in vivo have been investigated in B7-1 and/or B7-2 knockout mice, and the authors concluded that *CD86* could be more important for initiating T-cell responses, while *CD80* could be more significant for maintaining these immune responses [73]. Recently, [24] observed some important genes matching to important pathways involved in host immune response against parasites. The *PDGFRA* gene (OAR6: 67,950,121–69,892,816) was also observed by Benavides et al. [24], whose findings have pointed out as a key gene in cytokine signaling.

Conclusions

The traits indicating gastrointestinal parasites resistance shown adequate genetic variability to respond to selection in Santa Inês breed, and it is expected a higher genetic gain for famacha when compared to the others traits. The level of LD estimated for markers separated by less than 1 Mb indicates that the Ovine SNP12k BeadChip will likely be a suitable tool for identifying genomic regions associated with those traits related to gastrointestinal parasite resistance.

Several candidate regions related to immune system development and activation, inflammatory response, regulation of lymphocytes, and leukocytes proliferation were found in this study. These candidate regions and CG may help in the selection of animals with higher resistance to parasites, and consequently, reduce the anthelmintic drugs costs, also the production losses linked to the use of it. Furthermore, when reducing the use of anthelmintic drugs, a potential reduction in waste problems regarding meat and milk discharge due to drugs residues should be minimized, reducing the impact upon the environment.

Additional files

Additional file 1: Table S4. Genomic regions associated with egg counts per gram of feces (EPG$_{log}$) in Santa Inês sheep; **Table S5.** Genomic regions associated with famacha (FAM) in Santa Inês sheep; **Table S6.** Genomic regions associated with blood cell count (WBC) in Santa Inês sheep; **Table S7.** Genomic regions associated with red blood cells (RBC) in Santa Inês sheep; **Table S8.** Genomic regions associated with platelet (PLT) in Santa Inês sheep; **Table S9.** Genomic regions associated with hematocrit (HCT) in Santa Inês sheep; **Table S10.** Genomic regions associated with blood cell count (HGB) in Santa Inês sheep.

Additional file 2: Table S11. KEGG pathways ($P < 0.05$) for counting eggs per gram of feces (EPG$_{log}$), famacha (FAM), white blood cells (WBC), red blood cells (RBC), platelets (PLT), hematocrit (HCT), and hemoglobin (HGB) traits obtained from the DAVID software; **Table S12.** Gene Ontology terms for biological processes significantly ($P < 0.01$) related with counting eggs per gram of feces (EPG$_{log}$), famacha (FAM), white blood cells (WBC), red blood cells (RBC), platelets (PLT), and hematocrit (HCT) traits. (XLS 68 kb)

Abbreviations

BCS: Body condition score; EPG$_{log}$: Egg counts per gram of feces; FAM: Degree of anemia assessed by the famacha card; FC: Feces consistency; FS: Fur score; HCT: Hematocrit; HGB: Hemoglobin; LD: Linkage disequilibrium; PLT: Platelets; RBC: Red blood cell count; WBC: White blood cell count

Acknowledgements

To Fapesp, (Sao Paulo Research Foundation, grants # 2010/05516-7, #2011/00396-6 and #2014/07566-2). MP Berton, E Peripolli received scholarship from Coordination for the Improvement of Higher Education Personnel (CAPES; Coordenação de Aperfeiçoamento de Pessoal de Nível Superior) in conjunction with the Postgraduate Program on Genetics and Animal Breeding, Universidade Estadual Paulista, Faculdade de Ciências Agrárias e Veterinárias (FCAV, Unesp). F. B held productivity research fellowships from The Brazilian National Council for Scientific and Technological Development (CNPQ)

Funding

Sao Paulo Research Foundation – FAPESP grants # 2010/05516-7, #2011/00396-6 and #2014/07566-2.

Authors' contributions

MPB, RMOS, EP, NBS did the manuscript writing and discussion of the results. JFM, MSA, BVG, in Genomic modeling and genomic analysis. GB in concepts, writing and modeling. PSO in phenotypic data collection and genotyping. JPE in data management. FB JBSF in conception, funding, modeling, genomic analysis and coordination. All authors read and approved the final manuscript.

Competing interests

The authors declare that they have no competing interests.

Author details

[1]Departamento de Zootecnia, Faculdade de Ciências Agrárias e Veterinárias, Universidade Estadual Paulista, Via de acesso Prof. Paulo Donato Castellane, s/no, Jaboticabal, SP CEP 14884-900, Brazil. [2]Instituto Nacional de Investigación y Tecnología Agraria y Alimentaria INIA, Crta. de la Coruña, km 7,5 -, 28040 Madrid, Spain. [3]Departamento de Producción Agraria, School of Agricultural, Food and Byosystems Engineering, Universisdad Politécnica de Madrid, Campus Ciudad Universitaria Avda. Complutense 3 - Avda. Puerta Hierro, 28040 Madrid, Spain. [4]Instituto Nacional de Investigación Agropecuária (INIA), Ruta 50 Km. 12, Colonia, Uruguay. [5]Faculdade de Zootecnia e Engenharia de Alimentos, Nucleo de Apoio à Pesquisa em Melhoramento Animal, Biotecnologia e Transgenia, Universidade de São Paulo, Rua Duque de Caxias Norte, 225, Pirassununga, SP CEP 13635-900, Brazil.

References

1. Zygoyiannis D. Sheep production in the world and in Greece. Small Rumin Res. 2006;62:143-7.
2. Perry BD, Randolph TF, Mcdermott JJ, Sones KR, Thornton PK. Investing in Animal Health Research to Alleviate Poverty. ILRI (International Livestock Research Institute); 2002.
3. Liu J, Zhang L, Xu L, Ren H, Lu J, Zhang X, et al. Analysis of copy number variations in the sheep genome using 50K SNP BeadChip array. BMC Genomics. 2013;14:229.
4. Da Silva M V, Lopes PS, Guimarães SE, De R, Torres A. Utilização de marcadores genéticos em suínos. I. Características reprodutivas e de resistência a doenças The use of genetic markers in swine. I. Reproductive and disease resistance traits. 2002;
5. Amarante AFT, Bricarello PA, Rocha RA, Gennari SM. Resistance of Santa Ines, Suffolk and Ile de France sheep to naturally acquired gastrointestinal nematode infections. Vet Parasitol. 2004;120:91-106.
6. O'Connor LJ, Walkden-Brown SW, Kahn LP. Ecology of the free-living stages of major trichostrongylid parasites of sheep. Vet Parasitol. 2006;142:1-15.
7. Rocha RA, Amarante AFT, Bricarello PA. Comparison of the susceptibility of Santa Inês and Ile de France ewes to nematode parasitism around parturition and during lactation. Small Rumin Res. 2004;55:65-75.
8. Bricarello PA, Amarante AFT, Rocha RA, Cabral Filho SL, Huntley JF, Houdijk JGM, et al. Influence of dietary protein supply on resistance to experimental infections with Haemonchus contortus in Ile de France and Santa Ines lambs. Vet Parasitol. 2005;134:99-109.
9. Costa RLD, Bueno MS, Veríssimo CJ, Cunha EA, Santos LE, Oliveira SM, et al. Performance and nematode infection of ewe lambs on intensive rotational grazing with two different cultivars of Panicum maximum. Trop Anim Health Prod. 2007;39:255-63.
10. Woolaston RR, Piper LR. Selection of Merino sheep for resistance to Haemonchus contortus: genetic variation. Anim Semit. 2016;62:451-60.
11. Woolaston RR, Windon RG. Selection of sheep for response to Trichostrongylus colubriformis larvae: genetic parameters. Anim Sci. 2016;73:41-8.
12. Morris CA, Wheeler M, Watson TG, Hosking BC, Leathwick DM. Direct and correlated responses to selection for high or low faecal nematode egg count in Perendale sheep. New Zeal J Agric Res. 2005;48:10.
13. Karlsson LJE, Greeff JC. Selection response in fecal worm egg counts in the Rylington Merino parasite resistant flock. Aust J Exp Agric. 2006;46:809-11.
14. Mcewan JC. Developing genomic resources for whole genome selection. Proceedings of the New Zealand Society of Animal Production. 2007. p. 148-53.
15. Kemper KE. The implications for the host-parasite relationship when sheep are bred for enhanced resistance to gastrointestinal nematodes. PhD thesis. Melbourne: The University of Melbourne; 2010.
16. Pickering NK, Blair HT, Hickson RE, Dodds KG, Johnson PL, McEwan JC. Genetic relationships between dagginess, breech bareness, and wool traits in New Zealand dual-purpose sheep. J Anim Sci. 2013;91:4578-88.
17. Mcrae KM, Stear MJ, Good B, Keane OM. The host immune response to gastrointestinal nematode infection in sheep. Parasite Immunol. 2015;37:605-13.
18. Miller JE, Horohov DW. Immunological aspects of nematode parasite control in sheep. J Anim Sci. 2006;84(Suppl):124-32.
19. Bolormaa S, van der Werf JHJ, Walkden-Brown SW, Marshall K, Ruvinsky A. A quantitative trait locus for faecal worm egg and blood eosinophil counts on chromosome 23 in Australian goats. J Anim Breed Genet. 2010;127:207-14.
20. Alba-Hurtado F, Muñoz-Guzmán MA. Immune Responses Associated with Resistance to Haemonchosis in Sheep. Biomed Res. Int. 2013. doi: 10.1155/2013/162158.
21. Coltman DW, Wilson K, Pilkington JG, Stear MJ, Pemberton JM. A microsatellite polymorphism in the gamma interferon gene is associated

with resistance to gastrointestinal nematodes in a naturally-parasitized population of Soay sheep. Parasitology. 2001;122:571–82.

22. Crawford AM, Paterson KA, Dodds KG, Diez Tascon C, Williamson PA, Roberts Thomson M, et al. Discovery of quantitative trait loci for resistance to parasitic nematode infection in sheep: I. Analysis of outcross pedigrees. BMC Genomics. 2006;7:178.

23. McRae KM, McEwan JC, Dodds KG, Gemmell NJ. Signatures of selection in sheep bred for resistance or susceptibility to gastrointestinal nematodes. BMC Genomics. 2014;15:637.

24. Benavides MV, Sonstegard TS, Kemp S, Mugambi JM, Gibson JP, Baker RL, et al. Identification of novel loci associated with gastrointestinal parasite resistance in a red Maasai x Dorper backcross population. PLoS One. 2015;10:1–20.

25. Pickering NK, Auvray B, Dodds KG, Mcewan JC. Genomic prediction and genome-wide association study for dagginess and host internal parasite resistance in New Zealand sheep. BMC Genomics. 2011;16:1–11.

26. Atlija M, Arranz J-J, Martinez-Valladares M, Gutiérrez-Gil B. Detection and replication of QTL underlying resistance to gastrointestinal nematodes in adult sheep using the ovine 50K SNP array. Genet Sel Evol. 2016;48:4.

27. Ueno H, Gonçalves PC. Manual para diagnóstico das helmintoses de ruminantes. 1998. p. 149.

28. Van Wyk JA, Cabaret J, Michael LM. Morphological identification of nematode larvae of small ruminants and cattle simplified. Vet Parasitol. 2004;119:277–306.

29. Gordon H McL. The diagnosis of helminthosis in sheep. Med. Vet. Rev.1967; 67(140–168).

30. Russel AJF, Doney JM, Gunn RG. Subjective assessment of body fat in live sheep. J Agric Sci, Camb. DigiTop - USDA's Digital Desktop Library. 2016;72: 451–4.

31. Bath GF, Hansen JW, Krecek RC, Van Wyk JA, Vatta A. Sustainable approaches for managing haemonchosis in sheep and goats. Final Report of Food and Agricultural Organization (FAO) Technical Cooperation Project in South Africa. Food and Agriculture Organization of the United Nations; 2001.

32. Gianola D, Fernando RL. Bayesian methods in animal breeding theory. J Anim Sci. 1986;63:217–44.

33. Aguilar I, Misztal I, Johnson DL, Legarra A, Tsuruta S, Lawlor TJ. Hot topic: a unified approach to utilize phenotypic, full pedigree, and genomic information for genetic evaluation of Holstein final score. J Dairy Sci. 2010;93:743–52.

34. VanRaden PM, Van Tassell CP, Wiggans GR, Sonstegard TS, Schnabel RD, Taylor JF, et al. Invited review: reliability of genomic predictions for north American Holstein bulls. J Dairy Sci. 2009;92:16–24.

35. Vitezica ZG, Aguilar I, Misztal I, Legarra A. Bias in genomic predictions for populations under selection. Genet. Res. (Camb.) 2011;93:357–66.

36. Misztal I, Tsuruta S, Strabel T, Auvray B, Druet T, Lee DH. BLUPF90 and related programs (BGF90). Proc. 7th World Congr. Genet. Appl. to Livest. Prod. 2002;28:21–2

37. Geweke J. Evaluating the accuracy of sampling-based approaches to the calculation of posterior moments. In: Bernardo JM, Berger JO, David AP, Smith AFM, editors. Bayesian statistics. New York: Oxford University; 1992. p. 625-31. Cap. 4

38. Heidelberg P, Welch P. Simulation run length control in the presence of an initial transient. Oper Res. 1983;31:1109–14.

39. Raftery AE, Lewis S. How many iterations in the Gibbs sampler? Bayesian Stat. 1992:763–73.

40. R Development Core Team. R: A language and environment for statistical computing. Vienna: R Foundation for Statistical Computing; 2008. ISBN 3- 900051-07-0. URL http://www.R-project.org. Accessed 25 Aug 2016.

41. Aguilar I, Misztal I, Legarra A, Tsuruta S. Efficient computation of the genomic relationship matrix and other matrices used in single-step evaluation. J Anim Breed Genet. 2011;128:422–8.

42. Wang H, Misztal I, Aguilar I, Legarra A, Muir WM. Genome-wide association mapping including phenotypes from relatives without genotypes. Genet Res (Camb). 2012;94:73–83.

43. Strandén I, Garrick DJ. Technical note: derivation of equivalent computing algorithms for genomic predictions and reliabilities of animal merit. J Dairy Sci. 2009;92:2971–5.

44. Zhang Z, Liu J, Ding X, Bijma P, de Koning DJ, Zhang Q. Best linear unbiased prediction of genomic breeding values using a trait-specific marker-derived relationship matrix. PLoS One. 2010;5:1–8.

45. Wang H, Misztal I, Legarra A. Differences between genomic-based and pedigree- based relationships in a chicken population, as a function of quality control and pedigree links among individuals. J Anim Breed Genet. 2014;131:445–51.

46. Huang DW, Sherman BT, Lempicki RA. Bioinformatics enrichment tools: paths toward the comprehensive functional analysis of large gene lists. Nucleic Acids Res. 2009;37:1–13.

47. McRae AF, McEwan JC, Dodds KG, Wilson T, Crawford AM, Slate J. Linkage disequilibrium in domestic sheep. Genetics. 2002;160:1113–22.

48. McRae AF, Pemberton JM, Visscher PM. Modeling linkage disequilibrium in natural populations: the example of the Soay sheep population of St. Kilda, Scotland. Genetics. 2005;171:251–8.

49. Meadows JRS, Chan EKF, Kijas JW. Linkage disequilibrium compared between five populations of domestic sheep. BMC Genet. 2008;9:61.

50. García-Gámez E, Sahana G, Gutiérrez-Gil B, Arranz J-J. Linkage disequilibrium and inbreeding estimation in Spanish Churra sheep. BMC Genet. 2012;13:43.

51. Mastrangelo S, Di Gerlando R, Tolone M, Tortorici L, Sardina MT, Portolano B. Genome wide linkage disequilibrium and genetic structure in Sicilian dairy sheep breeds. BMC Genet. 2014;15:108.

52. Zhao F, Wang G, Zeng T, Wei C, Zhang L, Wang H, et al. Estimations of genomic linkage disequilibrium and effective population sizes in three sheep populations. Livest Sci. 2014;170:22–9.

53. Bisset SA, Van Wyk JA, Bath GF, Morris CA, Stenson MO, Malan FS. Phenotypic and genetic relationships amongst FAMACHA score, faecal egg count and performance data in Merino sheep exposed to Haemonchus contortus infection in South Africa. In: Proceedings of the 5th International Sheep Veterinary Congress, Stellenbosch, 2011.

54. Bishop SC, Jackson F, Coop RL, Stear MJ. Genetic parameters for resistance to nematode infections in Texel lambs and their utility in breeding programmes. Anim Sci. 2004;78:185–94.

55. Benavides MV, de Souza CJH, Moraes JCF, Berne MEA. Is it feasible to select humid sub-tropical Merino sheep for faecal egg counts? Small Rumin Res. 2016;137:73–80.

56. Riley DG, Van Wyk JA. Genetic parameters for FAMACHA© score and related traits for host resistance/resilience and production at differing severities of worm challenge in a Merino flock in South Africa. Vet Parasitol. 2009;164:44–52.

57. Hawkins PT, Stephens LR. PI3K signalling in inflammation. Biochim Biophys Acta - Mol Cell Biol Lipids. 1851;2015:882–97.

58. Prasad KVS, Ao Z, Yoon Y, Wu MX, Rizk M, Jacquot S, et al. CD27, a member of the tumor necrosis factor receptor family, induces apoptosis and binds to Siva, a proapoptotic protein. Immunology. 1997;94:6346–51.

59. Wang W, Soto H, Oldham ER, Buchanan ME, Homey B, Catron D, et al. Identification of a novel chemokine (CCL28), which binds CCR10 (GPR2). J Biol Chem. 2000;275:22313–23.

60. Junior Janeway CA, Travers P, Walport M, Shlomchik MJ. Immunobiology. 5th ed. New York: Garland Science; 2001.

61. Cooper MD, Herrin BR. How did our complex immune system evolve? Nat Rev Immunol. 2010;10:2–3.

62. Trzaskowski B, Latek D, Yuan S, Ghoshdastider U, Debinski A, Filipek S. Action of molecular switches in GPCRs–theoretical and experimental studies. Curr Med Chem. 2012;19:1090–109.

63. Teichmann SA, Chothia C. Immunoglobulin superfamily proteins in Caenorhabditis elegans. J Mol Biol. 2000;296:1367–83.

64. Lin M, Sutherland DR, Horsfall W, Totty N, Yeo E, Nayar R, et al. Cell surface antigen CD109 is a novel member of the alpha(2) macroglobulin/C3, C4, C5 family of thioester-containing proteins. Blood. 2002;99:1683–91.

65. Kondo M, Scherer DC, Miyamoto T, King AG, Akashi K, Sugamura K, et al. Cell-fate conversion of lymphoid-committed progenitors by instructive actions of cytokines. Nature. 2000;407:383–6.

66. Reymond N, Fabre S, Lecocq E, Adelaide J, Dubreuil P, Lopez M. Nectin4/PRR4, a new Afadin-associated member of the Nectin family that trans-interacts with Nectin1/PRR1 through V domain interaction. J Biol Chem. 2001;276:43205–15.

67. Sheridan C, Sadaria M, Bhat-Nakshatri P, Goulet R, Edenberg HJ, McCarthy BP, et al. Negative regulation of MHC class II gene expression by CXCR4. Exp Hematol. 2006;34:1085–92.

68. Contento RL, Molon B, Boularan C, Pozzan T, Manes S, Marullo S, et al. CXCR4-CCR5: a couple modulating T cell functions. Proc Natl Acad Sci. 2008;105:10101–6.

69. Zlotnik A, Yoshie O. Chemokines: a new classification system and their role in immunity. Immunity. 2000;12:121–7.

70. Shanaka WW, Rodrigo I, Jin X, Blackley SD, Rose RC, Schlesinger JJ. Differential enhancement of dengue virus immune complex infectivity mediated by signaling-competent and signaling-incompetent human Fc-gamma-RIA (CD64) or Fc-gamma-RIIA (CD32). J Virol. 2006;80:10128–38.

71. Williams MJ, Ando I, Hultmark D. Drosophila melanogaster Rac2 is necessary for a proper cellular immune response. Genes Cells. 2005;10:813–23.

10

Level of dietary energy and 2,4-thiazolidinedione alter molecular and systemic biomarkers of inflammation and liver function in Holstein cows

Afshin Hosseini[1], Mustafa Salman[2], Zheng Zhou[1], James K. Drackley[1], Erminio Trevisi[3] and Juan J. Loor[1]* (ID)

Abstract

Background: The objective of the study was to evaluate the effect of overfeeding a moderate energy diet and a 2,4-thiazolidinedione (TZD) injection on blood and hepatic tissue biomarkers of lipid metabolism, oxidative stress, and inflammation as it relates to insulin sensitivity.

Results: Fourteen dry non-pregnant cows were fed a control (CON) diet to meet 100% of NRC requirements for 3 wk, after which half of the cows were assigned to a moderate-energy diet (OVE) and half of the cows continued on CON for 6 wk. All cows received an intravenous injection of 4 mg TZD/kg of body weight (BW) daily from 2 wk after initiation of dietary treatments and for 2 additional week. Compared with CON cows and before TZD treatment, the OVE cows had lower concentration of total protein, urea and albumin over time. The concentration of cholesterol and tocopherol was greater after 2 wk of TZD regardless of diet. Before and after TZD, the OVE cows had greater concentrations of AST/GOT, while concentrations of paraoxonase, total protein, globulin, myeloperoxidase, and haptoglobin were lower compared with CON cows. Regardless of diet, TZD administration increased the concentration of ceruloplasmin, ROMt, cholesterol, tocopherol, total protein, globulin, myeloperoxidase and beta-carotene. In contrast, the concentration of haptoglobin decreased at the end of TZD injection regardless of diet. Prior to TZD injection, the mRNA expression of *PC*, *ANGPTL4*, *FGF21*, *INSR*, *ACOX1*, and *PPARD* in liver of OVE cows was lower compared with CON cows. In contrast, the expression of *HMGCS2* was greater in OVE compared with CON cows. After 1 wk of TZD administration the expression of *IRS1* decreased regardless of diet; whereas, expression of *INSR* increased after 2 wk of TZD injection. Cows fed OVE had lower overall expression of *TNF*, *INSR*, *PC*, *ACOX1*, *FGF21*, and *PPARD* but greater *HMGCS2* expression. These differences were most evident before and after 1 wk of TZD injection, and by 2 wk of TZD differences in expression for most genes disappeared.

Conclusions: Based on molecular and blood data, administration of TZD enhanced some aspects of insulin sensitivity while causing contradictory results in terms of inflammation and oxidative stress. The bovine liver is TZD-responsive and level of dietary energy can modify the effects of TZD. Because insulin sensitizers have been proposed as useful tools to manage dairy cows during the transition period, further studies are required to investigate the potential hepatotoxicity effect of TZD (or similar compounds) in dairy cattle.

Keywords: Inflammation, Insulin sensitivity, Nutrition, PPAR

* Correspondence: jloor@illinois.edu
[1]Department of Animal Sciences and Division of Nutritional Sciences, University of Illinois, 1207 West Gregory Drive, Urbana, IL 61801, USA
Full list of author information is available at the end of the article

Background

Thiazolidinediones (TZD) are agonists of peroxisome proliferator-activated receptor gamma (PPARG) that elicit insulin-sensitizing effects in non-ruminants [1–3] and dairy cows [4–6]. The administration of TZD (4 mg of TZD/kg of BW) during the late prepartum period appeared to alter the dynamics of plasma glucose, non-esterified fatty acids (NEFA), and hydroxybutyric acid (BHBA) concentrations and dry matter intake (DMI) during the periparturient period [5, 7]. Furthermore, a greater concentration of blood insulin was reported in TZD-treated cows, which likely accounted for the lower NEFA [6].

We previously reported that in dry and non-pregnant dairy cows fed a control lower-energy (CON) or higher-energy (OVE) diet receiving 4 mg of TZD/kg of BW daily for 2 wk [8] the concentrations of glucose (4.55 vs. 4.65 mmol/L), insulin (27 vs. 35 µU/mL), and (BHBA) (0.27 vs. 0.37) were increased during the TZD administration (2 to 4 wk after diet initiation). In contrast, the concentration of NEFA (0.18 vs. 0.15 mmol/L) and adiponectin (ADIPOQ; 34.6 vs. 30.3 µg/L) remained unchanged during TZD administration. More importantly, the ratios of glucose/insulin (0.51 vs. 0.54) and NEFA/insulin (0.22 vs. 0.18) decreased during TZD, suggesting an improvement in insulin sensitivity. The mRNA expression in subcutaneous adipose tissue of PPARG and its targets FASN and SREBF1, which are the main regulators of adipogenesis and lipogenesis, were upregulated by TZD [8]. Greater expression of the insulin sensitivity-related genes IRS1, SLC2A4, INSR, SCD, INSIG1, DGAT2, and ADIPOQ in subcutaneous adipose tissue of OVE cows indicated that greater energy intake did not impair insulin sensitivity. In skeletal muscle, TZD altered expression of carbohydrate- and fatty acid oxidation-related genes. The OVE cows had greater mRNA expression of PC and PCK1, which was indicative of increased glyceroneogenesis [8].

A comprehensive review on the topic insulin sensitivity in dairy cattle underscored the role of skeletal muscle, adipose tissue, and liver in the overall glucose and insulin relationship [9]. The level of dietary energy fed prepartum has long been known to alter fat deposition and other metabolic pathways in tissues like liver [10]. Cows suffering from fatty liver had higher serum concentrations of NEFA, greater serum TNF, and had signs of systemic insulin resistance [11]. Some evidence indicates that excess NEFA concentrations alter hepatic function and inflammatory status [12, 13]. Importantly, the severity of inflammation around parturition seems related to some inflammatory indices observed in the dry period [14]. Thus, the available data seem to suggest that NEFA could cause inflammation and that inflammation also could cause the greater release of NEFA.

Our hypothesis was that overfeeding and TZD administration lead to changes in the dynamics of biomarkers of

oxidative stress, inflammation, and metabolism in blood and liver tissue and might be related with systemic insulin sensitivity. To address the hypothesis, we measured in samples from Hosseini et al. [8] the plasma or serum concentrations of acute-phase proteins (haptoglobin, ceruloplasmin, albumin, cholesterol, and adiponectin), health and liver function biomarkers (total protein, myeloperoxidase, globulin, GGT, AST/GOT and total bilirubin), oxidative stress (paraoxonase and reactive oxygen metabolites (ROMt)), protein metabolism (urea), and vitamins (retinol, αtocopherol, beta-carotene). In addition, the mRNA expression of genes associated with glucose homeostasis and gluconeogenesis, lipid metabolism nuclear receptors and their targets, inflammation and ketogenesis was measured in liver tissue.

Methods

Experimental design, animals management and sampling

The Institutional Animal Care and Use Committee (IACUC) of the University of Illinois approved all procedures for this study (protocol #12134). Detailed materials and methods can be found elsewhere [8]. In brief, fourteen dry non-pregnant Holstein cows [initial BW (kg) = 717 ± 39; initial BCS = 3.31 ± 0.14] were assigned randomly to treatment groups. Cows were offered the TMR once daily at 0600 h and had unlimited access to fresh water. All cows were fed a control diet (CON; NE_L = 1.30 Mcal/kg) to meet 100% of NRC requirements for 3 wk, after which half of the cows were assigned to a moderate-energy diet (OVE; NE_L = 1.60 Mcal/kg) and half of the cows continued on CON for 6 wk. The OVE diet was fed ad libitum and resulted in cows consuming ~180% of NRC. Control cows were fed to consume only 100% of NRC. All cows received an intravenous injection of 4 mg TZD/kg of BW daily after the morning feeding into the jugular vein starting d 15 after the initiation of dietary treatments and until d 28. Blood samples were collected before the morning feeding from the coccygeal vein or artery every 5 ± 2 ds from –7 to 14 d of diet initiation (before TZD administration) and from 15 to 28 d of diet initiation (during TZD administration) for measurement of metabolites and hormones.

Biopsies, RNA isolation, primer design and evaluation, and quantitative PCR

Liver tissue was harvested via percutaneous biopsy as described previously [15] before the morning feeding at 14 d, before TZD administration, 21 and 28 d, during the TZD administration and 35 ds relative to diet initiation. Details of these procedures are reported in the supplemental material. Tissue was frozen immediately in liquid N until RNA extraction. The selection of the primers in liver was conducted base on core cellular functions, e.g. insulin signaling, glucose metabolism and synthesis, lipid metabolism, hepatokines, cytokines and inflammatory mediators (Additional file 1: Tables S1 and S2). The amplicons were sequenced and the fragment sequences were blasted and confirmed using the National Center for Biotechnology Information (NCBI). The sequences are shown in Additional file 1: Table S3. The geometric mean of internal control genes (ICG) *GAPDH*, ribosomal protein S9 (*RPS9*), and ubiquitously-expressed transcript (*UXT*) were used for data normalization (V2/3 ≤ 0.15) in liver tissue. qPCR perfromance is reported in Additional file 1: Table S4.

Statistical analysis

Data were analyzed with the Proc MIXED procedure of SAS 9.4 (SAS Institute Inc., Cary, NC). After normalization with the geometric mean of the ICG, the triplicate qPCR data for each gene were averaged and then $\log_2$ transformed prior to statistical analysis. Fixed effects in the model were diet, time, and diet × time. Cow within diet was designated as the random effect. Initial BW and BCS were included as covariates in the analysis for all variables, except when the covariate was

not significant ($P > 0.05$). Blood metabolites were log-scale transformed if needed to comply with normal distribution of residuals, and means subsequently were back-transformed. Significance was declared at $P \le 0.05$, while trend was declared at $P \le 0.10$.

Results

Results related to NE_L intake, BW, blood TZD as well as blood levels of BHBA, NEFA, glucose, insulin, adiponectin were reported previously [8].

Blood biomarkers before TZD administration

The OVE cows had lower ($P \le 0.05$) concentration over time of total protein, urea, and albumin (Fig. 1). The overall concentration of cholesterol and tocopherol increased ($P < 0.05$) at 14 d after diet initiation, whereas the serum concentration of total bilirubin tended ($P = 0.10$) to decrease at 14 d after diet initiation (Table 1). Prior to TZD, diet did not affect ($P > 0.05$) cholesterol, total bilirubin, or tocopherol. Serum concentrations of glucose, ceruloplasmin, ROMt, GGT, retinol, AST/GOT, paraoxonase, globulin, myeloperoxidase, haptoglobin, and beta-carotene

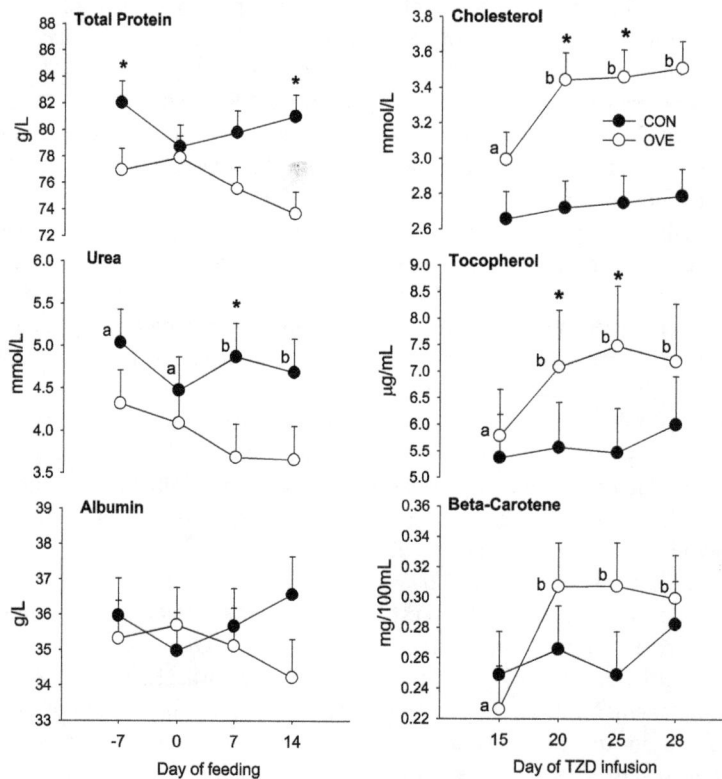

Fig. 1 Temporal concentrations [−7 to 14 d relative to diet initiation; before TZD administration and 15 to 28 d relative to diet initiation (during TZD administration)] of blood biomarkers in cows fed either a controlled-energy diet (CON, 1.30 Mcal/kg diet dry matter; n = 7) or a moderate-energy diet (OVE, 1.60 Mcal/kg diet dry matter; n = 7) before (*left panels*) and during (*right panels*) 2,4-thiazolidinedione (TZD) administration. *Significant (*P* < 0.05) difference due to diet or Diet × Time. a,bSignificant (*P* < 0.05) overall difference due to time

Table 1 Plasma concentrations of biomarkers from −7 to 14 d relative to diet initiation, before 2,4-thiazolidinedione (TZD) administration and 15 to 28 d relative to diet initiation and during TZD administration in cows fed either a controlled-energy diet (CON, 1.30 Mcal/kg; n = 7) or a moderate-energy diet (OVE, 1.60 Mcal/kg; n = 7)

Item	Diet			Day					P-value	
	CON	OVE	SEM	−7	0	7	14	SEM	Diet	Day
Before TZD				−	−	−	−			
Metabolism										
Cholesterol, mmol/L	2.64	2.68	0.15	2.62^a	2.54^a	2.64^a	2.84^b	0.11	0.86	< 0.01
Liver function										
Total Bilirubin, µU/mL	1.18	1.06	0.10	1.19^c	1.09^d	1.11^c	1.08^d	0.09	0.14	0.10
Antioxidants–anti-inflammation										
α-Tocopherol, mmol/L	5.22	5.28	0.46	5.19^a	5.02^a	5.13^a	5.65^b	0.34	0.94	< 0.01

Item	Diet			Day					P-value	
	CON	OVE	SEM	15	20	25	28	SEM	Diet	TZD
During TZD				+	+	+	+			
Health and liver function										
GGT, U/L	22.7	21.3	2.8	21.3^c	22.2^{cd}	21.6^{cd}	22.9^d	2.1	0.53	0.07
Globulin, g/L	43.6^b	38.8^a	1.0	39.8^a	42.0^b	41.5^b	41.5^b	1.0	0.03	0.02
Myeloperoxidase, U/L	491^b	419^a	21.5	399^a	474^b	484^b	462^b	24.1	0.03	0.05
Total Protein, g/L	78.7^b	73.0^a	1.2	74.1^a	77.2^b	76.0^b	76.2^b	1.0	< 0.01	0.021
Total Bilirubin, µmol/L	1.19^d	1.03^c	0.11	1.48	1.39	1.39	1.37	0.42	0.06	0.97
Acute-phase proteins										
Ceruloplasmin, µmol/L	3.30	3.14	0.11	2.95^a	3.30^b	3.30^b	3.34^b	0.10	0.33	< 0.01
Haptoglobin, g/L	0.65^b	0.51^a	0.05	0.62^a	$0.57^{a,b}$	0.61^a	0.51^b	0.068	< 0.01	0.05
Oxidative stress										
ROMt, mg H_2O_2/100 mL	13.8	13.4	0.57	12.1^a	14.0^b	14.1^b	14.2^b	0.51	0.62	< 0.01
Paraoxonase, U/mL	113^b	91.4^a	6.4	100	104	99.8	105	5.14	0.03	0.15
Liver injury										
AST/GOT, U/L	66.0^a	77.6^b	3.6	67.8	72.8	75.8	70.8	3.3	0.04	0.16
Antioxidants–anti-inflammation										
Retinol, µg/100 mL	37.8^c	47.6^d	3.7	42.6	43.5	42.9	41.8	2.77	0.09	0.56

[a,b]Means within a row with different superscripts differ (P ≤ 0.05) for TZD, while [c,d] denote trends (P ≤ 0.10)

also were not affected (P > 0.05, data not shown) by diet, time (14 wk) or their interaction.

Blood biomarkers and hormones during TZD administration

The Fig. 1 and Table 1 contain the main effects of diet, TZD, and interactions for plasma parameters. Although plasma cholesterol did not differ statistically prior to TZD (Table 1), during TZD there was greater (interaction P < 0.05) concentration of cholesterol and tocopherol in cows fed OVE (Fig. 1). Regardless of diet, the TZD increased (P < 0.05) GGT, ceruloplasmin, and ROMt concentrations at d 28 from diet initiation (Table 1). An interaction during TZD was detected for concentrations of globulin, myeloperoxidase, total protein, and haptoglobin due to lower overall concentrations in cows fed OVE. Independent of TZD, the concentrations of total bilirubin and paraoxonase were lower and AST/GOT greater in cows fed OVE (Table 1). No effect of diet, TZD or their interaction (P > 0.05) was detected for urea and albumin concentrations (Fig. 1).

Gene expression in response to level of diet and TZD

The Fig. 2 and Table 2 contain the main effects of diet, TZD, and interactions for genes related to insulin signaling, glucose metabolism, fatty acid oxidation, hepatokines, ketogenesis and inflammation. Prior to and during TZD, the expression of ACOX1 and

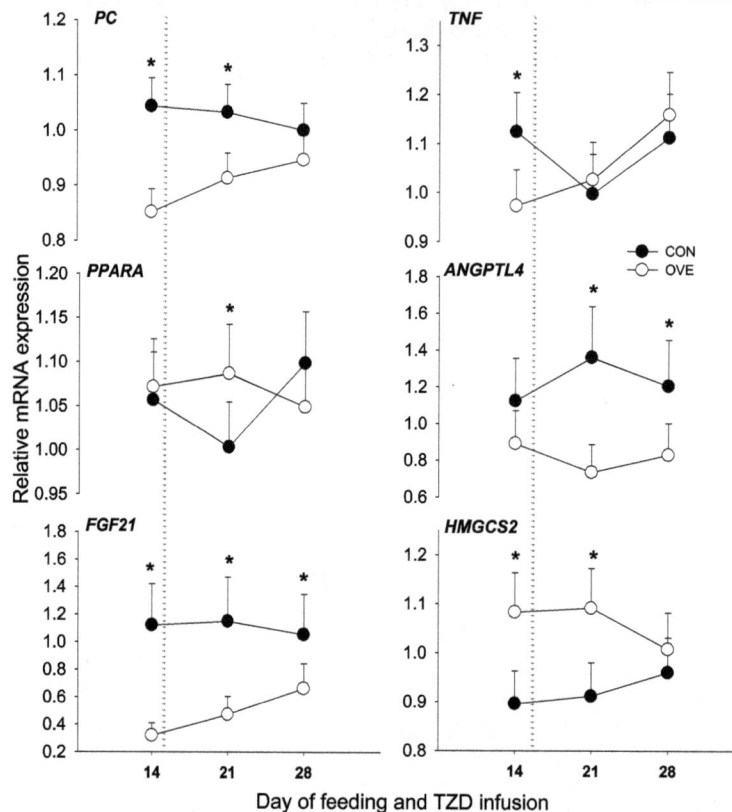

Fig. 2 Expression of genes associated with gluconeogenesis (*PC*), lipid metabolism nuclear receptors (*PPARA*), ketogenesis (*HMGCS2*), hepatokines (*ANGPTL4* and *FGF21*), and inflammation (*TNF*) in liver tissue of cows fed either a controlled-energy diet (CON, 1.30 Mcal/kg diet dry matter; n = 7) or a moderate-energy diet (OVE, 1.60 Mcal/kg diet dry matter; n = 7) before and during 2,4-thiazolidinedione (TZD) administration. The *dotted line* denotes the start of TZD infusion on d 15 relative to diet initiation. *Significant (P < 0.05) difference due to diet or Diet × Time. a,bSignificant (P < 0.05) overall difference due to time

PPARD was lower (*P* < 0.05) in cows fed OVE (Table 2). In OVE cows, the mRNA expression of *PC*, *FGF21*, and *INSR* was lower (*P* < 0.05), while the mRNA expression of *HMGCS2* was greater (*P* < 0.05) during TZD (Fig. 2 and Table 2). Regardless of diet, the TZD administration increased (*P* < 0.05) the mRNA of *IRS1* and *TNF* at 28 d after diet initiation (Fig. 2 and Table 2).

Discussion

The ruminant liver is the main site for gluconeogenesis [16], and a spike of insulin into the bloodstream suppresses not only gluconeogenesis but also glycogenolysis and AT tissue lipolysis [9]. In fact, elevated insulin concentrations can stimulate glycogenesis in liver and improve overall insulin sensitivity via

Table 2 Relative mRNA expression of target genes in liver tissue at 14 to 28 d relative to diet initiation and after a 7-day washout period (35 d) in cows fed a controlled-energy diet (CON, 1.30 Mcal/kg; n = 7) or a moderate-energy diet (OVE, 1.60 Mcal/kg; n = 7). All cows received an intravenous injection of 2,4-thiazolidinedione (TZD) (+) for 2 wk after biopsy collection at 14 d after initiation of dietary treatments, with the additional 2 wk of study serving as a washout period

Gene	Diet			Day					P-value	
	CON	OVE	SEM	14	21	28	35	SEM	Diet	TZD
TZD				−	+	+	−			
IRS1	1.03	1.02	0.03	1.04a	0.99b	1.03a	-c	0.029	0.77	0.04
INSR	1.12b	1.02a	0.05	1.06a	1.04a	1.11b	−	0.04	0.02	0.03
ACOX1	1.03b	0.96a	0.02	1.01	0.98	1.00	−	0.02	< 0.01	0.16
PPARD	1.07b	0.99a	0.04	1.04	1.01	1.03	−	0.03	0.01	0.21

a,bMeans within a row with different superscripts differ (P ≤ 0.05) due to diet or TZD
cTissue sample not available

glucose uptake in peripheral tissues such as AT and skeletal muscle [9]. In humans with type 2 diabetes, the injection of TZD decreased hepatic fat content and increased hepatic insulin sensitivity along with increased glucose uptake by peripheral tissues [17]. Thus, despite inherent differences in gastrointestinal tract anatomy, circulating insulin can alter insulin sensitivity in extrahepatic tissues in ruminants and non-ruminants.

Besides its effect on insulin sensitivity, previous data from non-ruminants also indicated that TZD changes the dynamics of blood biomarkers, e.g. adiponectin, insulin and glucose. Similarly, in dairy cows the administration of TZD (4 mg of TZD/kg of BW) during the late-prepartum period altered the dynamics of plasma glucose, NEFA, BHBA concentrations and also DMI during the periparturient period. The TZD helped cows to maintain BCS postpartum [5, 7] and achieve greater concentrations of insulin [6], which likely accounted for the lower NEFA, i.e. the drug probably induced greater insulin sensitivity in peripheral tissues such as AT. In the present study the TZD administration increased triacylglycerol concentration and altered expression of carbohydrate- and fatty acid oxidation-related genes in skeletal muscle [8]. Furthermore, the upregulation of *PPARG* with the TZD injection provided evidence that TZD could enhance adipogenesis and lipogenesis in SAT, while differentially regulating glucose homeostasis and fatty acid oxidation in skeletal muscle [8].

Diet, TZD and blood biomarkers
Overfeeding energy intake and hepatic function
The lower blood concentration of urea in OVE cows was in accordance with previous results [18]. A better liver function status in OVE cows is supported by the lower total bilirubin [12, 13]. In addition, the hyperinsulinemia in OVE cows [8] along with lower urea concentration also might be related to increased gluconeogenesis in liver along with amino acid uptake and protein synthesis in skeletal muscle [19].

In addition to its well-described role as a negative acute-phase protein (negAPP), the rate of albumin synthesis is affected by feed intake [20–22], i.e. greater concentrations are correlated with higher energy intake but in the absence of differences in energy intake a lower concentration is indicative of altered liver function. The difference between the total protein and albumin content in plasma provides an estimate of globulin concentration [13]. The dynamics of the change in globulin concentration is considered an index of immunoglobulin production [12]. Thus, as indicated by the present data, the mild decrease in total protein and albumin concentrations in OVE despite the greater DMI [8] is suggestive of a possible impairment in

synthesis at the liver level brought about by a mild inflammation. However, the more pronounced decrease globulin seems to argue against inflammation playing a role. These results are in accordance with some previous studies in energy-overfed dairy cows [18]. Whether excess energy intake caused a mild chronic degree of inflammation [23] remains to be determined.

The sustained decrease in plasma concentrations of cholesterol before calving and the weak increase after calving have been associated with an inflammatory-like status [12, 13], but also reflect the pattern of DMI, i.e. gradual decrease through calving followed by a gradual increase. Clearly, the response in the present study could not be compared with the periparturient period where negative energy balance is a consequence of higher energy requirements for milk production rather and lower DMI. Thus, the greater blood concentration of cholesterol in OVE cows is likely a reflection of the greater DMI, i.e. cholesterol synthesized by the intestine [24]. A review of the literature [25] concluded that inflammation could alter lipid transport in the circulation through changes in the dynamics of apolipoproteins.

Paraoxonase is considered an antioxidative stress enzyme and a negAPP that binds to plasma apolipoproteins [26]. An antioxidative function of paraoxonase has been suggested in dairy cows [27], as well as a link with the dynamics of positive acute-phase proteins (posAPP) and negAPP around parturition [12]. In humans, the serum paraoxonase activity is reduced by oxidized low density lipoprotein (LDL) and is maintained or increased by antioxidants [28]. Reactive oxygen metabolites (ROMt) are considered useful biomarkers of oxidative stress; the results of an in vivo study in mice and an in vitro study in human hepatocyte cell cultures revealed that concentrations of ROMt can be increased by feeding high-carbohydrate diets [29]. Those studies revealed that the increase in ROMt decreased the capacity of hepatocytes to respond to an oxidative challenge. In our study, the lower concentration of paraoxonase coupled with greater cholesterol in OVE cows indicated a reduction in the ability of the liver to synthesize this enzyme. In fact, work with dairy cows has revealed a positive relationship between increases in plasma cholesterol and paraoxonase, which underscores the idea that liver function in OVE cows experienced a negative effect for some functional aspects.

The liver produces the major portion of the systemic haptoglobin in mammals, but in humans its concentration has been used as a marker for adiposity [30]. In bovine, however, haptoglobin is a major posAPP and its concentration in plasma is increased during inflammation, infection, and often during the transition period [31]. Despite the marked increased in energy intake, the lower concentration of haptoglobin in OVE cows likely reflects a lower inflammatory and oxidative stress response.

Regardless of TZD, the overall blood concentrations of cholesterol, albumin, and total protein were within a non-pathologic range [32]. Along with the lower concentration of paraoxonase, haptoglobin, and myeloperoxidase, the greater concentrations of cholesterol in OVE cows indicated the absence of important impairment of the synthesis and release of the lipoproteins, and also of a severe acute-phase reaction in the liver. Furthermore, the lower bilirubin in OVE cows indicated normal hepatic bilirubin clearance capacity and agrees with the lower inflammatory status [32]. Despite all the positive systemic signs in OVE cows, they had greater concentration of AST/GOT, which in clinical studies is an indicator of liver tissue damage [33]. However, the concentration of AST/GOT can be affected by catabolism of nutrients such as amino acids via the citric acid cycle [34]. Thus, the greater AST/GOT in OVE cows might be related not only to "leakage" from liver cells but also to greater citric acid cycle flux, e.g. to produce glucose, as a result of the greater DMI and the greater absorption of amino acids from the gut.

The greater blood concentrations of α-tocopherol, retinol, and beta-carotene in OVE cows is suggestive of adequate or better vitamin status, and these molecules have been linked with the immune response. Beta-carotene and α-tocopherol are antioxidants that bind free radicals to prevent lipid peroxidation and improve the immune response [35, 36]. Retinol (vitamin A) circulates in the blood bound to retinol-binding protein (RBP), a protein that is synthesized in the liver and controls availability of the vitamin in times of insufficient dietary intake [37]. Inhibition of RBP transcription or function is one determinant of circulating levels of retinol. Similar to retinol, the liver is considered the "master regulator" of systemic vitamin E availability, i.e. it controls disposition, metabolism, and excretion [38]. In the context of lipoprotein metabolism, re-packing of α-tocopherol into the very-low density lipoproteins is central for extra-hepatic availability. Thus, hepatic metabolism of cholesterol and lipoproteins clearly has direct bearing on systemic α-tocopherol concentrations. In the context of insulin sensitivity, the greater overall concentration of retinol in cows fed OVE could have had an effect on peripheral insulin sensitivity as demonstrated recently [39]. These relationships likely explain the greater α-tocopherol concentration in OVE cows which also had greater cholesterol concentration (i.e. an index of lipoprotein synthesis). Besides dietary availability, in non-ruminants, the acute-phase response causes a decrease in the hepatic synthesis of vitamin A and of α-tocopherol namely due to alterations in hepatic function [40]. Clearly, the greater blood concentration of these vitamins especially retinol in OVE cows was in part due to the greater DMI and diet composition, and the resulting storage of these vitamins is not expected to impair liver function [41].

TZD, insulin sensitizing effect versus Hepatotoxicity

Historically, TZD has been used over the last two decades to improve insulin sensitivity in type II diabetic subjects. The results of in vivo and in vitro studies [42, 43] indicated that Troglitazone (TGZ), a thiazolidinedione, caused hepatotoxicity through non-metabolic and metabolic factors [44]. Some of the present data could reflect a mild hepatotoxic effect of the short-term TZD administration. For instance, in addition to the decrease in the negAPP (e.g. cholesterol), there seems to have been differential regulation of the posAPP by TZD, e.g. ceruloplasmin increased and haptoglobin decreased. Results from a previous study revealed that TZD can decrease haptoglobin expression in adipocytes in vitro and in vivo [45]. Because the cows were clinically healthy during the study, it is also possible that the greater concentration of ceruloplasmin after TZD regardless of diet was associated with a stressful effect also reflected by the gradual increase in ROMt with TZD regardless of diet.

There is evidence from rats [46] and humans [47] that the family of TZD compounds might cause oxidative stress and elicit a direct effect on mitochondrial physiology, hence, playing a role in TZD-mediated hepatotoxicity. In our study, the administration of TZD increased ROMt concomitantly with GGT, AST/GOT, myeloperoxidase, globulin and total protein. A study in mice reported that insulin-sensitizing compounds may act as antioxidants while increasing oxidative stress in the liver [42]. A recent study detected an increase in the synthesis of liver enzymes such as AST and ALT in Chinese subjects with type II diabetic using 4 mg of a TZD-like compound for three month. The mechanistic reasons for elevated liver enzymes upon TZD treatment remain to be determined.

The blood concentration of cholesterol has been used as an indicator of lipoprotein metabolism and is linked to the quantity of lipid mobilized and re-esterified as VLDL within the intestine and liver [32]. In the present study, the greater cholesterol concentration during TZD might indicate the increased export of VLDL to peripheral tissues such as adipose and skeletal muscle. We speculate that such response would have provided insulin-sensitive tissues with fatty acids for esterification, an idea supported by the greater concentration of triacylglycerol that was detected in muscle during TZD injection [8]. Similar to cholesterol, it is possible that the greater concentration of α-tocopherol is a reflection of an increase in VLDL export from the liver [48] to peripheral tissues during the TZD injection period.

TZD modulates hepatic metabolic pathways via PPARA and its targets

The state of systemic insulin sensitivity or glucose tolerance can be estimated using the euglycemic clamp [49] and glucose tolerance tests [4]. In our study, the OVE cows had a greater overall concentration of insulin [8], thus, the lower hepatic expression of *INSR* in the liver was somewhat expected [50]. Although it is well-established that the ruminant liver is not a net user of glucose, our results of insulin signaling-related genes are in accordance with previous results in rodents. For instance, a hyperinsulinemic state in obese mice led to a decrease in insulin binding and phosphorylation of the insulin receptor and the kinase IRS-1 in liver [51]. The data seem to agree with studies in sheep demonstrating differences in the binding of insulin to receptors in liver as a function of circulating concentration [50]. Furthermore, our data indicate that TZD could increase insulin sensitivity and signaling in cow liver [52].

Comprehensive reviews of available data have concluded that PPARα is the main regulator of hepatic fatty acid oxidation, ketogenesis, triglyceride turnover, and gluconeogenesis in non-ruminants and likely in ruminants [53, 54]. For example, the results of studies with *PPARA* knockout mice and bovine cells provided strong evidence that gluconeogenesis and glycerol metabolism are directly controlled by PPARA [55] via its targets PC and PCK1 [56, 57]. The temporal profile of *PC* in the present study seems to suggest the existence of an alternate mechanism whereby insulin sensitizers (over the long-term) could affect hepatic gluconeogenesis differently in response to dietary energy level, i.e. overfeeding moderate energy diets over a period of weeks seems to render the liver less responsive to TZD while feeding to meet energy requirements cannot prevent the expected negative effect of TZD on insulin sensitivity [58]. The exact mechanisms for the TZD effect merit further study.

In ruminants, there is indirect evidence that long-chain fatty acids (LCFA) can activate hepatic β-oxidation via PPARA and its downstream co-activators to decrease tissue TAG accumulation [53]. ACOX1 is one of the PPARA targets which is involved in peroxisomal fatty acid oxidation in non-ruminants; it is the first enzyme in peroxisomal LCFA oxidation [59, 60]. A lower *ACOX1* expression in OVE cows could be explained by the lower plasma NEFA compared with CON cows. In the present study, the different regulation of mRNA expression of proximal and outer-mitochondrial β-oxidation enzymes by TZD administration might have been related with the fact that this compound binds to mitochondrial membrane proteins to modulate mitochondrial metabolism [61].

Research in non-ruminants has demonstrated that ANGPTL4 [62] and FGF21 [63, 64] are the targets of PPARA, and are considered hepatokines, i.e. proteins secreted from liver that can regulate tissue adaptations to feed restriction [65, 66]. A study with mouse primary hepatocytes underscored the negative effect of insulin on mRNA expression of *ANGPTL4* [67]. In rodents, fasting and starvation activated ANGPTL4 in a PPARA-dependent manner [68]; the study of Loor et al. [65] uncovered a potential link between PPARA and ANGPTL4 in cows during nutrition-induced ketosis. ANGPTL4 is an endogenous inhibitor of lipoprotein lipase (LPL) that regulates peripheral tissue uptake of plasma TAG-derived fatty acids [69], hence, helping to regulate energy homeostasis.

In vitro studies in rodents have revealed that PPAR agonists such as TZD are potent inducers of FGF21 [70]. In diabetic rodents and monkey, FGF21 is considered an effective metabolic regulator of glucose and lipid homeostasis in the context of insulin resistance, glucose intolerance and dyslipidemia [68]. Additional studies indicated an increase in plasma concentration and expression of FGF21 during the transition period [66].

In the present study, the lower expression of *ANGPTL4* and *FGF21* in OVE agrees with the fact they were in positive energy balance. Furthermore, a lower *FGF21* expression would be expected in OVE cows because they had higher blood insulin concentration [8]. The upregulation of *FGF21* mRNA expression at two weeks. After TZD administration agrees with a mouse study in which the TZD induced the expression of this gene [70]. Our data indicate that bovine *PPARA* is TZD responsive, and may participate in the regulation of hepatic insulin sensitivity and energy expenditure via its targets. The lower *PPARD* mRNA expression in OVE cows might be related to the lower plasma NEFA concentration; in an in vitro study with rodents, an upregulation of *PPARD* was determined to be important in order to "sense" concentrations of LCFA as a way to control mitochondrial oxidation in order to protect against fatty acid-induced cellular dysfunction.

The HMGCS2 enzyme is rate-limiting for the production of ketone bodies in liver and the neonatal intestine [71]. In growing ruminants, the mRNA expression of *HMGCS2* also is associated with ketogenesis in ruminal epithelium, a process that helps the rumen develop [72]. Studies with rodents uncovered that the expression of *HMGCS2* is regulated by PPARA or by conditions that enhance LCFA influx into the liver (e.g. feed restriction) [73]. In contrast, insulin signaling in liver is a negative regulator of *HMGCS2* expression and activity. Thus, in the present study the greater mRNA expression of *HMGCS2* in OVE cows might have been linked with a lower plasma insulin concentration. In contrast, the gradual downregulation of *HMGCS2* expression during the TZD injection period could have been associated with an insulin-sensitizing effect.

Conclusions

Despite a marked increase in energy balance status in OVE cows in response to overfeeding energy there were modest alterations in the acute-phase response, but it should be kept in mind that the linear decrease in albumin from day 0 to 14 is an indication of a gradual impairment of some liver functions. Based on molecular and blood data it seems that administration of TZD enhanced some aspects of insulin sensitivity while causing contradictory results in terms of inflammation and oxidative stress. Overall, data indicated that bovine liver is TZD responsive and level of dietary energy can modify the effects of TZD. Because insulin sensitizers have been proposed as useful tools to manage dairy cows during the transition period, further studies are required to investigate the potential hepatotoxicity effect of TZD (or similar compounds) in dairy cattle.

Additional file

Additional file 1: Table S1. GenBank accession number, sequence and amplicon size of primers used to analyze gene expression by quantitative PCR. **Table S2** Gene symbol, gene name, and description of the main biological function and biological processes of the targets analyzed in subcutaneous adipose tissue and muscle. **Table S3** Sequencing results obtained from PCR product. **Table S4** qPCR performance among the genes measured in adipose tissue and muscle.

Abbreviations

ACD: Citric acid and dextrose; DIM: Days in milk; DMI: Dry matter intake; IACUC: Institutional animal care and use committee; PPAR: Peroxisome proliferator-activated receptors; ROMt: Reactive oxygen metabolites

Acknowledgements

Afshin Hosseini (HO 4596/1-1) received fellowship support from the German Research Foundation (DFG). Juan Loor (JJL) was supported by National Institute of Food and Agriculture (Grant: ILLU-538-914). The specific roles of the authors are articulated in the 'author's contributions' section.

Funding

Not applicable.

Authors' contributions

AH, MS, ZZ, and ET performed analyses and analyzed data. JJL and JKD conceived the animal experiment. AH and JJL wrote the manuscript. All authors approved the final version of the manuscript.

Competing interests

The authors declare that they have no competing interests.

Author details

[1]Department of Animal Sciences and Division of Nutritional Sciences, University of Illinois, 1207 West Gregory Drive, Urbana, IL 61801, USA. [2]Department of Animal Nutrition and Nutritional Diseases, University of Ondokuz Mayıs, 55139 Samsun, Turkey. [3]Istituto di Zootecnica, Facoltà di Scienze Agrarie Alimentari ed Ambientali, Università Cattolica del Sacro Cuore, 29122 Piacenza, Italy.

References

1. Ahmadian M, Suh JM, Hah N, Liddle C, Atkins AR, Downes M, et al. PPARgamma signaling and metabolism: the good, the bad and the future. Nat Med. 2013;19(5):557–66.
2. Jonas D, Van Scoyoc E, Gerrald K, Wines R, Amick H, Triplette M, et al. in Drug Class Review: Newer Diabetes Medications, TZDs, and Combinations. Final Original Report. 2011. Oregon Health & Science University, Portland (OR).
3. Ye J. Challenges in drug discovery for Thiazolidinedione substitute. Acta Pharmaceutica Sinica B. 2011;1(3):137–42.
4. Schoenberg KM, Overton TR. Effects of plane of nutrition and 2,4-thiazolidinedione on insulin responses and adipose tissue gene expression in dairy cattle during late gestation. J Dairy Sci. 2011;94(12):6021–35.
5. Smith KL, Butler WR, Overton TR. Effects of prepartum 2,4-thiazolidinedione on metabolism and performance in transition dairy cows. J Dairy Sci. 2009; 92(8):3623–33.
6. Smith KL, Stebulis SE, Waldron MR, Overton TR. Prepartum 2,4-thiazolidinedione alters metabolic dynamics and dry matter intake of dairy cows. J Dairy Sci. 2007;90(8):3660–70.
7. Schoenberg KM, Perfield KL, Farney JK, Bradford BJ, Boisclair YR, Overton TR. Effects of prepartum 2,4-thiazolidinedione on insulin sensitivity, plasma concentrations of tumor necrosis factor-alpha and leptin, and adipose tissue gene expression. J Dairy Sci. 2011;94(11):5523–32.
8. Hosseini A, Tariq MR, Trindade da Rosa F, Kesser J, Iqbal Z, Mora O, et al. Insulin sensitivity in adipose and skeletal muscle tissue of dairy cows in response to dietary energy level and 2,4-Thiazolidinedione (TZD). PLoS One. 2015;10(11):e0142633.
9. De Koster JD, Opsomer G. Insulin resistance in dairy cows. Vet Clin North Am Food Anim Pract. 2013;29(2):299–322.
10. Khan MJ, Jacometo CB, Graugnard DE, Corrêa MN, Schmitt E, Cardoso F, et al. Overfeeding dairy cattle during late-pregnancy alters hepatic PPARalpha-regulated pathways including Hepatokines: impact on metabolism and peripheral insulin sensitivity. Gene Regul Syst Bio. 2014;8:97–111.
11. Ohtsuka H, Koiwa M, Hatsugaya A, Kudo K, Hoshi F, Itoh N, et al. Relationship between serum TNF activity and insulin resistance in dairy cows affected with naturally occurring fatty liver. J Vet Med Sci. 2001;63(9):1021–5.
12. Bionaz M, Trevisi E, Calamari L, Librandi F, Ferrari A, Bertoni G. Plasma paraoxonase, health, inflammatory conditions, and liver function in transition dairy cows. J Dairy Sci. 2007;90(4):1740–50.
13. Bertoni G, Trevisi E, Han X, Bionaz M. Effects of inflammatory conditions on liver activity in puerperium period and consequences for performance in dairy cows. J Dairy Sci. 2008;91(9):3300–10.
14. Trevisi E, Zecconi A, Bertoni G, Piccinini R. Blood and milk immune and inflammatory profiles in periparturient dairy cows showing a different liver activity index. J Dairy Res. 2010;77(3):310–7.
15. Dann HM, Litherland NB, Underwood JP, Bionaz M, D'Angelo A, McFadden JW, et al. Diets during far-off and close-up dry periods affect periparturient metabolism and lactation in multiparous cows. J Dairy Sci. 2006;89(9):3563–77.
16. Bas P. Changes in activities of lipogenic enzymes in adipose tissue and liver of growing goats. J Anim Sci. 1992;70(12):3857–66.
17. Chang E, Park CY, Park SW. Role of thiazolidinediones, insulin sensitizers, in non-alcoholic fatty liver disease. J Diabetes Investig. 2013;4(6):517–24.
18. Graugnard DE, Bionaz M, Trevisi E, Moyes KM, Salak-Johnson JL, Wallace RL, et al. Blood immunometabolic indices and polymorphonuclear neutrophil function in peripartum dairy cows are altered by level of dietary energy prepartum. J Dairy Sci. 2012;95(4):1749–58.
19. Bolster DR, Jefferson LS, Kimball SR. Regulation of protein synthesis associated with skeletal muscle hypertrophy by insulin-, amino acid- and exercise-induced signalling. Proc Nutr Soc. 2004;63(2):351–6.

20. Kirsch R, Frith L, Black E, Hoffenberg R. Regulation of albumin synthesis and catabolism by alteration of dietary protein. Nature. 1968;217(5128):578–9.

21. Nicholson JP, Wolmarans MR, Park GR. The role of albumin in critical illness. Br J Anaesth. 2000;85(4):599–610.

22. Lunn PG, Austin S. Dietary manipulation of plasma albumin concentration. J Nutr. 1983;113(9):1791–802.

23. De Matteis L, Bertoni G, Lombardelli R, Wellnitz O, Van Dorland HA, Vernay MCMB, et al. Acute phase response in lactating dairy cows during hyperinsulinemic hypoglycaemic and hyperinsulinemic euglycaemic clamps and after intramammary LPS challenge. J Anim Physiol Anim Nutr (Berl). 2017;101(3):511–20.

24. Duske K, Hammon HM, Langhof AK, Bellmann O, Losand B, Nürnberg K, et al. Metabolism and lactation performance in dairy cows fed a diet containing rumen-protected fat during the last twelve weeks of gestation. J Dairy Sci. 2009;92(4):1670–84.

25. Katoh N. Relevance of apolipoproteins in the development of fatty liver and fatty liver-related peripartum diseases in dairy cows. J Vet Med Sci. 2002; 64(4):293–307.

26. Mackness B, Durrington PN, Mackness MI. Human serum paraoxonase. Gen Pharmacol. 1998;31(3):329–36.

27. Turk R, Juretić D, Geres D, Svetina A, Turk N, Flegar-Mestrić Z. Influence of oxidative stress and metabolic adaptation on PON1 activity and MDA level in transition dairy cows. Anim Reprod Sci. 2008;108(1–2):98–106.

28. Aviram M, Rosenblat M, Billecke S, Erogul J, Sorenson R, Bisgaier CL, et al. Human serum paraoxonase (PON 1) is inactivated by oxidized low density lipoprotein and preserved by antioxidants. Free Radic Biol Med. 1999;26(7–8):892–904.

29. Collison KS, Saleh SM, Bakheet RH, Al-Rabiah RK, Inglis AL, Makhoul NJ, et al. Diabetes of the liver: the link between nonalcoholic fatty liver disease and HFCS-55. Obesity (Silver Spring). 2009;17(11):2003–13.

30. Chiellini C, Santini F, Marsili A, Berti P, Bertacca A, Pelosini C, et al. Serum haptoglobin: a novel marker of adiposity in humans. J Clin Endocrinol Metab. 2004;89(6):2678–83.

31. Ceciliani F, Ceron JJ, Eckersall PD, Sauerwein H. Acute phase proteins in ruminants. J Proteome. 2012;75(14):4207–31.

32. Bertoni G, Trevisi E. Use of the liver activity index and other metabolic variables in the assessment of metabolic health in dairy herds. Vet Clin North Am Food Anim Pract. 2013;29(2):413–31.

33. Kaneko JJ, Harvey JW, Bruss ML. Clinical biochemistry of domestic animals. Fifth ed. London: Academic Press Limited; 1997.

34. Giannini EG, Testa R, Savarino V. Liver enzyme alteration: a guide for clinicians. CMAJ. 2005;172(3):367–79.

35. Bendich A. Physiological role of antioxidants in the immune system. J Dairy Sci. 1993;76(9):2789–94.

36. LeBlanc SJ, Herdt TH, Seymour WM, Duffield TF, Leslie KE. Peripartum serum vitamin E, retinol, and beta-carotene in dairy cattle and their associations with disease. J Dairy Sci. 2004;87(3):609–19.

37. Noy N. Vitamin a in regulation of insulin responsiveness: mini review. Proc Nutr Soc. 2016;75(2):212–5.

38. Traber MG. Vitamin E regulatory mechanisms. Annu Rev Nutr. 2007;27:347–62.

39. Berry DC, Noy N. Signaling by vitamin a and retinol-binding protein in regulation of insulin responses and lipid homeostasis. Biochim Biophys Acta. 2012;1821(1):168–7.

40. Gruys E, Toussaint MJ, Niewold TA, Koopmans SJ. Acute phase reaction and acute phase proteins. J Zhejiang Univ Sci B. 2005;6(11):1045–56.

41. Wolf G. Multiple functions of vitamin a. Physiol Rev. 1984;64(3):873–937.

42. Kassahun K, Pearson PG, Tang W, McIntosh I, Leung K, Elmore C, et al. Studies on the metabolism of Troglitazone to reactive intermediates in vitro and in vivo. Evidence for novel biotransformation pathways involving Quinone Methide formation and Thiazolidinedione ring scission. Chem Res in Toxicol. 2001;14(1):62–70.

43. Hu D, Wu CQ, Li ZJ, Liu Y, Fan X, Wang QJ, et al. Characterizing the mechanism of thiazolidinedione-induced hepatotoxicity: an in vitro model in mitochondria. Toxicol Appl Pharmacol. 2015;284(2):134–41.

44. Masubuchi Y. Metabolic and non-metabolic factors determining Troglitazone Hepatotoxicity: a review. Drug Metab Pharmacokinet. 2006; 21(5):347–56.

45. Vernochet C, Davis KE, Scherer PE, Farmer SR. Mechanisms regulating repression of haptoglobin production by peroxisome proliferator-activated receptor-gamma ligands in adipocytes. Endocrinology. 2010; 151(2):586–94.

46. Narayanan PK, Hart T, Elcock F, Zhang C, Hahn L, McFarland D, et al. Troglitazone-induced intracellular oxidative stress in rat hepatoma cells: a flow cytometric assessment. Cytometry A. 2003;52(1):28–35.

47. Garcia-Ruiz I, Rodríguez-Juan C, Díaz-Sanjuán T, Martínez MA, Muñoz-Yagüe T, Solís-Herruzo JA. Effects of rosiglitazone on the liver histology and mitochondrial function in ob/ob mice. Hepatology. 2007;46(2):414–23.

48. Cohn W, Loechleiter F, Weber F. Alpha-tocopherol is secreted from rat liver in very low energy lipoproteins. J Lipid Res. 1988;29(10):1359–66.

49. Petterson JA, Dunshea FR, Ehrhardt RA, Bell AW. Pregnancy and undernutrition alter glucose metabolic responses to insulin in sheep. J Nutr. 1993;123(7):1286–95.

50. Gill RD, Hart IC. Properties of insulin and glucagon receptors on sheep hepatocytes: a comparison of hormone binding and plasma hormones and metabolites in lactating and non-lactating ewes. J Endocrinol. 1980; 84(2):237–47.

51. Saad MJ, Saad MJ, Araki E, Miralpeix M, Rothenberg PL, White MF, et al. Regulation of insulin receptor substrate-1 in liver and muscle of animal models of insulin resistance. J Clin Invest. 1992;90(5):1839–49.

52. Smith DH, Palmquist DL, Schanbacher FL. Characterization of insulin binding to bovine liver and mammary microsomes. Comp Biochem Physiol A Comp Physiol. 1986;85(1):161–9.

53. Bionaz M, Chen S, Khan MJ, Loor JJ. Functional role of PPARs in ruminants: potential targets for fine-tuning metabolism during growth and lactation. PPAR Res. 2013;2013:684159.

54. Kersten S. Integrated physiology and systems biology of PPARalpha. Mol Metab. 2014;3(4):354–71.

55. Rakhshandehroo M, Sanderson LM, Matilainen M, Stienstra R, Carlberg C, de et al. Comprehensive analysis of PPARalpha-dependent regulation of hepatic lipid metabolism by expression profiling. PPAR Res. 2007;2007: 26839.

56. Patsouris D, Mandard S, Voshol PJ, Escher P, Tan NS, Havekes LM, et al. PPARalpha governs glycerol metabolism. J Clin Invest. 2004;114(1):94–103.

57. White HM, Koser SL, Donkin SS. Differential regulation of bovine pyruvate carboxylase promoters by fatty acids and peroxisome proliferator-activated receptor-alpha agonist. J Dairy Sci. 2011;94(7):3428–36.

58. Way JM, Harrington WW, Brown KK, Gottschalk WK, Sundseth SS, Mansfield TA, et al. Comprehensive messenger ribonucleic acid profiling reveals that peroxisome proliferator-activated receptor gamma activation has coordinate effects on gene expression in multiple insulin-sensitive tissues. Endocrinology. 2001;142(3):1269–77.

59. Dreyer C, Krey G, Keller H, Givel F, Helftenbein G, Wahli W. Control of the peroxisomal beta-oxidation pathway by a novel family of nuclear hormone receptors. Cell. 1992;68(5):879–87.

60. Tugwood JD, Issemann I, Anderson RG, Bundell KR, McPheat WL, Green S. The mouse peroxisome proliferator activated receptor recognizes a response element in the 5′ flanking sequence of the rat acyl CoA oxidase gene. EMBO J. 1992;11(2):433–9.

61. Colca JR, McDonald WG, Waldon DJ, Leone JW, Lull JM, Bannow CA, et al. Identification of a novel mitochondrial protein ("mitoNEET") cross-linked specifically by a thiazolidinedione photoprobe. Am J Physiol Endocrinol Metab. 2004;286(2):E252–60.

62. Kersten S. Regulation of lipid metabolism via angiopoietin-like proteins. Biochem Soc Trans. 2005;33(Pt 5):1059–62.

63. Kharitonenkov A, Shiyanova TL, Koester A, Ford AM, Micanovic R, Galbreath EJ, et al. FGF-21 as a novel metabolic regulator. J Clin Invest. 2005;115(6): 1627–35.

64. Hondares E, Rosell M, Gonzalez FJ, Giralt M, Iglesias R, Villarroya F. Hepatic FGF21 expression is induced at birth via PPARalpha in response to milk intake and contributes to thermogenic activation of neonatal brown fat. Cell Metab. 2010;11(3):206–12.

65. Loor JJ, Everts RE, Bionaz M, Dann HM, Morin DE, Oliveira R, et al. Nutrition-induced ketosis alters metabolic and signaling gene networks in liver of periparturient dairy cows. Physiol Genomics. 2007;32(1):105–16.

66. Schoenberg KM, Giesy SL, Harvatine KJ, Waldron MR, Cheng C, Kharitonenkov A, et al. Plasma FGF21 is elevated by the intense lipid mobilization of lactation. Endocrinology. 2011;152(12):4652–61.

67. Kuo T, Chen TC, Yan S, Foo F, Ching C, McQueen A, et al. Repression of glucocorticoid-stimulated angiopoietin-like 4 gene transcription by insulin. J Lipid Res. 2014;55(5):919–28.

68. Ryden M. Fibroblast growth factor 21: an overview from a clinical perspective. Cell Mol Life Sci. 2009;66(13):2067–73.

69. Mattijssen F, Kersten S. Regulation of triglyceride metabolism by Angiopoietin-like proteins. Biochim Biophys Acta. 2012;1821(5):782–9.

70. Oishi K, Tomita T. Thiazolidinediones are potent inducers of fibroblast growth factor 21 expression in the liver. Biol Pharm Bull. 2011;34(7):1120–1.

71. Hegardt FG. Mitochondrial 3-hydroxy-3-methylglutaryl-CoA synthase: a control enzyme in ketogenesis. Biochem J. 1999;338(Pt 3):569–82.

72. Connor EE, Li RW, Baldwin RL, Li C. Gene expression in the digestive tissues of ruminants and their relationships with feeding and digestive processes. Animal. 2010;4(7):993–1007.

73. Meertens LM, Miyata KS, Cechetto JD, Rachubinski RA. Capone JPA mitochondrial ketogenic enzyme regulates its gene expression by association with the nuclear hormone receptor PPARalpha. EMBO J. 1998;17(23):6972–8.

11

Candidate genes for male and female reproductive traits in Canchim beef cattle

Marcos Eli Buzanskas[1]* ⓘ, Daniela do Amaral Grossi[2], Ricardo Vieira Ventura[3], Flavio Schramm Schenkel[4], Tatiane Cristina Seleguim Chud[5], Nedenia Bonvino Stafuzza[5], Luciana Diniz Rola[6], Sarah Laguna Conceição Meirelles[7], Fabiana Barichello Mokry[8], Maurício de Alvarenga Mudadu[9], Roberto Hiroshi Higa[9], Marcos Vinícius Gualberto Barbosa da Silva[10], Maurício Mello de Alencar[11], Luciana Correia de Almeida Regitano[11] and Danísio Prado Munari[5]

Abstract

Background: Beef cattle breeding programs in Brazil have placed greater emphasis on the genomic study of reproductive traits of males and females due to their economic importance. In this study, genome-wide associations were assessed for scrotal circumference at 210 d of age, scrotal circumference at 420 d of age, age at first calving, and age at second calving, in Canchim beef cattle. Data quality control was conducted resulting in 672,778 SNPs and 392 animals.

Results: Associated SNPs were observed for scrotal circumference at 420 d of age (435 SNPs), followed by scrotal circumference at 210 d of age (12 SNPs), age at first calving (six SNPs), and age at second calving (four SNPs). We investigated whether significant SNPs were within genic or surrounding regions. Biological processes of genes were associated with immune system, multicellular organismal process, response to stimulus, apoptotic process, cellular component organization or biogenesis, biological adhesion, and reproduction.

Conclusions: Few associations were observed for scrotal circumference at 210 d of age, age at first calving, and age at second calving, reinforcing their polygenic inheritance and the complexity of understanding the genetic architecture of reproductive traits. Finding many associations for scrotal circumference at 420 d of age in various regions of the Canchim genome also reveals the difficulty of targeting specific candidate genes that could act on fertility; nonetheless, the high linkage disequilibrium between loci herein estimated could aid to overcome this issue. Therefore, all relevant information about genomic regions influencing reproductive traits may contribute to target candidate genes for further investigation of causal mutations and aid in future genomic studies in Canchim cattle to improve the breeding program.

Keywords: Animal breeding, Composite breed, Genome-wide association, Genomic data, Single nucleotide polymorphism

Background

Cattle breeding programs in Brazil have given greater emphasis to the study and selection of reproductive traits due to their economic importance for the production system. In males, scrotal circumference traits are related to the reproductive potential of bulls, because testis size is associated with the production and quality of sperm and the production of sex hormones [1].

In general, female reproductive traits are difficult to measure and, in some cases, are strongly influenced by environmental factors. The reproductive performance of heifers depends on the age at which they calve for the first time; the ones that calve earlier have a more productive life [2]. In addition to first calving, another important factor is that the cow continues producing calves regularly to maintain its productivity and diminish calving interval [3]. Studies have reported that indirect selection of females based on the performance of bulls is possible, considering the favorable genetic correlations

* Correspondence: marcosbuz@gmail.com
[1]Departamento de Zootecnia, Universidade Federal da Paraíba (UFPB), Areia, Paraíba 58397-000, Brazil
Full list of author information is available at the end of the article

between scrotal circumference measures and age at calving [2, 4, 5].

Many achievements in animal breeding were obtained based on the classical approach, using information from phenotypes and pedigree. Nowadays, molecular data analyses are bringing new insights to the genetic architecture of species. Regarding reproductive traits of beef cattle, genome-wide association studies (GWAS) are useful tools for the identification of candidate genes that could be used to classify precocious or more fertile individuals [6, 7].

The identification of candidate genes provides a better understanding of the distribution of genes that affect traits of economic interest, as well as a basis for further studies to identify causal mutations. Despite its potential, important observations are needed for this approach. According to Tabor et al. [8], the difficulty of replication over time or across populations in this approach might indicate that more detailed studies are needed to certify its causality. Therefore, the aim of this study was to perform a GWAS to identify genomic regions and candidate genes to uncover the genetic architecture of scrotal circumference at 210 and 420 d of age and age at first and second calving and aid in the breeding process in Canchim cattle.

Methods
Canchim breed

The Canchim is a composite breed (62.5% Charolais and 37.5% Zebu) developed in the 1940s by the Brazilian Agricultural Research Corporation (Embrapa), located in São Carlos city, SP, Brazil [9]. Different crossbreeding schemes have been studied to produce Canchim cattle with different Charolais-Zebu proportions and to achieve greater genetic variability in the population [10–13]. One of these schemes was used to produce animals of the "MA" genetic group, which was derived from mating Canchim-Zebu animals (F1) with Charolais animals, resulting in approximate contributions of 65.6% Charolais and 34.4% Zebu [14].

The Canchim cattle represents about 3% of the beef cattle produced in Brazil [15]. Indicine breeds are much more representative as purebreds or crossbred animals, being responsible for 80% of the beef cattle industry in the country [16].

Traits analyzed

The estimated breeding values (EBVs) used in our study considered the following traits: scrotal circumference at weaning adjusted for 210 d of age (SC210), scrotal circumference adjusted for 420 d of age (SC420), age at first calving (AFC), and age at second calving (ASC). Linear interpolation was previously used to adjust the scrotal circumferences for 210 and 420 d of age.

The EBVs were obtained from the genetic evaluation carried out by the Brazilian Agricultural Research Corporation (Embrapa) for the Canchim breed which considered 318,307 animals in the relationship matrix and 267,002 animals with valid records. In general, the traits analyzed by Embrapa for the genetic evaluation of Canchim that could be highlighted are body weight traits measured at various ages, reproductive traits of males and females, carcass quality, navel visual score of males and females, and hair coat. The EBVs were estimated by multi-trait analysis using the REMLF90 program [17], under an animal model. For the studied traits, the fixed effects considered in the contemporary groups were year and season of birth (January to March; April to September; and October to December), farm of birth, genetic group, and feeding system.

As described by Mokry et al. [18], the genotyped animals were chosen according to the EBV for some traits (ribeye area, back fat thickness, and productive and reproductive traits), accuracy, family size, and proportion of males and females. The mean EBV values for SC210, SC420, AFC, and ASC were 1.44 mm, 2.07 mm, −4.34 d, and −1.23 d, respectively. Minimum and maximum values for SC210, SC420, AFC, and ASC varied from −8.76 to 12.74, −17.41 to 21.80, −50.89 to 44.64, and −42.46 to 44.66, respectively. De-regressed proofs were not used due to limited data. Furthermore, de-regressed proofs calculated with low accuracies are expected to have a smaller genomic contribution due to Mendelian sampling and will have more noise added in de-regressed calculations [19].

Genotypes and quality control

The BovineHD BeadChip SNP panel (Illumina Inc., San Diego, CA) was used to genotype 194 males and 205 females: 285 animals of the Canchim breed and 114 animals of the MA genetic group, calves of 49 bulls and 355 cows. The animals were born between 1999 and 2005 and come from seven farms in the States of São Paulo and Goiás. A detailed description of these animals was previously presented by Buzanskas et al. [20] and Mokry et al. [18]. Genotype quality control excluded SNPs with significant deviations ($P < 10^{-5}$) from the Hardy-Weinberg equilibrium, SNPs with minor allele frequencies of less than 0.05, and call rate for SNPs and animals lower than 0.90. Only autosomal chromosomes and SNPs with known positions, according to the UMD_3.1 bovine genome assembly [21], were used for GWAS.

Genome-wide association study

The Generalized Quasi-Likelihood Score (GQLS) method [22] was used for GWAS. In this method, a logistic regression was used to associate the EBVs (treated as a covariate) to the genotypes (treated as the response variable), one SNP at a time. The EBVs were represented by $X_i = (X_1, \ldots, X_n)'$, in which Xi represents the EBVs for the ith animal. The genotype of the animals was coded as "0", "1" or "2" and, as the genotypes were represented by $Y_i = (Y_1, \ldots, Y_n)'$, in which $Y_i = \frac{1}{2}$ * (number of alleles for animal genotype i), the respective proportions would be equal to 0, ½ and 1. The expected SNP allele frequency is represented by μ, in which $\mu = (\mu_1, \ldots, \mu_n)' = E(X|\,Y)$, thus $0 < \mu_i < 1$.

To associate μ_i with X_i, the following logistic regression was defined:

$$\mu_i = E(Y_i, |X_i) = \frac{e^{\beta_0 + \beta_1 X_i}}{1 + e^{\beta_0 + \beta_1 X_i}}$$

in which β_0 is the constant term and β_1 is the slope.

Under the null hypothesis, μ_i can be interpreted by $\mu_i = \frac{e^{\beta_0}}{1+e^{\beta_0}}$, for all $i = 1, \ldots, n$. The mean vector of Y_i no longer depends on X_i, which becomes $\mu_i = E(Y) = \mu 1$, and "1" is a vector. The solution of the "quasi-likelihood score" results in an estimate for μ, such that $\hat{\mu} = \left(1'A^{-1}1\right)^{-1} 1'A^{-1} Y$, in which A^{-1} is the inverse of the relationship matrix of the individuals.

To obtain the GQLS, the statistic WG was calculated as in Feng et al. [22], as follows:

$$W_G = \frac{2}{\hat{\mu}(1-\mu)}\left[X'A^{-1}(Y-\mu 1)\right]' \times \left[X'A^{-1}X - (X'A^{-1}1)(1'A^{-1}1)^{-1}(1'A^{-1}X)\right]^{-1} \times \left[X'A^{-1}(Y-\mu 1)\right]$$

Under the null hypothesis, WG follows a Chi-squared distribution with one degree of freedom [23], resulting in P-values for each SNP. This method does not account for the genetic variance explained by the marker. The GQLS provides the advantage of accounting for the population structure by means of the pedigree-based relationship among animals (A^{-1}).

The false discovery rate (FDR) was used for multiple testing corrections [24] to verify significant SNPs. The P-values of each SNP were sorted in ascending order and the following formula applied:

$$q < mP(i)/i$$

in which q is the desired level of significance, m is the total number of SNPs, and P is the P-value of the ith SNP. To account for multiple comparisons, a genome-wide and chromosome-wise FDR of 5% (significant association) and 10% (suggestive association) was applied. The main reason for considering a minimum FDR of 10% was to obtain a comprehensive number of SNPs,

which might assist to comprehend the genetic architecture of the studied traits.

Pairwise linkage disequilibrium was estimated using the r^2 measure [25] for the SNPs with significant and suggestive associations. The software Haploview [26] was used to estimate r^2 and plot the results. Due to the high linkage phase consistency between Canchim and MA genetic group (above 0.80 up to 100 kb) [27], analyses were conducted considering Canchim and MA as one population.

Gene mapping and in silico functional analyses
The SNPs associated with the EBVs for SC210, SC420, AFC, and ASC traits were surveyed to their corresponding genes or to surrounding genes. The National Center for Biotechnology Information (NCBI) database [28] and Ensembl Genome Browser [29] were accessed to proceed with in silico functional analyses of the genes. If the SNP was located in the intergenic region (i.e. not assigned to any gene), we observed, through the integrated maps of the NCBI variation viewer, for the closest gene(s) and calculated the distance(s).

The PANTHER tool [30] was used to access and gather biological processes and pathway associations. The AnimalQTLdb database [31] was used to verify previous reports of quantitative trait loci (QTL) in the surroundings of significant SNPs. The main reason to use these tools was to validate our findings.

Results
A total of 672,778 SNPs and 392 animals were used for GWAS. The Manhattan plots for chromosome-wise (in blue) and genome-wide (in red) significantly associated SNPs after FDR correction for SC210, SC420, AFC, and ASC, respectively, are presented in Fig. 1. The greatest number of significant SNPs was observed for SC420 (435 SNPs), followed by SC210 (12 SNPs), AFC (six SNPs), and ASC (four SNPs) when considering FDR 10% for all traits.

In Table 1 are presented SNPs and genes identified for SC210, AFC, and ASC. For SC420, SNPs and genes are presented in Table 2. Due to the large number of SNPs associated with SC420, full information regarding genes, pseudogenes, and non-coding RNA are presented in Additional files 1, 2, and 3. For FDR 5%, a total of 249 (chromosome-wise) and 50 (genome-wide) SNPs were observed for SC420 (Additional file 1). Considering FDR 10%, the regions observed in Table 1 presented a suggestive association; therefore, the genes identified in these regions were surveyed.

For SC210, chromosome-wise significantly associated SNPs ($P \leq 0.00001$) were located on chromosomes 20 and 28. Five SNPs were located in the intergenic or intragenic regions of *SMIM23*, *PAPD7*, *ICE1*, and *EDARADD*

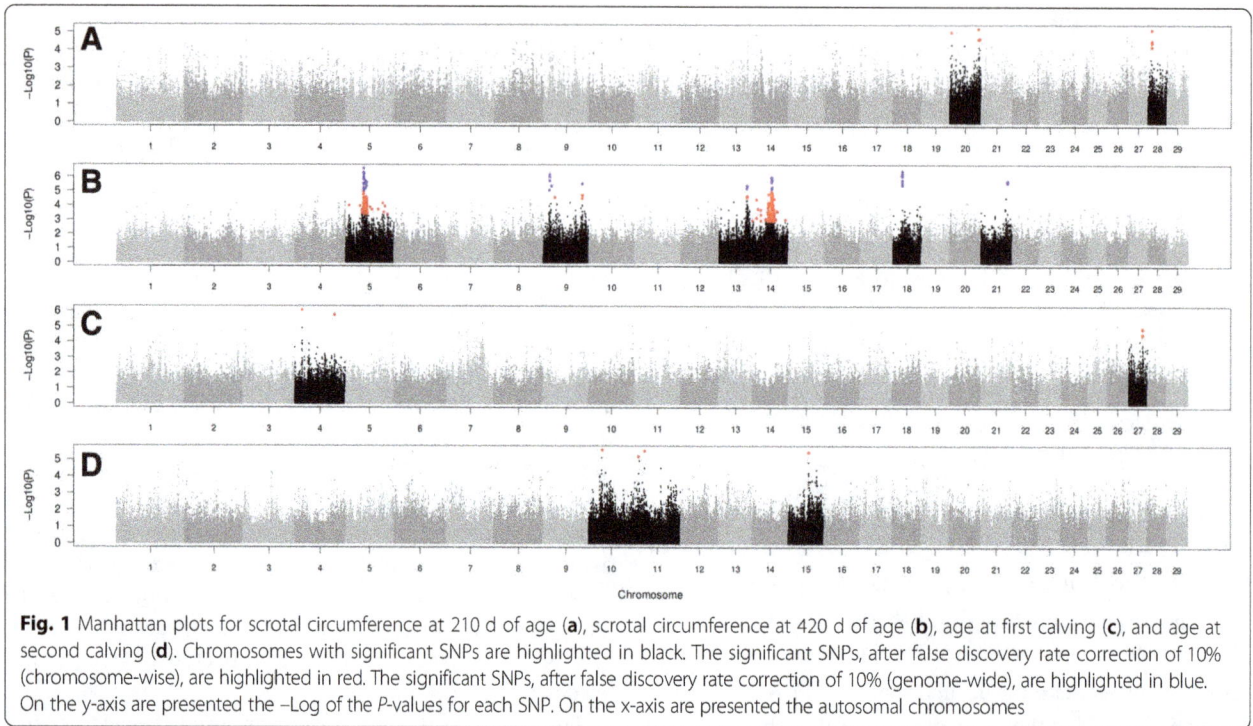

Fig. 1 Manhattan plots for scrotal circumference at 210 d of age (**a**), scrotal circumference at 420 d of age (**b**), age at first calving (**c**), and age at second calving (**d**). Chromosomes with significant SNPs are highlighted in black. The significant SNPs, after false discovery rate correction of 10% (chromosome-wise), are highlighted in red. The significant SNPs, after false discovery rate correction of 10% (genome-wide), are highlighted in blue. On the y-axis are presented the –Log of the *P*-values for each SNP. On the x-axis are presented the autosomal chromosomes

genes, and non-coding RNA *LOC101907249* (Table 1 and Fig. 1a). On chromosome 28, seven SNPs were located in the *EDARADD* gene.

For SC420, chromosome-wise and genome-wise associations were observed on chromosomes 5, 9, 13, 14, 18, and 21 (Fig. 1b). The SNPs were close or within a total

of 64 genes, 10 non-coding RNAs, 13 pseudogenes, and two transfer RNAs. Full information on SNPs, position, and genes are presented in the Additional files 1, 2, and 3.

Six significantly associated SNPs ($P \le 0.00001$) for AFC (Table 1 and Fig.1c) were located on chromosomes

Table 1 Chromosome-wise association for scrotal circumference (SC210), age at first (AFC) and second calving (ASC)

Trait	Genes	SNP Reference	Chr:Pos	Distance to gene	P-value (Min-Max)
SC210	SMIM23	rs135355728[a]	20:3.48	14.22	8.96E-06
SC210	PAPD7	rs137042056[a], rs136276163[a]	20:66.52..66.53	48.62..38.66	5.85E-06-p2.32E-05
SC210	ICE1	rs41582170[b]	20:67.81	0.00	2.13E-05
SC210	LOC101907249	rs136535499[a]	20:69.05	240.87	2.13E-05
SC210	EDARADD	rs110746860[b], rs110371081[b], rs109902875[b], rs110870694[b], rs110610232[b], rs134356559[b], rs210911576[b]	28:9.14..28:9.16	0.00	6.71E-06-5.76E-05
AFC	NXPH1	rs133411648[a]	4:17.46	45.77	8.26E-07
AFC	EXOC4	rs110606254[a]	4:98.31	0.00	1.74E-06
AFC	ZMAT4	rs134390082[a], rs137553882[a], rs133519327[a], rs135481346[a]	27:35.19..35.21	48.32..29.34	1.23E-05-3.18E-05
ASC	FMN1	rs134100268[b]	10:29.88	0.00	2.43E-06
ASC	TMEM182	rs43661848[a]	11:7.69	197.15	6.29E-06
ASC	LOC790871	rs136610615[a]	11:22.29	69.94	2.77E-06
ASC	UBQLN3	rs43031470[c]	15:48.70	0.00	3.39E-06

[a]intergenic region
[b]intron variant
[c]exon variant

Gene symbols, SNP reference number, and chromosomes (Chr) and positions (Pos, in megabase) were obtained from NCBI website. Distances to gene (kilobase) are presented from 5' to gene direction. If distance equals zero (0.00), the SNP is on intragenic region. *P*-values are presented as the minimum (Min) and maximum (Max) significance obtained from the generalized quasi-likelihood method

Table 2 Highlighted genes from genome-wide (in bold) and chromosome-wise associations for scrotal circumference at 420 d of age

Symbol	SNP Reference	Chr:Pos	Distances to gene	P-value (Min - Max)
RAP1B	**rs110520377**[a], **rs133990240**[a], rs109547215[a], **rs110160018**[b], **rs110261691**[c], **rs109099268**[b], **rs133124963**[b], **rs109023687**[b], **rs110034677**[b], **rs110091099**[b], **rs109950552**[b], **rs137658592**[b], rs133340933[b], rs134626455[b], rs110001336[b], **rs109288126**[b], rs137319832[b], **rs109506571**[b], **rs109248631**[b], **rs110027103**[b], **rs109561643**[b], **rs109210079**[b], **rs110160918**[b], rs134366426[b], rs109273768[b], rs109024096[b], rs110798702[b], **rs110625630**[b], **rs110219262**[b], **rs134311132**[b], **rs135497432**[b], **rs133173059**[b]	5:45.35..45.40	8.58..0.00	2.47E-07 - 7.64E-05
SRGAP1	**rs110268648**[b], **rs109748105**[b], **rs134621421**[b]	5:49.98..49.98	0.00	1.88E-06 - 6.55E-06
FOXM1	rs135705262[b]	5:107.39	0.00	4.00E-04
TOP1	**rs134822694**[b], **rs135287766**[b]	13:70.39..70.40	0.00	3.86E-06
STAU2	rs137465376[b], rs134711539[b], rs137821036[a]	14:38.89..38.97	0.00..14.40	8.66E-05 - 5.00E-04
PEX2	rs137442228[d], rs110035827[a], rs41730291[a]	14:42.33..42.37	0.00..40.77	4.10E-04 - 9.80E-04
FABP5	rs109480456[a], rs136045797[a], rs133930486[e], rs136613853[b], rs137684819[c], rs133054550[a], rs133483556[a], rs136236059[a], rs136236059[a], rs137344980[a], rs135804214[a], rs136785030[a], rs135727060[a], rs133436244[a], rs134708967[a], rs43103204[a]	14:46.63..46.73	15.48..0.00..83.31	1.55E-05 - 1.16E-03
FABP12	rs43765470[b], rs43765465[a], rs41730924[a]	14:46.89..46.92	0.00..28.04	3.48E-05 - 2.88E-04
MED30	rs135065691[c], rs41734435[b], rs135292147[a]	14:48.94..48.97	4.03..0.00..9.76	8.27E-04 - 9.48E-04
TRHR	rs133457508[e]	14:57.52	0.00	2.27E-04

[a]intergenic region [b]intron variant [c]downstream variant [d]5' UTR [e]upstream variant
Gene symbols, SNP reference number, and chromosomes (Chr) and positions (Pos, in megabase) were obtained from NCBI website. Distances to gene (kilobase) are presented from 5' to gene and 3' to gene directions. If distance equals zero (0.00), the SNP is on intragenic region. P-values are presented as the minimum (Min) and maximum (Max) significance obtained from the generalized quasi-likelihood method

4 (*NXPH1* and *EXOC4* genes) and 27 (*ZMAT4* gene). For ASC, four SNPs were significantly associated ($P \leq 0.000001$) and located on chromosomes 10 (*FMN1* gene), 11 (*TMEM182* gene and *LOC790871* pseudogene), and 15 (*UBQLN3* gene). Biological processes for SC210, AFC, and ASC are presented in Fig. 2.

For SC420, biological processes and pathway associations are presented in Figs. 3 and 4, respectively. PANTHER tool was able to account for 62 genes overrepresented among pathways when using *Bos taurus* as background. Most of the biological processes were

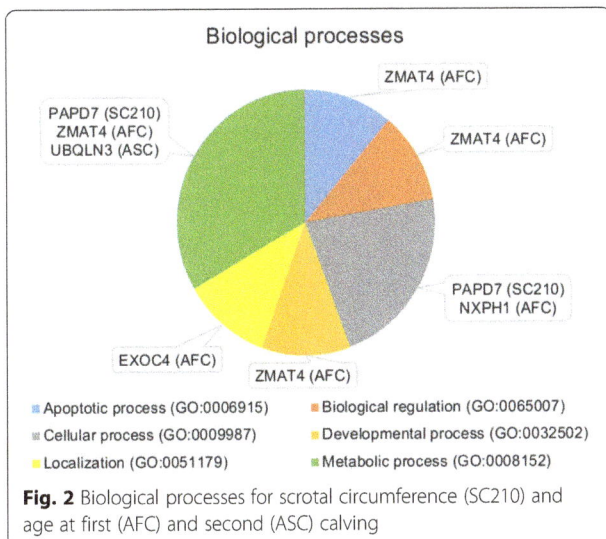

Fig. 2 Biological processes for scrotal circumference (SC210) and age at first (AFC) and second (ASC) calving

involved in metabolic (32 genes), cellular (22 genes), developmental (14 genes), biological regulation (13 genes), and localization (11 genes) processes. Biological processes were observed for the immune system (seven genes), multicellular organismal process (seven genes), response to stimulus (seven genes), apoptotic process (six genes), cellular component organization or biogenesis (six genes), biological adhesion (four genes), and reproduction (one gene). Pathway analysis showed that the *RAP1B* gene is involved in four pathways, while the *IFNG*, *COL12A1*, and *SRGAP1* genes are involved in two pathways.

The average linkage disequilibrium for each trait by chromosome was equal to 0.70 (AFC - BTA27), 0.94 (SC210 – BTA20), 0.89 (SC210 – BTA28), 0.64 (SC420 – BTA5), 0.69 (SC420 – BTA9), 0.90 (SC420 – BTA13), 0.56 (SC420 – BTA14), 0.85 (SC420 – BTA18), and 0.99 (SC420 – BTA21). Some significant or suggestive regions presented r2 equal to zero. We highlighted the SC420 trait due to the higher number of regions presenting significant or suggestive associations (Additional file 4: Figs. S1, S2, S3 and S4).

Discussion

The suggestive associations observed for SC210, AFC, and ASC, could provide an insight over potential regions responsible for the genetic variability of the traits. The main reason to consider FDR 10% was to study and describe, in a broader point of view, the biological

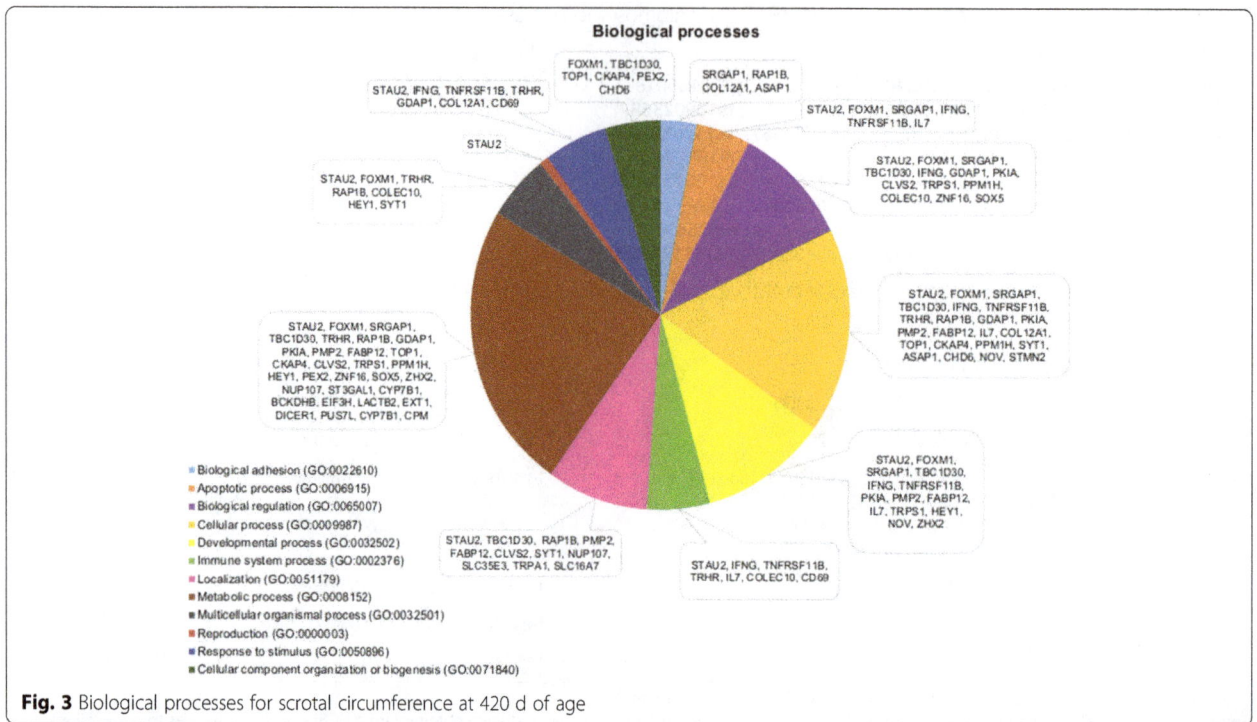

Fig. 3 Biological processes for scrotal circumference at 420 d of age

pathways to aid in the comprehension of these complex traits. Furthermore, due to the polygenic inheritance of these traits, it was expected that many loci with small effects were responsible for expressing the phenotype (or EBV). In some cases, we observed the lack of information regarding the SNP or gene identified.

We observed high average linkage disequilibrium among significantly or suggestively associated SNPs for most of the traits studied; therefore, there is evidence of direct or indirect associations that could be affecting the traits. Mokry et al. [27], when studying this Canchim population, observed that the linkage disequilibrium estimated could

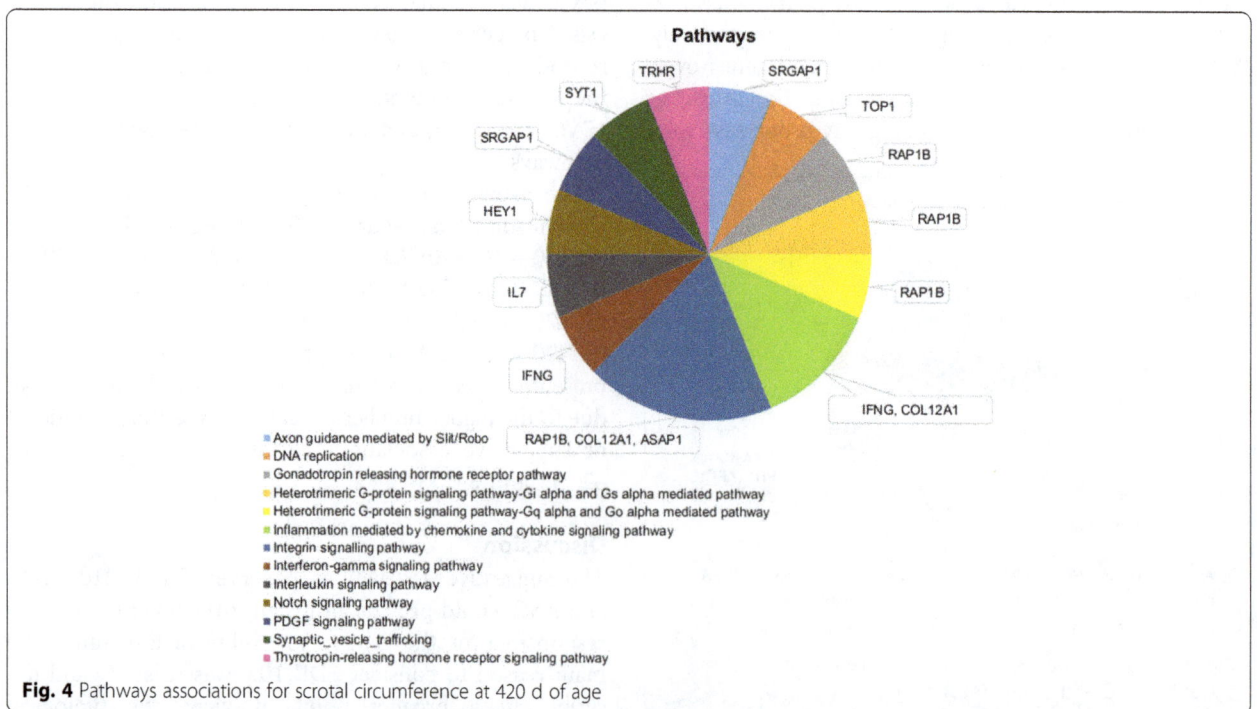

Fig. 4 Pathways associations for scrotal circumference at 420 d of age

be used for genomic selection and GWAS (minimum r^2 varying from 0.33 to 0.40 up to 2.5 kb).

For SC210, we have found that the *SMIM23* gene is located inside QTL regions associated with the rate of non-return to estrus in Swedish Red and Swedish Holstein breed animals [32]. In this same region, McClure et al. [33] found QTL for scrotal circumference in American Angus cattle. Studies reported a QTL in the *PAPD7*, *ICE1*, and *LOC101907249* gene regions associated with percentage abnormal sperm in Holstein bulls [34] and calving ease in Angus breed cows [33]. The *EDARADD* gene participates in cell differentiation and mutations in this gene can result in abnormal development of tissues and organs of ectodermal origin [35]. In this region, a QTL associated with pregnancy rate trait have been reported by Boichard et al. [36]. No specific biological process for *SMIM23*, *ICE1*, and *LOC101907249* genes was observed in the literature.

Few suggestive associations were observed for SC210 and none of the SNPs associated with SC210 presented pleiotropic effect with SNP associated with SC420 (discussed below) or other studied traits (discussed as follows), which could be due to low genetic correlation among EBVs for these traits in our dataset.

Regarding the SC420 trait, we have identified the STAU2 gene, which participates in multiple biological processes (Table 2 and Fig. 3), highlighting reproduction, developmental and immune system. This gene was involved in muscle development [37] in *Bos taurus indicus*, *Bos taurus taurus*, and *Bos taurus taurus* x *Bos taurus indicus*. According to Ramayo-Caldas et al. [37], the *STAU2* gene was present in two main networks of *PPARGC1A* and *HNF4G* genes, which acts as transcription factors that activate a variety of hormone receptors [38] and binding to fatty acids [39]. Peddinti et al. [40] observed, through genetic network analysis, that one of the main biological networks for bovine germinal vesicle and bovine cumulus granulosa cell proteomes contain the *SRGAP1* and *TOP1* genes, respectively.

The *RAP1B* gene is involved in four pathways (Fig. 4), including gonadotropin-releasing hormone receptor pathway. This gene was also observed associated with cell-to-cell signaling network (integrin, ephrin receptor, and mitogen-activated protein kinase signaling network) [40]. The *MED30* and *TRHR* genes are strongly related to thyroid hormone, triggering hormonal processes associated with reproductive systems in males and females [41, 42].

Genes related to fatty acid, cholesterol, and triacylglycerol, such as *FABP5* [43], *FABP12* [44], *PEX2* [45], and *MED30* [46] are critical for energy and hormone production in males and females. Cholesterol is a precursor of steroid hormone production, such as testosterone, and therefore it is involved in male growth and reproductive development.

Important genome-wide associations on chromosome 14 for scrotal circumference were observed by Fortes et al. [5], demonstrating that this chromosome presents regions of interest that could be explored to identify sexually precocious animals. These authors observed associations with age at first corpus luteum in the same regions as scrotal circumference and have attributed their results to the genetic correlation between these traits, thus genes and SNPs associated with puberty in heifers were likely to be relevant for puberty in bulls, and vice versa. On chromosome 14, a large number of SNPs associated with puberty were identified in both bulls and heifers. Urbinati et al. [47] found important selection signatures in Canchim cattle on chromosomes 5 and 14, which were related to pigmentation (strongly selected trait in Charolais and Canchim), productive and reproductive traits.

Reports of QTL associated with reproductive traits of interest may give support to the results found in our study; however, most of the QTL were reported for female traits. On chromosome 5, QTL associated with the concentration of follicle-stimulating hormone in Brahman and Hereford crosses [48], have been described. The QTL were observed as associated with interval to first estrus after calving [7] and dystocia in dairy cattle [49] on chromosome 13. On chromosome 14, QTL associated with gestation period [50], number of inseminations per conception [51], and ovulation rate [52] have been reported.

As favorably correlated responses between scrotal circumference measures and age at first calving traits are expected through selection, whereas the genes previously reported could be highlighted as a candidate to, directly and indirectly, improve the reproductive performance of males and females, respectively. Moreover, fat deposition in cattle could be reflected in sexual precocity and carcass finishing, traits that have become the main concern of beef cattle breeders.

Few genes were observed across the positions of significant SNPs for AFC (Table 1 and Fig. 1c). According to Blaschek et al. [53], an association of the SNP rs110984522 (178.73 kb apart from the SNP rs133411648) was observed for sire fertility in Holstein breed. The *EXOC4* gene plays a role in insulin processing, metabolism of proteins, and peptide hormone metabolism pathways. Reports of QTL associated with scrotal circumference [33] have been described in this region. The *ZMAT4* gene, located on chromosome 27, participates in the apoptotic, biological, developmental, and metabolic processes. QTL associated with dystocia [49] and calving ease [54] in Holstein cattle and non-return rate [55] in Angus cattle have been described.

Significant SNPs for ASC are presented in Table 1 and Fig. 1d. The *FMN1* gene located on chromosome 10 participates in actin cytoskeleton organization and is of great importance for cell and muscle movements [56]. QTL associated with calving ease have been observed in this region [33].

A QTL associated with scrotal circumference was observed in the region of the *TMEM182* gene [33]. No function or biological processes have been described for the *LOC790871* and *TMEM182* genes in the literature or databases consulted. In the region comprising the *LOC790871* gene, QTL regions have been reported associated with the following traits: scrotal circumference [33], subcutaneous fat [57], and sperm motility [34] in Angus, Holstein, and Charolais × Holstein crossbred cattle, respectively.

It has been verified that the *UBQLN3* gene is expressed in the testes, acts in spermatogenesis in humans and rats [58] and is conserved in mammals (*Homo sapiens*, *Mus musculus*, *Rattus norvegicus*, *Canis lupus familiaris*, and *Bos taurus*). This gene is involved in the protein processing in endoplasmic reticulum pathway. A QTL associated with weight and carcass traits has been described [33] in the region in which the UBQLN3 gene is located.

Conclusions

Few associations were observed for SC210, AFC, and ASC, reinforcing their polygenic inheritance and the complexity of understanding the genetic architecture of reproductive traits. Finding many associations for SC420 in various regions of the Canchim genome also reveals the difficulty of targeting specific candidate genes that could act on fertility; nonetheless, the high linkage disequilibrium between loci herein estimated could aid to overcome this issue. Therefore, all relevant information about genomic regions influencing reproductive traits may contribute to target candidate genes for further investigation of causal mutations and aid in future genomic studies in Canchim cattle to improve the breeding program.

Additional files

Additional file 1: Significantly associated single nucleotide polymorphism (SNP) with scrotal circumference at 420 days of age (SC420) after Bonferroni (B) and false discovery rate (FDR) correction at a chromosome-wise (CW) and genome-wide (GW) level.

Additional file 2: Genome-wide (in bold) and chromosome-wise associations for scrotal circumference at 420 days of age. Gene symbols, SNP reference number, and chromosomes (Chr) and positions (Pos, in megabase) were obtained from NCBI website. Distances to gene (kilobase) are presented from 5′ to gene and 3′ to gene directions. If distance equals zero (0.00), the SNP is on intragenic region. *P*-values are presented as the minimum (Min) and maximum (Max) significance obtained from the generalized quasi-likelihood method.

Additional file 3: Candidate genes identified for scrotal circumference at 420 days of age.

Additional file 4: Fig. S1. Linkage disequilibrium for scrotal circumference at 420 d of age on chromosome 5. **Fig. S2.** Linkage disequilibrium for scrotal circumference at 420 d of age on chromosome 9. **Fig. S3.** Linkage disequilibrium for scrotal circumference at 420 d of age on chromosome 13. **Fig. S4.** Linkage disequilibrium for scrotal circumference at 420 d of age on chromosome 14. **Fig. S5.** Linkage disequilibrium for scrotal circumference at 420 d of age on chromosome 18. **Fig. S6.** Linkage disequilibrium for scrotal circumference at 420 d of age on chromosome 21.

Abbreviations
AFC: Age at first calving; ASC: Age at second calving; BTA: *Bos taurus* chromosome; EBV: Estimated breeding value; FDR: False discovery rate; GQLS: Generalized Quasi-Likelihood Score; GWAS: Genome-wide association study; QC: Quality control; QTL: Quantitative trait loci; r^2: Linkage disequilibrium measure; SC210: Scrotal circumference at 210 days of age; SC420: Scrotal circumference at 420 days of age; SNP: Single nucleotide polymorphism

Acknowledgements
The authors would like to acknowledge the Universidade Estadual Paulista - Faculdade de Ciências Agrárias e Veterinárias (UNESP/FCAV), University of Guelph, and the Brazilian Agricultural Research Corporation (Embrapa). We also thank Dr. Mehdi Sargolzaei for assistance and providing the software used in our study.

Funding
This study was supported by the "Conselho Nacional de Desenvolvimento Científico e Tecnológico" (CNPq) - 449,564/2014–2. DPM, MVGBS, LCAR, and MMA were supported by a fellowship from CNPq. NBS received a Post-Doctoral fellowship from CAPES/PNPD. TCSC and LDR were supported by the São Paulo Research Foundation (Fapesp) fellowship (2015/08939-0 and 2013/13972-0).

Authors' contributions
MEB: contributed to the development of this research with data analysis, interpretation, figure compositions, manuscript writing, and revision. FSS, DAG, and RVV: data analysis, manuscript discussion, and revision. SLCM, FBM, MMA, LCAR, RHH, MAM, and MVGBS: experimental design. NBS, TCSC, LDR, and DPM: data interpretation. SLCM, LCAR, MMA, and FBM: experimental design, interpretation, and manuscript revision. All authors read and approved the final manuscript.

Competing interests
The authors declare that they have no competing interests.

Author details
[1]Departamento de Zootecnia, Universidade Federal da Paraíba (UFPB), Areia, Paraíba 58397-000, Brazil. [2]Fast Genetics, Saskatoon, SK S7K 2K6, Canada. [3]Beef Improvement Opportunities (BIO), Guelph, ON N1K 1E5, Canada. [4]Department of Animal and Poultry Science, University of Guelph, Centre for Genetic Improvement of Livestock (CGIL), Guelph, ON N1G 2W1, Canada. [5]Departamento de Ciências Exatas, Faculdade de Ciências Agrárias e Veterinárias, Universidade Estadual Paulista (Unesp), Jaboticabal, São Paulo 14884-900, Brazil. [6]Departamento de Zootecnia, Núcleo de Pesquisa e Conservação de Cervídeos, Faculdade de Ciências Agrárias e Veterinárias, Universidade Estadual Paulista (Unesp), Jaboticabal, São Paulo 14884-900, Brazil. [7]Department of Animal Science, Federal University of Lavras (UFLA), Lavras, Minas Gerais 37200-000, Brazil. [8]Department of Genetics and

Evolution, Federal University of São Carlos (UFSCar), São Carlos, São Paulo 13565-905, Brazil. [9]Embrapa Agricultural Informatics, Campinas, São Paulo 13083-886, Brazil. [10]Embrapa Dairy Cattle, Juiz de Fora, Minas Gerais 36038-330, Brazil. [11]Embrapa Southeast Livestock, São Carlos, São Paulo 13560-970, Brazil.

References

1. Trocóniz JF, Beltrán J, Bastidas H, Larreal H, Bastidas P. Testicular development, body weight changes, puberty and semen traits of growing Guzerat and Nellore bulls. Theriogenology. 1991;35:815–26.
2. Martin LC, Brinks JS, Bourdon RM, Cundiff LV. Genetic effects on and subsequent heifer puberty. J Anim Sci. 1992;70:4006–17.
3. Buzanskas ME, Savegnago RP, Grossi DA, Venturini GC, Queiroz SA, Silva LOC, et al. Genetic parameter estimates and principal component analysis of breeding values of reproduction and growth traits in female Canchim cattle. Reprod Fertil Dev. 2013;25:775–81.
4. Burns BM, Gazzola C, Holroyd RG, Crisp J, McGowan MR. Male reproductive traits and their relationship to reproductive traits in their female progeny: a systematic review. Reprod Domest Anim. 2011;46:534–53.
5. Fortes MRS, Lehnert SA, Bolormaa S, Reich C, Fordyce G, Corbet NJ, et al. Finding genes for economically important traits: Brahman cattle puberty. Anim Prod Sci. 2012;52:143–50.
6. Feugang JM, Kaya A, Page GP, Chen L, Mehta T, Hirani K, et al. Two-stage genome-wide association study identifies integrin beta 5 as having potential role in bull fertility. BMC Genomics. 2009;10:176.
7. Sahana G, Guldbrandtsen B, Bendixen C, Lund MS. Genome-wide association mapping for female fertility traits in Danish and Swedish Holstein cattle. Anim Genet. 2010;41:579–88.
8. Tabor HK, Risch NJ, Myers RM. Candidate-gene approaches for studying complex genetic traits: practical considerations. Nat Rev Genet. 2002;3:1–7.
9. Vianna AT, Pimentel-Gomes F, Santiago M. Formação do Gado Canchim pelo Cruzamento Charolês-Zebu. Nobel: São Paulo; 1978.
10. Alencar MM. Bovino-Raça Canchim: Origem e Desenvolvimento. Brasília: Embrapa-DMU; 1988.
11. Baldi F, Albuquerque LG, Alencar MM. Random regression models on Legendre polynomials to estimate genetic parameters for weights from birth to adult age in Canchim cattle. J Anim Breed Genet. 2010;127: 289–99.
12. Buzanskas ME, Grossi DA, Baldi F, Barrozo D, Silva LOC, Torres Júnior RAA, et al. Genetic associations between stayability and reproductive and growth traits in Canchim beef cattle. Livest Sci. 2010;132:107–12.
13. Barbosa PF. O Canchim na Embrapa Pecuária Sudeste. In: Convenção Nacional da Raça Canchim, São Carlos: Embrapa Pecuária Sudeste, 2000;55–69.
14. Andrade PC, Grossi DA, Paz CCP, Alencar MM, Regitano LCA, Munari DP. Association of an insulin-like growth factor 1 gene microsatellite with phenotypic variation and estimated breeding values of growth traits in Canchim cattle. Anim Genet. 2008;39:480–5.
15. ABCCAN. Associação Brasileira de Criadores de Canchim 2015. http://www.abccan.com.br. Accessed 05 Nov 2015.
16. Associação Brasileira dos Criadores de Zebu-ABCZ. Estatística total Brasil RGN + RGD - período 1939 a 2012 2013. http://www.abcz.org.br. Accessed 5 Nov 2013.
17. Misztal I. Reliable computing in estimation of variance components. J Anim Breed Genet. 2008;125:363–70.
18. Mokry FB, Higa RH, Mudadu MA, Lima AO, Meirelles SLC, MVG B d S, et al. Genome-wide association study for backfat thickness in Canchim beef cattle using random Forest approach. BMC Genet. 2013;14:47.
19. Boddhireddy P, Kelly MJ, Northcutt S, Prayaga KC, Rumph J, DeNise S. Genomic predictions in Angus cattle: comparisons of sample size, response variables, and clustering methods for cross-validation. J Anim Sci. 2014;92:485–97.
20. Buzanskas ME, Grossi DA, Ventura RV, Schenkel FS, Sargolzaei M, Meirelles SLC, et al. Genome-wide association for growth traits in canchim beef cattle. PLoS One. 2014;9:e94802.
21. Zimin A V, Delcher AL, Florea L, Kelley DR, Schatz MC, Puiu D, et al. A whole-genome assembly of the domestic cow, Bos Taurus. Genome Biol. 2009;10:r42.
22. Feng Z, Wong WWL, Gao X, Schenkel F. Generalized genetic association study with samples of related individuals. Ann Appl Stat. 2011;5:2109–30. doi:10.1214/11-AOAS465.
23. Heyde CC. Quasi-likelihood and its application: a general approach to optimal parameter estimation. Canberra: Springer; 1997.
24. Benjamini Y, Hochberg Y. Controlling the false discovery rate a practical and powerful approach to multiple testing. J R Stat Soc Ser B. 1995;57:289–300.
25. Hill W, Robertson A. Linkage disequilibrium in finite populations. Theor Appl Genet. 1968;38:226–31.
26. Barrett JC, Fry B, Maller J, Daly MJ. Haploview: analysis and visualization of LD and haplotype maps. Bioinformatics. 2005;21:263–365.
27. Mokry FB, Buzanskas ME, Mudadu MA, Grossi DA, Higa RH, Ventura RV, et al. Linkage disequilibrium and haplotype block structure in a composite beef cattle breed. BMC Genomics. 2014;15:S6.
28. National Center for Biotechnology Information. 2016. http://www.ncbi.nlm.nih.gov/snp/. Accessed 11 Nov 2016.
29. Ensembl Genome Browser. 2016. http://www.ensembl.org/index.html. Accessed 12 Nov 2016.
30. Mi H, Muruganujan A, Thomas PD. PANTHER in 2013: modeling the evolution of gene function, and other gene attributes, in the context of phylogenetic trees. Nucleic Acids Res. 2013;41:377–86.
31. AnimalQTLdb. 2016. http://www.animalgenome.org. Accessed 15 Nov 2016.
32. Holmberg M, Andersson-Eklund L. Quantitative trait loci affecting fertility and calving traits in Swedish dairy cattle. J Dairy Sci. 2006;89:3664–71.
33. McClure MC, Morsci NS, Schnabel RD, Kim JW, Yao P, Rolf MM, et al. A genome scan for quantitative trait loci influencing carcass, post-natal growth and reproductive traits in commercial Angus cattle. Anim Genet. 2010;41:597–607.
34. Druet T, Fritz S, Sellem E, Basso B, Gérard O, Salas-Cortes L, et al. Estimation of genetic parameters and genome scan for 15 semen characteristics traits of Holstein bulls. J Anim Breed Genet. 2009;126:269–77.
35. Gargani M, Valentini A, Pariset L. A novel point mutation within the EDA gene causes an exon dropping in mature RNA in Holstein Friesian cattle breed affected by X-linked anhidrotic ectodermal dysplasia. BMC Vet Res. 2011;7:1–7.
36. Boichard D, Grohs C, Bourgeois F, Cerqueira F, Faugeras R, Neau A, et al. Detection of genes influencing economic traits in three French dairy cattle breeds. Genet Sel Evol. 2003;35:77–101.
37. Ramayo-Caldas Y, Fortes MRS, Hudson NJ, Porto-Neto LR, Bolormaa S, Barendse W, et al. A marker-derived gene network reveals the regulatory role of PPARGC1A, HNF4G, and FOXP3 in intramuscular fat deposition of beef cattle. J Anim Sci. 2014;92:2832–45.
38. Summermatter S, Shui G, Maag D, Santos G, Wenk MR, Handschin C. PGC-1α improves glucose homeostasis in skeletal muscle in an activity-dependent manner. Diabetes. 2013;62:85–95.
39. Wisely GB, Miller AB, Davis RG, Thornquest AD, Johnson R, Spitzer T, et al. Hepatocyte nuclear factor 4 is a transcription factor that constitutively binds fatty acids. Structure. 2002;10:1225–34.
40. Peddinti D, Memili E, Burgess SC. Proteomics-based systems biology modeling of bovine germinal vesicle stage oocyte and cumulus cell interaction. Plos One. 2010;5(6):e11240.
41. Choksi NY, Jahnke GD, St Hilaire C, Shelby M. Role of thyroid hormones in human and laboratory animal reproductive health. Birth Defects Res B Dev Reprod Toxicol. 2003;68:479–91.
42. Krassas GE. Thyroid disease and female reproduction. Clin Endocrinol. 2007;66:309–21.
43. Michal JJ, Zhang ZW, Gaskins CT, Jiang Z. The bovine fatty acid binding protein 4 gene is significantly associated with marbling and subcutaneous fat depth in Wagyu x Limousin F2 crosses. Anim Genet. 2006;37:400–2.
44. Liu RZ, Li X, Godbout R. A novel fatty acid-binding protein (FABP) gene resulting from tandem gene duplication in mammals: transcription in rat retina and testis. Genomics. 2008;92:436–45.
45. Van Veldhoven PP. Biochemistry and genetics of inherited disorders of peroxisomal fatty acid metabolism. J Lipid Res. 2010;51:2863–95.
46. Wang Q, Sharma D, Ren Y, Fondell JD. A coregulatory role for the TRAP-mediator complex in androgen receptor-mediated gene expression. J Biol Chem Chem. 2002;277:42852–8.
47. Urbinati I, Stafuzza NB, Oliveira MT, TCS C, Higa RH, Regitano LC d A, et al. Selection signatures in Canchim beef cattle. J Anim Sci Biotechnol. 2016;7:29.
48. Casas E, Lunstra DD, Stone RT. Quantitative trait loci for male reproductive traits in beef cattle. Anim Genet. 2004;35:451–3.

49. Seidenspinner T, Bennewitz J, Reinhardt F, Thaller G. Need for sharp phenotypes in QTL detection for calving traits in dairy cattle. J Anim Breed Genet. 2009;126:455–62.

50. Maltecca C, Weigel KA, Khatib H, Cowan M, Bagnato A. Whole-genome scan for quantitative trait loci associated with birth weight, gestation length and passive immune transfer in a Holstein x Jersey crossbred population. Anim Genet. 2009;40:27–34.

51. Schulman NF, Sahana G, Iso-Touru T, McKay SD, Schnabel RD, Lund MS, et al. Mapping of fertility traits in Finnish Ayrshire by genome-wide association analysis. Anim Genet. 2011;42:263–9.

52. Gonda MG, Arias JA, Shook GE, Kirkpatrick BW. Identification of an ovulation rate QTL in cattle on BTA14 using selective DNA pooling and interval mapping. Anim Genet. 2004;35:298–304.

53. Blaschek M, Kaya A, Zwald N, Memili E, Kirkpatrick BW. A whole-genome association analysis of noncompensatory fertility in Holstein bulls. J Dairy Sci. 2011;94:4695–9.

54. Ashwell MS, Heyen DW, Weller JI, Ron M, Sonstegard TS, Van Tassell CP, et al. Detection of quantitative trait loci influencing conformation traits and calving ease in Holstein-Friesian cattle. J Dairy Sci. 2005;88:4111–9.

55. Ron M, Feldmesser E, Golik M, Tager-Cohen I, Kliger D, Reiss V, et al. A complete genome scan of the Israeli Holstein population for quantitative trait loci by a daughter design. J Dairy Sci. 2004;87:476–90.

56. Zhou F, Leder P, Martin SS. Formin-1 protein associates with microtubules through a peptide domain encoded by exon-2. Exp Cell Res. 2006;312: 1119–26.

57. Gutiérrez-Gil B, Williams JL, Homer D, Burton D, Haley CS, Wiener P. Search for quantitative trait loci affecting growth and carcass traits in a cross population of beef and dairy cattle. J Anim Sci. 2009;87:24–36.

58. Conklin D, Holderman S, Whitmore TE, Maurer M, Feldhaus AL. Molecular cloning, chromosome mapping and characterization of UBQLN3 a testis-specific gene that contains an ubiquitin-like domain. Gene. 2000;249:91–8.

12

Evaluation of fatty acid metabolism and innate immunity interactions between commercial broiler, F1 layer × broiler cross and commercial layer strains selected for different growth potentials

Nicky-Lee Willson[1,5]* [iD], Rebecca E. A. Forder[1], Rick G. Tearle[2], Greg S. Nattrass[3], Robert J. Hughes[1,4] and Philip I. Hynd[1]

Abstract

Background: The broiler industry has undergone intense genetic selection over the past 50 yr. resulting in improvements for growth and feed efficiency, however, significant variation remains for performance and growth traits. Production improvements have been coupled with unfavourable metabolic consequences, including immunological trade-offs for growth, and excess fat deposition. To determine whether interactions between fatty acid (FA) metabolism and innate immunity may be associated with performance variations commonly seen within commercial broiler flocks, total carcass lipid %, carcass and blood FA composition, as well as genes involved with FA metabolism, immunity and cellular stress were investigated in male birds of a broiler strain, layer strain and F1 layer × broiler cross at d 14 post hatch. Heterophil: lymphocyte ratios, relative organ weights and bodyweight data were also compared.

Results: Broiler bodyweight ($n = 12$) was four times that of layers ($n = 12$) by d 14 and had significantly higher carcass fat percentage compared to the cross ($n = 6$; $P = 0.002$) and layers ($P = 0.017$) which were not significantly different from each other ($P = 0.523$). The carcass and whole blood FA analysis revealed differences in the FA composition between the three groups indicating altered FA metabolism, despite all being raised on the same diet. Genes associated with FA synthesis and β-oxidation were upregulated in the broilers compared to the layers indicating a net overall increase in FA metabolism, which may be driven by the larger relative liver size as a percentage of bodyweight in the broilers. Genes involved in innate immunity such as *TLR2* and *TLR4*, as well as organelle stress indicators *ERN1* and *XBP1* were found to be non-significant, with the exception of high expression levels of *XBP1* in layers compared to the cross and broilers. Additionally there was no difference in heterophil: lymphocytes between any of the birds.

Conclusions: The results provide evidence that genetic selection may be associated with altered metabolic processes between broilers, layers and their F1 cross. Whilst there is no evidence of interactions between FA metabolism, innate immunity or cellular stress, further investigations at later time points as growth and fat deposition increase would provide useful information as to the effects of divergent selection on key metabolic and immunological processes.

Keywords: Broiler, Cellular stress, Fatty acid metabolism, Innate immunity, Layer, Selection

* Correspondence: nicky-lee.willson@adelaide.edu.au
[1]School of Animal and Veterinary Sciences, The University of Adelaide, Roseworthy, SA 5371, Australia
[5]The Australian Poultry and Cooperative Research Centre, University of New England, PO Box U242, Armidale, NSW 2351, Australia
Full list of author information is available at the end of the article

Background

Over the past 50 yr the intensification (improved housing, husbandry and nutrition) of the broiler industry and concurrent commercial genetic selection for growth, feed efficiency and yield has resulted broiler growth increases in excess of 400% [1], with broilers having the capacity to reach 2 kg of live weight within 35 d post hatch [2, 3]. At least 85% of production improvements has been attributed to genetic selection with meat production efficiency continually increasing by 2–3% per year through selective breeding programs alone [1, 4].

Selection for feed efficiency is largely measured as feed conversion ratio (FCR), the amount of feed intake (FI) per unit bodyweight gain. In poultry systems, feed accounts for approximately 70% of total production costs [5]. Selection for efficiency has resulted in an FCR decrease of over 50% over the past 5 decades, maintaining poultry as a cost efficient source of protein [1]. Despite continued improvements, there still remains significant (>10%) variation in performance traits, including feed efficiency, bodyweight and growth rate within broiler strains [6]. This performance variation can result in an economic cost to both the producer and industry [7]. For example, variation in live weight is problematic for modern automated processing plants which reject carcasses out of a relatively narrow weight range, thus requiring further handling and sorting, and hence can incur economic loss to the processor [7].

Maintenance of innate immunity and intestinal barrier function is one parameter thought to be nutritionally costly to the host, in which exasperated or diminished immune responses could lead to increased performance variation [8]. Our previous study (Willson N-L, Nattrass GS, Hughes RJ, Forder REA, and Hynd PI, unpublished) compared high and low performing broilers to determine whether or not innate immune function could be consistently linked to the phenotypic expression of FCR. A candidate gene approach was used to determine whether functional changes in innate immune parameters could be consistently associated with high or low FCR, the results of which, there was no association. Variable expression in the pathogen recognition receptor Toll-like receptor 2 (*TLR2*) and membrane protein *CD36* also known as *FAT/ CD36*, was however of interest, as both of which have been linked to each other and various roles in fatty acid metabolism. Lee and Hwang [9] have reported on links between fatty acids and TLR activation, with saturated fatty acids activating *TLR2* and *TLR4* signalling pathways and unsaturated fatty acids having an inhibitory effect on TLR-mediated signalling pathways and gene expression. *TLR2* is known to form complexes in lipid rafts with *CD36*, [10], and *CD36* has been described in facilitating *TLR2* signalling, although the mechanism remains somewhat unclear [11]. Furthermore *CD36*, is thought to promote the synthesis of triglycerides in adipocytes, the clearance of chylomicrons from plasma, as well as mediate lipid metabolism and fatty acid transport [12, 13]. Additionally, studies in broilers have found that *CD36* has a novel role in the visceral fat deposition of male broilers, and indicated that avian fat deposition has spatial and sex specific differences [14].

Fat deposition in broilers has been an unfavourable consequence of selection for growth, particularly up until the 1970s, however there has been reductions in body fat content from 26.9% in the 1970s to 15.3% in commercial breeds in the past decade (see Tallentire et al., [15] for review). Fat deposition is negatively linked to FCR, with observations that heavier chickens usually have a higher FCR and deposit a higher amount of fat [16]. The major site for fat deposition in broilers is the abdominal fat pad, which is highly correlated to total carcass fat [16, 17]. Fat has been demonstrated to account for 15–18% of the total broiler bodyweight and is considered the most variable body component, with a coefficient of variation for the total body fat content between 15 and 20%, and higher again for abdominal fat, varying between 25 and 30% [18–21]. Excess fat accumulation and the variation may be considered the net balance of dietary absorbed fat, the rate of fat synthesis (primarily hepatic lipogenesis), and fat catabolism [22]. As obesity is correlated with chronic low grade inflammation in humans [23], and that exasperated or diminished immune responses can result in inflammation potentially leading to decreased growth performance of the host, including chickens [24], it was hypothesised that interactions between fatty acid metabolism and innate immunity may be associated with performance variations commonly seen within commercial broiler flocks.

To investigate whether innate immunity and fatty acid metabolism are contributing to flock performance variation, we compared broiler and layer chicken strains that have been intensively selected for different traits; high carcass yield and growth efficiency for broilers, commercial egg production and egg efficiency for layers [25]. This selection over the years has seen the two strains diverge for these traits, with the bodyweight of broilers being five times that of layers by 6 wk of age [26]. The aim of the current experiment was to utilise broilers, layers, and a layer × broiler F1 cross to identify how genetic selection has influenced carcass lipid composition, key genes involved in fatty acid metabolism and select innate immune parameters to enable a better understanding of the biological factors underpinning feed efficiency, growth and performance variation.

Methods

All animal procedures were approved by the University of Adelaide Animal Ethics Committee (approval #S-2015-

171) and the PIRSA Animal Ethics Committee (approval #24/15).

Birds and management

In total, 150 newly hatched male chicks (50 broiler strain, 50 F1 layer × broiler cross, 50 layer strain) were obtained from the HiChick Breeding Company Pty Ltd., Bethel, South Australia. The cross progeny were produced by HiChick utilising their commercial breeding lines. Briefly, three Isa Brown roosters and 135 Isa Brown breeder hens were used to produce layer progeny, three broiler breeder roosters and 135 broiler breeder hens used to produce the broiler progeny, and three Isa Brown roosters and 135 broiler breeder hens used to produce the F1 layer × broiler cross. All progeny were produced via natural mating. The F1 cross was utilised as an intermediate growth phenotype against broiler and layer strain progeny. Chicks were separated by breed and placed 25 chicks/rearing pen in a temperature and climate controlled room at the SARDI PPPI Poultry Research Unit, Roseworthy Campus, The University of Adelaide.

All birds were fed ad libitum (standard commercially available broiler starter diet, no in-feed antimicrobials or coccidiostats added), and had unrestricted access to water via nipple drinker lines. The three experimental groups were selected for their growth potential: Fast growing (broilers; $n = 50$) moderate growing (F1 layer × broiler; $n = 50$) and slow growing (layer strain; $n = 50$). Bodyweight was recorded weekly. On d 0, −7, −14 and −28 post hatch, 36 birds ($n = 12$ birds/breed) were randomly selected and euthanised by cervical dislocation for subsequent sampling.

Total carcass lipid and total blood lipid composition

Eviscerated carcasses (fat pads left intact on carcass) were weighed and immediately frozen at −20 °C. Whole carcasses were submerged into liquid nitrogen for 3 min, shattered with a mallet in zip lock bags to contain all fragments, and homogenised in a 1700 W blender. Sub samples of homogenate were aliquoted (10 mL) and stored at −20 °C for analysis of total carcass lipid % and lipid composition. Total lipids were extracted at the Waite Lipid Analysis Service (WLAS), Waite Campus SA, using the methods of Folch [27]. Fatty acid composition of tissues was determined and quantified using a Hewlett-Packard 6890 GC (CA, USA) equipped with flame ionization detection and a capillary column (50 m × 0.32 mm internal diameter) coated with 70% cyanopropyl polysilphenylene-siloxane with a film thickness of 0.25 µm (BPX-70, SGE, Victoria, Australia). Fatty acid transmethylation for fatty acid methyl ester (FAME) extraction, and gas chromatography analysis of FAME were run by the methods of Folch [28]. Fatty acid peaks were identified by comparing the retention time of each

peak against the retention times of a fatty acid standard of known composition. Each peak from a trace was expressed as the relative percentage of the total FAME in the sample. The detection limit of each fatty acid was 0.05% of total fatty acids.

Total blood fatty acids were measuring using the PUFAcoat dried blood spot (DBS) card, developed by the Waite Lipid Analysis Service (WLAS), Waite Campus SA. Samples were prepared by placing a drop of blood on PUFAcoat DBS card and dried at room temperature for 5 h. See Liu, Mühlhäusler [29] for full methods and validation of the PUFAcoat DBS card. In brief, lipids were extracted using a modified Folch method and FAME were extracted into heptane for gas chromatography. A Hewlett-Packard 6890 GC (CA, USA) equipped with a BPX70 capillary column 50 m × 0.32 mm, film thickness 0.25 µm (SGC Pty Ltd., Victoria, Australia), programmed temperature vaporisation injector and a flame ionisation detector (FID) was used. The identification and quantification of FAME were achieved by comparing the retention times and peak area values of unknown samples to those of commercial lipid standards (Nu-Chek Prep Inc., Elysian, MN, USA) using the Hewlett-Packard Chemstation data system.

Heterophil: Lymphocyte ratios

Blood was collected by cardiac puncture immediately following cervical dislocation. Blood smears were made by placing 1 drop of whole blood on the end of a Starfrost frosted slide (ProSci Tech). Slides were air-dried and fixed in 100% methanol for 1 min, feather side down. Slides were stained with Geisma-Wright stain on a Hema-Tek 2000. A total of 100 cells (Cell types; lymphocytes, heterophils, eosinophils, basophils and monocytes) were counted at a 40 × magnification. Subsequent heterophil: lymphocyte ratios were determined.

RNA extraction, library preparation and sequencing

In total, 18 liver samples from d 14 post hatch (broiler $n = 6$, cross $n = 6$, layer $n = 6$) birds were randomly selected for RNA-sequencing. Total RNA was isolated using an RNeasy Plus Mini Kit (Qiagen, Hilden, Germany). Approximately 80 mg of frozen (−80 °C) liver tissue was homogenised in 2 mL of Trizol reagent (Invitrogen, Carlsbad, CA). Aliquots of the Trizol homogenate (1 mL) were combined with chloroform (200 µL) and centrifuged for 15 mins at 4 °C. The upper aqueous phase (350 µL) was transferred to a gDNA eliminator spin column and centrifuged at >8000 × g (14,000 rpm) for 30 s. The flow through (300 µL) was collected and combined with 70% ethanol (300 µL) for transfer onto RNeasy columns. The remaining collection and wash steps were performed to the manufacturer's specifications. RNA was eluted in 200 µL of RNA-free water.

Purity and concentration was determined using UV spectrophotometry (Nanodrop 1000; Thermo Scienfic, Wilmington, DE).

RNA-Seq was carried out by the ACRF Cancer Genomics Facility, Adelaide, SA. The sample quality was analysed on an Agilent Bio-analyser (minimum RIN requirement of 7) and sequencing libraries were made using 2 μL of total RNA. PolyA mRNA isolation was performed using oligo dT beads. Libraries were prepared using KAPA Library Quantification Kits for Ilumina platforms (KAPABiosystems, Massachusetts, USA). 2 × 100 nt sequencing was carried out on an Illumin HiSeq 2500 Sequencing System to generate a minimum depth of 25 million reads.

RNA sequence (RNA-seq) analysis

Reads were returned in fastq format. Low-quality base calls were trimmed from the 3′ end of reads with FastQC and adaptor sequences were trimmed from the 3′ end of reads with Cutadapt. Hisat2 [30] was used to map reads to the reference chicken genome Galgal5.0 (ftp://ftp.ncbi.nlm.nih.gov/genomes/Gallus_gallus). Duplicate reads were then removed. Stringtie [30] was used to define the transcripts from the read mappings for each sample, and to merge the transcript definitions for all samples. Transcripts were cleaned up using in-house scripts. The number of raw read counts were calculated for each transcript and sample using the function featureCounts of the R package Rsubread [31]. Another R package, edgeR [32] was used to analyse differential gene expression using normalised counts per million transcripts (CPM) to correct for varying depth of sequence among samples. Differential expression of genes was considered significant at $P < 0.05$, and a false discovery rate of <0.05, with any fold change considered. Transcript data were aggregated by gene. Genes where the maximum CPM was <1 were removed. Twenty two candidate genes primarily involved in fatty acid metabolism were selected from the RNA-seq analysis for inclusion in this study (Table 1).

Statistical analysis

Data were analysed by one-way ANOVA in SPSS (IBM SPSS Statistics 22). Any data not normally distributed were logged (Log_{10}) to normalise and analysed by one-way ANOVA. Statistical significance was accepted at $P < 0.05$ level after which Post Hoc tests were performed using Tukeys[HSD] to differentiate between the three groups of birds at each sampling time point.

Results

Bodyweight data

Bodyweights were recorded for a 28-day grow out period (Table 2). Starting bodyweights (mean ± SEM) at hatch

Table 1 Candidate genes involved with fatty acid metabolism and select parameters of innate immunity

Gene name	RNA target	Accession no.[a]
ACACA	Acetyl-CoA Carboxylase	NM_205505.1
ACADL	Acyl-CoA dehydrogenase	NM_001006511.2
ACLY	ATP-Citrate-lyase	NM_001030540.1
ACSL1	Acyl-CoA synthetase	NM_001012578.1
APOA1	Apolipoprotein A1	NM_205525.4
APOC3	Apolipoprotein cIII	NM_001302127.1
CD36	FATCD36	NM_001030731.1
CPT1A	Carnitine palmitoyltransferase 1	NM_001012898.1
CPT2	Carnitine palmitoyltransferase 12	NM_001031287.2
FABP1	fatty acid binding protein 1	NM_204192.3
FADS6	Δ6 desaturase	XM_426241.5
FASN	Fatty Acid Synthase	NM_205155.2
LPL	Lipoprotein Lipase	NM_205282.1
MDH1	Malate dehydrogenase	NM_001006395.2
ME1	Malic Enzyme 1	NM_204303.1
PPARA	peroxisome proliferator-activated receptor alpha	NM_001001464.1
RXRA	Retinoic X receptor-α	XM_003642291.3
SCD	Stearoyl-CoA desaturase	NM_204890.1
TLR2A	Toll-Like Receptor 2	NM_001161650
TLR4	Toll-Like Receptor-4	NM_001030693
XBP1	X-box binding protein	NM_001006192
ERN1	Inositol-requiring kinase 1	NM_001285501.1

[a]NCBI accession number

(d 0) were significantly different between broiler n= 56 (44.4 ± 0.4 g); cross n = 57 (42.5 ± .04 g; P = 0.008) and layer line n = 54 (38.5 ± 0.4 g; P < 0.001) males. Bodyweights remained significantly different between all three groups of birds for the remainder of the grow-out period ($P < 0.001$).

Organ weights

Organ weights were expressed as a percentage of total bodyweight to account for growth differences between broilers, layers and the F1 cross (Fig. 1). At d 0 and d 7 the layers had significantly lower relative liver weight percentages than the broiler and cross males (P = 0.006 and $P < 0.001$ respectively). Liver weight as a percentage

Table 2 Weekly bodyweights (grams) for broiler, cross, and layer line males for d 7, −14, −21 and −28 post hatch

	d 0	d 7	d 14	d 21	d 28
Broiler	44.4 ± 0.4[a]	195 ± 2[a]	560 ± 8[a]	1153 ± 22[a]	2102 ± 35[a]
Cross	42.5 ± 0.4[b]	137 ± 3[b]	311 ± 8[b]	603 ± 12[b]	1037 ± 31[b]
Layer	38.5 ± 0.4[c]	84 ± 1[c]	159 ± 2[c]	261 ± 3.82[c]	403 ± 6[c]

[a-c]Means (± SEM) within the same column with different superscripts are significantly different ($P < 0.05$)

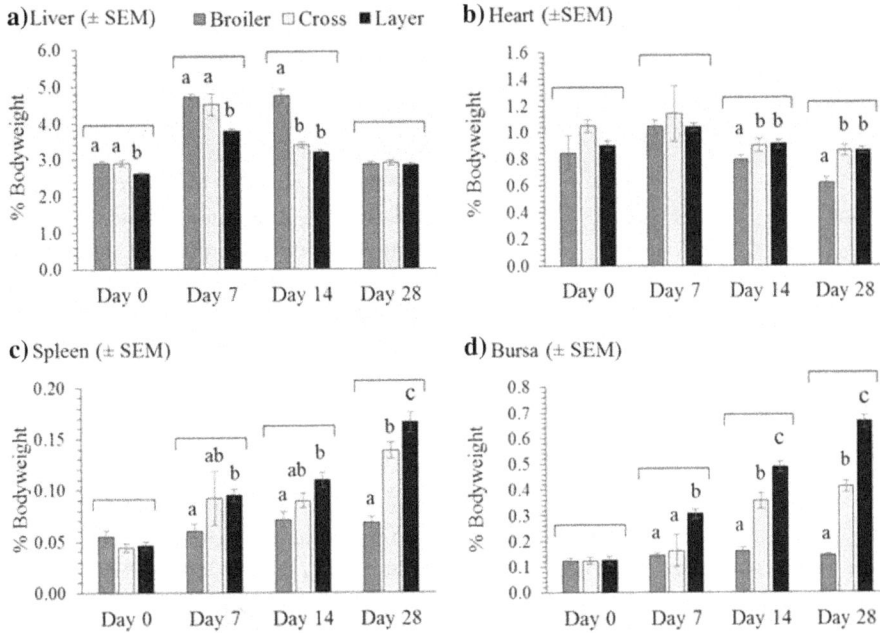

Fig. 1 Organ weights presented as a percentage of total bodyweight (± SEM) for broiler, cross and layer line males at d 0, d 7, d 14 and d 28 post hatch for: a) Liver, b) Heart, c) Spleen and d) Bursa. a-c Differing scripts within each time point are significantly different (P < 0.05)

of bodyweight peaked at d 14 in the broilers, which were significantly different from both the cross and layer birds (P < 0.001; Fig. 1a), whereas the cross and layer birds reached peak relative liver weights at d 7 post hatch. By d 28 post hatch there were no differences in relative liver weight (~2.9% of total bodyweight) between the three groups of birds (P = 0.852).

The heart accounted for 0.85–1.08% of total bodyweight at both d 0 and d 7 with no significant differences (P = 0.202 and P = 0.611) between broiler, cross and layers birds at each time point respectively (Fig.1b). The relative weight of the layer's hearts remained constant for the 28 d growth period, representing ~1% of total bodyweight. The broilers had significantly lower relative heart weights than the layer and cross birds at d 14 and d 28 post hatch (P < 0.001).

Relative spleen weights were not different between any of the three groups at d 0 (P = 0.233; Fig 1c). Layers had significantly heavier relative spleen weights than broilers from d 7 post hatch onwards (P = 0.004). The cross and layer spleen weights continued to increase in relative weight over the 28 d period, whereas the broilers reached their maximum relative spleen weight by d 14 post hatch. By d 28 post hatch broiler spleens accounted for 0.07% of total body weight whereas layer spleens accounted for 0.17% of total bodyweight (P < 0.001).

No significant differences were found in relative bursa weight between broilers, layers and cross birds at d 0 (P = 0.997; Fig. 1d). Relative bursa weights peaked in broilers at d 14 post hatch, exhibiting a 0.04% increase

from d 0-d 14 (0.12%–0.16%) then reducing slightly by d 28 to 0.14% of total bodyweight. Relative weights of the bursa increased in layer and cross birds at all sample time points. The increases were most pronounced in the layer birds with the bursa significantly different from both the crossed and layer birds at both d 14 (P < 0.001) and d 28 (P < 0.001). At d 28 post hatch the bursa weights were 0.14% and 0.67% of total bodyweight for broilers and layers respectively.

Total carcass and total blood lipids
Total carcass fat (%) and subsequent fatty acid composition was evaluated on eviscerated homogenised carcasses and blood samples at d 14 post hatch only. Broilers (n = 12) had significantly higher total carcass fat percentage (11.3%) than the cross (n = 6, 8.9%; P = 0.017) and layer line males (n = 12, 7.7%; P = 0.002; Fig. 2). The cross and layer total body fat percentages were not significantly different (P = 0.523).

The fatty acid composition of the carcasses varied indicating differential fatty acid metabolism (Table 3). The layers had higher levels of total saturated fatty acids (SFAs), followed by broilers, and then the cross, all significantly different (P = 0.001). The broilers had higher levels of palmitic acid (C16), whereas the layers had higher levels of stearic acid (C18), indicating increased elongation of SFAs in the layers. The same SFA pattern was seen in the blood (Table 4). Total carcass monounsaturated fatty acids (MUFAs) were higher in the broilers and cross relative to the layers (P < 0.001), indicating increased elongation of

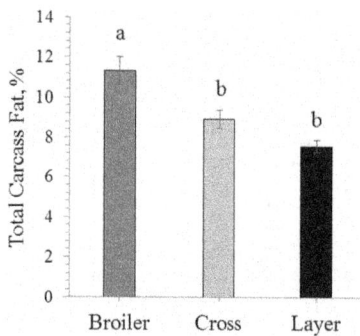

Fig. 2 Mean ± SEM Total carcass fat % for eviscerated homogenised carcasses for broilers ($n = 12$), cross ($n = 6$) and layer line ($n = 12$) males at d 14 post hatch. *a-b* Differing scripts are statistically different ($P < 0.05$)

Heterophil: lymphocyte ratios

The cross birds appeared to have a lower number of heterophils and a higher number of lymphocytes than the broiler and layer birds, however no statistical differences were detected in the heterophil; lymphocyte ratios between broilers, layers or the F1 cross (Fig. 3; $P = 0.203$). The differences were likely reflective of the high individual variation in cell frequencies, which is reflected by the large standard error. In addition to the heterophils and lymphocytes, basophils ($P = 0.094$), monocytes ($P = 0.773$) and eosinophils ($P = 0.561$) were also assessed however no significant differences were detected in the cell frequencies between any of the groups.

Gene expression

The 22 candidate genes selected (Table 5) revealed that broilers ($n = 6$) in comparison to layers ($n = 6$) had significant hepatic upregulation of genes involved in lipid transport; *APOA1* ($P = 0.019$), *APOC3* ($P = 0.003$), lipogenesis; *ACACA* ($P = 0.001$), *ME1* ($P = 0.022$), *FASN* ($P < 0.001$), *GPAM* ($P = 0.001$), *MDH1* ($P < 0.001$), *SCD* ($P < 0.001$), fatty acid transport; *FABP1* ($P = 0.001$), *ACLY* ($P < 0.001$), and fatty acid oxidation; *ACADL* ($P = 0.003$), *CPT-2* ($P < 0.001$), (Fig. 4). An exception was the downregulation of *FADS6* ($P = 0.054$) in broilers, a rate-limiting

MUFAs in the broilers and cross, this pattern also reflected in the blood. The cross and layers had significantly higher carcass percentages of polyunsaturated fatty acids (PUFAs), both omega-3 and omega-6. This was reflective both the n-6: n-3 ratio as well as the PUFA: SFA ratios between the three groups of birds. The composition of the serum and the composition of the carcass was generally the same for broilers, layers and the F1 crosses.

Table 3 Fatty acid composition (% of total identified fatty acids) in homogenised carcass samples for broiler ($n = 12$), cross ($n = 6$) and layer line males ($n = 12$) fed the same commercial broiler diet formulation at d 14 post hatch

Fatty acid	Broiler ($n = 12$)	Cross ($n = 6$)	Layer ($n = 12$)	P-value
Eviscerated carcass				
Total Carcass Fat %	11.3[a]	8.90[b]	7.56[b]	<0.001
Total SFA	37.7 ± 0.3[b]	36.8 ± 0.2[c]	38.6 ± 0.2[a]	0.001
Palmitic acid C_{16}	27.7 ± 0.24[a]	25.9 ± 0.19[b]	25.3 ± .25[b]	<0.001
Stearic acid C_{18}	7.8 ± 0.12[c]	8.4 ± 0.15[b]	10.0 ± 0.18[a]	<0.001
TFA	0.8 ± 0.03[b]	0.9 ± 0.05[ab]	1.0 ± 0.06[a]	0.038
Total MUFA	49.5 ± 0.27[a]	48.7 ± 0.34[a]	44.0 ± 0.41[b]	<0.001
Palmitoleic acid (C_{16}1n-7)	7.8 ± 0.17[a]	6.2 ± 0.27[b]	4.8 ± 0.19[c]	<0.001
Oleic acid (C_{18}1n-9)	38.6 ± .27[a]	38.9 ± 0.27[a]	35.8 ± 0.19[b]	<0.001
Vaccenic acid (C_{18}1n-7)	2.7 ± 0.07[b]	3.1 ± 0.09[a]	3.0 ± .06[a]	0.003
Total PUFAn-3	1.5 ± 0.01[b]	1.6 ± 0.02[b]	1.9 ± 0.05[a]	<0.001
α-Linolenic acid (C_{183}n-3)	1.1 ± 0.01	1.1 ± 0.00	1.1 ± 0.01	0.684
Eicosapentanoic acid (C_{22}5n-3)	0.1 ± 0.0	0.1 ± 0.0	0.1 ± 0.0	-
Docosahexanoic acid (C_{22}6n-3)	0.2 ± 0.01[c]	0.3 ± 0.02[b]	0.6 ± 0.02[a]	<0.001
Total PUFAn-6	10.4 ± 0.12[c]	12.0 ± .017[b]	14.5 ± 0.32[a]	<0.001
Linoleic acid (C_{18}2n-6)	9.8 ± 0.12[c]	11.0 ± 0.13[b]	12.8 ± 0.24[a]	<0.001
Arachidonic acid (C_{20}4n-6)	0.3 ± 0.02[c]	0.6 ± 0.03[b]	1.1 ± 0.07[a]	<0.001
n-6: n-3 ratio	6.88[c]	7.42[b]	7.68[a]	<0.001
(MUFA + PUFA): SFA	1.61[ab]	1.68[a]	1.57[b]	0.004
PUFA: SFA	0.31[b]	0.40[a]	0.43[a]	<0.001

[1]Data are expressed as the percentage of identified fatty acids ± Standard error of means (SEM)
[a-c]Means within the same row for each parameter with different superscripts are significantly different ($P < 0.05$)

Table 4 Fatty acid composition (% of total identified fatty acids) in PUFAcoat DBS blood spot samples for broiler, cross and layer line males fed the same commercial broiler diet formulation at d 14 post hatch

Fatty acid	Broiler ($n = 12$)	Cross ($n = 6$)	Layer ($n = 10$)	P-value
Total SFA	43.7 ± 0.7	43.05 ± 0.3	46.0 ± 1.2	0.107
Palmitic acid C_{16}	$24. \pm 0.52$	22.5 ± 0.18	23.9 ± 1.78	0.424
Stearic acid C_{18}	14.9 ± 0.39^b	16.16 ± 0.21^{ab}	17.07 ± 0.51^a	0.004
TFA	0.85 ± 0.03^b	0.93 ± 0.05^b	1.1 ± 0.06^a	0.004
Total MUFA	33.55 ± 0.33^a	28.55 ± 0.65^b	23.63 ± 0.60^c	<0.001
Palmitoleic acid (C_{16}1n-7)	4.19 ± 0.17^a	2.55 ± 0.07^b	1.69 ± 0.13^c	<0.001
Oleic acid (C_{18}1n-9)	$26.53 \pm .25^a$	23.08 ± 0.65^b	19.36 ± 0.50^c	<0.001
Vaccenic acid (C_{18}1n-7)	1.96 ± 0.05^{ab}	2.11 ± 0.06^a	1.78 ± 0.10^b	0.036
Total PUFAn-3	2.84 ± 0.13^b	3.58 ± 0.18^a	3.66 ± 0.25^a	0.007
α-Linolenic (C_{18}n-3)	0.69 ± 0.02	0.71 ± 0.03	0.63 ± 0.03	0.189
Eicosapentanoic (C_{22}5n-3)	0.133 ± 0.01	0.35 ± 0.04	0.31 ± 0.03	0.628
Docosahexanoic (C_{22}6n-3)	1.59 ± 0.09^b	2.2 ± 0.14^a	2.4 ± 0.18^a	0.001
Total PUFAn-6	19.06 ± 0.43^v	$23.86 \pm .059^b$	25.62 ± 1.1^a	<0.001
Linoleic (C_{18}2n-6)	16.38 ± 0.36^b	19.45 ± 0.33^a	19.72 ± 0.72^a	<0.001
Arachidonic (C_{20}4n-6)	1.26 ± 0.05^c	2.6 ± 0.23^b	4.06 ± 0.35^a	<0.001
n-6: n-3 ratio	6.82	6.68	7.14	0.418
(MUFA + PUFA): SFA	1.27	1.30	1.18	0.071
PUFA: SFA	0.51^b	0.64^a	0.65^a	0.002

[1]Data are expressed as the percentage of identified fatty acids ± Standard error of means (SEM)
[a-c]Means within the same row for each parameter with different superscripts are significantly different ($P < 0.05$)

enzyme involved in the elongation of PUFAs. Broilers when compared to the cross ($n = 6$) birds exhibited generalised upregulation of fatty acid metabolism, although not as pronounced as seen between broilers and layers. Significant hepatic upregulation for lipid transport; APOC3 ($P = 0.029$), lipogenesis; GPAM ($P = 0.035$), MDH1 ($P < 0.001$), fatty acid transport; FABP1 ($P = 0.008$), ACLY ($P < 0.001$), and fatty acid oxidation; ACADL ($P = 0.015$), CPT-2 ($P = 0.019$) were observed for broilers. Layers and cross comparisons indicated no real differential expression in fatty acid metabolism between the groups, with the exception of down regulation of lipogenic gene SCD1 ($P = 0.003$) and fatty acid oxidation CPT-2 ($P < 0.001$) and ACAA1 ($P < 0.001$) genes. Layers in comparison to the

cross also had upregulated expression of the transcription factor PPARA ($P = 0.047$), a difference not seen elsewhere.

Endoplasmic reticulum (ER) stress-related gene ERN1 was not differentially expressed between any of the three groups ($P = 0.67$). XBP1 was found to be significantly upregulated in layers in comparison to both broilers ($P = 0.002$) and crossed birds ($P = 0.007$). Toll-like receptors TLR2 and TLR4 were not found to be differentially expressed between any of the three groups ($P = 0.951$).

Pearson's two-tailed correlations with individual bird bodyweights (Table 5), revealed 15 of the 22 genes were highly correlated with bodyweight at $P < 0.01$, 2 genes correlated at $P < 0.05$ and 6 of the genes non-significant with bodyweight. The highest correlation was between malate dehydrogenase (MDH1) and bodyweight ($r = 0.902$; $P < 0.001$).

Discussion

The aim of this experiment was to elucidate how genetic selection has influenced carcass composition, fatty acid metabolism and select innate immune parameters. The objective was to further develop the understanding of factors which may be underpinning performance variation in modern broilers. Our previous experimental work did not provide sufficient phenotypic variation in feed conversion ratio within flock, thus it was decided to

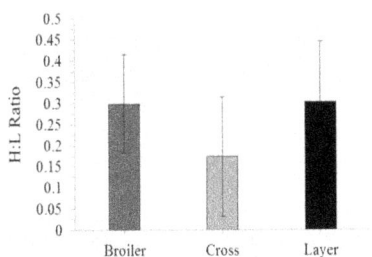

Fig. 3 Heterophil: Lymphocyte (H:L) ratios (±SD) for broilers ($n = 6$), Cross ($n = 6$) and layer line males ($n = 6$)

Table 5 Pearson correlation coefficient (r) of target gene against individual bodyweight (BW), and mean expression levels (CPM) of genes between broilers ($n = 6$), cross ($n = 6$) and layers ($n = 6$)

Gene name	r (Gene vs. BW)[1]	Broiler ($n = 6$)	Cross ($n = 6$)	Layer ($n = 6$)	Regulation[2]
ACACA	0.695**	3135.4 ± 118.6[a]	2918.5 ± 101.4[a]	2367.5 ± 135.4[b]	↑
ACADL	0.734**	751.3 ± 27.8[a]	648.1 ± 15.3[b]	620.9 ± 23.4[b]	↑
ACLY	0.855**	3605.7 ± 201.6[a]	2386.9 ± 117.4[b]	1956.9 ± 163.3[b]	↑
ACSL1	0.336	693.5 ± 61.9	553.0 ± 21.6	612.8 ± 27.1	
APOA1	0.639**	1519.0 ± 107.2[a]	1348.5 ± 42.2[ab]	1194.1 ± 57.3[b]	↑
APOC3	0.736**	1859.5 ± 131.2 [a]	1472.7 ± 63.2[b]	1307.6 ± 77.4[b]	↑
CD36	0.593**	517.8 ± 24.6[b]	580.4 ± 15.1 [ab]	596.5 ± 15.8[a]	↓
CPT1A	0.044	244.5 ± 37.1	247.6 ± 15.5	233.5 ± 10.0	
CPT2	0.853**	224.7 ± 8.4 [a]	195.4 ± 6.7[b]	151.8 ± 4.3[c]	↑
FABP1	0.722**	998.8 ± 96.5[a]	687.5 ± 30.3[b]	606.7 ± 38.4[b]	↑
FADS6	0.547*	109.6 ± 8.3	130.4 ± 11.7	145.1 ± 9.1	↓
FASN	0.769**	10,794 ± 755.5[b]	8475.9 ± 480.1[a]	6486.9 ± 559.1[a]	↑
LPL	0.600**	48.2 ± 22.3[b]	90.1 ± 9.1 [ab]	117.9 ± 9.6[a]	↓
MDH1	0.902**	667.0 ± 28.0[a]	462.6 ± 18.8[b]	386.7 ± 12.3[b]	↑
ME1	0.601**	1045.0 ± 127.5[a]	963.7 ± 75.0[ab]	600.6 ± 92.9[b]	↑
PPARA	0.376	447.1 ± 17.3 [ab]	434.6 ± 17.1[b]	496.8 ± 15.3[a]	
RXRA	0.012	65.9 ± 2.9	63.4 ± 2.8	64.6 ± 3.0	
SCD	0.817**	2785.2 ± 130.0[a]	2322.6 ± 81.9[a]	1413.6 ± 233.1[b]	↑
TLR2A	0.041	21.1 ± 1.3	20.9 ± 1.9	20.4 ± 1.7	
TLR4	0.360	10.2 ± 0.9	9.3 ± 0.4	12.6 ± 1.2	
XBP1	0.620**	225.6 ± 9.1[b]	231.4 ± 9.9[b]	281.8 ± 12.4[a]	↓
ERN1	0.578*	28.9 ± 2.5	23.9 ± 1.2	23.2 ± 1.1	↑

[1]Pearson's correlation coefficient of target gene against individual bodyweight of all three groups of birds (BW); *Sig at $P < 0.05$, **Sig at $P < 0.01$
[2]Relative direction of regulation: ↑ Broiler upregulated (broiler > cross > layer); ↓Broiler downregulated (broiler < cross < layer)
a–cMeans (± SEM) within the same row for each parameter with different superscripts are significantly different ($P < 0.05$). Means values are counts per million (CPM) transcripts, to correct for varying sequence depth between individual samples

investigate birds with grossly different growth potentials; namely, broilers, layers and a layer × broiler F1 cross. Although samples were taken at multiple time points, d 14 was selected as the primary sampling date due to the rapid growth acceleration seen in broilers from 2 to 3 wk of age. By sampling at this time point it was hoped to capture physiological changes at the beginning of the growth acceleration to further understand broiler growth rates.

As expected the growth rates of the broiler progeny well exceeded those of the layer strain progeny. By d 14 the broilers were four times the weight of the layer strain males and twice the weight of the F1 cross. The total lipid carcass percentage of the broilers was higher than both the layers and the cross, which weren't significantly different from each other, despite the cross being twice the weight of the layers. Interestingly multiple studies have shown that the dietary fatty acid composition is reflected in the fatty acid composition of the tissues and serum of broilers [33, 34]. Despite being raised in the same environmental conditions and fed the same diet,

the fatty acid composition of the carcasses and blood spots differed between the three groups in this study, suggesting difference existed in fatty acid metabolism. The broilers had increased overall MUFA percentages, which would correlate with the significant upregulation of SCD1 which encodes the rate-limiting enzyme converting SFAs into MUFAs [35]. Comparisons of the total SFA, MUFA and PUFAs revealed layers had higher n-6 and n-3 levels, indicating two possibilities, layer strains either have a higher physiological requirement for long chain PUFAs, or, layers are more efficient at converting available dietary linoleic and alpha-linolenic fatty acids to their long chain derivatives. The gene encoding the enzyme FADS6, which is rate limiting in the elongation of PUFAs, was found to be upregulated in the layers in comparison to the broilers which may support this concept.

Whilst it may be anticipated that the increased fat deposition is due to either increased lipogenesis and/or a decrease in fatty acid β-oxidation, we saw a net overall increase in both lipogenesis and fatty acid β-oxidation

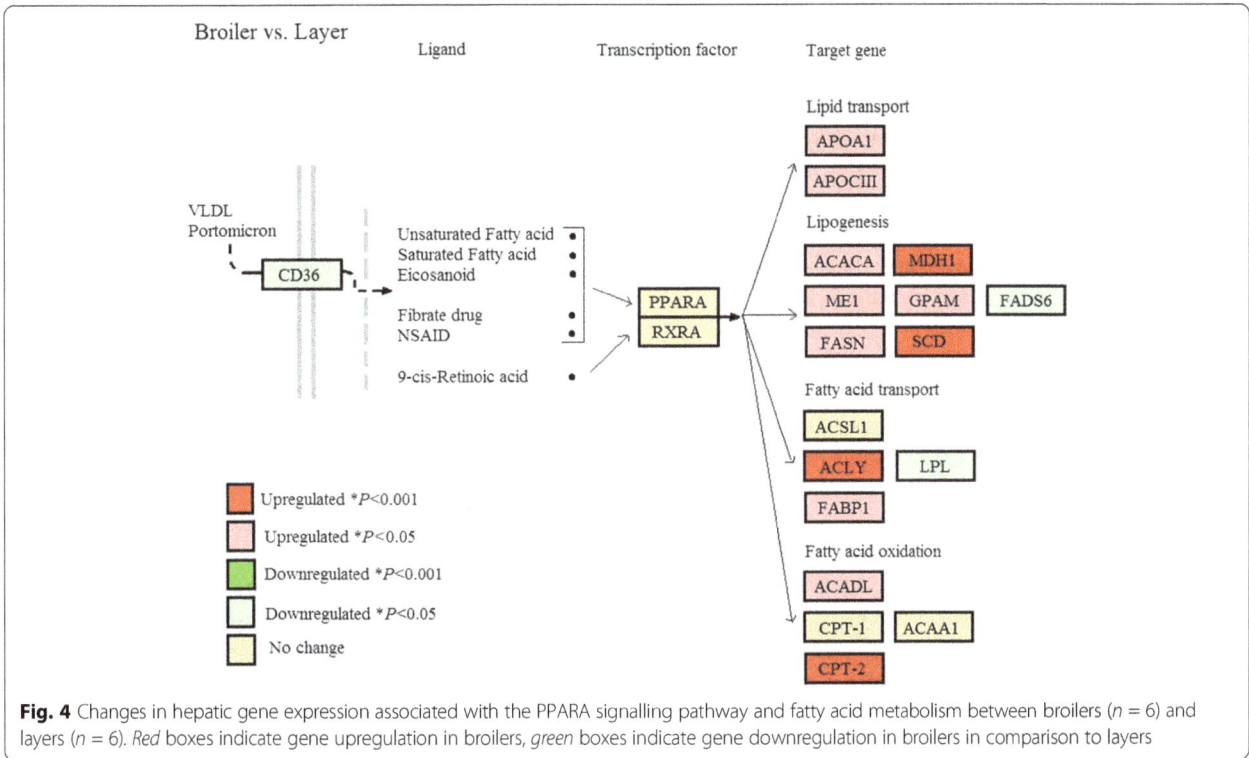

Fig. 4 Changes in hepatic gene expression associated with the PPARA signalling pathway and fatty acid metabolism between broilers ($n = 6$) and layers ($n = 6$). *Red* boxes indicate gene upregulation in broilers, *green* boxes indicate gene downregulation in broilers in comparison to layers

genes in the broilers compared to layers or the F1 cross. Although this could be controlled by transcription factors regulating FA metabolism, such as the nuclear receptor *PPARA*, we found no evidence to support this. The higher metabolic activity may therefore be reflective of the weight of the liver at d 14 which was relatively larger than that of the layers expressed as a percentage of bodyweight. An early increase in liver mass has also been observed in multiple studies, including comparisons of modern broilers and heritage lines [2]. In the current study the layer and crossed birds had reached their peak relative liver mass by d 7, however the broilers had higher relative weights at d 7 and reached their relative maximum weights at d 14 post hatch. By d 28 there were no differences in relative liver mass between the broilers, layers and their F1 cross. Schmidt et al., [2] propose this early increase in liver mass could correspond to increased liver capacity required in early post hatch, and that a possible effect of selection may have shifted earlier maturation of the liver in modern broiler lines. The relative heart weights followed a similar pattern to the liver in that they were at their maximums in the first 2 wks post hatch. From d 14 onwards the broiler relative heart weights had significantly reduced when compared to the cross and layers. These findings are not surprising as the reduction in cardiac relative size and capacity has been well documented in broilers due to genetic selection for increased growth [2, 18, 36].

Additional to differential fatty acid metabolism, it was hypothesised that innate immune parameters may also be interacting with fatty acid metabolism ultimately influencing performance variation. Modern broilers are now considered obese relative to layer strains, so obesity-related pathologies such as inflammation and cellular stress may be anticipated to be increased in broilers. To test this hypothesis immune organ weights (spleen and bursa), heterophil: lymphocyte ratios, as well as Toll-like Receptors (*TLR2a, TLR4*), fatty acid translocase (*CD36*) and endoplasmic reticulum stress indicator genes (*ERN1, XBP1*) were included in the current study.

The relative weight of both the spleen and bursa continued to increase in the cross and layer birds from d 0 until d 28 post hatch. The broilers reached maximum relative spleen and bursa weights at d 14 and then decreased from there on in. There has been conflicting interpretation as to whether relative increased immune organ size equates to a better immune defence system. Once such study found that the size of the spleen for example was correlated with changes in body condition, and that size was elevated in individual birds in prime body condition [37]. It could be argued that all of our birds were in good body condition, as there was no disease, parasite infection or mortality. Body condition as measure of fatness v leanness however, as used by Møller et al., [37], would assume the layers and the F1 cross were in better relative condition than the broilers, and potentially reflective of the smaller immune organs. Additionally broilers have repeatedly been shown to be less responsive to immune challenges experimentally, and this has been attributed to a negative consequence of genetic selection [24]. Although relative decreased lymphoid organ weights (% of bodyweight) were observed in the broilers compared to the cross and layers, there was no evidence to suggest that the broilers were compromised immunologically due to increases in fat deposition in an unchallenged experimental setting. Heterophil: lymphocyte ratios were not significantly different between any of the birds although there was a high level of variation between the individuals. The cross did appear to have a lower ratio, however this is more likely attributed to a lower number of samples and the high variation in individual birds than a significant trend.

Whilst short-term stress is of minimal consequence to broilers, long-term stress results in increased serum corticosterone, increased heterophil: lymphocyte ratios and altered protein, carbohydrate and lipid metabolism, and increased deposition of abdominal fat [38]. This poses a question, could a broiler be chronically stressed at a cellular level, particularly with the reduction of organ weights relative to overall bodyweight as growth increases? To investigate whether there was any evidence of organelle stress occurring, two key ER stress indicators which initiate the unfolded protein response (UPR) were assessed, inositol-requiring kinase 1 (ERN1) and x-box binding protein (XBP1). Saturated fatty acids have been shown to trigger the UPR response in hepatocytes and the UPR has been linked to lipid synthesis and breakdown [39]. Despite the broiler, layer and F1 cross birds having differing SFA levels, no differences were found in the expression levels of ERN1, XBP1 however was found to be upregulated in the layers in comparison to both the broiler and cross. Given that ERN1 levels are showing no indication of ER stress, the differential expression of XBP1 may align with the suggestion that XBP1 functions as a mediator of hepatic lipogenesis, distinct from its function in ER stress and the UPR [40]. It is thought to regulate the transcription of genes involved with fatty acid synthesis, including SCD1 and ACACA, with deletion of XBP1 resulting in decreased triglyceride, cholesterol and free fatty acids [40]. It is difficult to conclude whether XBP1 is exhibiting a regulatory effect on lipogenesis in the layers however the aforementioned genes are not seen to be increased in the layers compared to the broilers or the cross.

Additional to organelle stress, Toll-Like receptors, including TLR2 and TLR4 have received attention for their roles in the development of obesity and insulin resistance, although the mechanisms by which they contribute still remain unclear. Mice lacking TLR2 and TLR4 genes do show however that TLRs are involved in the development of obesity [41]. In macrophage cell cultures, saturated fatty acids, such as stearic acid and palmitic acid, have been shown to activate TLR2 and TLR4 signalling pathways, which consequently activates down steam pro-inflammatory pathways, Conversely, PUFAs, particularly

n-3 s, have been shown to inhibit TLR2/4 expression, activation and downstream signalling [42]. In our current study we found no differential expression of TLR2a in the avian liver in any of the three types of birds. Additionally we found no evidence in the expression levels of TLR4 to suggest that the differing fatty acid profiles of the birds was having an effect or interaction with the expression of TLR4 at d 14 post hatch. This was also the case for CD36, with the exception of a down regulation in the broilers in comparison to the layers. Given the biological diversity for the role of CD36, this likely does not translate into down regulated facilitation of fatty acid transport given the overall upregulation of fatty acid metabolism seen in the broilers.

Conclusion

The results indicate a total upregulation of fatty acid metabolism in broiler chickens when compared to an F1 cross and commercial layer strain. This increase is most likely as a result of genetic selection for growth, with the overall increase resulting in increased FA synthesis as well as β-oxidation in the liver. There was no evidence to suggest that at d 14 post hatch that the broilers are in a state of cellular hepatic stress or demonstrating changes in innate immunity parameters such as TLR2 and TLR4 expression. This is despite the broilers growing at four times the rate of the layers and with significant increases in fat %. Day 14 post hatch was selected to capture the physiological changes as the broiler growth acceleration begins. It is possible that the d 14 sample time point was too early in relation to fatty acid metabolism and innate immunity/cellular stress interactions to capture changes that may ultimately be driving performance. Analysis at additional time points in the grow out phase could better revel indicators of chronic stress as the organ weights continue to decrease by relative weight, contributing to metabolic stress and altering metabolism.

Acknowledgements
This research was conducted within the Australian Poultry Cooperative Research Centre, established and supported under the Australian Government's Cooperative Research Centres Program. N-LW received a stipend from the University of Adelaide (F J Sandoz Scholarship) and the Australian Poultry Cooperative Research Centre for PhD studies. We would like to thank D. Schultz, M. Bowling and N. Heberle for assistance during the trial and the HiChick Breeding Company Pty Ltd. for providing the birds.

Funding
This work was financially supported by the Australian Poultry Cooperative Research Centre. The funder had no role in study design, data collection, and analysis. The funder read and approved the manuscript for publication.

Authors' contributions
N-LW, PH, RF, RH and GN designed the study. N-LW and RF were involved in performing the experiment and laboratory analysis. RT performed RNA-seq data analysis. N-LW wrote the manuscript, PH, RF and RH revised the manuscript. All authors read and approved the final manuscript.

Competing interests

The authors declare they have no competing interests.

Author details

[1]School of Animal and Veterinary Sciences, The University of Adelaide, Roseworthy, SA 5371, Australia. [2]Davies Research Centre, School of Animal and Veterinary Sciences, The University of Adelaide, Roseworthy, SA 5371, Australia. [3]South Australian Research and Development Institute (SARDI), Livestock and Farming Systems, Roseworthy, SA 5371, Australia. [4]South Australian Research and Development Institute (SARDI), Pig and Poultry Production Institute, Roseworthy, SA 5371, Australia. [5]The Australian Poultry and Cooperative Research Centre, University of New England, PO Box U242, Armidale, NSW 2351, Australia.

References

1. Zuidhof MJ, Schneider BL, Carney VL, Korver DR, Robinson FE. Growth, efficiency, and yield of commercial broilers from 1957, 1978 and 2005. Poult Sci. 2014;93:2970–82.
2. Schmidt CJ, Persia ME, Feierstein E, Kingham B, Saylor WW. Comparison of a modern broiler line and a heritage line unselected since the 1950s. Poult Sci. 2009;88:2610–9.
3. Robins A, Phillips CJC. International approaches to the welfare of meat chickens. Worlds Poult Sci J. 2011;67:351–69.
4. Gous RM. Nutritional limitations on growth and development in poultry. Livest Sci. 2010;130:25–32.
5. Aggrey SE, Karnuah AB, Sebastian B, Anthony NB. Genetic properties of feed efficiency parameters in meat type chickens. Genet Sel Evol. 2010;42:25.
6. Emmerson DA. Commercial approaches to genetic selection for growth and feed conversion in domestic poultry. Poult Sci. 1997;76:1121–5.
7. Hughes RJ, Heberle ND, Barekatin R, Edwards NM, Hynd PI. Flock uniformity- is it important and how is it assessed? Annual Australian Poultry Science Symposium. 2017;28:93–6.
8. Kohl KD. Diversity and function of the avian gut microbiota. J Comp Physiol B. 2012;182:591–602.
9. Lee JY, Hwang DH. The modulation of inflammatory gene expression by lipids: mediation through toll-like receptors. Mol Cells. 2006;21:174–85.
10. Hoebe K, Georgel P, Rutschmann S, Du X, Mudd S, Crozat K, et al. CD36 is a sensor of diacylglycerides. Nature. 2005;433:523–7.
11. Wolowczuk I, Verwaerde C, Viltart O, Delanoye A, Delacre M, Pot B, et al. Feeding our immune system: impact on metabolism. Clin Dev Immunol. 2008;8
12. Drover VA, Ajmal M, Nassir F, Davidson FNO, Nauli AM, Sahoo D, et al. CD36 deficiency impairs intestinal lipid secretion and clearance of chylomicrons from the blood. J Clin Invest. 2005;115:1290–7.
13. Silverstein RL, Febbraio M. CD36, a scavanger receptor involved in immunity, metabolism, angiogenesis, and behaviour. Sci Signal. 2009;2:re3.
14. Shu G, Liao WY, Feng JY, Yu KF, Zhai YF, Wang SB, et al. Active immunization of fatty acid translocase specifically decreased visceral fat deposition in male broilers. Poult Sci. 2011;90:2557–64.
15. Tallentire CW, Leinonen I, Kyriazakis I. Breeding for efficiency in the broiler chicken: a review. Agron Sustain Dev. 2016;36:66. doi:10.1007/s13593-016-0398-2.
16. Gaya LG, Ferraz JBS, Rezende FM, Mourao GB, Mattos EC, Eler JP, et al. Heritability and genetic correlation estimates for performance and carcass and body composition traits in a male broiler line. Poult Sci. 2006;85:837–43.
17. Zerehdaran S, Vereijken ALJ, van Arendonk AM, van der Waaij EH. Estimation of genetic parameters for fat deposition and carcass traits in broilers. Poult Sci. 2004;83:521–5.
18. Havenstein GB, Ferket PR, Qureshi MA. Carcass composition and yield of 1957 versus 2001 broilers when fed representative 1957 and 2001 broiler diets. Poult Sci. 2003;82:1509–18.
19. Leenstra FR. Effect of age, sex, genotype and environment on fat deposition in broiler chickens-a review. Worlds Poult Sci J. 1986;42:12–25.
20. Daval S, Lagarrigue S, Douaire M. Messenger RNA levels and transcription rates of hepatic lipogenesis genes in genetically lean and fat chickens. Genet Sel Evol. 2000;32:521–31.
21. Choct M, Naylor A, Hutton O, Nolan J. A report for the Rural Industries Research and Development Corporation. 2000. Publication No 98/123. https://rirdc.infoservices.com.au/downloads/98-123
22. Sanz M, Lopez-Bote CJ, Menoyo D, Bautista JM. Abdominal fat deposition and fatty acid synthesis are lower and beta-oxidation is higher in broiler chickens fed diets containing unsaturated rather than saturated fat. J Nutr. 2000;130:3034–7.
23. Lumeng CN, Saltiel AR. Inflammatory links between obesity and metabolic disease. J Clin Invest. 2011;121:2111–7.
24. Lochmiller RL, Deerenberg C. Trade-offs in evolutionary immunology: just what is the cost of immunity. Oikos. 2000;88:87–98.
25. Druyan S. The effects of genetic line (broilers vs. layers) on embryo development. Poult Sci. 2010;89:1457–67.
26. Zhao R, Muehlbauer E, Decuypere E, Grossman R. Effect of genotype-nutrition interaction on growth and somatotropic gene expression in the chicken. Gen Comp Endocrinol. 2004;136:2–11.
27. Folch J, Lees M, Sloane Stanley GH. A simple method for the isolation and purification of total lipids from animal tissues. J Biol Chem. 1957; 226:497–509.
28. Kartikasari LR, Hughes RJ, Geier MS, Makrides M, Gibson RA. Dietary alpha-linolenic acid enhances omega-3 long chain polyunsaturated fatty acid levels in chicken tissues. Prostaglandins Leukot Essent Fatty Acids. 2012;4-5:103–9.
29. Liu G, Mühlhäusler BS, Gibson RA. A method for long term stabilisation of long chain polyunsaturated fatty acids in dried blood spots and its clinical application. Prostaglandins Leukot Essent Fatty Acids. 2014;91:251–60.
30. Pertea M, Kim D, Pertea GM, Leek JT, Salzberg SL. Transcript-level expression analysis of RNA-seq experiments with HISAT, StringTie and Ballgown. Nat Protoc. 2016;11:1650–67.
31. Liao Y, Smith GK, Shi W. The subread aligner: fast, accurate and scalable read mapping by seed-and-vote. Nucleic Acids Res. 2013;41:10.e108.
32. Robinson MD, McCarthy DJ, Smyth GK. edgeR: a bioconductor package for differential expression analysis of digital gene expression data. Bioinformatics. 2010;26:139–40.
33. Frttsche KL, Cassity NA, Huang SC. Effect of dietary fats on the fatty acid compositions of serum and immune tissues in chickens. Poult Sci. 1991;70:1213–22.
34. Newman RE, Bryden WL, Fleck E, Ashes JR, Buttemer WA, Storlien LH, et al. Dietary n-3 and n-6 fatty acids alter avian metabolism: metabolism and abdominal fat deposition. Br J Nutr. 2002;88:11–8.
35. Ntambi JM. Regulation of stearoyl-CoA desaturase by polyunsaturated fatty acids and cholesterol. J Lipid Res. 1999;40:1549–58.
36. Collins KE, Kiepper BH, Ritz CW, McLendon BL, Wilson JL. Growth, livability, feed consumption, and carcass composition of the Athens Canadian random bred 1955 meat-type chicken versus the 2012 high-yielding Cobb 500 broiler. Poult Sci. 2014;93:2953–62.
37. Møller AP, Christe P, Erritzøe CJ, Mavarez J. Condition, disease and immune defense. Oikos. 1998;83:301–6.
38. Virden WS, Kidd MT. Physiological stress in broilers: ramifications on nutrient digestibility and responses. J Appl Poult Res. 2009;18:338–47.
39. Hotamisligil GS, Erbay E. Nutrient sensing and inflammation in metabolic diseases. Nat Rev Immunol. 2008;8:923–34.
40. Lee A-H, Scapa EF, Cohen DE, Glimcher LH. Regulation of hepatic lipogenesis by the transcription factor XBP1. Science. 2008;320:1492–6.
41. Fresno M, Alvarez R, Cuesta N. Toll-like receptors, inflammation, metabolism and obesity. Arch Physiol Biochem. 2011;111:151–64.
42. Wahli W, Michalik L. PPARs at the crossroads of lipid signaling and inflammation. Trends Endocrinol Metab. 2012;23:351–63.

13

Circulating leptin and its muscle gene expression in Nellore cattle with divergent feed efficiency

Lúcio Flávio Macedo Mota[1,2], Cristina Moreira Bonafé[1], Pâmela Almeida Alexandre[3], Miguel Henrique Santana[3], Francisco José Novais[3], Erika Toriyama[4], Aldrin Vieira Pires[1], Saulo da Luz Silva[5], Paulo Roberto Leme[5], José Bento Sterman Ferraz[3] and Heidge Fukumasu[3*]

Abstract

Background: Leptin has a strong relation to important traits in animal production, such as carcass composition, feed intake, and reproduction. It is mainly produced by adipose cells and acts predominantly in the hypothalamus. In this study, circulating leptin and its gene expression in muscle were evaluated in two groups of young Nellore bulls with divergent feed efficiency. Individual dry matter intake (DMI) and average daily gain (ADG) of 98 Nellore bulls were evaluated in feedlot for 70 d to determinate the residual feed intake (RFI) and select 20 animals for the high feed efficient (LRFI) and 20 for the low feed efficient (HRFI) groups. Blood samples were collected on d 56 and at slaughter (80 d) to determine circulating plasma leptin. Samples of *Longissimus dorsi* were taken at slaughter for leptin gene expression levels.

Results: DMI and RFI were different between groups and LRFI animals showed less back fat and rump fat thickness, as well as less pelvic and kidney fat weight. Circulating leptin increased over time in all animals. Plasma leptin was greater in LRFI on 56 d and at slaughter ($P = 0.0049$). Gene expression of leptin were greater in LRFI animals ($P = 0.0022$) in accordance with the plasma levels. The animals of the LRFI group were leaner, ate less, and had more circulating leptin and its gene expression.

Conclusion: These findings demonstrated that leptin plays its physiological role in young Nellore bulls, probably controlling food intake because feed efficient animals have more leptin and lower residual feed intake.

Keywords: Beef cattle, Energy homeostasis, Fat depositon, Residual feed intake

Background

Leptin is a polypeptide hormone produced mainly by adipose cells [1], being also expressed by other tissues such as skeletal muscle [2], mammary gland [3], and others [4]. The main physiological function of leptin is energy homeostasis by controlling feed intake by inhibiting hunger [5–7]. It acts on a specific receptor (LEPR) in the brain, most specific to the hypothalamus in ventromedial, dorsomedial, and arcuate nuclei [8]. Although the physiological role of leptin is to control food intake, obese individuals generally present greater circulating levels of the hormone due to their greater percentage of body fat [9]. Obese people usually present resistance to leptin by alterations in the *LEPR* signaling and/or diminished crossing into the blood brain barrier [10].

Fat metabolism and deposition is important for livestock, affecting the quality of the products (beef, milk, etc) and providing energy reserves for reproduction and lactation [11]. Soon after the discovery of the obese gene in 1994 [1], its product, leptin, was identified for potential applications in animal production [12]. Indeed, research has been done on different aspects of leptin in livestock, including the use of its polymorphisms in animal breeding programs [13–17] and the use of circulating plasma leptin as a predictor of body composition [18, 19].

* Correspondence: fukumasu@usp.br
[3]Departamento de Medicina Veterinária, Faculdade de Zootecnia e Engenharia de Alimentos, Universidade de São Paulo, Av. Duque de Caxias Norte n°225, Pirassununga 13635-900, SP, Brazil
Full list of author information is available at the end of the article

In sheep, plasma concentration of leptin is more related to variation in body fatness (35%) than to nutritional status (17%) [20]. Corroborating this result, leptin levels in beef cattle (*Bos taurus*) were positively associated with adipose cell size [21], body fatness [22], and with the 12th rib and rump fat thickness, explaining 16.8% of the variation in the 12th rib fat thickness [19]. In the same work, the authors [19] showed the circulating leptin was positively associated with residual feed intake (RFI), a measure of feed efficiency, however, they explained very little of the variation in RFI (<3.2% of the variance). A moderate association ($r = 0.31$) of leptin with RFI was also showed by Richardson et al. [23] in Angus steers.

In *Bos indicus*, the majority of beef cattle grown in Brazil, *Leptin* gene polymorphisms were associated mostly with growth and carcass traits [17, 24, 25], but no study was found on the circulating plasma leptin and/or muscle gene expression association with feed efficiency. Thus, the aim of this work was to characterize the circulating plasma leptin and its muscle gene expression in young Nellore bulls with divergent residual feed intake.

Methods
Animals and experimental design
A feeding trial of 98 Nellore bulls (16 to 20 months old and 376 ± 29 kg BW) was conducted at the Faculdade de Zootecnia e Engenharia de Alimentos, Universidade de São Paulo (FZEA / USP), Pirassununga, SP, Brazil. All the details regarding animals, traits, and diet can be found in Alexandre et al. [26]. Briefly, the data collection period consisted of 70 d preceeded by 21 d of adaptation to diet and environment. Animals were weighed in the beginning, the end, and every 14 d of the experimental period. Additionally, daily dry matter intake (DMI) for each animal was measured by weighting the orts every day. Residual feed intake (RFI) was calculated [27], and two groups were formed: low RFI (LRFI) and high RFI (HRFI), each composed of 20 extreme animals.

Plasmatic leptin quantification
Plasma samples were collected at 56 d of feeding trial and at slaughter (80 d) by jugular venipuncture using vacutainer tubes containing sodium heparin as anticoagulant (BD Vacutainer Plus, BD, Brazil). All blood samples were centrifuged at 3,500×g for 15 min at 4 °C, and plasma was collected and stored at −20 °C until assayed for leptin. The concentration was determined in duplicate 100 μL aliquots of plasma samples, using the leptin RIA kit (Multi-species leptin RIA kit, Cat. # XL-85 K, Millipore, St. Charles, MO, USA) and following the manufacturer's instructions. Intra and interassay CVs for the leptin assay were less than 10% as described by Delavaud et al. [20].

Slaughter and tissue sample collection
Animals were slaughtered on d 80 at the Experimental slaughterhouse of University of São Paulo, after fasting from feed for 16 h. Immediately after slaughter, samples of the medial portion from the right *Longissimus lumborum* muscle (between the 12th and 13th ribs) were taken from 10 animals of each RFI group. The samples were collected and immediately immersed in RNA stabilization solution (RNAholder – Bioagency, São Paulo/SP, Brazil). The sample were maintained overnight at 4 °C and then stored at −80 °C until RNA extraction.

Gene expression of *Leptin*
Total RNA was extracted from 300 mg of powdered tissue samples using Trizol reagent following the manufacturer's protocol (Invitrogen, Carlsbad, CA, USA). After quantification in a spectrophotometer, 1 μg of total RNA was treated with DNAse (Invitrogen, Carlsbad, CA, USA) and reverse transcribed into cDNA using Go Script ™ Reverse Transcription System kit (Promega Corporation, Madison, WI, USA). Real time-PCR was performed on a CFX ConnectTM Real-Time PCR detection system (Bio-Rad, Hercules, CA, USA), using SYBR Green RT-PCR kit (Applied Biosystems, Foster City, CA, USA) with the following cycle parameters: 95 °C for 3 min and 40 cycles at 95 °C for 10 s and 60 °C for 30 s. Primers utilized for PCR were leptin (F) 5'GGGCACGTCAGCATCTATTA3' and leptin (R) 5'CCTGTCTGCTGTTATGGTCTTA3', and for the endogenous control, the ribosomal 18S (F) 5'CCTGC GGCTTAATTTGACTC3' and 18S (R) 5'AACTAAG-AACGGCCATGCAC3' were used. The amplification efficiency was 0.90 to 0.99 for leptin and 18S, respectively. All reactions were performed in triplicate, and the method of Livak and Schimttgen was used for gene expression analysis [28].

Statistical analysis
All analyses were performed using GraphPad Prism version 6.0 for Mac (GraphPad Sotware, La Jolla California, USA). The phenotypic measures assessed in the LRFI and HRFI groups were first tested for Gaussian distribution by the Shapiro-Wilk test, and later, they were tested for difference between the means of the groups by Student's t-test for normal distributed data and Mann–Whitney-Wilcoxon test for nonparametric data. The correlation between the two leptin measures was performed with the Pearson test. A two-way ANOVA followed by the Fisher post-test was used for leptin plasmatic concentration on d 56 and d 80. Statistically significant results were considered when $P < 0.05$.

Results

Characterization of feed efficiency groups

Phenotypic traits (related to body weight, feed efficiency, muscle and fat deposition, liver weight, and cascass yeld), are already published and no significant difference of sire or age on RFI was found [26]. Animals of LRFI and HRFI groups showed no difference in body weight (initial, final, or average daily gain) and rib eye area (initial, final, and gain). On the other hand, DMI and RFI were different between groups, wherein LRFI animals presented less DMI [26]. In addition, LRFI animals presented less fat deposition in both back fat and rump fat and less pelvic and kidney fat relative weight in comparison with HRFI. Thus, LRFI animals eat less and are leaner than HRFI animals [26].

Circulating plasma leptin and muscle gene expression

The correlation between those two measures was moderate to strong and significant (r = 0.5817 and P = 0.0071). In this experiment, leptin concentration in plasma increased over time in both groups (Fig. 1a), accounting for 21.3% of the total variation and was considered significant (P < 0.0001). In addition, leptin concentration was different between groups (P = 0.0049). However, no interaction between time and groups was noted though (P = 0.513).

Leptin gene expression was performed in muscle samples collected at slaughter. LRFI animals showed greater expression of leptin (P = 0.0022) compared to HRFI animals, reflecting the increased leptin plasma concentration of animals at slaughter (Fig. 1b).

Discussion

In this study, it was shown that circulating plasma leptin and muscle gene expression were greater in high feed efficiency group (LRFI) than in low feed efficiency group (HRFI) of Nellore bulls. In addition, feed efficient animals showed less fat deposition, probably a consequence of significant less DMI. In a previous work from the group, it was demonstrated that LRFI animals have less circulating plasma cholesterol at the end of the feeding trial [26]. Since the main role of leptin is regulate food intake in different species [29–31], these results support the physiological role of this hormone in Nellore cattle, since leptin is controlling DMI, plasma cholesterol, and fat deposition. Supporting this concept, less efficient animals have less muscle gene expression and circulating leptin which increases DMI, plasmatic cholesterol, and fat deposition.

Nonetheless, in other experiments on cattle, leptin concentration was positively associated with DMI, body fat, and RFI ([19, 23, 32]. In these cases, leptin seems to be related to body fat mass but not with the control of food intake. Interestingly, this phenotype resembles the resistance to leptin in obese humans where even elevated levels

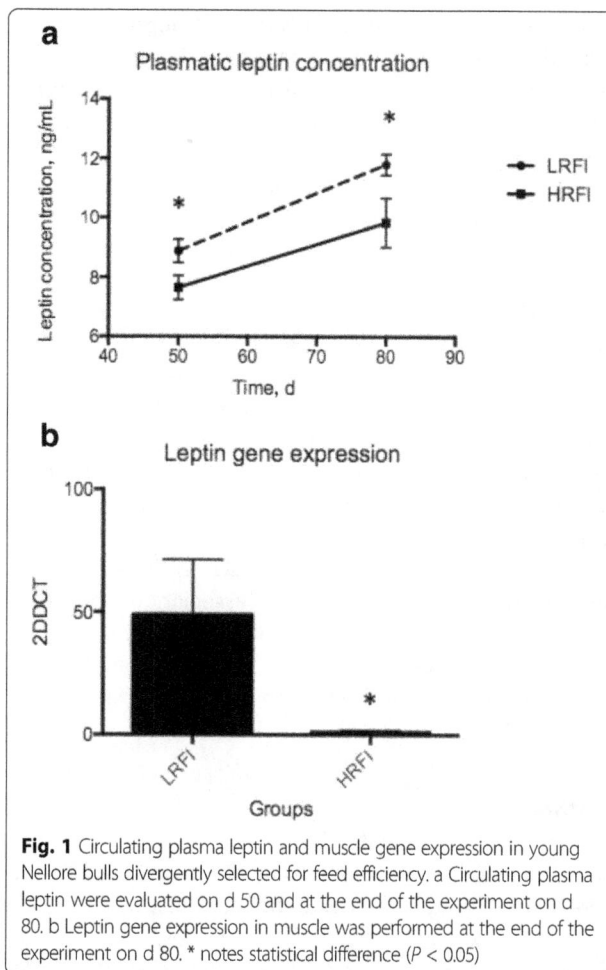

Fig. 1 Circulating plasma leptin and muscle gene expression in young Nellore bulls divergently selected for feed efficiency. a Circulating plasma leptin were evaluated on d 50 and at the end of the experiment on d 80. b Leptin gene expression in muscle was performed at the end of the experiment on d 80. * notes statistical difference (P < 0.05)

of leptin produced by greater percentage of body fat failed to control hunger and modulated body fat [33]. In accordance with the hypothesis, a recently published work from Foote et al. [34] suggested that beef cattle on a high grain finishing ration could become slightly leptin resistant.

The opposing results observed in the current study and others suggest the hypothesis that animals in this experiment are not in an "obese" phenotype, being instead where leptin reflects fat body mass. This could be explained by the conjunction of: (1) having used young bulls while others used steers [18, 19, 23, 34] or heifers [34, 35] and/or; (2) Nellore is an indicine breed and most experiments on cattle are performed in taurine breeds. The first possibility is probably the most relevant since the effect of sex on fat deposition has long been described. Total rate of fat deposition relative to muscle is similar for heifers and steers but significantly lower in bulls [36]. These authors concluded that differences in fattening patterns among sexes result from a combination of fattening at a lighter weight of carcass muscle in heifers than steers and in steers than bulls, in addition to a more rapid rate of fat deposition relative to muscle in heifersand steers compared to bulls. In addition, the effect of sex is well known to be related to feed efficiency, where bulls are more efficient than steers or heifers [37, 38]. In humans, females have greater circulating leptin as compared to males, even after the correction for differences in body fat mass [39]. In vitro experiments with human adypocytes in primary culture showed that both testosterone, and its biologically active metabolite dihydrotestosterone inhibited leptin secretion up to 62%, supporting that sex is clearly a relevant factor on leptin metabolism [39].

One should also consider the possible effect of genetic differences between Nellore and taurine breeds for the observed effect in this experiment. Indicine cattle are known to be more adapted to tropical climate but with less growth performance than taurine breeds. Corroborating this idea, Marcondes et al. [40] demonstrated the dry matter intake and performance of steers were higher in Nellore crossbreds (Nellore-Simmental and Nellore-Red Angus) than that of pure Nellore, however, no difference in feed efficiency was noted between groups. Paschal and colleagues [41] studying the postweaning and feedlot growth and carcass characteristics of five indicine breeds and one taurine (Angus) showed that Zebu crosses (including Nellore) grew faster postweaning and were heavier and taller than Angus crosses. However, the Angus cross was more desirable in marbling score and quality grade. Beef products derived from Nellore are recognized by the international market as very lean meat due to lack of intramuscular fat [42]. The same authors demonstrated the difference in intramuscular fat content in skeletal muscle of Nellore and Angus cattle is due to a slightly enhanced muscle adipogenesis in Angus cattle. Altogether, these results support the concept that Nellore have diferent growth curves in comparison to taurine breeds taking more time for fat deposition.

Therefore, it is plausible that the increased muscle gene expression of leptin and circulating leptin levels in feed efficient young Nellore bulls in comparison to less efficient animals, although, others demonstrated the opposite. This result also calls attention for the use of leptin as a possible predictor for body fat mass in cattle, especially if considered for use in Nellore cattle.

Conclusion
To the best of our knowledge, this work demonstrated for the first time that leptin plays its physiological role in young Nellore bulls, rather its pathological role, controlling food intake because feed efficient animals have more circulating leptin and lower residual feed intake.

Abbreviations
ADG: Average daily gain; BWf: Final body weight; BWg: Body weight gain; BWi: Initial body weight; DMI: Dry matter intake; HRFI: Low feed efficient; RBg: Rib eye gain; REAf: Final rib eye area; REAi: Initial rib eye area; RFI: Residual feed intake; RIA: Radioimmunoassay; LEPR: Leptin receptor; LRFI: High feed efficient

Acknowledgements
We would also like to thank Michael James Stablein of the University of Illinois Urbana-Champaign for his translation services and review of this work.

Funding
This work was financially supported by "Fundação de Amparo à Pesquisa do Estado de São Paulo" (FAPESP 2010/05650–5; 2014/02493–7, 2014/07566–2).

Authors' contributions
PAA, CMB, SLS, PRL, and HF designed the experimental study; LFMM and CMB performed the leptin circulating analysis; PAA, MHS, LFMM, and HF conducted the in vivo experiment and sample collection; ET helped LFMM on gene expression analysis; LFMM, PAA, FJN, and HF drafted the manuscript. All authors read, made critical revisions, and approved the final manuscript for submission.

Competing interests
The authors declare that they have no competing interests.

Author details
[1]Departamento de Zootecnia, Universidade Federal dos Vales do Jequitinhonha e Mucuri, Diamantina, MG 39100-000, Brazil. [2]Present adress: Faculdade de Ciências Agrárias e Veterinárias, Universidade Estadual Paulista, Jaboticabal, SP 14884-900, Brazil. [3]Departamento de Medicina Veterinária, Faculdade de Zootecnia e Engenharia de Alimentos, Universidade de São Paulo, Av. Duque de Caxias Norte n°225, Pirassununga 13635-900, SP, Brazil. [4]Departamento de Zootecnia e Desenvolvimento Agrossocioambiental Sustentável, Faculdade de Veterinária, Universidade Federal Fluminense, Niteroi, RJ 24230-340, Brazil. [5]Departamento de Zootecnia, Faculdade de Zootecnia e Engenharia de Alimentos, Universidade de São Paulo, Pirassununga 13635-900, SP, Brazil.

References
1. Zhang Y, Proenca R, Maffei M, Barone M, Leopold L, Friedman JM. Positional cloning of the mouse obese gene and its human homologue. Nature. 1994;372:425–32.
2. Wang T, Hartzell DL, Flatt WP, Martin RJ, Baile CA. Responses of lean and obese Zucker rats to centrally administered leptin. Physiol Behav. 1998;65:333–41.
3. Bonnet M, Gourdou I, Leroux C, Chilliard Y, Djiane J. Leptin expression in the ovine mammary gland: putative sequential involvement of adipose, epithelial, and myoepithelial cells during pregnancy and lactation. J Anim Sci. 2002;80:723–8.
4. Margetic S, Gazzola C, Pegg G, Hill R. Leptin: a review of its peripheral actions and interactions. Int J Obes. 2002;26:1407–33.
5. Elias CF, Aschkenasi C, Lee C, Kelly J, Ahima RS, Bjorbaek C, et al. Leptin differentially regulates NPY and POMC neurons projecting to the lateral hypothalamic area. Neuron. 1999;23:775–86.
6. Fekete C, Légrádi G, Mihály E, Huang QH, Tatro JB, Rand WM, et al. Alpha-Melanocyte-stimulating hormone is contained in nerve terminals

innervating thyrotropin-releasing hormone-synthesizing neurons in the hypothalamic paraventricular nucleus and prevents fasting-induced suppression of prothyrotropin-releasing hormone ge. J Neurosci. 2000;20:1550–8.

7. Houseknecht KL, Portocarrero CP. Leptin and its receptors: regulators of whole-body energy homeostasis. Domest Anim Endocrinol. 1998;15:457–75.

8. Fei H, Okano HJ, Li C, Lee GH, Zhao C, Darnell R, et al. Anatomic localization of alternatively spliced leptin receptors (Ob-R) in mouse brain and other tissues. Proc Natl Acad Sci U S A. 1997;94:7001–5.

9. Considine RV, Sinha MK, Heiman ML, Kriauciunas A, Stephens TW, Nyce MR, et al. Serum immunoreactive-leptin concentrations in normal-weight and obese humans. N Engl J Med. 1996;334:292–5.

10. Lubis AR, Widia F, Soegondo S, Setiawati A. The role of SOCS-3 protein in leptin resistance and obesity. Acta Med Indones. 2008;40:89–95.

11. Wylie ARG. Leptin in farm animals: where are we and where can we go? Animal. 2011;5:246–67.

12. Hossner KL. Cellular, molecular and physiological aspects of leptin: potential application in animal production. Can J Anim Sci. 1998;78:463–72.

13. Buchanan FC, Fitzsimmons CJ, Van Kessel AG, Thue TD, Winkelman-Sim DC, Schmutz SM. Association of a missense mutation in the bovine leptin gene with carcass fat content and leptin mRNA levels. Genet Sel Evol. 2002;34:105–16.

14. Lagonigro R, Wiener P, Pilla F, Woolliams JA, Williams JL. A new mutation in the coding region of the bovine leptin gene associated with feed intake. Anim Genet. 2003;34:371–4.

15. Schenkel FS, Miller SP, Ye X, Moore SS, Nkrumah JD, Li C, et al. Association of single nucleotide polymorphisms in the leptin gene with carcass and meat quality traits of beef cattle. J Anim Sci. 2005;83:2009–20.

16. Nkrumah JD, Li C, Yu J, Hansen C, Keisler DH, Moore SS. Polymorphisms in the bovine leptin promoter associated with serum leptin concentration, growth, feed intake, feeding behavior, and measures of carcass merit. J Anim Sci. 2005;83:20–8.

17. Silva DBS, Crispim BA, Silva LE, Oliveira JA, Siqueira F, Seno LO, et al. Genetic variations in the leptin gene associated with growth and carcass traits in Nellore cattle. Genet Mol Res. 2014;13:3002–12.

18. Nkrumah JD, Keisler DH, Crews DH, Basarab JA, Wang Z, Li C, et al. Genetic and phenotypic relationships of serum leptin concentration with performance, efficiency of gain, and carcass merit of feedlot cattle. J Anim Sci. 2007;85:2147–55.

19. Foote AP, Hales KE, Kuehn LA, Keisler DH, King DA, Shackelford SD, et al. Relationship of leptin concentrations with feed intake, growth, and efficiency in finishing beef steers. J Anim Sci. 2015;93:4401–7.

20. Delavaud C, Bocquier F, Chilliard Y, Keisler DH, Gertler A, Kann G. Plasma leptin determination in ruminants: effect of nutritional status and body fatness on plasma leptin concentration assessed by a specific RIA in sheep. J Endocrinol. 2000;165:519–26.

21. Delavaud C, Ferlay A, Faulconnier Y, Bocquier F, Kann G, Chilliard Y. Plasma leptin concentration in adult cattle: effects of breed, adiposity, feeding level, and meal intake. J Anim Sci. 2002;80:1317–28.

22. Brandt MM, Keisler DH, Meyer DL, Schmidt TB, Berg EP. Serum hormone concentrations relative to carcass composition of a random allotment of commercial-fed beef cattle. J Anim Sci. 2007;85:267–75.

23. Richardson EC, Herd RM, Archer JA, Arthur PF. Metabolic differences in Angus steers divergently selected for residual feed intake. Aust J Exp Agric. 2004;44:441.

24. da Silva RCG, Ferraz JBS, Meirelles FV, Eler JP, Balieiro JCC, Cucco DC, et al. Association of single nucleotide polymorphisms in the bovine leptin and leptin receptor genes with growth and ultrasound carcass traits in Nellore cattle. Genet Mol Res. 2012;11:3721–8.

25. de Oliveira JA, da Cunha CM, do A CB, de O SL, ARM F, de P NG, et al. Association of the leptin gene with carcass characteristics in Nellore cattle. Anim Biotechnol. 2013;24:229–42.

26. Alexandre PA, Kogelman LJA, Santana MHA, Passarelli D, Pulz LH, Fantinato-Neto P, et al. Liver transcriptomic networks reveal main biological processes associated with feed efficiency in beef cattle. BMC Genomics. 2015;16:1073.

27. Koch RM, Swiger LA, Chambers D, Gregory KE. Efficiency of feed use in beef cattle. J Anim Sci. 1963;22:486–94.

28. Livak KJ, Schmittgen TD. Analysis of relative gene expression data using real-time quantitative PCR and the 2(−Delta Delta C(T)) method. Methods. 2001;25:402–8.

29. Halaas JL, Gajiwala KS, Maffei M, Cohen SL, Chait BT, Rabinowitz D, et al. Weight-reducing effects of the plasma protein encoded by the obese gene. Science. 1995;269:543–6.

30. Larsson H, Elmståhl S, Berglund G, Ahrén B. Evidence for leptin regulation of food intake in humans. J Clin Endocrinol Metab. 1998;83:4382–5.

31. Henry BA, Goding JW, Alexander WS, Tilbrook AJ, Canny BJ, Dunshea F, et al. Central administration of leptin to ovariectomized ewes inhibits food intake without affecting the secretion of hormones from the pituitary gland: evidence for a dissociation of effects on appetite and neuroendocrine function. Endocrinology. 1999;140:1175–82.

32. Minton JE, Bindel DJ, Droullard JS, Titgemeyer EC, Grieger DM, Hill CM. Serum leptin is associated with carcass traits in finishing cattle. J Anim Sci. 1998;76:231.

33. Myers MG, Leibel RL, Seeley RJ, Schwartz MW. Obesity and leptin resistance: distinguishing cause from effect. Trends Endocrinol Metab. 2010;21:643–51.

34. Foote AP, Tait RG, Keisler DH, Hales KE, Freetly HC. Leptin concentrations in finishing beef steers and heifers and their association with dry matter intake, average daily gain, feed efficiency, and body composition. Domest Anim Endocrinol. 2016;55:136–41.

35. Kelly AK, McGee M, Crews DH, Fahey AG, Wylie AR, Kenny DA. Effect of divergence in residual feed intake on feeding behavior, blood metabolic variables, and body composition traits in growing beef heifers. J Anim Sci. 2010;88:109–23.

36. Berg RT, Jones SDM, Price MA, Hardin RT, Fukuhara R, Butterfield RM. Patterns of carcass fat deposition in heifers, steers and bulls. Can J Anim Sci. 1979;59:359–66.

37. Hedrick H. Bovine growth and composition. Research Bulletin Monograph 928. Washington, DC: Agricultural Experimental Station, 1968; pp. 56.

38. Bailey CM, Probert CL, Bohman VR. Growth rate, feed utilization and body composition of young bulls and steers. J Anim Sci. 1966;25:132–7.

39. Wabitsch M, Blum WF, Muche R, Braun M, Hube F, Rascher W, et al. Contribution of androgens to the gender difference in leptin production in obese children and adolescents. J Clin Invest. 1997;100:808–13.

40. Marcondes M, Valadares Filho S, Oliveira I, Veiga P, Paulino R, Ferreira R, et al. Eficiência alimentar de bovinos puros e mestiços recebendo alto ou baixo nível de concentrado. Brazilian J Anim Sci. 2011;406:1313–24.

41. Paschal JC, Sanders JO, Kerr JL, Lunt DK, Herring AD. Postweaning and feedlot growth and carcass characteristics of Angus-, gray Brahman-, Gir-, Indu-Brazil-, Nellore-, and red Brahman-sired F1 calves. J Anim Sci. 1995;73:373–80.

42. Martins TS, Sanglard LMP, Silva W, Chizzotti ML, Rennó LN, Serão NVL, et al. Molecular factors underlying the deposition of intramuscular fat and collagen in skeletal muscle of Nellore and Angus cattle. PLoS One. 2015;10:e0139943.

14

Plasma and cerebrospinal fluid interleukin- 1β during lipopolysaccharide-induced systemic inflammation in ewes implanted or not with slow-release melatonin

Janina Skipor[1][*]iD, Marta Kowalewska[1]iD, Aleksandra Szczepkowska[1]iD, Anna Majewska[1], Tomasz Misztal[2], Marek Jalynski[3], Andrzej P. Herman[2] and Katarzyna Zabek[4]

Abstract

Background: Interleukin-1β (IL-1β) is important mediator of inflammatory-induced suppression of reproductive axis at the hypothalamic level. At the beginning of inflammation, the main source of cytokines in the cerebrospinal fluid (CSF) is peripheral circulation, while over time, cytokines produced in the brain are more important. Melatonin has been shown to decrease pro-inflammatory cytokines concentration in the brain. In ewes, melatonin is used to advance the onset of a breading season. Little is known about CSF concentration of IL-1β in ewes and its correlation with plasma during inflammation as well as melatonin action on the concentration of IL-1β in blood plasma and the CSF, and brain barriers permeability in early stage of lipopolysaccharide (LPS)-induced inflammation.

Methods: Systemic inflammation was induced through LPS administration in melatonin- and sham-implanted ewes. Blood and CSF samples were collected before and after LPS administration and IL-1β and albumin concentration were measured. To assess the functions of brain barriers albumin quotient (QAlb) was used. Expression of IL-1β (*Il1B*) and its receptor type I (*Il1r1*) and type II (*Il1r2*) and matrix metalloproteinase (*Mmp*) 3 and 9 was evaluated in the choroid plexus (CP).

Results: Before LPS administration, IL-1β was on the level of 62.0 ± 29.7 pg/mL and 66.4 ± 32.1 pg/mL in plasma and 26.2 ± 5.4 pg/mL and 21.3 ± 8.7 pg/mL in the CSF in sham- and melatonin-implanted group, respectively. Following LPS it increased to 159.3 ± 53.1 pg/mL and 197.8 ± 42.8 pg/mL in plasma and 129.8 ± 54.2 pg/mL and 139.6 ± 51.5 pg/mL in the CSF. No correlations was found between plasma and CSF IL-1β concentration after LPS in both groups. The QAlb calculated before LPS and 6 h after was similar in all groups. Melatonin did not affected mRNA expression of *Il1B*, *Il1r1* and *Il1r2* in the CP. The mRNA expression of *Mmp3* and *Mmp9* was not detected.

Conclusions: The lack of correlation between plasma and CSF IL-1β concentration indicates that at the beginning of inflammation the local synthesis of IL-1β in the CP is an important source of IL-1β in the CSF. Melatonin from slow-release implants does not affect IL-1β concentration in plasma and CSF in early stage of systemic inflammation.

Keywords: Albumin, Cerebrospinal fluid, Ewes, Interleukin −1β, LPS, Melatonin

* Correspondence: j.skipor@pan.olsztyn.pl
[1]Institute of Animal Reproduction and Food Research, Polish Academy of Sciences, Olsztyn, Poland
Full list of author information is available at the end of the article

Background

It is well established that immune stress caused by infections and inflammatory diseases reduces animals productivity and is a powerful modulator of mechanisms regulating reproduction at all levels of the hypothalamic–pituitary–gonadal axis [1]. A few pathways have been suggested to be responsible for the immune-mediated inhibition of reproductive activity at the level of hypothalamus and one of these involves pro-inflammatory mediators such as cytokines [2]. Hypothalamic interleukin (IL)-1β and tumor necrosis factor (TNF)-α mediates the lipopolysaccharide (LPS)-induced suppression of gonadotropin releasing hormone (GnRH) and luteinizing hormone (LH) release in female rats [3]. In ewes, the central IL-1β is an important modulator of the GnRH biosynthesis and release during immune/inflammatory challenge. The thesis about the crucial role of these cytokines in the transmission of signals from the immune to neuroendocrine systems seems to be supported by the presence of IL-1β and its receptors in the hypothalamus in LPS-treated ewes [4].

In general, passage of molecules from the periphery to the brain is restricted by brain barriers: blood-brain barrier (BBB) located in cerebromicrovascular endothelial cells and blood-cerebrospinal fluid barrier (BCSFB) located in the epithelial cells of choroid plexus (CP) [5]. The origin of central pro-inflammatory cytokines is differentiated. At the beginning of inflammation, the main source of cytokines present in the cerebrospinal fluid (CSF) is peripheral circulation, while over time of inflammation, endogenous cytokines produced in the brain seems to be more important. In rats treated intravenously with IL-1β two waves of cellular activation at the brain appears, the first one at the blood side of the BBB 30 min after IL-1β administration and then after 3 h at the parenchymal side of the BBB [6]. It has been demonstrated in rats, that early after (5 h) injection of IL-1β to the brain BBB becomes permeable to intravenously administered contrast [7]. In this process matrix metalloproteinases (MMPs) as enzymes that catalyze the proteolytic cleavage of basal lamina components and thus remodeling of the extracellular matrix and brain barriers permeability play an important role [8]. Interestingly, melatonin attenuated BBB hyperpermeability in IL-1β stimulated rat brain microvessels endothelial cells in vitro as well as in vivo in mouse traumatic brain injury model [9]. From the periphery, IL-1β is transported throughout the brain barriers [10]. Transport of IL-1β has been suggested to occur via a type II IL-1 receptor [11]. This receptor may also be released from cells and function as decoy receptor to block IL-1β action in contrast to type I IL-1β receptor that transduce IL-1 signals after binding with IL-1β [12, 13]. Expression of type II IL-1 receptor mRNA (*Il1r2*) and I IL-1β receptor (*Il1r1*)

was detected in the brain endothelial cells [14] and in the CP [15]. In ewes, intravenous injection of LPS is one model of systemic inflammation which has been used to study mechanisms responsible for the immune-mediated inhibition of reproductive activity [4]. So far little is known about CSF concentration of IL-1β in ewes and its correlation with plasma concentration during LPS-induced systemic inflammation. Melatonin receptors MT1 and MT2 has been demonstrated in the ovine CP what unable direct melatonin action on the CP [16]. Melatonin action on the concentration of IL-1β in blood plasma and the CSF is particularly interesting due to the use of melatonin from continuous slow-release implants to advance the onset of a breeding season in sheep and goats [17].

The present study aimed at evaluating effect of LPS alone and with melatonin slow-release implants on the concentration of IL-1β in blood plasma and the CSF as well as on the BBB permeability in ewes early after LPS administration. Additionally, we evaluated effect of melatonin implantation on mRNA expression of *Il1B* and its receptors *Il1r1 and Il1r2* as well as *Mmp3* and *Mmp9* in the CP under the influence of IL-1β.

Methods

Animals and experimental design

All animal procedures were conducted in accordance with the Polish Guide for the Care and Use of Animals and approved by the Local Ethics Committee (No. 25/2012). Female adult sheep (4–5 years old, 50–60 kg body weight) of the Blackheaded Mutton breed (n = 14) were maintained indoors under natural lighting conditions (latitude 52°N, 21°E) and fed a constant diet of hay, straw and commercial concentrates according to the recommendations proposed by the National Research Institute of Animal Production for adult ewes. Water and mineral licks were available ad libitum. On the beginning of May, ewes were ovariectomized under general anaesthesia and then (middle of May) subcutaneously implanted with an oestradiol (E2) implant, which maintained plasma E2 concentrations of 2–4 pg/L [18]. After 2 wk of recovery ewes were implanted under general anaesthesia with stainless steel guide cannulae (1.2 mm o.d.) into the third ventricle as described earlier [19]. In the middle of May, ewes were first sampled for blood and then one- half of the animals was randomly melatonin-implanted (n = 7, slow-release implant of 18 mg, Melovine Ceva Sante Animale, France) and the second half of the animals was sham-implanted (n = 7). Approximately 40 d later, ewes were implanted with the jugular vein catheter early on the morning (7:00 am) and have been placed in individual cages where they could lie down and have access to hay and water. To prevent the stress of social isolation, all ewes had visual contact.

After that stainless catheter was introduced into the third ventricle and control blood and CSF samples were collected. Then immune stress was induced by intravenously injection of LPS from *Escherichia coli* 055:B5 (Sigma, USA), at the dose of 400 ng/kg of body weight, dissolved in saline (0.9% *w/v* NaCl) at a concentration of 10 mg/L (10 µg/mL) as it was used previously in ewes [20]. The individual body mass of experimental ewes were at the range of 52 kg to 63 kg, therefore injection volume of LPS solution/saline was at the range of 2.1 to 2.5 mL. Body temperature was measured before and after LPS administration. Blood plasma was collected just before and after melatonin implantation and then every hour after LPS administration (Fig. 1). Blood samples were collected through a catheter inserted into the jugular vein. First 2 mL of blood samples were removed then 10 mL were collected into the tubes with stabilizer (EDTA). Immediately after centrifuging, the blood plasma were divided into separate (1 mL) aliquots and stored at −40 °C until assayed for IL-1β and albumin. To collect the CSF samples, a stainless steel catheter (1.0 mm o.d., 0.8 mm i.d.) was carefully introduced into the guide cannula and connected to a special cannula-Eppendorf tube system joined to a PHD 2000 infuse/withdrawal pump (Hugo Sachs Elektronik Harvard Apparatus, Germany). The CSF collection from the third ventricle of conscious ewes was performed during a 7 h period (1 h before and 6 h after LPS administration) at a rate of 20 µL/min. The tubes of CSF samples were kept in an ice bath during sampling, and the volume of one sample collected during the 30 min period was about 500 µL. Immediately after filling, the tubes were stored at −80 °C until assayed for IL-1β and albumin. The ewes were euthanized 6 h after LPS administration. After decapitation, the brains were dissected, CP were removed from their anchoring to the Galien's vein and the split was made along the mid-line, separating the CP from each lateral ventricle. CPs were then immediately frozen in liquid nitrogen and stored at −80 °C until use.

Hormones and IL-1β concentration measurement

The ability of the melatonin implants to maintain permanently high blood concentrations of melatonin was monitored by determining melatonin concentrations in blood plasma samples obtained before and 40 d after melatonin-implantation using radioimmunoassay (RIA) described by Misztal et al. [21]. The assay sensitivity for melatonin was 16.8 ± 8.0 pg/mL, and the intra- and interassay coefficients of variations were 10.5 and 13.2%, respectively. Blood plasma and CSF IL-1β concentrations were determined by commercially available enzyme-linked immunosorbent assay (ELISA Kit for ovine Interleukin 1β, Cloud-Clone Corp., USA), following the manufacturer's instructions. The optical density of individual wells was measured by a spectrophotometric microplate reader (Epoch, BioTek, Switzerland) at a wavelength of 450 ± 10 nm. The concentration of IL-1β in samples were determined by comparing the optical density of the samples to the standard curve. The detection limit of the assay was less than 5.9 pg/mL.

Western blotting and CSF/blood plasma quotient

To assess the functions of BCSFB albumin quotient (QAlb) was used. The CSF and blood plasma samples (equal volume 16 µL CSF and diluted blood plasma (1:160) with addition of 4 µL of loading buffer) from the sham- and melatonin-implanted group (one animal from each group) were loaded onto 10% sodium dodecyl sulfate (SDS) polyacrylamide gels together with serial dilutions of sheep serum albumin (SSA, Biorbyt, USA,

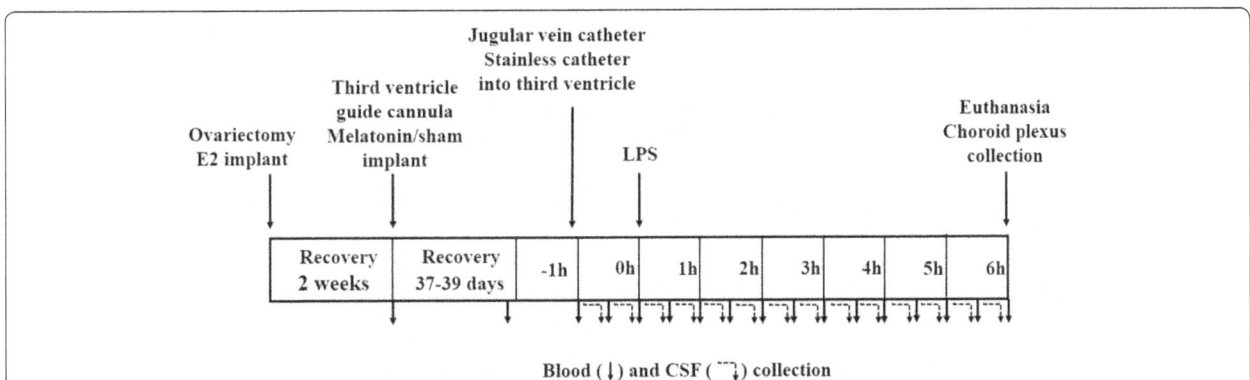

Fig. 1 Schematic diagram of the experimental design. At the beginning of May all the ewes (*n* = 14) were ovariectomized and implanted with an oestradiol (E2). Two weeks later, animals were implanted under general anaesthesia with stainless steel guide cannulae into the third ventricle and ewes were melatonin- (*n* = 7) or sham-implanted (*n* = 7). Approximately 40 d later, melatonin- and sham-implanted ewes were treated with lipopolysaccharide (LPS). Blood samples were collected for melatonin, cortisol albumin and IL-1β concentrations measurement. Cerebrospinal fluid (CSF) samples were collected for albumin and IL-1β concentrations measurement. The choroid plexuses were collected 6 h after LPS administration

range 0.25 to 5 µg/µL). Electrophoresis was performed using the MiniProtean II electrophoretic apparatus (BioRad, USA) at 60 mV constant voltage. Thereafter, proteins were transferred onto 0.2 µm thick nitrocellulose membranes (Whatman Inc., Germany) at 30 V for 1.5 h in a semi-dry transfer system (BioRad, USA). After 1.5 h blocking with block buffer (TBST, 50 g/L nonfat milk in 10 mL Tris buffer saline containing 0.5% Tween 20) at room temperature, the membranes were extensively washed in TBST and incubated overnight at 4 °C with rabbit polyclonal antibodies against sheep albumin at 1:300 dilution (Anitbodies-online, Germany). After final wash, membranes were developed using chemiluminescence SuperSignal® West Dura Kit (Thermo Scientific, USA) and visualized by VersaDoc 4000 MP Imaging System (BioRad, USA). Based on a SSA dilution curve, the albumin concentration was calculated in both CSF and blood plasma samples by measuring optical densities of the bands (Image Lab 5.2.1, Software, BioRad, USA). The integrity of BCSFB was estimated by the ratio of albumin concentrations in CSF and blood plasma. The albumin quotient was evaluate as follows: QAlb = Alb (CSF)/ Alb (blood plasma).

Relative gene expression assays

The total RNA from the CP was isolated using NucleoSpin RNA II Kit (Marcherey-Nagel, Germany). All steps of the isolation were performed according to the manufacturer's protocol. The purity and concentration of the isolated RNA were quantified spectrophotometrically using a NanoDrop 1000 instrument (Thermo Fisher Scientific, USA). The integrity of RNA was verified by electrophoresis using 1.2% agarose gel stained with ethidium bromide (Sigma Aldrich, USA). To synthesize cDNA, the DyNAmo cDNA Synthesis Kit (Thermo Fisher Scientific, USA) and 1 µg of total RNA were used. Expression of interleukin 1β (*Il1B*) and its receptor type I (*Il1r1*) and type II (*Il1r2*) and matrix metalloproteinases (*Mmp*) 3 and 9 in the ovine CP was determined by real-time PCR. Specific primer pairs for the different genes were used according to the literature or were designed using Primer-BLAST (National Center for Biotechnology Information) and were synthesized by Genomed (Poland) and are presented in Table 1. One reaction mixture for real-time PCR reaction (10 µL) contained 3 µL of diluted (1:14 reference genes, 1:10 *Il1B*, *Il1r1*, *Il1r2* and 1:8 *Mmp3* and *Mmp9*) cDNA, 0.2 µmol/L of the forward and reverse primers and 5 µL of mastermix from a DyNAmo SYBR Green qPCR Kit with ROX (Thermo Fisher Scientific, USA). The following protocol was used: 95 °C for 10 min for Hot Start modified Tbr DNA polymerase, followed by 35 cycles of 15 s of denaturation at 95 °C, 30 s of annealing at X °C (see Table 1) and 30 s of extension at 72 °C. After the cycles, a final melting curve analysis under continuous fluorescence measurement was performed to evaluate the specific amplification. The results were analyzed using Real-time PCR Miner (on-line available: http://www.miner.ewindup.info/version2), based on the algorithm developed by Zhao and Fernald [22].

Table 1 Sequences of oligonucleotide primers used for real time-PCR

GenBank Acc. No.	Gene	Amplicon size, bp	Temp. of primers annealing, °C	Forward/ reverse	Sequence 5' → 3'	Reference
X54796.1	*Il1B*	137	59	forward	CAGCCGTGCAGTCAGTAAAA	[35]
				reverse	GAAGCTCATGCAGAACACCA	
NM_001206735.1	*Il1r1*	124	59	forward	GGGAAGGGTCCACCTGTAAC	[35]
				reverse	ACAATGCTTTCCCCAACGTA	
NM_001046210	*Il1r2*	161	59	forward	CGCCAGGCATACTCAGAAA	[36]
				reverse	GAGAACGTGGCAGCTTCTTT	
XM_004015970.1	*Mmp3*	112	60	forward	AAGGCAGACATTTTTGGCGG	Originally designed
				reverse	ATGCCTCTTGGGGAACCTGC	
XM_004014614.1	*Mmp9*	115	60	forward	CTTCCGATGGAAAGAACGGGC	Originally designed
				reverse	GGGATCACAACGCCTTTGC	
NM_001034034	*Gapdh*	143	60	forward	TGACCCCTTCATTGACCTTC	[27]
				reverse	GATCTCGCTCCTGGAAGATG	
NM_001009784.1	*Actb*	122	60	forward	GCCAACCGTGAGAAGATGAC	[27]
				reverse	TCCATCACGATGCCAGTG	
BC_108088.1	*Hdac1*	115	60	forward	CTGGGGACCTACGGGATATT	[35]
				reverse	GACATGACCGGCTTGAAAAT	

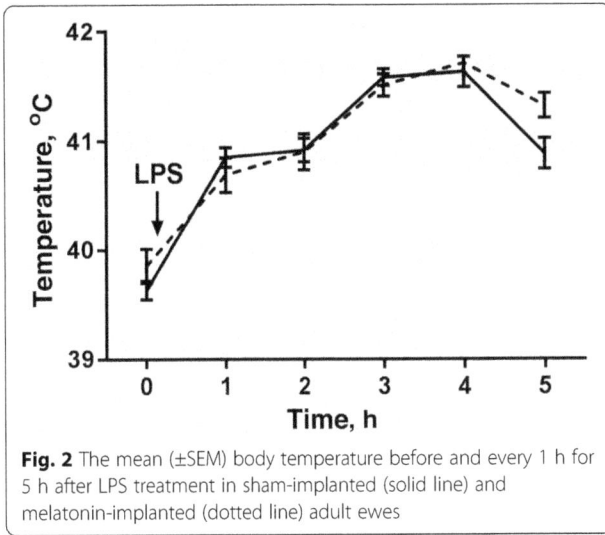

Fig. 2 The mean (±SEM) body temperature before and every 1 h for 5 h after LPS treatment in sham-implanted (solid line) and melatonin-implanted (dotted line) adult ewes

Statistical analysis

All data are presented as the mean ± standard error of the mean [SEM]. The real-time PCR results are presented as the relative gene expression of the target gene vs. the mean of 3 reference genes (*Gapdh*, *Actb*, *Hdac1*). The body temperature, melatonin and IL-1β concentrations, were analyzed by a one-way ANOVA for repeated measures and relative gene expression by t-test using PRISM 6 GraphPad Software (San Diego, USA). For statistical analysis, percentage data of QAlb were multiplied by 0.1 and then arcsin transformed according to the eq. $(Y = deg.(arcsin(sqrt(Y/100))))$ using PRISM 6 GraphPad Software. The transformed data were subjected to one-way ANOVA. Additionally, the differences between QAlb in sham-and melatonin-implanted ewes before LPS administration were analyzed by Welch test. The relationship between variables was analyzed using Pearson's correlation coefficient. Statistical significance was assumed at $P < 0.05$.

Results

The mean body temperature before the LPS administration was 39.6 ± 0.1 °C in the sham-implanted group and 39.9 ± 0.2 °C in the melatonin-implanted group and increased significantly ($P \leq 0.05$) to 41.6 ± 0.1 °C and 41.7 ± 0.1 °C 4 h after LPS administration, respectively (Fig. 2). Additionally, in all ewes LPS administration induced rapid breathing and shortness of breath, sneezing, and stopped feed intake, anhedonia and reduced social interactions which are collectively termed 'sickness behavior'. The plasma melatonin concentration in melatonin-implanted ewes increased significantly ($P < 0.05$) from 8.2 ± 2.2 pg/mL (before implantation) to 84.6 ± 15.0 pg/mL (one mo after implantation) while in sham-implanted stayed on the level of 9.6 ± 1.1 pg/mL. As shown on Table 2 immunoreactive IL-1β was not detected in blood plasma collected 1 h before LPS in 2 out of 6 ewes in sham-implanted group and in 4 out of 7 ewes in melatonin-implanted group. In other ewes IL-1β concentration ranged from 49.4 to 197.2 pg/mL and 128.6 to 192.7 pg/mL in sham- and melatonin-implanted ewes, respectively. Two and a half h after LPS administration the concentration of IL-1β increased in all investigated animals and reached the mean level of 159.3 ± 53.1 pg/mL and 197.8 ± 42.8 pg/mL in sham- and melatonin-implanted ewes, respectively. Melatonin did not affect IL-1β concentration in blood plasma. In CSF collected before LPS, IL-1β was not detected in 2 out of 7 ewes in melatonin group. In sham-implanted ewes IL-1β ranged from 13.5 to 51.7 pg/mL while in other melatonin-implanted ewes ranged from 11.8 to 67.4 pg/mL. After LPS treatment the mean concentration of IL-1β increased in all investigated animals in sham-implanted (129.8 ± 54.2 pg/mL) and melatonin-implanted (139.6 ± 51.5 pg/mL) ewes. There was no effect ($P > 0.05$) of melatonin on CSF IL-1β concentration as well as no differences ($P > 0.05$) between IL-1β in

Table 2 Individual measurements of IL-1β concentration (pg/mL) in blood plasma and cerebrospinal fluid (CSF) in sham-implanted and melatonin implanted ewes before and 3 h after lipopolysaccharide (LPS, 400 ng/kg) administration

Sham-implanted					Melatonin-implanted				
	Plasma, pg/mL		CSF, pg/mL			Plasma, pg/mL		CSF, pg/mL	
Ewe #	Before	After LPS	Before	After LPS	Ewe #	Before	After LPS	Before	After LPS
1	0.0	65.3	27.8	48.0	1	143.3	211.0	26.4	410.8
2	0.0	105.9	13.5	35.5	2	128.	134.7	67.4	146.5
3	53.0	58.5	21.0	50.4	3	0.0	210.7	19,2	39.6
4	76.1	113.0	20.3	365.6	4	0.0	298.4	0.0	198.5
5	197.2	399.0	22.8	66.7	5	0.0	80.3	0.0	12.3
6	49.4	214.2	51.7	213.7	6	192.7	378.7	24.2	54.0
					7	0.0	71.0	11,8	115.4
Mean (±SEM)	62.0 (±29.7)	159.3[a] (±53.1)	26.2 (±5.4)	129.8[a] (±54.2)		66.4 (±32.1)	197.8[a] (±42.8)	21.3 (±8.7)	139.6[a] (±51.5)

[a]significant vs. concentration before LPS administration at $P < 0.05$

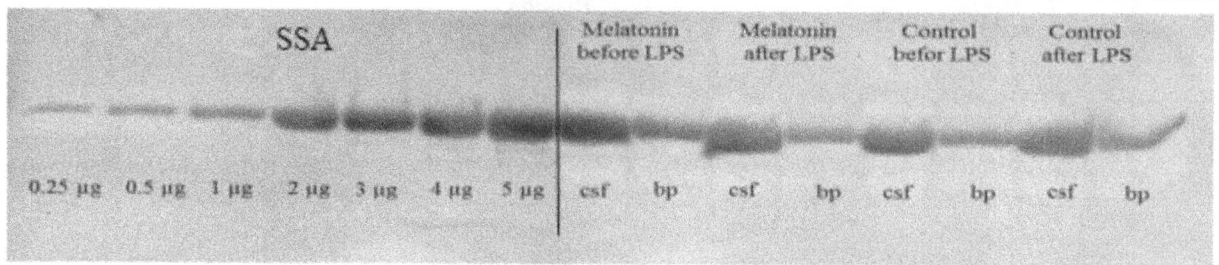

Fig. 3 Determination of albumin levels by western blot analysis in sheep cerebrospinal fluid (csf) and blood plasma samples (bp) collected before and 6 h after LPS administration. The densities of the bands were based on sheep serum albumin (SSA) dilution curve

blood plasma and CSF after LPS administration in both groups. No correlations was found between plasma and CSF IL-1β concentration after LPS administration in both sham- and melatonin-implanted groups ($r^2 = 0.08$; $P < 0.29$ vs. $r^2 = 0.01$; $P < 0.4$).

The albumin concentrations in the CSF collected from the third brain ventricle and blood plasma were calculated on the base of linear dilution curve of sheep serum albumin detected by western blot method (Fig. 3) and then the integrity of BCSFB was estimated by ratio of albumin concentrations in the CSF and blood plasma (QAlb). Six hours after LPS administration mean QAlb was on the level of 0.20% ± 0.05 and 0.14% ± 0.02 ($P > 0.05$) in sham- and melatonin-implanted ewes and was similar to that observed before LPS administration (0.18% ± 0.02 vs. 0.12% ± 0.03, $P = 0.0572$, Fig. 4). mRNA expression of *Il1B* and its receptors *Il1r1* and *Il1r2* in the CP collected 6 h after LPS administration were similar ($P > 0.05$) in both sham- and melatonin-implanted group (Fig. 5). Within 35 amplification cycles mRNA expression of *Mmp3* and *Mmp9* was not detected.

Discussion

Our results show that in ovariectomized and E2 treated ewes LPS in a dose of 400 ng/mL increased IL-1β concentration in blood plasma and the CSF early (2.5 h)

after LPS administration. In our previous study LH secretion as well as the GnRH release was observed to be suppressed 3 h after intravenous LPS administration in ewes [23]. The peripheral administration of LPS increases pro-inflammatory cytokine level IL-1β, TNFα and IL-6 in blood plasma in many animals, however, there are species differences in a dose of LPS necessary to trigger the response [24–26]. Ewes are much more sensitive to LPS in comparison to mice and rats in which the range of the applied LPS doses is from 5 μg/kg to 5 mg/kg. In our study the mean concentration of IL-1β reach the level of 159.3 ± 53.1 pg/mL in plasma and 129.8 ± 54.2 pg/mL in CSF in control and 197.8 ± 42.8 pg/mL in plasma and 139.6 ± 51.5 pg/mL in CSF of melatonin-implanted ewes. Individual plasma IL-1β levels were differentiated in both groups but means were similar to these observed in pigs 3 h after continuous LPS infusion in a dose of 250 ng/kg/h [26]. We did not find any significant differences in IL-1β mean concentration between groups (sham- and melatonin-implanted) as well as between compartments (blood plasma and the CSF). This indicates the lack of melatonin slow-release implants effect on IL-1β concentration in both plasma and CSF early (2.5 h) after LPS administration. In our study, daytime plasma melatonin concentration in melatonin-implanted ewes were similar to this observed in our previous study [27] and reported by Skinner and Malpaux [28] for plasma collected at night. We did not observed any correlation between plasma and CSF concentration of IL-1β in both groups, what may suggests that transport of IL-1β from blood to the CSF is not only the source of IL-1β in the CSF early after LPS administration, and that CSF IL-1β originates from other sources. These findings are in line with results described by Qann and colleagues [29], who observed that in rats subseptic doses of LPS (0.01–10 μg/kg) that are in range of the dose used in ewes induced *Il1B* mRNA expression only in the CP, the circumventricular organs and meninges. Moreover, secretion of cytokines to the brain by activated cells of the BBB has been described as an additional, to saturable transport system, pathway of cytokine access to the brain [10].

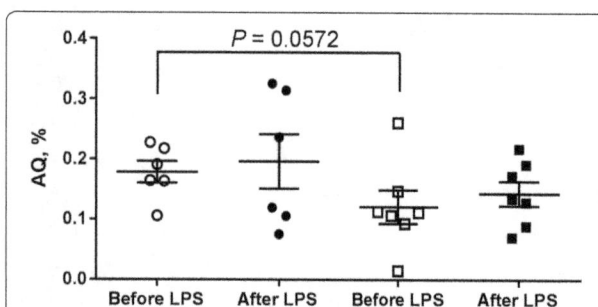

Fig. 4 Mean (±SEM) cerebrospinal fluid albumin quotient (QAlb) before and after lipopolysaccharide (LPS, 400 ng/kg) administration in sham- (circles) and melatonin-implanted (squares) adult ewes

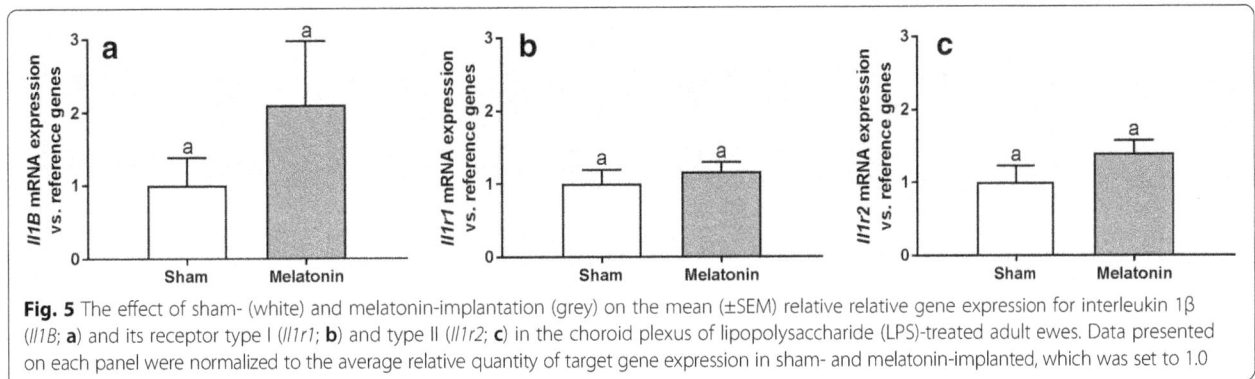

Fig. 5 The effect of sham- (white) and melatonin-implantation (grey) on the mean (±SEM) relative relative gene expression for interleukin 1β (*Il1B*; **a**) and its receptor type I (*Il1r1*; **b**) and type II (*Il1r2*; **c**) in the choroid plexus of lipopolysaccharide (LPS)-treated adult ewes. Data presented on each panel were normalized to the average relative quantity of target gene expression in sham- and melatonin-implanted, which was set to 1.0

In addition to evaluating the effect of LPS alone and with melatonin from slow-release implants on the concentration of IL-1β in blood plasma and the CSF, the second aim of this study was to investigate the BBB permeability in ewes early after LPS administration. The integrity of brain barriers was estimated by the ratio of albumin concentrations in CSF and blood plasma. The QAlb, calculated in our study before LPS administration was similar to that obtained by Chen et al. [30] for young (1–2 yr) and middle aged (3–5 yr) ewes. Moreover, the QAlb on the level of 0.18% ± 0.02 in sham-implanted and 0.12% ± 0.03 in melatonin-implanted ewes observed in ewes before LPS administration, despite the difference at the very edge of significance ($P = 0.0572$), seem to confirm previous observation related with melatonin as possible modulator of molecule passage throughout the brain barriers in sheep. These include: 1) higher steroids access to the CSF during long than short days in female sheep [18, 31], 2) higher passage of leptin from the periphery to the CSF in rams during long days than short days [32], 3) higher expression of tight junction proteins in the CP in ewes during short than long days [33] and 4) photoperiod-dependent change in CSF proteome composition in ewes [34]. The lack of differences in QAlb before and after LPS administration, found in our studies, indicates that despite the high level of IL-1β in blood and the CSF the integrity of BBB and BCSFB was not damaged 6 h after LPS administration. Indeed, in the CP collected 6 h after LPS administration the expression of mRNA for *Mmp*3 and 9, that are responsible for degradation of extracellular matrix and therefore for increase of BCSFB permeability was very weak.

Conclusions

In summary we demonstrated that intravenous LPS administration in ewes induces rapid increase of IL-1β in blood plasma and the CSF that is not modulated by melatonin from slow-release implants. The lack of changes in the brain barrier permeability early

after LPS administration, at the time when LPS-dependent suppression of GnRH secretion was observed in ewe, indicates that LPS acts mainly at the BBB and BCSFB which are a place for elaboration of signal molecules that communicate peripheral immune status to the brain.

Abbreviations
Actb: β-actin; BBB: Blood-brain barrier; BCSFB: Brain-cerebrospinal fluid barrier; bp: Blood plasma; CP: Choroid plexus; CSF: Cerebrospinal fluid; E2: Oestradiol; Gapdh: Glycer-aldehyde-3-phosphate dehydro-genase; GnRH: Gonadotropin releasing hormone; Hdac1: Histone deacetylase 1; IL: Interleukin; IL-1R1: Interleukin 1 receptor type I; IL-1R2: Interleukin 1 receptor type II; LH: Luteinizing hormone; LPS: Lipopolysaccharide; Mmp: Matrix metalloproteinase; MT: Melatonin receptor; QAlb: Albumin quotient; RIA: Radioimmunoassay; SDS: Sodium dodecyl sulfate; SEM: Standard error of the mean; SSA: Sheep serum albumin; TBST: Tris buffer saline with Tween; TNF: Tumor necrosis factor

Acknowledgements
We would like to thank Dr. Przemysław Gilun and Joanna Winnicka for expert technical assistance.
Marta Kowalewska participated as a part of PhD thesis.

Funding
This work was supported by a project funded by the National Science Centre allocated on the basis of decision-DEC 2011/03/B/NZ9/00118. Participation of Dr. T. Misztal, Dr. M. Jalynski and Dr. K. Zabek was supported by Ministry of Science and High Education.

Authors' contributions
SJ, HAP and JM conceived and designed the experiment; SJ, KM, SA, MT, JM, HAP and ZK performed animal experiment; KM, SA, MA and ZK performed laboratory analysis; SJ, KM and MA analyzed data; SJ, MK and MT prepared the manuscript. All authors read and approved the final manuscript.

Competing interests
The authors declare that they have no competing interests.

Author details
[1]Institute of Animal Reproduction and Food Research, Polish Academy of Sciences, Olsztyn, Poland. [2]The Kielanowski Institute of Animal Physiology and Nutrition, Polish Academy of Sciences, Jablonna n/Warsaw, Olsztyn,

Poland. [3]Veterinary Medicine Faculty, University of Warmia and Mazury, Olsztyn, Poland. [4]Department of Sheep and Goat Breeding, Animal Bioengineering Faculty, University of Warmia and Mazury in Olsztyn, Olsztyn, Poland.

References

1. Tomaszewska-Zaremba D, Herman A. The role of immunological system in the regulation of gonadoliberin and gonadotropin secretion. Reprod Biol. 2009;9:11–23.

2. Tremellen K, Pearce K, editors. Nutrition, fertility, and human reproductive function. Boca Raton: CRC Press; 2015.

3. Watanobe H, HayakawaY. Hypothalamic interleukin-1β and tumor necrosis factor-α, but not interleukin-6, mediate the endotoxin-induced suppression of the reproductive axis in rats. Endocrinology. 2003;144:4868–75.

4. Herman AP, Misztal T, Herman A, Tomaszewska-Zaremba D. Expression of interleukin (IL)-1b and IL-1 receptors genes in the hypothalamus of anoestrous ewes after lipopolysaccharide treatment. Reprod Domest Anim. 2010;45:e426–33.

5. Skipor J, Thiery JC. The choroid plexus-cerebrospinal fluid system: under valuated pathway of neuroendocrine signaling into the brain. Acta Neurobiol Exp. 2008;68:414–28.

6. Vitkovic L, Konsman JP, Bockaert J, Dantzer R, Homburger V, Jacque C. Cytokine signals propagate through the brain. Mol Psychiatry. 2000;5:604–15.

7. Blamire AM, Anthony DC, Rajagopalan B, Sibson NR, Perry VH, Styles P. Interleukin-1beta -induced changes in blood-brain barrier permeability, apparent diffusion coefficient, and cerebral blood volume in the rat brain: a magnetic resonance study. J Neurosci. 2000;20:8153–9.

8. Strazielle N, Khuth ST, Murat A, Chalon A, Giraudon P, Belin MF, et al. Pro-inflammatory cytokines modulate matrix metalloproteinase secretion and organic anion transport at the blood-cerebrospinal fluid barrier. J Neuropathol Exp Neurol. 2003;62:1254–64.

9. Alluri H, Wilson RL, Shaji CA, Wiggins-Dohlvik K, Patel S, Liu Y, et al. Melatonin preserves blood-brain barrier integrity and permeability via matrix metalloproteinase-9 inhibition. PLoS One. 2016; https://doi.org/10.1371/journal.pone.0154427.

10. Banks WA, Erickson MA. The blood-brain barrier and immune function and dysfunction. Neurobiol Dis. 2010;37:26–32.

11. Skinner RA, Gibson RM, Rothwell HJ, Pinteaux E, Penny JI. Transport of interleukin-1 across cerebromicrovascular endothelial cells. Br J Pharmacol. 2009;156:1115–23.

12. Colotta F, Re F, Muzio M, Bertini R, Polentarutti N, Sironi M, et al. Interleukin-1 type II receptor: a decoy target for IL-1 that is regulated by IL-4. Science. 1993;261:472–5.

13. Sims JE, Gayle MA, Slack JL, Alderson MR, Bird TA, Giri JG, et al. Interleukin 1 signaling occurs exclusively via the type I receptor. Proc Natl Acad Sci U S A. 1993;90:6155–9.

14. Daneman R, Zhou L, Agalliu D, Cahoy JD, Kaushal A, Barres BA. The mouse blood-brain barrier transcriptome: a new resource for understanding the development and function of brain endothelial cells. PLoS One 2010;5: e13741.

15. Liddelow SA, Temple S, Møllgård K, Gehwolf R, Wagner A, Bauer H, et al. Molecular characterisation of transport mechanisms at the developing mouse blood-CSF interface: a transcriptome approach. PLoS One. 2012;7:1–18.

16. Cogé SP, Guenin I, Fery I, Migaud M, Devavry S, Slugocki C, et al. The end of a myth: cloning and characterization of the ovine melatonin MT(2) receptor. Br J Pharmacol. 2009;158:1248–62.

17. Chemineau P, Malpaux B, Delgadillo JA, Guerin Y, Ravault JP, Thimonier J, et al. Control of sheep and goat reproduction: use of light and melatonin. Anim Reprod Sci. 1992;30:157–84.

18. Thiéry JC, Lomet D, Schumacher M, Liere P, Tricoire H, Locatelli A, et al. Concentrations of estradiol in ewe cerebrospinal fluid are modulated by photoperiod through pineal-dependent mechanisms. J Pineal Res. 2006;41: 306–12.

19. Skipor J, Misztal T, Kaczmarek MM. Independent changes of thyroid hormones in blood plasma and cerebrospinal fluid after melatonin treatment in ewes. Theriogenology. 2010;74:236–45.

20. Herman AP, Herman A, Haziak K, Tomaszewska-Zaremba D. Immune stress up regulates TLR4 and Tollip gene expression in the hypothalamus of ewes. J Anim Feed Sci. 2013;22:13–8.

21. Misztal T, Romanowicz K, Barcikowski B. Seasonal changes of melatonin secretion in relation to the reproductive cycle in sheep. J Anim Feed Sci. 1996;5:35–48.

22. Zhao S, Fernald RD. Comprehensive algorithm for quantitative real-time polymerase chain reaction. J Comput Biol. 2005;12:1047–64.

23. Herman AP, Krawczyńska A, Bochenek J, Haziak K, Romanowicz K, Misztal T, et al. The effect of rivastigmine on the LPS-induced suppression of GnRH/LH secretion during the follicular phase of the estrous cycle in ewes. Anim Reprod Sci. 2013;138:203–12.

24. Creasey AA, Stevens P, Kenney J, Allison AC, Warren K, Catlett R, et al. Endotoxin and cytokine profile in plasma of baboons challenged with lethal and sublethal Escherichia Coli. Circ Shock. 1991;33:84–91.

25. Givalois L, Dornand J, Mekaouche M, Solier MD, Bristow AF, Ixart G, et al. Temporal cascade of plasma level surges in ACTH, corticosterone, and cytokines in endotoxin-challenged rats. Am J Phys. 1994;267:R164–R70.

26. Aberg AM, Abrahamsson P, Johansson G, Haney M, Winsö O, Larsson JE. Does carbon monoxide treatment alter cytokine levels after endotoxin infusion in pigs? A randomized controlled study J Inflamm. 2008;5:13.

27. Kowalewska M, Szczepkowska A, Herman AP, Pellicer-Rubio MT, Jalynski M, Skipor J. Melatonin from slow-release implants did not influence the gene expression of the lipopolysaccharide receptor complex in the choroid plexus of seasonally anoestrous adult ewes subjected or not to a systemic inflammatory stimulus. Small Rumi Res. 2017;147:1–7.

28. Skinner DC, Malpaux B. High melatonin concentrations in third ventricular cerebrospinal fluid are not due to Galen vein blood recirculating through the choroid plexus. Endocrinology. 1999;140:4399–405.

29. Quan N, Stern EL, Whiteside M, Herkenham M. Induction of pro-inflammatory cytokine mRNAs in the brain after peripheral injection of subseptic doses of lipopolysaccharide in the rat. J Neuroimmunol. 1999;93: 72–80.

30. Chen Rl, Chen CPC, Preston JE. Elevation of CSF albumin in old sheep: relations to CSF turnover and albumin extraction at blood-CSF barrier. J Neurochem. 2010;113:1230–9.

31. Thiéry JC, Robel P, Canepa S, Delaleu B, Gayrard V, Picard-Hagen N, et al. Passage of progesterone into the brain changes with photoperiod in the ewe. Eur J Neurosci. 2003;18:895–901.

32. Adam CL, Findlay PA, Miller DW. Blood-brain leptin transport and appetite and reproductive neuroendocrine responses to intracerebroventricular leptin injection in sheep: influence of photoperiod. Endocrinology. 2006;147: 4589–98.

33. Lagaraine C, Skipor J, Szczepkowska A, Dufourny L, Thiéry JC. Tight junction proteins vary in the choroid plexus of ewes according to photoperiod. Brain Res. 2011;1393:44–51.

34. Teixeira-Gomes AP, Harichaux G, Gennetay D, Skipor J, Thiery JC, Labas V. Al. Photoperiod affects the cerebrospinal fluid proteome: a comparison between short day– and long day–treated ewes. Domest Anim Endocrinol. 2015;53:1–8.

35. Herman AP, Krawczyńska A, Bochenek J, Antushevich H, Herman A, Tomaszewska-Zaremba D. Peripheral injection of SB203580 inhibits the inflammatory-dependent synthesis of proinflammatory cytokines in the hypothalamus. BioMed Res Int. 2014;doi: https://doi.org/10.1155/2014/475152.

36. Herman AP, Bochenek J, Król K, Krawczyńska A, Antushevich H, Pawlina B, et al. Central interleukin-1β suppresses the nocturnal secretion of melatonin Mediat Inflamm. 2016. https://doi.org/10.1155/2016/2589483.

15

Energetic-protein supplementation in the last 60 days of gestation improves performance of beef cows grazing tropical pastures

Aline Gomes da Silva[1,2*] [iD], Mário Fonseca Paulino[1], Edenio Detmann[1], Henrique Jorge Fernandes[3], Lincoln da Silva Amorim[1,4], Román Enrique Maza Ortega[1], Victor Valério de Carvalho[1], Josilaine Aparecida da Costa Lima[1], Felipe Henrique de Moura[1], Mariana Benevides Monteiro[5] and Jéssika Almeida Bitencourt[1]

Abstract

Background: Nutrition is one of the most important factors that affect animal performance, and it therefore also impacts on financial results in beef systems. In this way, finding the best strategy for feeding supplements is of paramount importance. Aiming to evaluate the effect of supplement feeding strategies for beef cows in the last third of gestation, two experiments were conducted. In Experiment 1, 35 pregnant Nellore cows were assigned to a completely randomized design with four treatments: control, which received no supplement; supplementation for the last 30 d of gestation (30-d; 3.0 kg/d); supplementation for the last 60 d of gestation (60-d; 1.5 kg/d); or supplementation for the last 90 d of gestation (90-d; 1.0 kg/d). All supplemented treatments received the same total amount of supplement throughout the experiment: 90 kg (20% of crude protein). A second experiment (Experiment 2) was delineated to evaluate the effects of the amounts offered in Experiment 1 on intake and metabolism. Four multiparous pregnant Nellore cows were assigned to a 4 × 4 Latin square design, with periods of 15 d each.

Results: There was a linear effect of the number of days of supplementation on calving body weight (BW; $P < 0.05$) and a quadratic effect on BW change from parturition to d 31 post-calving ($P < 0.05$), with cows on the 60-d strategy losing less BW post-calving. No difference was found in offspring birth BW ($P > 0.10$). A significant linear effect on interval from parturition to conception ($P < 0.05$) was observed, with the highest calving to conception interval being observed in the 90-d strategy. The level of supplementation did not affect forage intake or neutral detergent fiber digestibility ($P > 0.10$). Nitrogen excreted through urine tended to increase linearly with the level of supplementation ($P < 0.10$).

Conclusion: Providing 1.5 kg of supplement during the last 60 d of gestation improves cow performance after calving, reducing the magnitude of BW lost, and reduces the number of days from calving to re-conception in the following breeding season compared to the usually recommended period of supplementation of 90 d pre-partum.

Keywords: Flushing, Nutrition, Parturition, Reproduction

* Correspondence: alinegomesdasilva@rocketmail.com; aline.g.silva@ufms.br
[1]Universidade Federal de Viçosa, Viçosa, Minas Gerais 36570–000, Brazil
[2]Present Address: Universidade Federal de Mato Grosso do Sul, Campo Grande, Mato Grosso do Sul 79074–460, Brazil
Full list of author information is available at the end of the article

Background

It is estimated that 50% of the cows in extensive beef systems do not receive adequate nutritional management, and this is a main reason for low fertility rates in tropical herds [1].

Among the factors affecting the reproductive performance of beef cattle, nutrition has perhaps the highest impact [2]. Supplementation to grazing animals is a practice that can be adopted under tropical conditions to increase animal performance. Because the last third of gestation usually coincides with the dry season and, consequently, with low quantity and quality of forage, producers in tropical conditions, as in Brazil, are usually oriented to supplement pregnant cows for the last 90 d of gestation.

The objective with this experiment was to study whether this period of supplementation could be reduced to avoid extra labor and, consequently, reduce feeding costs. Therefore, we conducted a study to evaluate the effects of different supplementation strategies for pregnant beef cows in the last third of gestation, receiving supplementation for 90, 60 or 30 d pre-partum.

Methods

The two experiments were conducted at the Department of Animal Science – Universidade Federal de Viçosa, Brazil, from July to December 2012. All animal care and handling procedures were ethically standardized and approved by the Animal Care and Use Committee of the Universidade Federal de Viçosa, Brazil.

Experiment 1 – Performance

Thirty-five multiparous, at an average age of five years old, single pregnant Nellore cows with 491.88 ± 55 kg of body weight (BW), a body condition score (BCS) of 4.7 ± 0.58 and 200 ± 15 d of gestation were used. Cows were housed in an experimental area of *Brachiaria decumbens* divided into four paddocks of 5.0 ha each, and had unlimited access to water, feeders and mineral salt (8.7% calcium, 9.0% phosphorous, 18.7% sodium, 9.0% sulfur, 2,400 mg/kg of zinc, 800 mg/kg of copper, 1,600 mg/kg of manganese, 40.0 mg/kg of iodine, 8.00 mg/kg of cobalt and 8.16 mg/kg of selenium). To avoid any effects of paddock on the responses, treatments were rotated among paddocks every 10 d.

Four strategies were evaluated: 30-d – cows received 3.0 kg/d of supplement beginning 30 d prior to calving; 60-d – cows received 1.5 kg/d of supplement beginning 60 d prior to calving; 90-d – cows received 1.0 kg/d of supplement beginning 90 d prior to calving; and control – no concentrate supplement was fed. All supplemented treatments received the same total amount of supplement throughout the experiment, 90 kg/d per head.

Animals were assigned to a completely randomized design with four treatments. There were nine cows each in the control group, 60-d and 90-d supplemented groups, and eight cows in the 30-d supplemented group. Supplement was fed in a collective feeder, which is experimental handling closer to what is normally observed in beef production systems due to cattle gregarious behavior. As the evaluations in Experiment 1 were focused on individual performance, and these measurements were collected individually, the animal was considered the experimental unit, as recommended by Detmann et al. [3].

Supplement was composed of corn, sorghum and soybean meal, and formulated to contain approximately 20% crude protein (CP; Table 1). The level of supplementation adopted for the 90-d treatment corresponds to the daily supply of 1 kg of supplement to approximately 23% of the requirements of CP and 17% of the daily energy requirements of a pregnant Nellore cow, with average BW of 465 kg and expected calf birth BW of 32 kg [4]. The other supplemented treatments were based on the supply of the same total amount of energy and CP, but offered in a reduced period of feeding.

Cow BW was recorded at the beginning of the experiment, approximately 90 d prior to the day of calving, at the week before the expected date of parturition (calving BW), and 31 d after parturition. Cow BCS was recorded at the beginning of experiment, prior to calving and in the first day of the breeding season on a scale ranging from 1 to 9, as recommended by NRC [5] by two experienced technicians. Calf BW was also recorded at birth. Shrunk BW (SBW) was calculated using adjustments proposed by Gionbelli et al. [6] for Nellore cows as follows:

$$SBW = 0.8084 \times BW^{1.0303}$$

After calving, cows were managed as a single herd until after pregnancy was diagnosed. During this period,

Table 1 Ingredients and chemical composition of supplements

Item[a]	Supplement
Ingredients, % as-fed basis	
Corn	33
Sorghum	33
Soybean meal	34
Chemical composition, g/kg	
OM	971
CP	208
apNDF	164

[a]OM – organic matter; CP – crude protein; apNDF – neutral detergent fiber corrected for ash and protein residue

all cows grazed the same pasture and received mineral supplement ad libitum.

Twenty-one and 31 d after calving, blood samples were taken from the jugular vein using vacuum tubes with clot accelerator and gel for serum separation (BD Vacuntainer® SST II Plus, São Paulo, Brazil). Immediately after collection, the samples were centrifuged at 3,600×g for 20 min. Then, the serum was frozen at −20 °C and subsequently analyzed for progesterone by the chemiluminescent method using Access Progesterone Reagent Kit (Ref. Number 33550, Beckman Coulter®, Brea, USA) in the Access 2 Immunoassay System (Beckman Coulter Inc., Brea, USA).

Pasture chemical composition (Table 2) was assessed by hand-plucked samples every two weeks. In the middle of every experimental month, a second pasture sample was also collected to estimate forage potentially digestible dry matter (pdDM) as proposed by Detmann et al. [5]. Four subsamples were randomly collected in each plot by cutting it close to the ground using a metal square (0.5 m × 0.5 m). Samples were weighed and oven dried at 60 °C for 72 h. After that, samples were mill grinded to pass through a 2-mm screen for indigestible neutral detergent fiber (iNDF) analysis [7]. A sub portion of 20 g of each sample was grinded to pass through a 1-mm screen for analyses of dry matter (DM), ash, crude protein (CP) and neutral detergent fiber (NDF).

Samples of forage were analyzed following procedures described by Detmann et al. [7] for DM (index INCT-CA G-003/1), CP (index INCT-CA N-001/1), ash (index INCT-CA M-001/1), NDF corrected for contaminant ash and protein (apNDF; index INCT-CA F-002/1, INCT-CA M-002/1, and INCT-CA N-004/1). The iNDF was evaluated using F57 filter bags (Ankom®) by a 288-h in situ incubation procedure [7].

The potentially digestible dry matter (pdDM) was estimated using the second sample collected in each month, as described previously, using the following eq. [5]:

$$pdDM\,(\%;\,dry\ matter\ basis) = 0.98 \times (100 - NDF) + (NDF - iNDF)$$

where: 0.98 is the true digestibility coefficient of intracellular content; NDF is forage content of neutral detergent fiber; and iNDF is forage content of indigestible neutral detergent fiber.

In the breeding season, cows were synchronized using the following protocol: on d 0, an intravaginal device of progesterone release (Tecnopec Primer®, São Paulo, Brazil) was inserted, and an i.m. injection of 2.0 mg of estradiol benzoate (Tecnopec RIC-BE®, São Paulo, Brazil) was performed. On day seven, the intravaginal device was removed and cows received a 2-mL injection of cloprosterol sodium (MSD Saúde Animal Ciosin®, São Paulo, Brazil). Finally, on day eight, cows received 0.5 mL of estradiol cypionate i.m. (Zoetis-Pfizer E.C.P.®, Campinas, Brazil). Fixed time artificial insemination (FTAI) was performed 46–52 h following intravaginal device removal (day nine). Semen from five Nellore sires were randomly assigned to each cow. The protocol was repeated once more in a way that cows that did not conceive were inseminated again 32 d after the first FTAI. Pregnancy diagnosis was determined via trans-rectal ultrasonography 30 d after each FTAI. The number of days from parturition to re-conception was calculated for each cow.

Response variables were analyzed using GLIMMIX in SAS 9.4. Initial BW of cows was used as a covariate for data analysis. Treatments were compared using orthogonal contrasts, contrasts were constructed in order to evaluate the effects of supplementation, and the linear and quadratic effects of days receiving supplementation (30, 60 and 90 d). For the variables which didn't present supplementation effect but a linear or quadratic effect was significant, a Dunnett's test was performed to identify whether a supplemented treatment differed from the control. Significance was considered at $P < 0.05$.

Experiment 2 – Intake and metabolism

In order to evaluate the effects of the amounts of supplement offered daily in Experiment 1 on intake and metabolism, a second experiment was conducted simultaneously. Four multiparous, five years old, single pregnant Nellore cows with 488 ± 22 kg of BW, BCS of 4.7 ± 0.3, and 210 ± 10 d of gestation were assigned to a 4 × 4 Latin square design, with experimental periods of 15 d each.

Cows were individually housed in an experimental area of *Brachiaria decumbens* divided into four paddocks of

Table 2 Potentially digestible forage mass and chemical composition of forage in Experiment 1

Item[a,b]	Experimental month		
	1	2	3
pdDM, kg/hm²	4820	4050	3410
OM, g/kg	912	913	922
CP, g/kg	77.4	66.6	65.2
NDIN, % of total N	6.94	7.54	9.18
apNDF, g/kg	618	628	679
iNDF, g/kg	225	251	272

[a]pdDM – potentially digestible forage dry matter; OM – organic matter; CP – crude protein; NDIN – neutral detergent insoluble N; apNDF – neutral detergent fiber corrected for ash and protein residue; iNDF – indigestible neutral detergent fiber
[b]pdDM was estimated for forage sampled in the area delimited by a metal square 0.5 m × 0.5 m ; chemical composition was evaluated in the hand-plucked forage sample

0.34 ha each, with free access to water, mineral salt and feeders. Experiment 2 started 15 d later than Experiment 1.

The experimental treatments evaluated were: 3.0 kg – cows received 3.0 kg of supplement daily; 1.5 kg – cows received 1.5 kg of supplement daily; 1.0 kg – cows received 1.0 kg of supplement daily; and 0.0 kg – no concentrate supplement was fed. The supplement used was the same used in Experiment 1 (Table 1).

Pasture chemical composition (Table 3) was assessed by hand-plucked samples, collected on the eighth day of each experimental period. On the same day, a second pasture sample was also collected to estimate forage pdDM. Samples were collected and processed as described in Experiment 1.

After six days of adaptation to treatments in each period, a nine-day intake trial was carried out. To estimate fecal excretion, chromium oxide (Cr_2O_3) was used as an external marker in the amount of 15 g per animal. The chromium oxide was packed in paper cartridges and delivered via the esophagus with a metal probe once daily, at 10:00. To estimate DM intake, iNDF was used as an internal marker. Once the intake trial had started, six days were allowed for stabilization of Cr_2O_3 excretion; and, after that, fecal samples were collected at 1500 h on the seventh day, at 1100 h on the eighth day, and at 0700 h on the ninth day of the intake trial (13th, 14th and 15th d of each experimental period, respectively).

Feces samples were collected directly from the rectum of cows, at amounts of approximately 200 g, dried (60 °C/72 h) and mill grinded as described for forage samples. Ground samples were proportionally combined to a pooled three-day sample per animal per period.

Samples of forage, feces and supplement were analyzed for DM, CP, ash, apNDF, and iNDF following procedures

Table 3 Potentially digestible forage mass and chemical composition of forage in Experiment 2

Item[a,b]	Experimental period			
	1	2	3	4
pdDM, kg/hm^2	4320	3740	3245	1640
OM, g/kg	925	917	922	938
CP, g/kg	68.1	75.7	59.8	52.3
NDIN, % of total N	10.4	10.4	12.3	10.9
apNDF, g/kg	652	644	712	719
iNDF, g/kg	257	239	312	345

[a]pdDM – potentially digestible forage dry matter; OM – organic matter; CP – crude protein; NDIN – neutral detergent insoluble N; apNDF – neutral detergent fiber corrected for ash and protein residue; iNDF – indigestible neutral detergent fiber
[b]pdDM was estimated for forage sampled in the area delimited by a metal square 0.5 m × 0.5 m; chemical composition was evaluated in the hand-plucked forage sample

previously described for Experiment 1. Fecal samples were also analyzed for levels of chromium by atomic absorption spectrophotometry (index INCT-CA M-005/1) and titanium dioxide by colorimetry (index INCT-CA M-007/1), as recommended by Detmann et al. [8].

Fecal excretion (FE) was estimated by the ratio of chromium oxide and its concentration in the feces. Dry matter intake (DMI) was estimated by using iNDF as an internal marker and calculated by the following equation:

$$DMI~(kg/d) = [(FE \times iNDF~feces - iNDF~supplement)$$
$$\div iNDF~forage] + SI$$

where FE is the fecal excretion (kg/d); iNDF feces is the concentration of iNDF in the feces (kg/kg); iNDF supplement is the iNDF in the supplement (kg/d); iNDF forage is the concentration of iNDF in forage (kg/kg) and SI is DM supplement intake.

At the 15th d of each period, two blood samples were collected immediately before and 4 h after supplementation, to estimate insulin levels pre- and post-supplementation, respectively. Blood was collected in tubes with clot activator and gel for serum separation (BD Vacuntainer® SST II Plus, São Paulo, Brazil), centrifuged at 3,600×g for 20 min and serum was immediately frozen at −20 °C in duplicate until further analysis. The same blood collected 4 h after supplementation was used for quantification of serum urea concentrations.

Insulin was analyzed by the chemiluminescent method using Access Ultrasensitive Insulin Reagent (Ref. Number 33410, Beckman Coulter®, Brea, USA) in the Access 2 Immunoassay System (Beckman Coulter Inc., Brea, USA). Urea was quantified by an enzymatic-colorimetric method using reagents provided by commercial kits (Ref. Number K056, Bioclin® Quibasa, Belo Horizonte, Brazil) in an automatic biochemistry analyzer (Mindray BS200E, Shenzhen, China). Serum urea N (SUN) was estimated as 46.67% of total blood urea.

Spot urine sampling at the 15th day of each experimental period (collected immediately before the 4 h after supplementation blood sampling) was used to assess the excretion of urinary nitrogenous compounds [9]. Urine volume was estimated using creatinine concentration as a marker and assuming a daily creatinine excretion (mg/d) of $34.5 \times SBW^{0.9491}$ [10]. Microbial N synthesis was estimated by using the technique of the purine derivatives in urine. Allantoin was estimated by colorimetry [11]. The urinary concentrations of creatinine and uric acid were obtained by colorimetric and enzymatic-colorimetric methods, respectively. The analyses of creatinine and uric acid were performed in an automatic biochemistry analyzer (Mindray BS200E, Shenzhen, China) using commercial kits (Ref. Number K067 for creatinine and K139 for uric acid, Bioclin® Quibasa, Belo Horizonte, Brazil).

Excretion of the purine derivatives in urine was calculated by the sum of the allantoin and uric acid excretions, which were obtained by the product between their concentrations in urine by the daily urinary volume. Absorbed purines were calculated from the excretion of purine derivatives [12], as follows:

$$Y = \frac{x - 0.301 \times BW^{0.75}}{0.8}$$

where Y = absorbed purines (mmol/d), x = excretion of purine derivatives (mmol/d), 0.8 = recovered absorbed purines. The $0.301 \times BW^{0.75}$ value = endogenous excretion of purine derivates.

Ruminal synthesis of nitrogen compounds was calculated as a function of the absorbed purines [12]:

$$Z = \frac{70 \times Y}{0.93 \times 0.137 \times 1,000}$$

where Z = ruminal synthesis of nitrogen compounds (g/d), Y = absorbed purines (mmol/d), 70 = purine N content (mg/mol), 0.93 = purine digestibility and 0.137 = relation of purine N:total N of microorganisms.

Linear, quadratic and cubic effects of amount of supplement fed daily were analyzed using GLIMMIX in SAS 9.4. Animal and period were considered as random effects. Significance was assumed at P < 0.05.

Results
Experiment 1
There was a linear effect of number of days of supplementation on cow BW at calving (Table 4; P < 0.05). Body weight at calving ranged from 508 kg for cows in the 30-d strategy to 531 kg for cows in the 90-d strategy. Supplemented treatments did not differ from the control by Dunnett's test as well. Looking at the individual means of each treatment, can be observed that, actually, control cows had a BW at calving intermediate compared to the supplemented cows. It possibly reflects the small numbers of animals used in this study coupled with the variability of the variable in question. Birth BW of calves averaged 34.4 kg and was not different among supplementation strategies used for their mothers in the last third of gestation (P > 0.10).

Thirty-one days after calving, cows that received supplementation pre-calving tended to be heavier (480 kg, P < 0.10) compared to control cows (465 kg). Supplemented cows also lost less BW from parturition to 31 d post-calving (P < 0.05); within the supplemented treatments, there was a quadratic effect of the number of days of supplementation on post-partum BW change (P < 0.05): cows from the 60-d strategy lost less BW (−25 kg) compared to cows in the 30-d and 90-d strategies (−50 kg on average). Nevertheless, supplementation or level of supplementation did not affect BCS at any time (P > 0.10).

Table 4 Cow BW and BCS, calf BW, cow progesterone concentrations and reproductive performance

Item[a]	Treatment[b]				SEM	P-value[c]		
	30-d	60-d	90-d	Control		S	L	Q
Supplement fed, kg/d	3.00	1.50	1.00	–				
Cow BW, kg, cow BCS and calf BW, kg								
Initial BW	494	517	503	503	18.0	0.95	0.74	0.43
Initial BCS	4.66	4.65	4.54	4.87	0.20	0.28	0.65	0.85
Calving BW	508	515	531	522	6.79	0.56	0.02	0.55
Calving BCS	4.74	4.83	4.83	4.82	0.15	0.95	0.70	0.82
Calf birth BW	33.8	31.7	35.8	36.2	1.93	0.28	0.43	0.10
Cow BW 31 d after calving	468	490	482	465	7.73	0.09	0.21	0.15
BW change from parturition to day 31 post-calving	−46.7	−24.9	−53.5	−62.1	8.24	0.04	0.58	0.03
Breeding season BCS	5.00	5.11	4.83	5.03	0.21	0.84	0.57	0.43
Progesterone concentration, ng/dL								
21 d after calving	0.27*	0.21	0.12	0.14	0.06	0.34	0.05	0.82
31 d after calving	1.24	0.70	0.69	0.14	0.37	0.08	0.29	0.55
Cow reproductive performance, %								
Calving to conception, d	57.2	63.1	84.2*	64.1	5.85	0.53	0.01	0.30

[a]BW– body weight; BCS – body condition score; FTAI – fixed time artificial insemination
[b]Treatments: 30-d – cows received 3.0 kg of concentrate supplement beginning 30 d prior to calving; 60-d – cows received 1.5 kg of concentrate supplement beginning 60 d prior to calving; 90-d – cows received 1.0 kg of concentrate supplement beginning 90 d prior to calving; and control – no concentrate supplement was fed
[c]S – effect of supplementation, supplemented treatments compared to the control; L and Q – effects of linear and quadratic order of supplement delivery strategy (30, 60 or 90 d). * Means statistically different from the control by Dunnett's test

At 21 d post-calving, progesterone concentration linearly reduced with an increase in days of supplementation (P = 0.05), cows in the 30-d strategy had higher progesterone concentration compared to control cows. Ten days later, at 31 d post-calving, supplemented cows tended to have greater progesterone concentration compared to cows receiving no supplement (P < 0.10).

A significant linear effect on interval from parturition to conception (P < 0.05) was observed, with the highest calving to conception interval being observed for cows in the 90-d strategy.

Experiment 2

Level of supplementation linearly increased DM intake (kg/d and g/kg of BW), DM digested (kg/d and g/kg of BW), OM (kg/d and g/kg of BW) and OM digested (kg/d), and crude protein (g/d – Table 5; P < 0.05). There was also a cubic effect of the level of supplementation on intake of crude protein (g/d; P < 0.05). There was no

effect of the level of supplementation on intake (g/d and g/kg BW) of forage DM, forage OM, forage apNDF, apNDF and iNDF.

Level of supplementation increased the digestibility of OM and CP linearly (P < 0.05; Table 6). There was also a cubic effect of the level of supplementation on digestibility of OM (P < 0.05).

The microbial N produced (g/d), the efficiency of microbial N produced in relation to N ingested and the efficiency of microbial N produced in relation to OM digested were not different among levels of supplementation (P > 0.10; Table 7). Level of SUN was also similar among treatments (P > 0.10), but ureic nitrogen excreted (g/d, UUN) tended to increase linearly with level of supplementation (P < 0.10).

Pre-supplementation levels of insulin were not different according to the level of supplementation (P > 0.10; Table 8), but insulin levels 4 h after supplementation increased linearly with the amount of supplement fed (P < 0.05).

Table 5 Intake according to amount of supplement fed to cows in the last third of gestation

Item[a]	Treatment[b]				SEM	P-value[c]		
	3.0 kg	1.5 kg	1.0 kg	0.0 kg		L	Q	C
Intake per day, kg/d								
Forage DM	5.56	5.69	6.00	5.61	0.72	0.96	0.73	0.86
DM	8.03	6.92	6.82	5.61	0.72	0.05	0.94	0.15
DM digested	4.00	2.97	2.76	2.08	0.45	0.01	0.67	0.08
Forage OM	5.11	5.26	5.55	5.22	0.65	0.90	0.72	0.89
OM	7.51	6.46	6.35	5.22	0.65	0.05	0.95	0.15
OM digested	4.03	3.01	2.83	2.13	0.41	0.01	0.68	0.06
Forage apNDF	3.70	3.89	4.11	3.85	0.48	0.83	0.66	0.94
apNDF	4.11	4.09	4.25	3.85	0.48	0.72	0.71	0.68
iNDF	1.56	1.69	1.71	1.60	0.19	0.88	0.53	0.95
CP, g/d	882	613	556	359	58.8	<0.01	0.48	<0.01
Intake per kg BW, g/kg BW								
Forage DM	9.95	9.78	10.15	9.58	1.22	0.84	0.88	0.77
DM	14.35	11.92	11.53	9.58	1.23	0.03	0.85	0.12
DM digested	7.16	5.09	4.68	3.55	0.77	0.01	0.54	0.07
Forage OM	9.14	9.05	9.39	8.92	1.12	0.89	0.87	0.81
OM	13.42	11.13	10.74	8.92	1.13	0.03	0.84	0.12
OM digested	5.06	4.74	4.75	6.25	0.95	0.24	0.20	0.39
Forage apNDF	6.64	6.70	6.94	6.59	0.84	0.97	0.81	0.87
apNDF	7.36	7.05	7.17	6.59	0.84	0.54	0.88	0.61
iNDF	2.81	2.93	2.87	2.74	0.33	0.89	0.70	0.99

[a]DM – dry matter; OM – organic matter; apNDF – neutral detergent fiber corrected for ash and protein residue; iNDF – indigestible neutral detergent fiber; CP – crude protein
[b]Treatments: 3.0 kg – cows received 3.0 kg of concentrate daily; 1.5 kg – cows received 1.5 kg of concentrate daily; 1.0 kg – cows received 1.0 kg of concentrate daily; and 0.0 kg – no concentrate supplement was fed
[c]L, Q and C – effects of linear, quadratic and cubic order of level of supplementation

Table 6 Coefficients of digestibility (%) according to amount of supplement fed

Item[a]	Treatment[b]				SEM	P-value[c]		
	3.0 kg	1.5 kg	1.0 kg	0.0 kg		L	Q	C
OM	53.87	45.94	44.44	39.57	3.30	<0.01	0.42	0.01
apNDF	52.39	49.52	50.22	48.77	2.55	0.18	0.69	0.25
CP	48.12	39.14	36.54	17.70	8.98	0.03	0.53	0.11

[a]DM – dry matter; OM – organic matter; apNDF – neutral detergent fiber corrected for ash and protein residue; CP – crude protein
[b]Treatments: 3.0 kg – cows received 3.0 kg of concentrate daily; 1.5 kg – cows received 1.5 kg of concentrate daily; 1.0 kg – cows received 1.0 kg of concentrate daily; and 0.0 kg – no concentrate supplement was fed
[c]L, Q and C – effects of linear, quadratic and cubic order of level of supplementation

Discussion

Nutritional status at calving is the most important factor that influences the interval from parturition to conception in beef cows. Postpartum nutrient intake can modulate the duration of the postpartum anestrous interval; however, if thin cows gain great amounts of weight after calving, ovulation occurs later than for cows that calve in good body condition and maintain body weight [2].

Cabral et al. [13] supplemented pregnant cows grazing pastures in similar conditions to the present study, but with different amounts of supplement per day, and observed a quadratic pattern in performance, with cows receiving 1.0 kg of supplement daily gaining more BW. The total amount of supplement provided for the 1.0 kg treatment in Cabral's work (84 kg) was similar to the present study (90 kg).

Based on the magnitude of post-partum BW change, supplementing cows with 1.5 kg of supplement during the last 60 d prior to calving was the most efficient strategy to lessen the post-partum negative energy balance, since cows on this strategy lost about half (–25 kg) of the BW lost in the 30-d and 90-d strategies (–50 kg). The differences found in BW changes are related to supplementing proper amounts of nutrients at key times and the effects when this supplement is removed from cow diet and its impact on cow metabolism, in addition

to changes that naturally occur due to calving and lactation. Therefore, supplementing adequate amounts of nutrients at times of higher requirements (last 60 d of gestation) [4, 5] seems to metabolically prepare the cow for the post-calving period, not only due to accretion in body reserves, but also due to its effects on cow metabolism later on, when supplements are no longer provided. These results provide evidence that adopting a strategy to deliver the same amount of supplement 60 d prior to calving, instead of the usually recommended 90 d, may not only be economical, but also improve cow post-partum performance.

Previous studies in cattle during late gestation [14] have provided evidence that feeding systems during the last third of gestation can alter the subsequent birth weight of the progeny, suggesting that the maternal dietary energy source may affect fetal growth [15]. Although the maternal intake of protein has also been shown to be an important factor for fetal growth, Summers et al. [16] found no difference in calf birth BW according to the supplementation strategy applied to their mothers. Similarly, no difference was observed for calf birth BW in the present study, probably because throughout the experiment forage presented median quality (Table 2).

In tropical pastures, the use of energetic-protein supplements can impact forage intake in different ways, depending largely on forage quality, amount and quality of supplement provided [17–19]. Linear increase of DM, OM and CP intake with levels of supplementation observed in the present work was simply due to the increase in supplement intake, as no difference in forage DM intake was observed in the present study.

During the intake and metabolism experiment (Table 3), average forage CP content was adequate or slightly below the minimum required by ruminal microorganisms [17]. In this way, the level of supplementation did not affect apNDF digestibility or intake.

Level of supplementation linearly increased insulin levels post-supplementation. For animals in an exclusive

Table 7 Nitrogen utilization according to amount of supplement fed

Item[a]	Treatment[b]				SEM	P-value[c]		
	3.0 kg	1.5 kg	1.0 kg	0.0 kg		L	Q	C
Nmic, g/d	72.87	77.99	48.02	36.14	16.4	0.17	0.58	0.82
Nmic, g/g N ingested	0.747	0.788	0.605	0.533	0.20	0.45	0.76	0.93
Nmic, g/kg OMD	26.71	26.21	19.99	14.87	7.19	0.25	0.72	0.67
SUN, mg/dL	13.77	12.72	12.37	10.85	1.70	0.20	0.88	0.41
UUN, g/d	61.36	47.59	33.35	32.39	9.85	0.09	0.53	0.48

[a]Nmic – microbial N; OMD – organic matter digested; SUN – Serum urea nitrogen; UUN – Urine urea N
[b]Treatments: 3.0 kg – cows received 3.0 kg of concentrate daily; 1.5 kg – cows received 1.5 kg of concentrate daily; 1.0 kg – cows received 1.0 kg of concentrate daily; and 0.0 kg – no concentrate supplement was fed
[c]L, Q and C – effects of linear, quadratic and cubic order of level of supplementation

Table 8 Insulin levels (μIU/mL) according to amount of supplement fed

Item	Treatment[a]				SEM	P-value[b]		
	3.0 kg	1.5 kg	1.0 kg	0.0 kg		L	Q	C
Pre-supplementation	1.68	1.50	1.18	1.13	0.43	0.13	0.79	0.63
Post-supplementation	2.20	1.48	1.33	1.25	0.38	0.03	0.21	0.13

[a]Treatments: 3.0 kg – cows received 3.0 kg of concentrate daily; 1.5 kg – cows received 1.5 kg of concentrate daily; 1.0 kg – cows received 1.0 kg of concentrate daily; and 0.0 kg – no concentrate supplement was fed
[b]L, Q and C – effects of linear, quadratic and cubic order of level of supplementation

forage diet, the relative amount of propionate, a glyco-genic precursor, available for metabolism is low, and supplementation significantly increases the proportion of propionate produced in the rumen [20].

Hawkins et al. [21] have suggested that an increase in insulin, concomitant with a decrease in growth hormone (GH), is an important relationship to consider for evaluating the impact of nutrition on reproduction. Insulin is an important mediator of nutritional effects on follicular dynamics in cattle, and can stimulate the release of gonadotropin-releasing hormone (GnRH) from the hypothalamus. In the ovaries, insulin may also stimulate cell proliferation and steroidogenesis [22]. In accordance with these findings, in the present study, supplementing with higher amounts of supplement daily linearly increased progesterone concentrations after calving.

The interval from calving to conception greatly influences the profitability of beef production. Hence, in beef systems, it is recommended that the calving interval is no longer than 85 d in order to assure the cow will produce a calf per year. Cows that conceive early calve early, and have a better opportunity to start reproductive cycles in time to re-conceive in the next breeding season. The calving date also affects the value of offspring, demonstrating the importance of supplementation strategies to improve early re-conception.

Cushman et al. [23] and Funston et al. [24] reported that heifer calves born early tend to conceive early in their first breeding season and remain in the herd. The calving date can also impact male offspring performance; steer calves born earlier in the calving season have greater weaning BW, hot carcass weight and marbling scores [24]. In this way, increasing early calving by early conception may increase progeny value at weaning, enhance carcass value of the steers and increase heifer pregnancy rates in their first breeding season.

All treatments in the present study presented acceptable calving intervals, but cows receiving higher amounts of supplement per day for a reduced number of days had lower calving intervals (30-d and 60-d vs. 90-d).

Conclusions

Providing 1.5 kg of supplement during the last 60 d of gestation, instead of 90 d that are usually recommended, is a nutritional management strategy that can be adopted to improve cow performance, reducing the magnitude of BW lost after calving and reducing the number of days from calving to re-conception in the following breeding season, with no negative effect on forage intake or digestibility.

Abbreviations

apNDF: Neutral detergent fiber corrected for ash and protein residue; BCS: Body condition score; BW: Body weight; CP: Crude protein; DM: Dry matter; FTAI: Fixed time artificial insemination; iNDF: Indigestible neutral detergent fiber; NDF: Neutral detergent fiber; OM: Organic matter; pdDM: potentially digestible dry matter; SUN: Serum urea N; UUN: Urine urea N

Acknowledgements

The authors thank to Fapemig and CNPq for financial support and CNPq and Capes for the scholarships provided to the first author.

Funding

This research was supported by funding from Fapemig – Fundação de Amparo à Pesquisa de MG, CNPq – Conselho Nacional de Desenvolvimento Científico e Tecnológico and Capes – Coordenação de Aperfeiçoamento de Pessoal de Nível Superior.

Authors' contributions

AGS, conceived the study, carried out the experimental trial, performed the statistical analysis and wrote the manuscript. MFP, contributed to draft the manuscript, and coordinate the research group. ED and HJF, contributed to the statistical analysis and to draft the manuscript. LSA, REMO, contributed to designing the experiment and to draft the manuscript. VVC, JACL, FHM, MBM, JAB, carried out the experimental trial, performed the chemical analysis and contributed to draft the manuscript. All authors read and approved the final manuscript.

Competing interests

The authors declare that they have no competing interests.

Author details

[1]Universidade Federal de Viçosa, Viçosa, Minas Gerais 36570–000, Brazil. [2]Present Address: Universidade Federal de Mato Grosso do Sul, Campo Grande, Mato Grosso do Sul 79074–460, Brazil. [3]Universidade Estadual de Mato Grosso do Sul, Aquidauana, Mato Grosso do Sul 79200–000, Brazil. [4]Biotran - Biotecnologia e Treinamento em Reprodução Animal, Alfenas, Minas Gerais 37130–000, Brazil. [5]Universidade Federal do Acre, Rio Branco, Acre 69920–900, Brazil.

References

1. Madureira ED, Maturana Filho M, Lemes KM, Silva JCB, Santini T. Análise crítica de fatores que interferem na fertilidade de vacas zebuínas. In: Proc of 9th symposium of beef cattle production. Viçosa, Brazil: Universidade Federal de Viçosa; 2014. p. 367–400.
2. Wettemann RP, Lents CA, Ciccioli NH, White FJ, Rubio I. Nutritional- and pre-weaning-mediated anovulation in beef cows. J Anim Sci. 2002;81(Suppl 2):E48–59.
3. Detmann E, Gionbelli MP, Paulino MF, Valadares Filho SC, Rennó LN. Considerations on research methods applied to ruminants under grazing. Nutritime. 2016;13:4711–31.
4. Valadares Filho SC, Costa e-Silva LF, Gionbelli MP, Rotta PP, Marcondes MI, Chizzotti ML. BR-CORTE 3.0 - Nutrient requirements of zebu and crossbred cattle. 3rd Ed. Viçosa: Universidade Federal de Viçosa; 2016.
5. National Research Council – NRC. Nutrient requirements of beef cattle. 8th ed. Washington: Academic Press; 2016.
6. Gionbelli MP, Duarte MS, Valadares Filho SC, Detmann E, Chizzotti ML, Rodrigues FC, et al. Achieving body weight adjustments for feeding status and pregnant or non-pregnant condition in beef cows. PLoS One. 2015;10:e0112111.
7. Valente TNP, Detmann E, Queiroz AC, Valadares Filho SC, Gomes DI, Figueiras JF. Evaluation of ruminal degradation profiles of forages using bags made from different textiles. Rev Bras Zootec. 2011;40:2565–73.
8. Detmann E, Souza MA, Valadares Filho SC, Queiroz AC, Berchielli TT, Saliba EOS, et al. Métodos para análise de alimentos – INCT – Ciência Animal. Visconde do Rio Branco. Brazil: Suprema; 2012. 214p.
9. Valadares RFD, Broderick GA, Valadares Filho SC, Clayton MK. Effect of replacing alfafa silage with high moisture corn on ruminal protein synthesis estimated from excretion of total purine derivatives. J Dairy Sci. 1999;82:2686–96.
10. Silva LFC, Valadares Filho SC, Chizzotti ML, Rotta PP, Prados LF, Valadares RFD, et al. Creatinine excretion and relationship with body weight of Nellore cattle. Rev Bras Zootec. 2012;41:807–10.
11. Chen XB, Gomes MJ. Estimation of microbial protein supply to sheep and cattle basid on urinary excretion of purine derivatives-an overview of the technical details. Ocasional publication. Ed. Rowett Research Institute: Buchsburnd Aberdeen, UK; 1992. 21p.
12. Barbosa AM, Valadares RFD, Valadares Filho SC, Pina DS, Detmann E, Leão MI. Endogenous fraction and urinary recovery of purine derivatives obtained by different methods in Nellore cattle. J Anim Sci. 2011;89:510–9.
13. Cabral CHA, Paulino MF, de Paula NF, Valadares RFD, de Araújo FL. Levels of supplementation for grazing pregnant beef cows during the dry season. Rev Bras Zootec. 2012;41:2441–9.
14. Radunz AE, Fluharty FL, Day ML, Zerby HN, Loerch SC. Prepartum dietary energy source fed to beef cows: I. Effects on pre- and postpartum cow performance. J Anim Sci. 2010;88:2717–28.
15. Radunz AE, Fluharty FL, Zerby HN, Loerch SC. Winter-feeding systems for gestating sheep I. Effects on pre- and postpartum ewe performance and lamb progeny preweaning performance. J Anim Sci. 2011;89:467–77.
16. Summers AF, Meyer TL, Funston RN. Impact of supplemental protein source offered to primiparous heifers during gestation on I. Average daily gain, feed intake, calf birth body weight, and rebreeding in pregnant beef heifers. J Anim Sci. 2015;93:1865–70.
17. Lazzarini I, Detmann E, Sampaio CB, Paulino MF, Valadares Filho SC, de Souza MA, et al. Intake and digestibility in cattle fed low-quality tropical forage and supplemented with nitrogenous compounds. Rev Bras Zootec. 2009;38:2021–30.
18. Costa VAC, Detmann E, Valadares Filho SC, Paulino MF, Henriques LT, Mantovni HC. Degradação in vitro da fibra em detergente neutro de forragem tropical de baixa qualidade em função de suplementação com proteína e/ou carboidratos. Rev Bras Zootec. 2008;37:494–503.
19. Costa VAC, Detmann E, Paulino MF, Valadares Filho SC, Carvalho IPC, Monteiro LP. Consumo e digestibilidade em bovinos em pastejo durante o período das águas sob suplementação com fontes de compostos nitrogenados e de carboidratos. Rev Bras Zootec. 2011;40:1788–98.
20. Huntington GB, Harmon DL, Richards CJ. Sites, rates, and limits of starch digestion and glucose metabolism in growing cattle. J Anim Sci. 2006;84(Suppl 1):E14–24.
21. Hawkins DE, Petersen MK, Thomas MG, Sawyer JE, Waterman RC. Can beef heifers and young postpartum cows be physiologically and nutritionally manipulated to optimize reproductive efficiency? J Anim Sci. 2000;77(Suppl 1):1–10.
22. Wettemann RP, Bossis I. Energy intake regulates ovarian function in beef cattle. J Anim Sci. 2000;77(Suppl 1):E1–E10.
23. Cushman RA, Kill LK, Funston RN, Mousel EM, Perry GA. Heifer calving date positively influences calf weaning weights through six parturitions. J Anim Sci. 2013;91:4486–91.
24. Funston RN, Musgrave JA, Meyer TL, Larson DM. Effect of calving distribution on beef cattle progeny performance. J Anim Sci. 2012;90:5118–21.

16

Mycotoxin binder improves growth rate in piglets associated with reduction of toll-like receptor-4 and increase of tight junction protein gene expression in gut mucosa

Linghong Jin[1,2], Wei Wang[1,2], Jeroen Degroote[1,2], Noémie Van Noten[1,2], Honglin Yan[1,2], Maryam Majdeddin[1,2], Mario Van Poucke[3], Luc Peelman[3], Anne Goderis[4], Kurt Van De Mierop[4], Ronny Mombaerts[4], Stefaan De Smet[2] and Joris Michiels[1]*

Abstract

Background: Deoxynivalenol (DON) is a mycotoxin produced by *Fusarium* species in the field, commonly found in cereal grains, which negatively affects performances and health of animals. Mycotoxin binders are supposed to reduce the toxicity of mycotoxins.

Method: The effect of a mycotoxin binder (containing acid-activated bentonite, clinoptilolite, yeast cell walls and organic acids) on growth performance and gut health was studied. Hundred and twenty weaning piglets were allocated to 4 treatments, with 5 pens of 6 piglets each, arranged in a 2×2 factorial design: control diet; control diet with 1 kg/t binder; control diet with DON; and control diet with DON and 1 kg/t binder. From d0–14, the diet of DON-challenged groups was artificially contaminated with a mixture of DON (2.6 mg/kg), 3-acetyl-deoxynivalenol (0.1 mg/kg) and 15-acetyl-deoxynivalenol (0.3 mg/kg), after which the total contamination level was reduced to 1 mg/kg, until d37. On d14, one pig from each pen was euthanized and distal small intestinal mucosa samples were collected for the assessment of intestinal permeability, and gene expression of tight junction proteins, toll-like receptor 4, inflammatory cytokines and intestinal alkaline phosphatase.

Results: After 37 d, there were no differences in growth performance between control and DON-challenged groups ($P > 0.05$). Nevertheless, groups that received diets with binder had a significantly higher average daily gain (ADG) and average daily feed intake (ADFI) for the first 14 d as well as for the whole period, compared to groups without binder ($P \leq 0.05$). Groups with binder in the diet also exhibited lower expression of toll-like receptor 4 in distal small intestinal mucosa at d14, compared to groups without binder ($P \leq 0.05$). Interestingly, comparing the two DON treatments, piglets fed DON and binder had significantly higher ADFI and ADG compared to those with only DON for the first 14-d ($P \leq 0.05$). Addition of binder to DON contaminated diets, also down-regulated the gene expression of toll-like receptor 4 ($P \leq 0.05$) and increased mRNA level zona occludens 1 ($P \leq 0.10$) as compared to DON.

Conclusions: The present data provide evidence that the binder improves growth rate in piglets associated with reduction of toll-like receptor-4 and increase of tight junction protein gene expression. However, the current study does not allow to assess whether the effects of the binder are mediated by alterations in the toxicokinetics of the mycotoxin.

Keywords: Binder, Deoxynivalenol, Gut barrier, Gut health, Mycotoxin, Pigs

* Correspondence: joris.michiels@ugent.be
[1]Department of Applied Biosciences, Ghent University, Valentin Vaerwyckweg 1, 9000 Ghent, Belgium
Full list of author information is available at the end of the article

Background

The contamination of feedstuffs with mycotoxins is a worldwide issue. Mycotoxins are harmful secondary metabolites of fungi which can cause intoxications at very low dosage. Deoxynivalenol (DON) is a type B trichothecene mycotoxin, produced by *Fusarium* species. DON is noted for two typical toxicological effects: reduced feed intake and induction of emesis in swine. Under natural conditions, DON is present along with its two major acetylated forms, 3-acetyl-deoxynivalenol (3A–DON) and 15-acetyl-deoxynivalenol (15A–DON) at lower concentrations than DON in cereals [1]. DON is physically stable and can easily enter the food chain [2]. Humans and all animal species can exhibit toxic effects after exposure to DON [3], with pigs being the most susceptible species [4].

The intestinal mucosa is constantly challenged by various chemical and biological contaminants, and functions as a vital barrier between the intestinal milieu and the luminal content [5, 6]. After consumption of feed contaminated with DON, the intestine can be exposed to high levels of the toxin [7]. Studies show that DON can affect the intestinal histology and morphology by affecting intestinal cell viability and proliferation. DON is also able to regulate the production of pro-inflammatory cytokines, increasing the expression of interleukin 1 beta (IL-1β), interleukin 2 (IL-2) and interleukin 6 (IL-6) in the jejunum, and IL-1β, IL-6 and tumor necrosis factor alpha (TNF-α) in the ileum in pigs [8]. An in vitro study indicated that DON is able to decrease transepithelial electrical resistance (TEER) and increase the permeability of IPEC-1 cells in a dose and time dependent manner [7]. The alterations in these two parameters are related to the decreased expression of specific tight junction proteins (TJPs) 9 [7, 9]. In the intestinal epithelium, the activation of mitogen-activated protein kinase (MAPK) by DON and its acetylated derivatives suppresses the expression of TJPs, which is responsible for the loss of barrier function [10, 11].

In order to solve the problem caused by mycotoxicosis, various strategies have been developed, including physical, chemical, and biological methods [12, 13]. The most common approach is the addition of mycotoxin binders to feeds [14]. Mycotoxin binders are large weight molecules, capable of binding to mycotoxins in animal feeds. These binder-mycotoxin complexes pass through the gastrointestinal tract without dissociating, preventing mycotoxin uptake [15]. The complex passes through the GIT and is excreted via the faeces, thereby helping to minimize absorption of mycotoxins by target organs and alleviating the adverse effects of mycotoxins.

DON was the most prevalent single mycotoxin found in all feedstuffs all over the world in 2015 [16]. However, there are few publications about the effects of toxin binders on gut health in piglets. This study was conducted to assess the effect of addition of a mycotoxin binder to the feed on gut health and performance in pigs following a 37-d dietary exposure to DON.

Methods

Animals and dietary treatments

An animal feeding experiment in a 2 × 2 factorial design with either or not addition of DON, and either or not addition of binder to the feed was performed. A total of 120 weaning (24 d of suckling period) piglets with an average weight of 7.3 kg were used in this study. They were provided with water and feed ad libitum throughout the experiment. Animal experimental procedures were in accordance to the guidelines of the Ethical Committee of the Faculty of Veterinary Sciences, Ghent University, Belgium.

Piglets were randomly allocated to 4 dietary treatments. Each treatment contained 5 pens with 6 piglets per pen. The 4 treatments were as follows: CON, negative control diet (uncontaminated basal diet); CON + BIN, negative control diet with 1 kg/t mycotoxin binder; DON, negative control diet with DON; DON + BIN, negative control diet with DON and 1 kg/t mycotoxin binder. The mycotoxin binder was a blend of indigestible adsorbents that bind mycotoxins in the GIT (Free-Tox, Nutrex NV, Belgium). It contains acid-activated bentonite, clinoptilolite, yeast cell walls and organic acids. From d 0 until d 14 (sampling on d 14) of the experiment, the diet of DON and DON + BIN was artificially contaminated with a mixture of DON (2.6 mg/kg), 3A–DON (0.1 mg/kg) and 15A–DON (0.3 mg/kg), after which the DON contamination level was reduced to 1 mg/kg from d 14 until d 37. The composition of the negative control wheat-barley-soybean based diets is given in Table 1. The DON-challenge diets were artificially contaminated with a fungal culture containing DON and its metabolites. DON was produced in vitro by *F. graminearum*. After growing up of the mold, the amount of DON was quantified by ELISA assay on the medium. The result was verified by LC/MS/MS on the certified standard blank wheat [17]. Results showed that the medium contained 240 mg/kg total DON metabolites (87.5% DON, 2.7% 3A–DON and 9.8% 15A–DON). All other mycotoxins were under the detection limit. Based on the concentration in the medium, the amount of medium needed for 3 mg/kg and 1 mg/kg in pre-starter and starter diets, respectively, was calculated. After homogenization, the medium was mixed into the basal diet. Three mg/kg for pre-starter period (from d 0 until d 14) was chosen because typically feed contamination with 2–5 mg/kg DON is required to induce reduction of feed intake and decrease of body weight gain [18]. It was further reduced to 1 mg/kg for the starter period (from d 14 until d 37) as model for chronic exposure.

Table 1 Composition of the negative control diet (CON) for pre-starter (d0-d14) and starter (d14-d37) periods

Pre-starter		Starter	
Ingredients	%	Ingredients	%
Wheat	22.57	Barley	25.00
Barley	22.50	Wheat	22.93
Whey	7.00	Corn	15.00
Extruded soybeans	2.40	Toasted soybeans	12.00
Calcium formiate and lactic acid	1.00	Whey	4.20
Potato protein	2.00	Extruded soybeans	1.70
Toasted soybeans	12.00	Calcium formiate and lactic acid	0.50
Extruded oats and barley	10.00	Coconut	0.20
Corn	7.50	Soybean meal, CP49	9.07
Soybean meal, CP49	4.04	Wheat gluten feed	2.69
Fat, > 88% triglycerides	0.50	Beet pulp, sugar 72%	2.00
Sodium bicarbonate	0.30	Fat, > 88% triglycerides	0.95
Organic acids mixture[a]	0.30	Organic acids mixture[a]	0.30
Lime fine	0.29	Salt	0.05
Premix[b]	7.60	Premix[c]	3.40
Composition		Composition	
NE, kcal/kg	2350	NE, kcal/kg	2350
Crude protein, g/kg	173	Crude protein, g/kg	178
Crude fibre, g/kg	37	Crude fibre, g/kg	38
Calcium, g/kg	5.2	Calcium, g/kg	7.9
Phosphorus, g/kg	4.5	Phosphorus, g/kg	4.8
Sugar + Starch, g/kg	457	Sugar + Starch, g/kg	438

[a]Organic acids mixture: contains formic acid, phosphoric acid and citric acid

[b]Providing per kg of complete diet: Vitamin A, 15,000 IU/kg; Vitamin D_3 2000 IU/kg; Vitamin E, 200 IU/kg; Vitamin K_3, 4.0 mg; Vitamin B_1, 3.0 mg; Vitamin B_2, 8.0 mg; Vitamin B_3, 20 mg; Vitamin B_6, 6.0 mg; Vitamin B_{12}, 50.0 μg; niacinamide, 40.0 mg; folic acid, 2.0 mg; biotin, 0.3 mg; Cu, 155 mg/kg; Fe, 150 mg/kg; Mn, 49 mg/kg; Zn, 104 mg/kg; I, 1.55 mg/kg; Se, 0.40 mg/kg

[c]Providing per kg of complete diet: Vitamin A, 15,000 IU/kg; Vitamin D_3 2000 IU/kg; Vitamin E, 102 IU/kg; Vitamin K_3, 4.0 mg; Vitamin B_1, 3.0 mg; Vitamin B_2, 8.0 mg; Vitamin B_3, 20 mg; Vitamin B_6, 6.0 mg; Vitamin B_{12}, 50.0 μg; niacinamide, 40.0 mg; folic acid, 2.0 mg; biotin, 0.3 mg; Cu, 155 mg/kg; Fe, 150 mg/kg; Mn, 49 mg/kg; Zn, 80 mg/kg; I, 1.49 mg/kg; Se, 0.40 mg/kg

Sampling

After 14 d of feeding, one pig out of each pen was euthanized by intra-peritoneal pentobarbiturate overdose. The GIT was removed and the small intestine was exposed for sample collection. A 10-cm segment from the 75% length of the small intestine (distal small intestine) was collected for Ussing chamber measurements following flushing with saline to remove residual content. Another 20 cm segment from the same region was flushed with saline, placed on a cold plate and slit longitudinally. Then, mucosa was harvested by scraping with a glass-slide followed by snap freezing and storage at –80 °C pending gene expression analysis.

Growth performance

The weight of piglets as well as the feed intake per pen were determined at d0, d14 and d37. Average Daily Gain (ADG, g/d), Average Daily Feed Intake (ADFI, g/d) and Feed to Gain ratio (F:G) were calculated for periods d0-

d14, d14-d37 and d0-d37. Diarrhoea and mortality were daily checked and recorded.

Ex vivo measurement of intestinal permeability

Permeability was assessed ex vivo in Ussing chambers by measuring the permeability for the macromolecular marker fluorescein isothiocyanate-dextran 4 (FD4, molecular weight 4 kDa) across sheets of mucosa as described by Wang et al. [19]. Briefly, fresh segments of mucosa samples from the distal small intestine (75% of the total small intestinal length) were separated from the seromuscular layer and mounted in the Ussing chamber system. Intestinal sheets were bathed in 6.5 mL Ringer buffer solution with 6 mmol/L glucose and 6 mmol/L mannitol in the serosal and mucosal sides, respectively. The system was maintained at 37 °C and oxygenated (95% O_2 and 5% CO_2). After a 20-min equilibration period, 0.8 mg/mL FD4 (Sigma-Aldrich, Bornem, Belgium) was added to the mucosal side. Samples from the serosal compartment were taken at 20 min intervals for 80 min to monitor mucosal-to-serosal fluxes of FD4. Fluorescence intensity of FD4 was determined by fluorescence spectrophotometry (Thermo Fisher Scientific, Marietta, OH, USA). The flux over the 100 min period was calculated and expressed as an apparent permeability coefficient as described before [19].

RNA isolation and reverse-transcription quantitative real-time PCR

Relative mRNA expression of TJPs (*ZO-1, ZO-2, OCLN, CLDN-1, CLDN-2, CLDN-5, CLDN-7*) and pro-inflammatory cytokines (*TNF-α, IFN-γ, IL-1β, IL-8*), toll-like receptor 4 (*TLR-4*) and a brush border enzyme intestinal alkaline phosphatase (*IAP*) were determined by reverse transcription quantitative real-time PCR (RT-qPCR) and performed according to the MIQE guidelines. Briefly, mucosal total RNA was extracted using the Bio-Rad Aurum Total RNA Fatty and Fibrous Tissue Kit (Bio-Rad Laboratories, Inc., Hercules, CA, USA) according to the manufacturer's instructions, including an on-column DNase I treatment to remove genomic DNA (gDNA). The concentration and purity ($OD_{260/280}$) of RNA were measured with the NanoDrop ND-1000 (NanoDrop Technologies, Thermo Scientific, Wilmington, DE, USA). 1 μg RNA was analyzed by 1% agarose gel electrophoresis to check RNA integrity (28S and 18S rRNA bands). In addition to this assessment, a minus-RT control PCR was performed using *YWHAZ* as primer to verify the absence

of any gDNA contamination. Following this,1 μg of high quality DNA-free RNA was reverse transcribed in the 20 μL reverse-transcription reaction with the ImProm-II cDNA synthesis kit (Promega, Madison, WI, USA), containing both oligo dT and random primers. The obtained cDNA was diluted 10 times with molecular grade water and a control PCR using 2 μL cDNA was performed to verify the reverse-transcription reaction.

Primers (Table 2) used for genes in the study were designed with Primer3Plus. The repeats, the secondary structure and single nucleotide polymorphism in target sequence were checked with RepeatMarker, mfold and dbSNP, respectively. All these primer sequences were gene isoform specific as they were designed based on certain exon-exon boundaries of published pig gene sequences corresponding to the accession number. Primers were then purchased from IDT (Integrated DNA Technologies, Leuven, Belgium).

The RT-qPCR was performed on the CFX96 Touch Real-Time PCR Detection System (Bio-Rad Laboratories, Inc.). Briefly, 2 μL cDNA template, 5 μL 2× KAPA SYBR FAST qPCR Kit Master Mix (Kapa Biosystems, Inc., Wilmington, MA, USA), 2 μL molecular grade water, 0.5 μL forward primer and 0.5 μL reverse primer (5 μmol/L each) were added to a total volume of 10 μL. The amplification conditions were as follows:1) enzyme activation and initial denaturation (95 °C for 3 min); 2) denaturation (95 °C for 20 s) and annealing/extension and data acquisition (annealing temperature depending

Table 2 Primer sequences used for reverse-transcription quantitative real-time PCR

Gene symbol[a]	Accession number	Nucleotide sequence of primers, 5'-3'		Product length, bp	Tm, °C
		Forward	Reverse		
CLDN-1	NM_001244539.1	TATGACCCCATGACCCCAGT	GCAGCAAAGTAGGGCACCTC	108	59
CLDN-2	NM_001161638.1	TTCCTCCCTGTTCTCCCTGA	CACTCTTGGCTTTGGGTGGT	152	62
CLDN-5	NM_001161636.1	GTGGTCCGCGAGTTCTACGA	CTTGACAGGGAAGCCGAGGT	171	60
CLDN-7	NM_001160076.1	GGTCCCCACAAACGTGAAGTA	TCACTCCCAGGCACAAGAGCA	114	60
HPRT-1	DQ178126	CCGAGGATTTGGAAAAGGT	CTATTTCTGTTCAGTGCTTTGATGT	181	60
IAP	XM_003133729.3	GGCCAACTACCAGACCATCG	CCGACTTCCCTGCTTTCTTG	116	60
IFN-γ	NM_213948.1	GCTTTTCAGCTTTGCGTGACT	CACTCTCCTCTTTCCAATTCTTCA	166	58
IL-1β	NM_214055.1	GCACCCAAAACCTGGACCT	CTGGGAGGAGGGATTCTTCA	143	58
IL-8	XM_003361958.3	TGTCAATGGAAAAGAGGTCTGC	CTGCTGTTGTTGTTGCTTCTCA	100	60
OCLD	NM_001163647.2	CATGGCTGCCTTCTGCTTCATTGC	ACCATCACACCCAGGATAGCACTCA	129	65
PPIA	NM_214353	CTGAAGCATACGGGTCCTGG	TGCCCTCTTTCACTTTGCCA	139	65
TBP	DQ178129	GATGGACGTTCGGTTTAGG	AGCAGCACAGTACGAGCAA	124	59
TLR-4	NM_001113039.2	TTCTTGCAGTGGGTCAAGGA	GACGGCCTCGCTTATCTGAC	135	58
TNF-α	NM_214022.1	CATGATCCGAGACGTGGAGC	AACCTCGAAGTGCAGTAGGC	151	62
ZO-1	XM_003480423.3	ATCTCGGAAAAGTGCCAGGA	CCCCTCAGAAACCCATACCA	172	61
ZO-2	XM_005660148.2	CCAGGAAGCACAGAATGCAA	AAGTCTGGCGGGACCTCTCT	148	61

[a]*CLDN-1* claudin-1, *CLDN-2* claudin-2, *CLDN-5* claudin-5, *CLDN-7* claudin-7, *HPRT-1* hypoxanthine phosphoribosyltransferase 1, *IAP* intestinal alkaline phosphatase, *IFN-γ* interferon gamma, *IL-1β* interleukin 1 beta, *IL-8* interleukin 8, *OCLN* occludin, *PPIA* peptidylprolyl isomerase A, *TBP* TATA-binding protein, *TLR-4* toll-like receptor 4, *TNF-α* tumor necrosis factor alpha, *ZO-1* zona occludens 1, *ZO-2* zona occludens 2

on primer for 40 s) repeated 40 cycles; and 3) dissociation (melt curve analysis from 70 to 90 °C with 0.5 °C increment every 5 s).

Primers used in this study were first optimized by gradient quantitative real-time PCR. A 5-fold dilution series (5 points, from 1 times to 625 times dilution) of cDNA as standard curve was included at 3 gradient temperatures to determine PCR amplification efficiency and specificity. The standard curve was also included in each run to determine PCR efficiency. In this study, PCR amplification efficiencies were consistently between 90% and 110%. Gene-specific amplification was verified by agarose gel electrophoresis and melting curve analysis. Efficiency was used to convert the Cq value into raw data with the highest expressed samples (lowest Cq value) as a calibrator for the normalization of raw data. The relative expression was expressed as a ratio of the target gene to the geometric mean of three stable expressed reference genes (PPIA, HPRT1 and TBP) [19].

Statistical analysis

After determination of normality and variance homogeneity, a general linear model with the fixed effects of mycotoxin and binder, and the interaction term was used with Tukey's test as a multiple comparison test in SAS Enterprise Guide 7 (SAS Institute, Cary, NC, USA). $P \leq 0.05$ was considered as significant. All data are expressed as mean ± standard errors.

Principal component analysis (PCA) as described by Montagne et al. was conducted to work out the variables that contributed most to the variation between subjects [20, 21]. In brief, the data of 17 variables were standardised before the application of PCA. At first, a scree plot was carried out to fix the number of principal components to be maintained. Five principal components were retained with the eigenvalues >1.0. In addition, variables that had a correlation coefficient between variable and all principal components ≤0.5 were excluded. Then, retained variables were grouped into families to check the correlation. Only the main representative variable with highest principal component loading, together with high correlation ($r > 0.55$; $P \leq 0.05$) within family was retained for the final analysis. Finally, 11 variables entered the final PCA.

Results
Growth performances

Only few pigs from different groups had diarrhoea problems in the first week of the experiment, likely following weaning stress. No case of emesis or mortality were observed. Overall, no clinical signs of toxicity were found. There were no significant differences between control groups (CON and CON + BIN) and DON-challenged groups (DON and DON + BIN) regarding growth performances (Table 3). In contrast, pigs supplemented with binder (CON + BIN and DON + BIN) consumed more feed (265 g/d vs. 242 g/d) and had a higher growth (197 g/d vs. 170 g/d) for the first 14-d when compared to pigs that received diets with no binder (CON and DON) ($P \leq 0.05$). Similarly, for the whole experimental period d0-d37, groups receiving diets with binder (CON + BIN and DON + BIN) showed an improved ADG (368 g/d vs. 341 g/d) and ADFI (548 g/d vs. 519 g/d) compared to groups that received diets without binder (CON and DON) ($P \leq 0.05$). Meanwhile, There was a trend that groups that received diets with binder (CON + BIN and DON + BIN) had a lower F:G compared to groups that received diets without binder (CON and DON) from d 1 until d 14 of the experiment ($P \leq 0.10$). Interestingly, within DON-challenged piglets, addition of the binder improved performance in the first 14-d of the experiment; DON contaminated diet supplemented with binder (DON + BIN) showed higher ADFI compared to diet only contaminated with DON (DON) (272 g/d vs. 227 g/d) ($P \leq 0.05$). This again resulted in a higher growth rate for treatment DON + BIN than treatment DON for period d0–14 (205 g/d vs. 159 g/d) ($P \leq 0.05$). Also, pigs supplemented with binder (DON + BIN) showed higher body weight at d14 compared to pigs that received diets with no binder (DON) (10.18 kg vs. 9.63 kg) ($P \leq 0.05$).

Permeability measurements in distal small intestine

Neither mycotoxin level, nor binder addition affected FD4 fluxes across distal small intestinal sheets ($P > 0.05$) (Fig. 1). The mean of control groups (CON and CON + BIN) was 7.5×10^{-7} cm/s; while the average value of DON-challenged groups (DON and DON + BIN) was 7.6×10^{-7} cm/s. On binder level, the difference of FD4 flux between groups that received diets with the addition of binder (CON + BIN and DON + BIN) and groups that received diets without the addition of binder (CON and DON) was larger compared to the difference between DON-challenged and DON-control groups but still not significant.

mRNA expression of tight junction proteins, inflammatory cytokines and brush border enzyme in distal small intestine

The gene expressions of TJPs (ZO-1, ZO-2, OCLN, CLDN-1, CLDN-2, CLDN-5, CLDN-7), pro-inflammatory cytokines (TNF-α, IFN-γ, IL-1β, IL-8), TLR-4 and IAP in distal small intestine are described in Table 4. Ingestion of diets contaminated with or without DON, did not change the gene expression of TJPs, pro-inflammatory cytokines, and IAP in distal small intestine ($P > 0.05$), whereas adding the binder to the diets down-regulated the expression of TLR-4 (0.72 for CON + BIN and DON + BIN vs 1.00 for CON and DON; $P \leq 0.05$). At the same time, there

Table 3 Growth performance (body weight, BW; average daily gain ADG; average daily feed intake, ADFI, and feed:gain ratio, F:G) for periods d0-d14 (pre-starter), d14-d37 (starter) and d0-d37 (total) of piglets fed diets with or without mycotoxins, and with or without binder ($n = 5$)

Item	Treatment[1]								P-value		
	CON		CON + BIN		DON		DON + BIN		Mycotoxin	Binder	Mycotoxin × Binder
	Mean	SE	Mean	SE	Mean	SE	Mean	SE			
Day 0–14 (pre-starter)											
BW d0, g	7.30	0.01	7.31	0.01	7.32	0.01	7.30	0.01	0.52	0.85	0.49
BW d14, g	9.82[ab]	0.13	9.96[ab]	0.15	9.63[b]	0.13	10.18[a]	0.12	0.97	0.02	0.05
ADG, g/d	181[ab]	10	188[ab]	11	159[b]	10	205[a]	9	0.80	0.02	0.03
ADFI, g/d	256[ab]	10	257[ab]	12	227[b]	10	272[a]	10	0.50	0.05	0.05
F:G	1.43	0.04	1.38	0.05	1.43	0.04	1.33	0.04	0.64	0.08	0.25
Day 14–37 (starter)											
BW d37, g	20.3	0.43	20.4	0.49	19.9	0.43	21.0	0.40	0.95	0.21	0.35
ADG, g/d	446	17	469	17	444	15	466	13	0.86	0.17	0.55
ADFI, g/d	724	21	760	23	716	21	750	19	0.68	0.11	0.44
F:G	1.59	0.03	1.62	0.03	1.62	0.03	1.62	0.03	0.68	0.55	0.83
Day 0–37 (total)											
ADG, g/d	344	14	366	14	337	12	370	11	0.90	0.05	0.21
ADFI, g/d	527	14	548	16	511	14	549	13	0.59	0.05	0.19
F:G	1.50	0.03	1.50	0.03	1.51	0.03	1.49	0.02	0.84	0.63	0.92

Means with different superscripts (a, b) within row represent differences among treatments ($P \leq 0.05$)

[1]CON negative control diet (uncontaminated basal diet), CON + BIN negative control diet with 1 kg/t mycotoxin binder, DON negative control diet with DON, DON + BIN negative control diet with DON and 1 kg/ton mycotoxin binder

was a trend that groups that received the diet with the addition of binder (CON + BIN and DON + BIN) up-regulated the expression of CLDN-1 compared to groups that received diet without the addition of binder (CON and DON) ($P \leq 0.10$). More specifically, *TLR-4* gene expression was down-regulated in the DON contaminated diet supplemented with mycotoxin binder (DON + BIN)

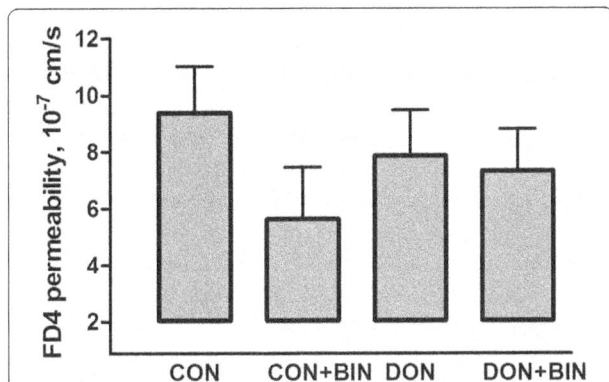

Fig. 1 Intestinal permeability for FD4 in distal small intestinal mucosa of piglets fed diets at d14 post-weaning. Data are means ± SE ($n = 5$). P for factor mycotoxin is 0.96, for factor binder 0.21, and for mycotoxin × binder 0.51. CON, negative control diet (uncontaminated basal diet); CON + BIN, negative control diet with 1 kg/t mycotoxin binder; DON, negative control diet with DON; DON + BIN, negative control diet with DON and 1 kg/t mycotoxin binder

compared to DON (0.57 vs 1.11; $P \leq 0.05$). At the same time, there was a tendency that DON + BIN up-regulated the expression of *ZO-1* compared to DON ($P \leq 0.10$).

Principal component analysis

BWd0, BWd14, ADG d0-d14, FD4 permeability, ZO-1, ZO-2, OCLN, CLDN-1, CLDN-2, CLDN-5, CLDN-7, TLR4, TNF-α, IFN-γ, IL-1β, IL-8 and IAP were the 17 variables used in the PCA. After application of a first PCA, 5 principal components were retained following a scree plot. BWd0 was the only variable that did not show high correlation on any principal component and was excluded. Then, BWd14 and ADG d0-d14 were grouped into growth performance family, ZO-1, ZO-2, OCLN, CLDN-1, CLDN-2, CLDN-5 and CLDN-7 were grouped into TJPs family, and TNF-α, IFN-γ, IL-1β and IL-8 were grouped into inflammatory cytokines family. FD permeability, TLR-4 and IAP were considered single representatives and were retained for final analysis. Some variables were highly correlated within family. In growth performance family, ADG d0-d14 was highly correlated with BWd14 ($r = 0.966$, $P \leq 0.01$), yet ADG d0–14 was not retained. Within the family of TJPS, OCLN was correlated with ZO-1 ($r = 0.762$, $P \leq 0.01$), ZO-2 ($r = 0.811$, $P \leq 0.01$) and CLDN-7 ($r = 0.683$, $P \leq 0.01$). Then, OCLD, CLDN-1, CLDN-2 and CLDN-5 were retained for the final PCA. For the family of

Table 4 Relative mRNA expression of tight junction proteins, pro-inflammatory cytokines and intestinal alkaline phosphatase in distal small intestinal mucosa of piglets fed diets with or without mycotoxin contamination, and with or without binder at d14 post-weaning (n = 5)

| Item[1] | Treatment[2] | | | | | | | | P-value | | |
| | CON | | CON + BIN | | DON | | DON + BIN | | Mycotoxin | Binder | Mycotoxin × Binder |
	Mean	SE	Mean	SE	Mean	SE	Mean	SE			
ZO-1	1.50	0.16	1.45	0.18	1.29	0.16	1.88	0.15	0.50	0.12	0.08
ZO-2	1.40	0.17	1.18	0.17	1.13	0.17	1.42	0.15	0.95	0.82	0.50
OCLN	1.09	0.24	0.97	0.24	0.83	0.24	1.60	0.22	0.45	0.18	0.13
CLDN-1	1.10	0.17	1.30	0.15	1.03	0.15	1.36	0.15	0.96	0.10	0.36
CLDN-2	1.68	0.28	1.17	0.28	0.87	0.28	1.25	0.25	0.19	0.81	0.26
CLDN-5	0.95	0.23	1.04	0.23	0.76	0.23	0.66	0.21	0.22	0.98	0.61
CLDN-7	1.83	0.23	1.36	0.23	1.20	0.23	1.62	0.21	0.42	0.92	0.26
TLR-4	0.88[ab]	0.12	0.87[ab]	0.12	1.11[a]	0.14	0.57[b]	0.11	0.77	0.04	0.05
TNF-α	0.98	0.20	0.84	0.20	0.85	0.20	0.98	0.18	0.99	0.98	0.92
IFN-γ	0.38	0.18	0.30	0.18	0.73	0.18	0.45	0.17	0.18	0.34	0.41
IL-1β	0.97	0.24	0.92	0.24	1.20	0.24	0.79	0.22	0.83	0.34	0.65
IL-8	1.48	0.31	1.00	0.31	0.89	0.31	1.20	0.28	0.54	0.78	0.57
IAP	0.91	0.22	1.00	0.22	0.73	0.22	1.03	0.20	0.72	0.37	0.65

Means with different superscripts (a, b) within row represent differences among treatments (P ≤ 0.05)

[1]CLDN-1 Claudin 1, CLDN-2 Claudin 2, CLDN-5 Claudin 5, CLDN-7 Claudin 7, IAP Intestinal alkaline phosphatase, IFN-γ Interferon, gamma, IL-1β Interleukin 1 beta, IL-8 Interleukin 8, OCLN Occludin, TLR-4 Toll like receptor 4, TNF-α Tumor necrosis factor alpha, ZO-1 Zona occludens 1, ZO-2 Zona occludens 2

[2]CON negative control diet (uncontaminated basal diet), CON + BIN negative control diet with 1 kg/t mycotoxin binder, DON negative control diet with DON, DON + BIN negative control diet with DON and 1 kg/t mycotoxin binder

inflammatory cytokines, only IFN-γ was excluded. Finally, 11 variables were kept for this final PCA (Table 5). The 5 principal components explained 85.5% of the variance, of which the first principal component contributing 22.6% and the second principal component contributing 22.2%. The first principal component grouped the TJPs family members OCLN, CLDN-1, CLDN-2 as well as TNF-α and brush border enzyme IAP together. Principal component 1 had higher principal component score in groups with addition of binder (CON + BIN and DON + BIN) compared to groups without addition of binder (CON and DON) (0.330 vs. −0.398) (P ≤ 0.10). In other words, ingestion of diets supplemented with binder tended to be associated with higher gene expression of OCLN, CLDN-1, CLDN-2, TNF-α and IAP as compared to diet without binder. This finding is consistent with the gene expression result of CLDN-1 in Table 4. It supports the finding that binder may also co-up-regulate the expression of other TJPs (OCLN, CLDN-2) and IAP. The second principal component indicates that the high expression of TLR-4 was associated with high expression of pro-inflammatory cytokines TNF-α and IL-1β. The third principal component denotes that high expression of CLDN-1, CLDN-2 and CLDN-5 was related to higher weight at d14. However, principal components 2 and 3 were not discriminatory for treatments.

Discussion

In the current experiment, mycotoxin contamination of piglet diets exhibited no effect on growth and gut health parameters. In contrast, the addition of a mycotoxin binder showed beneficial effects, in particular when diets were contaminated with 3 mg/kg of a mixture of DON and acetylated metabolites. Growth and feed intake were enhanced, in line with improvements of some selected gut health parameters.

Lack of effect of DON addition to feed on performance and gut health

In addition to lack of effect on growth performance, our results did not show an effect of DON on gut health in the distal small intestine, regarding intestinal permeability and mRNA expression of TJPs and pro-inflammatory cytokines, as well as IAP, after 14 d of feeding 3 mg/kg total DON. The lack of effect in the distal small intestine might be associated with the toxicokinetic properties of DON. In vivo and in vitro studies demonstrated that DON and its acetylated forms are rapidly absorbed from the upper GIT, involving stomach until proximal jejunum. After chronic exposure to DON in pigs, a fast and almost complete absorption (> 90%) occurs, with DON appearing within 15 min in the blood and reaching maximal concentrations 1.65 h after oral exposure [22]. Danicke et al. revealed that

Table 5 Loadings and principal component scores for 5 principal components obtained by principal component analysis [a] (PCA) of 11 variables from piglets fed diets with or without mycotoxin contamination, and with or without binder at d14 post-weaning

Treatment[b]	Principal component				
	1	2	3	4	5
	22.6%	22.2%	16.0%	15.6%	10.1%
CON	−0.225	0.119	0.119	0.719	0.517
CON + BIN	−0.019	0.006	0.390	−0.327	−0.455
DON	−0.571	0.630	−0.366	−0.477	0.119
DON + BIN	0.679	−0.630	−0.120	0.071	−0.151
P-value					
Mycotoxin	0.68	0.88	0.29	0.36	0.92
Binder	0.10	0.12	0.58	0.56	0.18
Mycot. × Binder	0.23	0.19	0.98	0.08	0.44
BW d14[c]			0.889		
FD4 permeability					0.974
OCLN	0.813				
CLDN-1	0.767			0.401	
CLDN-2	0.343		0.335	0.754	
CLDN-5		0.546	0.713		
TLR-4		0.891			
TNF-α	0.498	0.666			−0.300
IL-1β		0.866			
IL-8				0.910	
IAP	0.856				

[a]Rotation method: varimax with Kaiser normalisation; only correlations with |r| > 0.3 are indicated
[b]*CON* negative control diet (uncontaminated basal diet), *CON + BIN* negative control diet with 1 kg/t mycotoxin binder, *DON* negative control diet with DON, *DON + BIN* negative control diet with DON and 1 kg/t mycotoxin binder. Principal components scores of subjects were analysed by General Linear model with fixed factor mycotoxin addition, binder addition and the interaction
[c]*BW* body weight, *CLDN-1* Claudin 1, *CLDN-2* Claudin 2, *CLDN-5* Claudin 5, *FD4* FITC-dextran 4, *IAP* Intestinal alkaline phosphatase, *IL-1β* Interleukin 1 beta, *IL-8* Interleukin 8, *OCLN* Occludin, *TLR-4* Toll like receptor 4, *TNF-α* Tumor necrosis factor alpha

88.5% of the DON dose was detected in the stomach whereas only 1.5% in the small intestine [23]. Also, the acetylated derivatives of DON are rapidly hydrolysed to DON in vivo and then absorbed. Thus, the distal small intestine might be less susceptible to DON as the majority of DON is already absorbed in the proximal parts of the GIT [24]. All animal species can exhibit toxic effects when exposed to DON [3], with pigs being the most susceptible species [4]. However, the severity depends on various factors, including type and dose of DON, the route and duration of application, as well as the animal status [25]. After application of DON on the apical side or on the basolateral side of IPEC-J2 cells, Diesing et al. found that the apical epithelium seems to be more resistant to DON application while the same

concentration of DON from basolateral side severely impairs barrier integrity [9]. In our case, it can be assumed that most DON was absorbed in the upper GIT, reaching the more susceptible basolateral side whereby only a small part of DON was left in the less susceptible apical side. That's probably the reason why little effect of DON administration on gut health was seen in samples from the distal small intestine. Also, it should be taken into account that generally cytokine induction upon DON exposure occurs within hours of exposure [26], and thus differences after chronic exposure as in our study might not always be present.

So far, data about the effects of DON are incomplete, especially due to the lack of in vivo data. At first, feed naturally contaminated with DON was used in studies in vivo and in vitro. However, interpretation of data from naturally contaminated diets is complicated as co-occurrence with other mycotoxins is commonly found in cereals. Mycotoxicoses may be caused by multiple toxins, making it difficult to unravel the separate effects of the target mycotoxin DON. At present, it is challenging to study the effect of DON and its masked mycotoxins ADONs. Purified mycotoxins are generally used in in vitro experiments. In addition, it is difficult to correlate in vitro exposure with in vivo dosage as the amount of mycotoxin that can be absorbed in vivo does not necessarily correspond to the amount absorbed by cells in culture. Taken together, it's difficult to conclude what dose DON will exhibit toxic properties, as the dose, the type of mycotoxin, the route and duration of exposure can all influence the mode of action.

Binder addition to the feed improves performance and some parameters of gut health

The mycotoxin binder used in this study was a combination of acid-activated bentonite, clinoptilolite, yeast cell walls and organic acids and salt. Acid-activated bentonite increases the adsorption capacity by using a specific acid activation process to increase the surface area and to enlarge pores [27, 28]. Clinoptilolite, with a honeycomb like structure, serves to bind a broad range of mycotoxins [29]. Yeast cell walls, which contain α-D-mannans and β-D-glucans, have an active role in reducing mycotoxins in animal feed [30, 31]. Yeast cell walls used in the binder are extracted and harvested in the early stage of the fermentation, during which the network of covalent bonds is less dense, which offers more flexibility and a maximal accessibility of the mycotoxin binding sites [32–34].

In the current study, the ingestion of diets supplemented with binder reduced the expression of *TLR-4* compared to diets with no binder. TLR-4 plays an important role in recognizing Gram-negative bacteria and activation of the innate immune system. Activation of TLR-4 leads to the release of its downstream inflammatory modulators, including TNF-α and IL-1 [35]. This mechanism is well known and supported in the present study by the positive association between *TLR-4*, *TNF-α* and *IL-1β* mRNA expression that was evident in principal component 2 from the PCA. TLR-4 is most well-known for recognizing lipopolysaccharides (LPS), a structural component of the outer membrane of Gram-negative bacteria. LPS induce strong inflammatory responses in vivo, and are released when the cell is lysed or during bacterial cell division. Supplementation of toxin binders has shown reduced expression of pro-inflammatory cytokines such as IL-1β and IL-6 and other immune responses in LPS-induced pigs [36]. In other words, the mycotoxin binder might not only bind to mycotoxins, but might also bind other toxins such as bacterial endotoxins. This was not specifically investigated in the present study but might have occurred.

At the same time, there was a tendency that the supplementation of binder up-regulated the expression of *CLDN-1*. Claudins function as major components of the tight junction strands that regulate the permeability of epithelia. CLDN-1, a member of the claudin family, is an integral membrane protein. Enhanced *CLDN-1* expression can decrease paracellular permeability and tighten the tight junctions. As the mycotoxin binder in the current study may have adsorbed a range of toxins, it could be hypothesized that it adsorbed other xenobiotics which might impair the barrier function by deregulation of TJ assembly.

From the result of PCA, we know that groups with the supplementation of binder tended to have higher scores for principal component 1, which is positively associated with gene expression of *OCLN*, *CLDN-1*, *CLDN-2* and *IAP*, as compared to diets without binder. This finding is consistent with the gene expression result of *CLDN-1*. The PCA suggests that binder may also up-regulate the expression of other TJPs (OCLN, CLDN-2) and IAP. OCLD, together with the claudin group of proteins, is an important component of the tight junctions. Studies have shown that rather than being important in assembly and maintenance of tight junctions, OCLN is important in stability and barrier function of tight junctions. As the *OCLN* gene is essential and plays a fundamental role in modulating the epithelial tight junctions, OCLN is a respective marker of epithelial barrier and its presence or absence could reflect the permeability of intestinal epithelium [37]. Taken the correlation into consideration, the binder may stimulate the expression of other TJPs as OCLN is highly correlated to ZO-1, ZO-2 and CLDN-7. The family of ZO is a part of the cytoplasmic plaque of the TJPs. The importance of maintenance of gut barrier integrity is further illustrated by the grouping of weight

at d14 and the mRNA levels of claudins within principal component 3.

IAP is a brush border enzyme which is a component of the gut mucosal defence system. IAP is involved in regulating secretion of bicarbonate in the duodenum. Failure to neutralize acid environment can lead to acidified chyme injuring epithelial cells, finally increasing inflammation and intestinal permeability. IAP is also known to detoxify LPS and prevent bacterial translocation in the gut [38]. As discussed, LPS will induce strong inflammatory responses in vivo. In other words, IAP can inhibit the inflammatory responses by detoxification of LPS. So, IAP is an important indicator to gut health.

Taken together, the addition of mycotoxin binder could improve the gut health by decreasing the expression of TLR-4 as well as increasing the expression of TJPs and IAP.

Conclusions

The addition of a mycotoxin binder showed beneficial effects for weaned piglets, especially when diets were contaminated with 3 mg/kg of a mixture of DON in the pre-starter period. Growth and feed intake were enhanced. In line with this, reduced toll-like receptor-4 and increase of tight junction protein gene expression might shed light on the mode of action of the binder. However, the current study does not allow to assess whether the effects of the binder are mediated by alterations in the toxicokinetics of the mycotoxin.

Abbreviations
15A–DON: 15-acetyl- deoxynivalenol; 3A–DON: 3-acetyl- deoxynivalenol; AC: Activated carbon; ADFI: Average daily feed intake; ADG: Average daily gain; ADONs: Acetylated-deoxynivalenols; AFB1: Aflatoxin B1; CLDN-1: Claudin 1; CLDN-2: Claudin 2; CLDN-5: Claudin 5; CLDN-7: Claudin 7; DON: Deoxynivalenol; F:G: Feed to Gain Ratio; FB1: Fumonisin B1; FD4: FITC-dextran 4; GIT: Gastrointestinal tract; HCK: Hematopoietic cell kinase; HPRT-1: Hypoxanthine Phosphoribosyltransferase 1; HSCAS: Hydrated sodium calcium aluminosilicate; IAP: Intestinal alkaline phosphatase; IFN-γ: Interferon gamma; Ig: Immunoglobulins; IL-1β: Interleukin 1 beta; IL-8: Interleukin 8; LPS: Lipopolysaccharide; MAPK: Mitogen-activated protein kinase; NIV: Nivalenon; OCLN: Occludin; OTA: Ochratoxin A; PCA: Principal component analysis; PPIA: Peptidylprolyl isomerase A; SAPK/JNK: Stress-activated protein kinases/cJun N-terminal kinases; TBP: TATA-binding protein; TJP: Tight junction protein; TLR-4: Toll like receptor 4; TNF-α: Tumor necrosis factor alpha; ZEA: Zearalenone; ZO-1: Zona occludens 1; ZO-2: Zona occludens 2

Acknowledgements
We thank Ms. Tessa Van Der Eecken for her technical support and animal care.

Funding
The research was supported by Nutrex company. The funders had no role in data collection and analysis, or drafting of the manuscript.

Authors' contributions
LJ, SDS and JM conceived and designed the experiments. LJ, WW, JD, NVN, HY and MM conducted the piglets study, collected and analysed the samples. LJ,

WW, MVP and LP were involved in the gene expression analysis. LJ analysed the data and wrote the first draft of the manuscript. All authors read and approved the final manuscript.

Competing interests

The authors declare that they have no competing interests, except AG, KVDM and RM. AG, KVDM and RM are employed by the funding company, however they had no role in data collection and analysis, or drafting of the manuscript. They approved the final manuscript.

Author details

[1]Department of Applied Biosciences, Ghent University, Valentin Vaerwyckweg 1, 9000 Ghent, Belgium. [2]Laboratory for Animal Nutrition and Animal Product Quality, Department of Animal Production, Ghent University, Coupure Links 653, 9000 Ghent, Belgium. [3]Department of Nutrition, Genetics and Ethology, Faculty of Veterinary Medicine, Ghent University, Heidestraat 19, 9820 Merelbeke, Belgium. [4]Nutrex, Achterstenhoek 5, 2275 Lille, Belgium.

References

1. EFSA. Opinion of the Scientific Panel on contaminants in the food chain on a request from the commission related to Deoxynivalenol (DON) as undesirable substance in animal feed. EFSA J. 2004;73:1–42.
2. Turner PC, Burley VJ, Rothwell JA, White KL, Cade JE, Wild CP. Deoxynivalenol: rationale for development and application of a urinary biomarker. Food Add Contam. 2008;25:864–71.
3. Pestka JJ, Smolinski AT. Deoxynivalenol: toxicology and potential effects on humans. J Toxicol Environ Health-Part B-Crit Rev. 2005;8:39–69.
4. Eriksen GS, Pettersson H. Toxicological evaluation of trichothecenes in animal feed. Anim Feed Sci Technol. 2004;114:205–39.
5. Hecht G. Innate mechanisms of epithelial host defense: spotlight on intestine. Amer J Physiology-Cell Physiol. 1999;277:C351–8.
6. Podolsky D. Review article: healing after inflammatory injury–coordination of a regulatory peptide network. Alim Pharmacol Therap. 2000;14(Suppl1):87–93.
7. Pinton P, Nougayrède JP, Del Rio JC, Moreno C, Marin DE, Ferrier L, et al. The food contaminant deoxynivalenol, decreases intestinal barrier permeability and reduces claudin expression. Toxicol Appl Pharmacol. 2009;237:41–8.
8. Bracarense AP, Lucioli J, Grenier B, Drociunas Pacheco G, Moll WD, Schatzmayr G, et al. Chronic ingestion of deoxynivalenol and fumonisin, alone or in interaction, induces morphological and immunological changes in the intestine of piglets. Brit J Nutr. 2012;107:1776–86.
9. Diesing AK, Nossol C, Dänicke S, Walk N, Post A, Kahlert S, et al. Vulnerability of polarised intestinal porcine epithelial cells to mycotoxin deoxynivalenol depends on the route of application. PLoS One. 2011;6
10. Pestka JJ. Deoxynivalenol: mechanisms of action, human exposure, and toxicological relevance. Arch Toxicol. 2010;84:663–79.
11. Pinton P1, Braicu C, Nougayrede JP, Laffitte J, Taranu I, Oswald IP. Deoxynivalenol impairs porcine intestinal barrier function and decreases the protein expression of claudin-4 through a mitogen-activated protein kinase-dependent mechanism. J Nutr. 2010;140:1956–62.
12. Doyle MP, Applebaum RS, Brackett RE, Marth EH. Physical, chemical and biological degradation of mycotoxins in foods and agricultural commodities. J Food Prot. 1982;45:964–71.
13. Ramos AJ, Hernandez E. Prevention of aflatoxicosis in farm animals by means of hydrated sodium calcium aluminosilicate addition to feedstuffs: a review. Anim Feed Sci Technol. 1997;65:197–206.
14. Huwig A, Freimund S, Käppeli O, Dutler H. Mycotoxin detoxication of animal feed by different adsorbents. Toxicol Letters. 2001;122:179–88.
15. Döll S, Dänicke S. In vivo detoxification of Fusarium toxins. Arch Anim Nutr. 2004;58:419–41.
16. Biomin, Biomin mycotoxin survey 2016. 2017. https://www.biomin.net/ru/blog-posts/2016-biomin-mycotoxin-survey-results-for-swine-feed. Accessed 5 March 2017.
17. Monbaliu S, Van Poucke C, Detavernier C, Dumoulin F, Van De Velde M, Schoeters E, et al. Occurrence of mycotoxins in feed as analyzed by a multi-mycotoxin LC-MS/MS method. J Agri Food Chem. 2010;58:66–71.
18. Haschek WM, Rousseaux CG. Mycotoxins. Chapter 39 In Haschek and Rousseaux's Handbook Toxicol Pathol, 3rd Ed. 2013;645:699.
19. Wang W, Degroote J, Van Ginneken C, Van Poucke M, Vergauwen H, Dam TM, et al. Intrauterine growth restriction in neonatal piglets affects small intestinal mucosal permeability and mRNA expression of redox-sensitive genes. FASEB J. 2016;30:863–73.
20. Montagne L, Boudry G, Favier C, Le Huërou-Luron I, Lallès JP, Sève B. Main intestinal markers associated with the changes in gut architecture and function in piglets after weaning. Brit J Nutr. 2007;97:45–57.
21. Michiels J, De Vos M, Missotten J, Ovyn A, De Smet S, Van Ginneken C. Maturation of digestive function is retarded and plasma antioxidant capacity lowered in fully weaned low birth weight piglets. Brit J Nutr. 2013;109:65–75.
22. Goyarts T, Danicke S. Bioavailability of the Fusarium toxin deoxynivalenol (DON) from naturally contaminated wheat for the pig. Toxicol Letters. 2006;163:171–82.
23. Dänicke S, Valenta H, Klobasa F, Döll S, Ganter M, Flachowsky G. Effects of graded levels of Fusarium toxin contaminated wheat in diets for fattening pigs on growth performance, nutrient digestibility, deoxynivalenol balance and clinical serum characteristics. Arch Anim Nutr. 2004;58:1–17.
24. Awad WA, Aschenbach JR, Setyabudi FM, Razzazi-Fazeli E, Böhm J, Zentek J. In vitro effects of deoxynivalenol on small intestinal D-glucose uptake and absorption of deoxynivalenol across the isolated jejunal epithelium of laying hens. Poultry Sci. 2007;86:15–20.
25. Bondy GS, Pestka JJ. Immunomodulation by fungal toxins. J Toxicol Environ Health-Part B-Crit Rev. 2000;3:109–43.
26. Zhou HR, Islam Z, Pestka JJ. Rapid, sequential activation of mitogen-activated protein kinases and transcription factors precedes proinflammatory cytokine mRNA expression in spleens of mice exposed to the trichothecene vomitoxin. Toxicol Sci. 2003;72:130–42.
27. Motlagh MK, Youzbashi AA, Rigi ZA. Effect of acid activation on structural and bleaching properties of a bentonite. Iran J Mat Sci Engineer. 2011;8:50–6.
28. Salem A, Karimi L. Physico-chemical variation in bentonite by sulfuric acid activation. Kor J Chem Engineer. 2009;26:980–4.
29. Sabet FA, Libre NA, Shekarchi M. Mechanical and durability properties of self consolidating high performance concrete incorporating natural zeolite, silica fume and fly ash. Construct Building Mat. 2013;44:175–84.
30. Yiannikouris A, André G, Poughon L, François J, Dussap CG, Jeminet G, et al. Chemical and conformational study of the interactions involved in mycotoxin complexation with β-D-glucans. Biomacromolecules. 2006;7: 1147–55.
31. Kogan G, Kocher A. Role of yeast cell wall polysaccharides in pig nutrition and health protection. Livest Sci. 2007;109:161–5.
32. Aguilar-Uscanga B, Francois J. A study of the yeast cell wall composition and structure in response to growth conditions and mode of cultivation. Letters Appl Microbiol. 2003;37:268–74.
33. Klis FM, Boorsma A, De Groot PW. Cell wall construction in Saccharomyces cerevisiae. Yeast. 2006;23:185–202.
34. Klis FM, Mol P, Hellingwerf K, Brul S. Dynamics of cell wall structure in Saccharomyces cerevisiae. FEMS Microbiol Rev. 2002;26:239–56.
35. Telepnev M, Golovliov I, Grundström T, Tärnvik A, Sjöstedt A. Francisella tularensis inhibits toll-like receptor-mediated activation of intracellular signalling and secretion of TNF-α and IL-1 from murine macrophages. Cell Microbiol. 2003;5:41–51.
36. Collier CT, Carroll JA, Ballou MA, Starkey JD, Sparks JC. Oral administration of Saccharomyces cerevisiae boulardii reduces mortality associated with immune and cortisol responses to Escherichia coli endotoxin in pigs. J Anim Sci. 2011;89:52–8.
37. Saitou M, Furuse M, Sasaki H, Schulzke JD, Fromm M, Takano H, et al. Complex phenotype of mice lacking occludin, a component of tight junction strands. Mol Biol Cell. 2000;11:4131–42.
38. Bates JM, Akerlund J, Mittge E, Guillemin K. Intestinal alkaline phosphatase detoxifies lipopolysaccharide and prevents inflammation in zebrafish in response to the gut microbiota. Cell Host Microbe. 2007;2:371–82.

17

Impact of hormonal modulation at proestrus on ovarian responses and uterine gene expression of suckled anestrous beef cows

Manoel Francisco de Sá Filho[1], Angela Maria Gonella-Diaza[1], Mariana Sponchiado[1], Marcio Ferreira Mendanha[1], Guilherme Pugliesi[1], Roney dos Santos Ramos[1], Sónia Cristina da Silva Andrade[2], Gustavo Gasparin[3], Luiz Lehmann Coutinho[3], Marcelo Demarchi Goissis[1], Fernando Silveira Mesquita[4], Pietro Sampaio Baruselli[1] and Mario Binelli[1,5*] (iD)

Abstract

Background: This study evaluated the impact of hormonal modulation at the onset of proestrus on ovarian response and uterine gene expression of beef cows.

Methods: A total of 172 anestrous beef cows were assigned to one of four groups according to the treatment with estradiol cypionate (ECP) and/or equine chorionic gonadotropin (eCG) [CON (n = 43), ECP (n = 43), eCG (n = 44) and ECP + eCG (n = 42)].

Results: ECP-treated cows (ECP and ECP + eCG groups) presented greater occurrence of estrus (44.6% vs. 65.4%; P = 0. 01) and pregnancy per AI [47.1% vs. 33.3%; P = 0.07], but similar progesterone (P4) concentration at subsequent diestrus than cows not treated with ECP (CON and eCG groups). Nonetheless, eCG-treated cows (eCG and ECP + eCG groups) presented larger follicle at timed AI (12.6 ± 0.3 vs. 13.5 ± 0.3 mm; P = 0.03), greater ovulation rate (96.5% vs. 82.6%; P = 0.008) and greater P4 concentration at d 6 (3.9 ± 0.2 vs. 4.8 ± 0.2 ng/mL; P = 0.001) than cows not treated with eCG (CON and ECP groups). Next, cows with a new corpus luteum 6 d after TAI were submitted to uterine biopsy procedure. Uterine fragments [CON (n = 6), ECP (n = 6)] were analyzed by RNA-Seq and a total of 135 transcripts were differentially expressed between groups (73 genes up-regulated by ECP treatment). Subsequently, uterine samples were analyzed by qPCR (genes associated with cell proliferation). ECP treatment induced greater abundance of PTCH2 (P = 0.07) and COL4A1 (P = 0.02), whereas suppressed EGFR (P = 0.09) expression. Conversely, eCG treatment increased abundance of HB-EGF (P = 0.06), ESR2 (P = 0.09), and ITGB3 (P = 0.05), whereas it reduced transcription of ESR1 (P = 0.05). Collectively, supplementation with ECP or eCG at the onset of proestrous of anestrous beef cows influenced ovarian responses, global and specific endometrial gene expression.

Conclusion: Proestrus estradiol regulate the endometrial transcriptome, particularly stimulating proliferative activity in the endometrium.

Keywords: Cattle, eCG, Endometrium, Estradiol, Transcriptome

* Correspondence: binelli@usp.br
[1]Departamento de Reprodução Animal, FMVZ-USP, São Paulo, SP, Brazil
[5]Universidade de São Paulo, Faculdade de Medicina Veterinária e Zootecnia, Departamento de Reprodução Animal, Avenida Duque de Caxias Norte, 225, Pirassununga, SP Zip Code 13635900, Brazil
Full list of author information is available at the end of the article

Background

Synchronization of estrus and ovulation programs for timed artificial insemination (TAI) has been constantly incorporated on modern reproductive management of beef farms [1, 2]. These programs can induce the first postpartum ovulation and, consequently, hasten the establishment of pregnancy of suckled beef cows [1, 3–5]. However, a significant proportion of ovulated and inseminated cows are detected not-pregnant 30 d after insemination despite the satisfactory ovulation rate (~85%) following protocols for synchronization of ovulation [4, 6, 7]. The uterine environment plays a relevant role among factors that are likely to contribute to the observed failures [8–10].

Early classic studies demonstrated the significant impact of a coordinated and sequential exposure to ovarian steroids on uterine function [11–13]. Gene expression of bovine endometrium changes according to the phase of the estrous cycle and is closely controlled by circulating concentrations of estradiol (E2), progesterone (P4) and the expression ratio of their specific receptors [8, 14–17]. In this regard, proestrus E2 concentration is fundamental in modulation of the uterus for the subsequent luteal phase [8, 14, 18, 19]. This E2 priming may be important for induction of endometrial P4 receptors [20, 21] to avoid premature luteolysis and short cycles in beef cattle [22]. In cyclic dairy heifers, elevated E2 concentrations during proestrus, induce changes in uterine gene expression of E2 and P4 receptors (*ESR1* and *PGR*, respectively), oxytocin receptors, and expression of cyclooxygenase-2, and beta subunit inhibin serpin-14 throughout the subsequent estrus cycle [23]. Also, cyclic beef heifers that are exposed to a longer proestrus period exhibit alterations in the pattern of steroids receptors expression in the uterus and other proteins associated with uterine receptivity to pregnancy [19]. Therefore, it is reasonable to hypothesize that the modulation of E2 concentration during the synchronized proestrus by means of exogenous E2 supplementation could also alter the uterine gene expression of suckled anestrous beef cows.

Two pharmacological strategies to manipulate the proestrus phase have been extensively evaluated in cattle breeding programs; exogenous E2 supplementation or equine chorionic gonadotropin (eCG) administration. Firstly, exogenous E2 supplementation using E2 esters enhances the proportion of cows that display estrus [24–26], increases endometrial thickness in lactating dairy cows [27] and improves the pregnancy success of suckled beef cows [6, 25, 26]. Furthermore, Jinks et al. [28] demonstrated that, recipients beef cows with lower E2 concentration at periovulatory phase, receiving in vivo-produced embryo, presented a dramatic reduction on pregnancy establishment (45% vs. 65% of pregnancy rate). Secondly, administration of eCG at onset of the proestrus is an efficient alternative to increase final follicular growth, ovulation rate and plasma P4 concentration on subsequent diestrus [5, 26, 29, 30]. Such changes may be responsible for the increase in pregnancy rates of anestrous beef cows stimulated with eCG [5, 26, 29, 30]. Altogether, both pharmacological strategies to manipulate the proestrus are capable of altering the periovulatory steroidal endocrine profiles, potentially modulating the expression of genes associated with uterine receptivity and ultimately positively influencing pregnancy establishment of suckled anestrous beef cows.

Therefore, based on the importance of the proestrus hormonal milieu on fertility, we hypothesized that supplementation with estradiol cypionate (ECP) and/or eCG at the onset of proestrus alters the ovarian response and the uterine transcriptome of suckled anestrous beef cows. To assess the above mentioned hypothesis, we chose the following approaches. First, taking a comprehensive approach, RNA extracts from endometrial fragments were submitted to Next Generation RNA sequencing followed by functional enrichment analysis to potentially identify and characterize other ECP-regulated biological and molecular processes and pathways. Secondly, following a candidate gene approach, we tested the effect of ECP and/or eCG supplementation on the expression of selected molecules with relevant biological functions in the context of uterine biology, specifically associated with cell proliferation.

Methods

Animals

Animal procedures were approved by the Ethics and Animal Handling Committee of the Faculdade de Medicina Veterinária e Zootecnia, Universidade de São Paulo (CEUA-FMVZ/USP, No. 2287/2011). This experiment was conducted during the 2012/2013 spring-summer breeding seasons. A total of 172 suckled anestrous Nelore (*Bos indicus*) beef cows at 30–60 d postpartum from a commercial farm in the state of Parana, Brazil, were enrolled in this study. Cows were maintained on *Brachiaria brizantha* pasture with water and mineral supplementation ad libitum. Immediately prior to the initiation of the TAI protocol, information about body condition score from each cow were recorded (BCS; range, 1 = emaciated to 5 = obese; with 0.5 scale) [31].

Reproductive management and experimental design

After calving, cows were allocated into breeding groups according to calving date. At 30 to 60 d post-partum, females were synchronized using an E2-plus-P4-based TAI protocol. Briefly, suckled cows received an intravaginal P4-releasing insert previously used for 8 d (1 g of P4; DIB®, MSD Animal Health, São Paulo, Brazil) on D −10 along with an intramuscular (IM) administration of 2 mg estradiol benzoate (EB; Gonadiol®, MSD Animal Health, São Paulo, Brazil; Fig. 1). The P4 insert were removed eight day later (D −2). All cows received an intramuscular administration of 500 mg of cloprostenol (Ciosin®, MSD Animal Health, São Paulo, Brazil) at the moment of the P4 insert removal. At this moment, cows were blocked by BCS, parity (multiparous vs. primiparous) and the diameter of the largest follicle and then randomly assigned into one of four experimental groups [Control (CON): n = 43, Estradiol cypionate (ECP): n = 43, eCG: n = 44, and ECP + eCG: n = 42], in a 2 × 2 factorial arrangement. Cows from ECP group received an IM injection of 1 mg of ECP (E.C.P.; Zoetis, São Paulo, Brazil), cows from eCG group received an IM injection of 400 IU of eCG (Folligon®, MSD Animal Health), while cows from ECP + eCG group received both treatments and cows from CON group did not receive any treatment. In all groups, ovulation was induced by 10 μg of buserelin acetate (GnRH, Sincroforte, Ourofino Saúde Animal, Cravinhos, São Paulo, Brazil) IM administration 48 h after the P4 insert removal (D 0). Cows were artificially inseminated immediately after GnRH treatment. Inseminations were performed by a single technician using frozen-thawed semen from single Angus sire with proven fertility. The sire used had been previously used in TAI programs and had satisfactory (~50%) pregnancy results.

Estrus was determined based on the tail-head mark. At the time of the removal of the P4 insert, the tail-head was marked with chalk (Raidl-Maxi, RAIDEX GmbH, Dettingen/Erms, Germany). Estrus was deemed to have occurred in cattle without a tail-head mark at TAI.

Cows presenting a corpus luteum (CL) on D 6 (6 d after GnRH treatment) had the body of the uterus biopsied as previously described [32]. Fragments obtained from uterine biopsies were individually allocated in cryotubes and immediately immersed into liquid nitrogen. Day 6 was strategically selected as the moment in which an early embryo is expected to have recently accessed the uterine environments. Pregnancy was diagnosed by transrectal ultrasonography through the detection of a viable embryo (presence of heartbeat) on d 42 post-AI.

Blood sampling and hormone measurements

Blood sampling for determination of P4 concentrations was performed on D 6, concurrently with the uterine biopsy. Blood samples were collected by coccygeal venipuncture using evacuated tubes containing EDTA (BD, São Paulo, SP, Brazil) and immediately stored in ice. Plasma was separated by centrifugation at room temperature, 1,500 × g for 15 min, and stored at −20 °C. Progesterone concentrations were measured in all samples using a solid-phase radioimmunoassay (Coat-a-count, Siemens, Los Angeles, USA), as validated previously [33]. The P4 assay sensitivity was 0.08 ng/mL and the intra-assay coefficient of variation was 8.7%.

Ultrasound examinations

Transrectal ultrasound examinations were carried out on D −10, D −2, D 0 and D 6 to assess cyclic status, growth of the dominant follicle (DF), ovulation, and the presence of CL. Ultrasonography was performed with the aid of a B-mode (gray-scale) ultrasound instrument (8100, Chison Medical Imaging, Co, China), equipped with a multi-frequency linear-array transducer. The anestrous status was defined as the absence of CL in two consecutive ultrasound examinations performed on

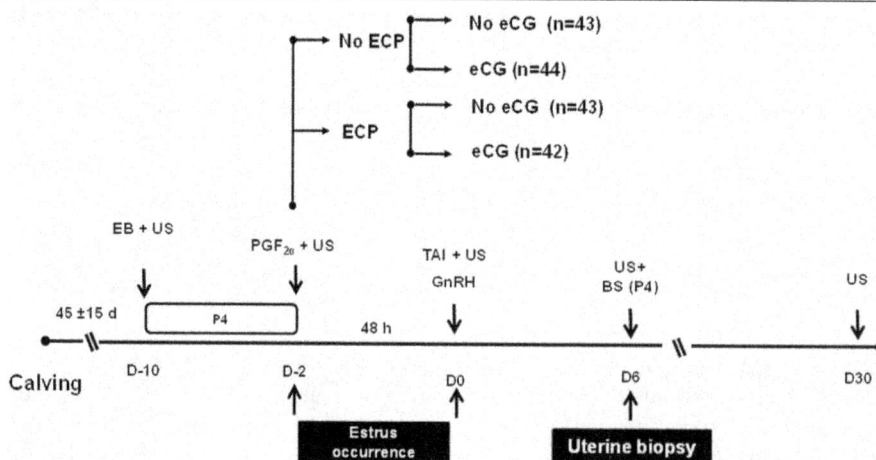

Fig. 1 Schematic diagram of the synchronization of ovulation protocol in suckled anestrous beef cows. EB = 2 mg of estradiol benzoate; P4 = progesterone; P4 insert = previously used intravaginal P4 insert containing 1.0 g of P4; GnRH = 100 mg of gonadorelin; ECP = 1 mg of estradiol cypionate; eCG = 400 IU of equine chorionic gonadotropin; PGF2α = 0.25 mg of cloprostenol; US = ultrasound examination; BS = blood sample. Cows from ECP group received ECP and cows from eCG group received eCG, while cows from CON did not receive any further treatment and cows from ECP + ECG received both treatments

D – 10 and D –2. Ovulation was defined as the presence of a recently formed CL on D 6 on the same ovary that the DF was observed on D –2 and D 0. The diameter of the DF at the time of P4 insert removal and at TAI, in addition to the diameter of new CL formed, was calculated as the average between measurements of two perpendicular axes of each structure.

Tissue processing, RNA isolation and cDNA synthesis

Approximately 30 mg of endometrial tissue was macerated in liquid nitrogen using a stainless steel mortar and pestle and immediately mixed with buffer RLT from the PureLink®, RNA Mini kit (Thermo Fisher Scientific, São Paulo, SP, Brazil), as per manufacturer's instructions. To maximize lysis, tissue suspension was passed at least ten times through a 21-ga needle, and centrifuged at $13,000 \times g$ for 3 min for removal of debris, prior to supernatant loading and processing on RNeasy columns. Columns were eluted with 40 µL of RNase free water. Concentration of total RNA on extracts was measured by a spectrophotometer (Nanovue™ Plus, Spectrophotometer, GE Healthcare, UK, by the absorbance at 260 nm). Subsequently, samples were treated with 80 µL of DNAse I solution (Life Technologies, São Paulo, SP, Brazil) for 15 min at room temperature during RNA extraction protocol, according to manufacturer's instructions. RNA samples were stored at –80 °C until cDNA synthesis. The cDNA was synthesized by reverse-transcription using High Capacity cDNA Reverse Transcription Kit (Life Technologies) according to manufacturer's instructions. Briefly, 10 µL of master mix containing RT buffer, dNTP mix, random primers, RNase inhibitor and reverse transcriptase were mixed to 1 µg of total RNA and final volume of the reaction was adjusted to 20 µL. Immediately, reactions were incubated at 25 °C for 10 min, followed by incubation at 37 °C for 2 h, and reverse-transcriptase inactivation at 85 °C for 5 min. Samples were stored at –20 °C.

RNAseq

Prior to the RNA-seq analyses, 12 samples (n = 6/group; ECP and CON) were selected according to previously established criteria by ovarian, occurrence of estrus, pregnancy and endocrine responses. Cows having similar DF diameter at the time of P4 insert removal [ECP (12.1 ± 0.7 mm) and CON (12.1 ± 0.6 mm)] and similar circulating P4 concentration at the time of uterine biopsy [ECP (3.8 ± 0.2 ng/mL) and CON (3.6 ± 0.2 ng/mL)] were considered suitable to further analysis. Additionally, cows were also selected based on pregnancy status 30 d after TAI in order to have both pregnant and non-pregnant cows represented in both experimental groups. Finally, only cows displaying estrus were selected in ECP treated group, whereas only cows that did not display estrus were chosen in the control group. The latter criterion was applied aiming to increase the distinction between two different E2 pre-ovulatory endocrine environments, as cows that display estrus present greater E2 concentration than those not displaying estrus [42].

Integrity of total RNA extracts was assessed using the Agilent RNA 6000 Nano chip (Bioanalyzer, Agilent Technologies). RNA Integrity Number (RIN) of extracts submitted to RNA sequencing analysis ranged from 8.3 to 8.7. Next, 4 µg of RNA were used with the TruSeq RNA Sample Preparation kit (Illumina, San Diego, CA) to prepare the libraries for RNA-Seq. The insert sizes were estimated through the Agilent DNA 1000 chip (Agilent Technologies) and the libraries concentration were measured through Quantitative Real-Time PCR (qPCR) with a KAPA Library Quantification kit (KAPA Biosystems). Samples were diluted, pooled in equimolar amounts and then sequenced at the Centro Genômico Funcional Aplicado a Agropecuária e Agroenergia at the University of São Paulo using a HiScanSQ sequencer (Illumina, San Diego, CA).

Bioinformatics analyses

Raw sequences were trimmed for adaptors and low quality using SeqyClean v1.3.12. (https://github.com/ibest/seqyclean) using 26 Phred quality parameter for maximum average error and a fasta file with contaminant sequences from the Univec database (https://www.ncbi.nlm.nih.gov/tools/vecscreen/univec/). Only high quality paired-end sequences were kept for further analyses The reads were mapped with Bowtie2 v2.1.0 [34] on the masked bovine genome assembly (*Bos taurus* UMD 3.1, NCBI). The mapping file was sorted using SAMTools v 0.1.18 [35] and read counts were obtained using the script from HTSeq-count v0.5.4p2 (http://htseq.readthedocs.io/en/release_0.9.1/). The differential expression analysis was performed with package DESeq2 [36] from R [37]. Using the function estimateSizeFactors, the normalized counts were obtained (baseMean values, which are the number of reads divided by the size factor or normalization constant). The standard deviation along the baseMean values was also calculated for each gene. In order to avoid artifacts caused by low expression profiles and high expression variance, only transcripts that had an average of baseMean >5 and the mean greater than the standard variation were analyzed. The threshold for evaluating significance was obtained by applying an alpha ≤0.10, considering the FDR-Benjamini-Hochberg P-value [38]. Integrated analysis of different functional databases was done using the functional annotation tool of the Database for Annotation, Visualization, and Integrated Discovery using as background the genes (DAVID) [39] using as background the set of genes that passed through the differential expression analysis filter.

qPCR

The samples employed in qPCR analysis were selected mirroring the general results obtained in regard to ovarian and endocrine responses. Cows receiving ECP should present greater occurrence of estrus, while cows from eCG treatment group should present greater circulation of P4 concentration at the moment of the uterine biopsy. Step-One Plus thermocycler (Life Technologies, Carlsbad, CA) and SYBR Green chemistry were used for quantitative PCR analysis. Primers were designed based on the mRNA sequence of target genes obtained from the RefSeq database, on Genbank (http://www.ncbi.nlm.nih.gov/genbank/). Sequences were masked to remove repetitive sequences with RepeatMasker (http://www.repeatmasker.org/) [40] and then, the masked sequences were used for primer design using the PrimerQuest software (IDT1, http://www.idtdna.com/primerquest/Home/Index). The characteristics of the primers were checked in Oligo Analyzer 3.1 software (IDT1, http://www.idtdna.com/analyzer/Applications/OligoAnalyzer/), while the specificity was compared by BLAST (NCBI, http://blast.ncbi.nlm.nih.gov). The qPCR products obtained from reactions performed with primers not previously validated were submitted to agarose gel electrophoresis and SANGER-DNA sequencing, and identities of target genes were confirmed. Details of primers are provided on Table 1. In order to select reference genes, the GeNorm Microsoft Excel applet was used, as this applet provides a measure of gene expression stability (M) [41]. The Glyceraldehyde-3-Phosphate Dehydrogenase (GAPDH), Actin, Beta (ACTB) and Ribosomal Protein S18 (RPS18)

were the most stable genes and were, therefore, selected as reference genes. Determination of qPCR efficiency and Cq (quantification cycle) values per sample were performed with LinRegPCR software (V2014.2; http://www.hartfaalcentrum.nl/index.php?main=files&fileName=LinRegPCR.zip&description=LinRegPCR:%20qPCR%20data%20analysis&sub=LinRegPCR). Quantification was obtained after normalization of the target genes expression values (Cq values) by the geometric mean of the endogenous control expression values. The following genes, associated with regulation of cell proliferation in the uterus, were selected: ovarian steroid receptors [Estrogen Receptor alpha (ESR1), Estrogen Receptor beta (ESR2), P4 receptor (PGR)], growth factors that regulate cellular proliferation [epidermal growth factor receptor (EGFR), heparin-binding EGF-like growth factor (HB-EGF) and patched homolog 2 (PTCH2)], and extracellular matrix [collagen, type IV, alpha 1 (COL4A1) and integrin, beta 3 (ITGB3)].

Statistical analyses from ovarian, endocrine and gene expression responses

The statistical analyses for ovarian responses were performed using the PROC GLIMMIX of SAS for Windows (SAS 9.3 Institute Inc., Cary, NC, USA, 2003). Continuous variables were presented as mean ± standard error of the mean (mean ± SEM) and percentage (%) for frequency of occurrence for binomial variables. The continuous response variables were subjected to response scaling test through the solution Guided Data Analysis of SAS. Variables that did not follow these assumptions were transformed accordingly. Binomial variables (i.e.

Table 1 Gene name, accession number, forward and reverse primer sequences used for qPCR analysis

Gene Name	Gene ID	Sequence ID	Forward primer sequence (5'–3')	Reverse sequence (5'–3')	Primer efficiency, %	Amplicon length, bp
Progesterone receptor	PGR	NM_001205356.1	GCCGCAGGTCTACCAGCCCTA	GTTATGCTGTCCTTCCATTGCCCTT	96.9	199
Estrogen receptor 1	ESR1	NM_001001443.1	CAGGCACATGAGCAACAAAG	TCCAGCAGCAGGTCGTAGAG	99.1	82
Estrogen receptor 2	ESR2	NM_174051.3	TCACGTCAGGCACGCCAGTAAC	CACCAGGTTGCGCTCAGACCC	99.5	155
Patched 2[a]	PTCH2	XM_005197904.1	CATCCTGCTGCTGTGTACTT	ATCGCCAGGACCAGTACTAT	99.9	87
Epidermal growth factor receptor	EGFR	XM_002696890.3	ATGCTCTATGACCCTACCAC	TTCCGTTACAAACTTTGCCA	97.6	178
Heparin-binding EGF-like growth factor	HB-EGF	NM_001144090.1	CATCCACGGAGAATGCAAATAC	CAGCAGACAGACGGATGATAG	98.6	181
Collagen, type IV, alpha 1	COL4A1	NM_001166511.1	CACGGCTACTCTTTGCTCTAC	GAAGGGCATGGTACTGAACTT	96.48	102
Integrin, beta 3 (platelet glycoprotein IIIa, antigen CD61)	ITGB3	NM_001206490.1	GGGAGAGTGCTATGGTTAGA	CTTCACAAGACACCCAAGAG	92.09	142
Actin Beta	ACTB	NM_173979.3	GGATGAGGCTCAGAGCAAGAGA	TCGTCCCAGTTGGTGACGAT	93.7	77
Glyceraldehyde-3-Phosphate Dehydrogenase	GAPDH	NM_001034034.2	GCCATCAATGACCCCTTCAT	TGCCGTGGGTGGAATCA	99.99	69
Ribosomal Protein S18	RPS18	AY786141.1	TGGAGAGTATTGCGCCTTCTC	CACAAGTTCCACCACACTATTGG	97.9	79

[a]Transcript variants X1 to X7

occurrence of estrus and ovulation rate) were analyzed by logistic regression using the SAS GLIMMIX procedure with models fitted to binomial distributions. The explanatory variables considered for inclusion in the models were the treatment with ECP, eCG and interaction of ECP and eCG. The effect of cow within each replicate was included as a random effect.

The qPCR data were tested for normality of residuals and homogeneity of variances followed by ANOVA using the GLIMMIX procedure of SAS fitting log normal distribution. The explanatory variables considered for inclusion in the models were the treatment with ECP, eCG and interaction between ECP and eCG. Final results are presented in natural log (Ln) scale (because of the log normal distribution considered) as normalized values of a specific gene transcript by the mean level of the transcript from Control (No-ECP and No-eCG treated animals). Down-regulation of expression in a specific experimental group may be represented by negative values relative to control because of Ln scale. To avoid negative values, the mean used for data normalization was divided by the fifth negative exponent. All data were compared with the relative mean expression level of the control group.

Statistical difference was considered when $P < 0.10$. Graphs were plotted with Sigmaplot (version 11.0; Systat Software, Inc. San Jose, CA, USA).

Results

Ovarian, pregnancy and endocrine responses

Animals receiving different hormonal therapies at the proestrus presented different rates of occurrence of estrus between P4-releasing device removal and TAI, final follicular growth, ovulatory responses and subsequent CL function (Table 2). There were no interactions between ECP and eCG treatment on response variables, except for the CL diameter 6 d after the TAI ($P = 0.06$). Larger CLs were observed in cows treated with eCG, especially in cows not treated with ECP. The ECP treated cows presented a greater frequency of occurrence of estrus [ECP = 64.7% (55/85) vs. No-ECP = 44.8% (39/87); $P = 0.008$] and presented greater pregnancy per TAI [ECP = 47.1% (40/85) vs. No-ECP = 33.3% (29/87); $P = 0.07$]. Cows treated with eCG presented greater rate of final follicular growth [eCG = 1.2 ± 0.1 mm/d vs. No-eCG = 0.9 ± 0.1 mm/d; $P = 0.01$], resulting in a greater DF diameter at TAI [eCG = 13.5 ± 0.3 mm vs. No-eCG = 12.6 ± 0.3 mm; $P = 0.03$]. Also, a greater proportion of cows receiving eCG displayed estrus [eCG = 62.8% (54/86) vs. No-eCG = 46.5% (40/86); $P = 0.03$] and ovulated [eCG = 96.5% (83/85) vs. No-eCG = 82.6% (71/86); $P = 0.008$]. A greater P4 concentration at the moment of uterine biopsy (D 6) was observed in cows receiving eCG at the onset of the proestrus [eCG = 4.8 ± 0.2 ng/mL vs.

Table 2 Overall occurrence and effects of treatment with estradiol cypionate (ECP) and/or equine chorionic gonadotropin (eCG) at onset of the proestrus on follicular and luteal development in an estradiol/progesterone-based synchronization protocol on anestrous suckled beef cows

Itens	Treatments[1]				Pvalue		
	No ECP		ECP				
	No eCG	eCG	No eCG	eCG	ECP	eCG	ECP × eCG
Number of cows	43	44	43	42	–	–	–
BCS at onset of the synchronization[2]	3.1 ± 0.1	3.0 ± 0.1	2.9 ± 0.1	3.0 ± 0.1	0.12	0.71	0.43
DF diameter at insert removal, mm[3]	11.0 ± 0.4	11.2 ± 0.4	10.8 ± 0.4	11.3 ± 0.4	0.90	0.38	0.77
DF diameter at TAI, mm[4]	12.6 ± 0.4	13.6 ± 0.4	12.7 ± 0.4	13.4 ± 0.4	0.90	0.03	0.68
Daily DF growth, mm/d[5]	0.9 ± 0.1	1.3 ± 0.1	0.9 ± 0.1	1.1 ± 0.1	0.52	0.01	0.25
Occurrence of estrus, %[6]	37.2	52.3	55.8	73.8	0.008	0.03	0.77
Ovulation rate, %[7]	81.4	95.5	83.7	97.6	0.54	0.008	0.71
CL diameter at d 6 after TAI, mm	17.8 ± 0.6[b]	20.1 ± 0.5[a]	18.6 ± 0.6[ab]	18.7 ± 0.6[ab]	0.54	0.04	0.06
Plasma P4 at d 6 after TAI, ng/mL	3.8 ± 0.3	5.1 ± 0.3	4.1 ± 0.3	4.6 ± 0.3	0.94	0.001	0.18
Pregnancy per TAI, %	30.2	36.4	44.2	50.0	0.07	0.42	0.95

[1]Suckled anestrous beef cows received an previously used intravaginal insert containing 1.0 g of progesterone (P4) and 2.0 mg of estradiol benzoate on the first day of the estrus/ovulation synchronization protocol (D −10). The P4 insert was removed eight days later (D −2), and cows from ECP group received an IM treatment of 1 mg of ECP, cows from eCG group received an IM injection of 400 IU of eCG, while cows from ECP + ECG received both treatments and cows from CON did not receive any treatment at this moment. All cows received GnRH IM and were timed artificially inseminated (TAI) 48 h after the P4 insert removal (D 0). Different letters within the same row indicate the presence of difference between groups (P <0.05) when an interaction between eCG and ECP was observed
[2]BCS = Body condition score collected at insertion of the P4 insert
[3]DF = Dominant follicle
[4]TAI = timed artificial insemination
[5]DF growth between the P4 insert removal and TAI divided by two
[6]Estrus determined based on the tail-head mark
[7]Number of cows with a new CL formed 6 d after the TAI divided by the number of animal synchronized

No-eCG = 3.9 ± 0.2 ng/mL; P = 0.001]. However, there was no influence of eCG treatment on the pregnancy per TAI [eCG = 43.0% (37/86) vs. No-eCG = 37.2% (32/86); P = 0.42].

RNA-seq

RNA sequencing produced a total of ~334 million reads with an average of 27.5 million reads for each group. Six biological replicates were analyzed for each phenotype (please see Statistical Analyses section above) with the reads ranging from 17 to 26 million per sample after filtering (Additional file 1: Table S1). Approximately ~65% of the total reads uniquely mapped to the UMD 3.1 reference genome (https://www.ncbi.nlm.nih.gov/genome?term=bos%20taurus). Only the uniquely mapped reads were considered in the analysis. From the remaining, approximately 20% of the reads were not uniquely mapped, and 15% unmapped reads. After applying the variance and minimal value of baseMean filtering, a total of 15,161 genes were included on the differential expression analysis. A total of 310 out of the 15,161 analyzed genes showed differential expression (adjusted P-value <0.1), of which 73 and 62 were upregulated in the endometrium of ECP and CON samples, respectively (see Volcano plot, Fig. 2 and Additional file 2: Table S2). Differentially expressed genes (DEG) with the greatest expression profiles were RPS2 [ribosomal protein S2], GABARAP [GABA (A) receptor-associated protein], up-regulated in the CON endometrium, and PEPD [peptidase D], SG100g [calcium binding protein G] and CEACAM1 [carcinoembryonic

antigen-related cell adhesion molecule 1], up-regulated in the ECP group. Heatmap on Fig. 3 shows the 50 genes with the lowest p-adjusted values. It is possible to observe the similarity of gene expression patterns among individuals within each group, as indicated by the shades of green (for low expression) or red color (high expression).

Sequences of all reads were deposited in the Sequence Read Archive (SRA) of the NCBI (http://www.ncbi.nlm.nih.gov/sra/; Additional file 3: Table S3) and, an overview of these data has been deposited in NCBI's Gene Expression Omnibus (GEO) and is accessible through GEO Series accession number GSE67807.

Functional enrichment analysis of RNA-seq data - DAVID results

KEGG pathway and Gene ontology (GO) term analyses were performed with DAVID (Table 3). Functional enrichment analysis using DAVID revealed two KEGG pathways overrepresented by the ECP-upregulated transcripts: pathways in cancer (5 genes; $P < 0.01$) and small cell lung cancer (3 genes; $P < 0.05$). On the other hand, ECP downregulated transcripts indicated the enrichment of three pathways: Parkinson's disease (3 genes; $P = 0.06$), oxidative phosphorylation (3 genes; $P = 0.06$) and Alzheimer's disease (3 genes; $P = 0.09$). More specifically, ECP-upregulated transcripts associated with pathways in cancer were [gene symbol (fold change; adjusted P value on RNA-seq); respectively]: $LAMC3$ (1.55; $P = 0.10$), $PTCH1$ (1.51; $P = 0.09$), $PTCH2$ (1.52; $P = 0.03$), $PIK3R3$ (1.22; $P = 0.10$), and $PIAS1$ (1.18; $P = 0.09$), whereas ECP

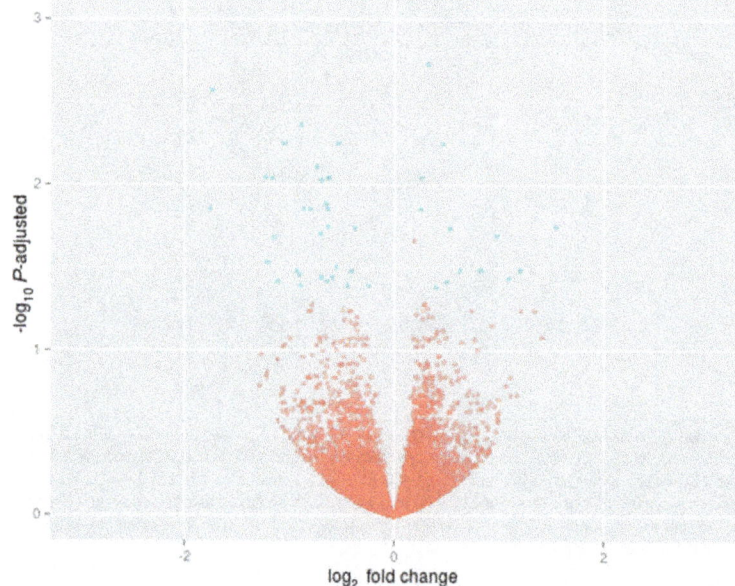

Fig. 2 Volcano plot obtained from DESeq analysis. Volcano plot shows that the vertical lines axe is $\log_2$-fold change and the horizontal axis is the statistical significance (P value ≤ 0.10). Genes with P value ≤0.10 are marked with blue dots

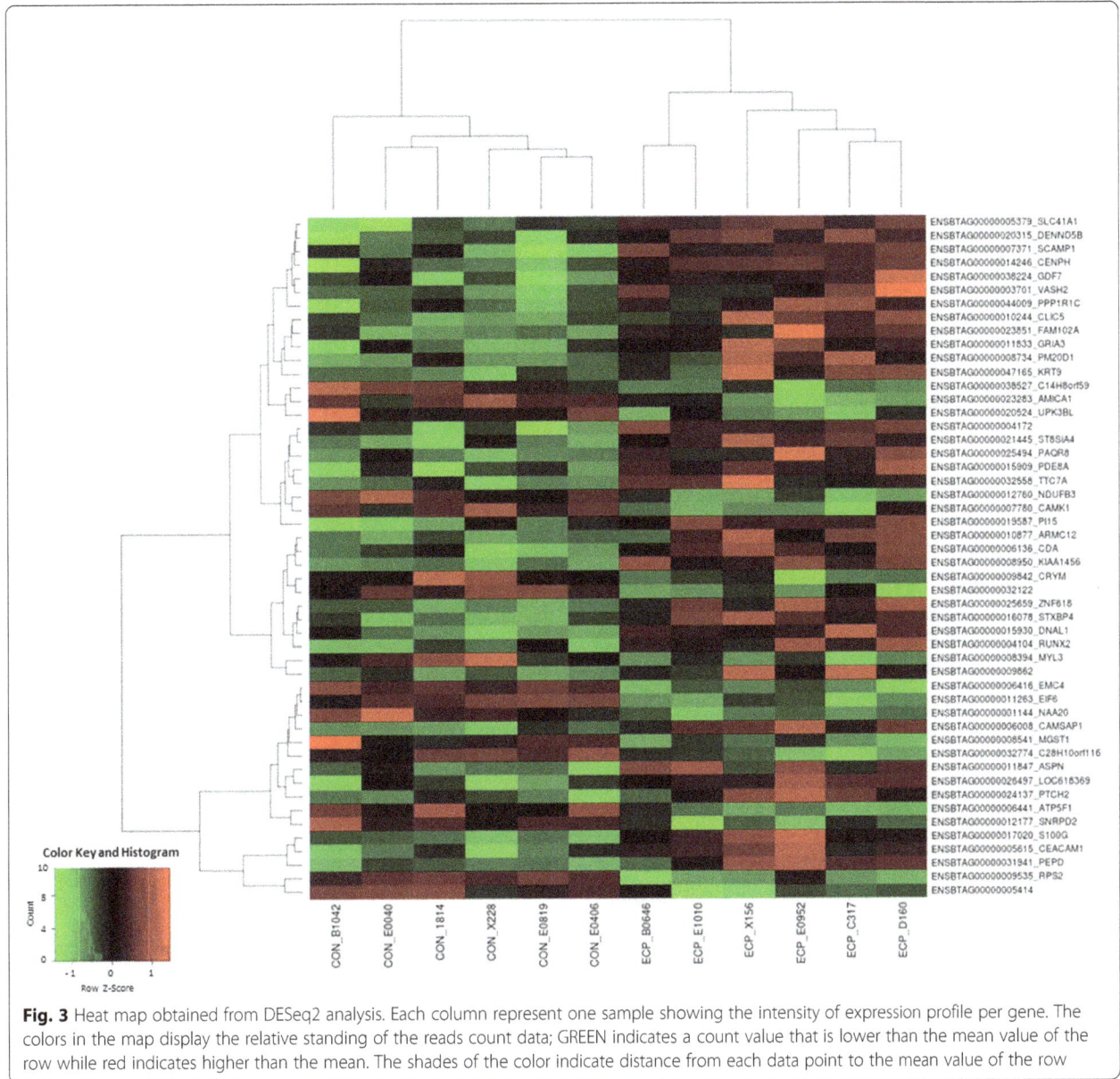

Fig. 3 Heat map obtained from DESeq2 analysis. Each column represent one sample showing the intensity of expression profile per gene. The colors in the map display the relative standing of the reads count data; GREEN indicates a count value that is lower than the mean value of the row while red indicates higher than the mean. The shades of the color indicate distance from each data point to the mean value of the row

downregulated transcripts associated with oxidative phosphorylation were *ATP5F1* (1.18; $P = 0.01$), *ATP5J* (1.24; $P = 0.06$), and *NDUFB3* (1.37; $P = 0.01$). Additionally, analysis of GO terms identified that ECP upregulated transcripts over represented epidermis development [*ADAM9* and ENSBTAG00000017455 (uncharacterized protein)]. On the other hand, ECP downregulated GO terms indicated the enrichment of five biological processes: generation of metabolic precursors and energy (*GPI, NDUFB3, ATP5F1, IDH3B, ATP5J*), Translation (*RPS2, EEF1D,* ENSBTAG00000013866, ENSBTAG00000011263), and mRNA processing, mRNA metabolic process and RNA splicing with 3 common genes (*GEMIN7, SNRPD2, STRAP*).

Cell proliferation-related gene expression
According to selection criteria described previously, uterine tissue used in the qPCR analysis derived from cows that presented different estrus responses [CON (10.1%), ECP (90.9%), eCG (66.7%) and ECP + ECG (83.3%)] and P4 concentration at uterine biopsy [CON (3.4 ± 0.2 ng/mL), ECP (3.7 ± 0.2 ng/mL), eCG (5.3 ± 0.4 ng/mL) and ECP + ECG (5.0 ± 0.6 ng/mL)].

There were no interactions ($P > 0.10$) between ECP and eCG on the expression of the transcripts evaluated. ECP treatment induced greater endometrial abundance of *PTCH2* ($P = 0.07$) and *COL4A1* ($P = 0.02$) genes, whereas it reduced *EGFR* ($P = 0.09$) gene expression (Figs. 4 and 5). The ECP treatment did not affect gene

Table 3 KEGG Pathways and Gene ontologies (GO category) of mRNA transcripts differentially expressed in cows treated with estradiol cypionate (ECP).

Category	Term	Count	P value	Genes
Upregulated in ECP				
KEGG PATHWAY	Pathways in cancer	5	0.008	*PIK3R3, PTCH1, PTCH2, LAMC3, PIAS1*
KEGG PATHWAY	Small cell lung cancer	3	0.020	*PIK3R3, LAMC3, PIAS1*
GO TERM BP_FAT	Epidermis development	2	0.095	*ADAM9, ENSBTAG00000017455*
Downregulated in ECP				
KEGG PATHWAY	Parkinson's disease	3	0.062	*NDUFB3, ATP5F1, ATP5J*
KEGG PATHWAY	Oxidative phosphorylation	3	0.064	*NDUFB3, ATP5F1, ATP5J*
KEGG PATHWAY	Alzheimer's disease	3	0.090	*NDUFB3, ATP5F1, ATP5J*
GO TERM BP_FAT	Generation of precursor metabolites and energy	5	0.004	*GPI, NDUFB3, ATP5F1, IDH3B, ATP5J*
GO TERM BP_FAT	Translation	4	0.059	*RPS2, EEF1D, ENSBTAG00000013866, ENSBTAG00000011263*
GO TERM BP_FAT	mRNA processing	3	0.071	*GEMIN7, SNRPD2, STRAP*
GO TERM BP_FAT	mRNA metabolic process	3	0.087	*GEMIN7, SNRPD2, STRAP*
GO TERM BP_FAT	RNA splicing	3	0.041	*GEMIN7, SNRPD2, STRAP*

Enrichment analysis was performed with DAVID tools (https://david.ncifcrf.gov/tools.jsp)

expression of *ESR1* ($P = 0.90$), *ESR2* ($P = 0.61$), *HB-EGF* ($P = 0.80$) and *ITGB3* ($P = 0.57$). On the other hand, eCG treatment induced greater endometrial abundance of *HB-EGF* ($P = 0.06$), *ESR2* ($P = 0.09$), and *ITGB3* ($P = 0.05$) genes, whereas reduced the gene expression of *ESR1* ($P = 0.05$). Supplementation with eCG did not alter expression of *EGFR* ($P = 0.34$), *PTCH2* ($P = 0.31$) and *COL4A1* ($P = 0.19$). Additionally, expression of *PGR* was not altered by either ECP ($P = 0.51$) or eCG ($P = 0.25$) treatments.

Discussion

The present study investigated the impact of hormonal manipulation of proestrus on ovarian response and on uterine transcriptome 6 d post-TAI. The most relevant observations from this study are: 1) ECP treatment improves occurrence of estrus and pregnancy per AI, whereas eCG treatment enhances final follicular growth, size of ovulatory follicle, ovulation rate and subsequent P4 concentration, 2) the endometrial transcriptional profile is regulated by ECP supplementation and cell proliferation was one of the overrepresented gene ontology terms; 3) selected candidate genes with altered expression further support an ECP effect on cellular proliferation and tissue morphology.

Synchronized cows displaying estrus before TAI exhibited larger dominant follicles, greater E2 concentration during the proestrus/estrus, greater luteal function on the subsequent estrus cycle, and greater conception rate when compared to cows that did not display estrus [6, 25, 42–44]. In agreement, exogenous ECP treatment at the onset of proestrus improved the proportion of suckled beef cows displaying estrus, determining greater pregnancy outcomes following TAI than non-ECP treated cows [25, 26], similar to what was observed in the present

study. Furthermore, the eCG treatment at onset of the proestrus was effective to increase conception rates in suckled beef cows [1, 5, 29, 30]. Also corroborating with the present results, eCG-treated cows presented greater final follicular growth, follicular diameter at TAI, ovulation rate and plasma P4 concentration on subsequent diestrus [5, 26, 29, 30]. Therefore, the hormonal therapies established in the present study may be considered a pro-fertility model for suckled anestrous beef cows and potentially allow the establishment of two distinct periovulatory endocrine milieus, that are associated with an uterine environment of better receptivity. Specifically, it was expected that ECP-treated cows present greater periovulatory E2 concentration due to the exogenous estradiol administration. Additionally, those cows treated with eCG also presented greater concentrations of E2 during proestrus/estrus due to endogenous estradiol from a healthy larger DF at TAI, in addition to presenting greater concentrations of P4 during early diestrus.

Unexpectedly, transcriptome analysis of D 6 endometrium from cows treated or not with ECP did not reveal dramatic differences of gene expression patterns. Our model was unique in selecting for cows displaying estrus behavior in ECP-treated group versus not displaying estrus behavior in the control group. Estrus behavior is associated with higher pregnancy rates [6, 25, 43, 44]. Global transcriptome analysis of D 14 endometrium from high fertility heifers compared to low fertility ones did not reveal substantial differences [45]. Another study using a similar criterion for high and low fertility revealed that D 7 endometrium presented 417 DEG, however, most of the DEG exhibited fold change between 1.0 and 2.0 [46]. These results are in agreement with our

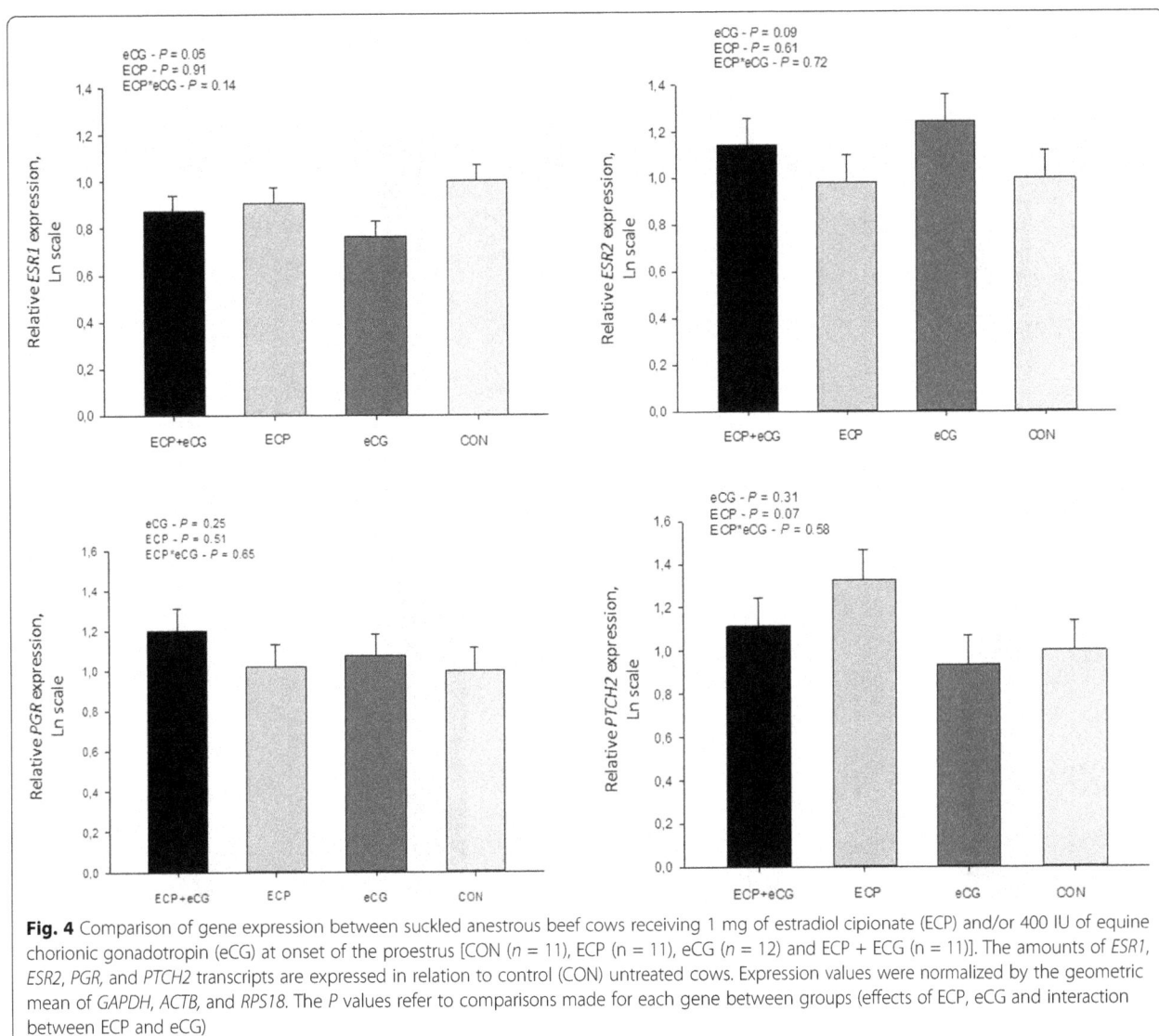

Fig. 4 Comparison of gene expression between suckled anestrous beef cows receiving 1 mg of estradiol cipionate (ECP) and/or 400 IU of equine chorionic gonadotropin (eCG) at onset of the proestrus [CON (n = 11), ECP (n = 11), eCG (n = 12) and ECP + ECG (n = 11)]. The amounts of *ESR1*, *ESR2*, *PGR*, and *PTCH2* transcripts are expressed in relation to control (CON) untreated cows. Expression values were normalized by the geometric mean of *GAPDH*, *ACTB*, and *RPS18*. The P values refer to comparisons made for each gene between groups (effects of ECP, eCG and interaction between ECP and eCG)

data showing that endometrial gene expression is not dramatically different between groups with contrasting fertility; however, it is important to point out that half of the samples came from pregnant animals, whereas the other half came from non-pregnant cows in both ECP or control groups.

Estradiol levels are higher after ECP administration [28, 42, 47] and estrus behavior is correlated with estradiol levels [42, 48]; however, we did not quantify estradiol plasma concentrations. It was observed in ovariectomized cows that estradiol benzoate injection alters global gene expression of the endometrium when compared to a control group or progesterone treatment; whereas a combined estradiol and progesterone group shows data closer to estradiol treatment, suggesting that estradiol counteracts progesterone effects [49]. In our model, progesterone is the dominant steroid hormone at the time of sample collection; however

its impact on gene expression is likely influenced by the previous exposure to estradiol.

We have observed previously that the endometrial tissue of cows ovulating larger follicles expressed markers of proliferative activity earlier than cows ovulating smaller follicles [8]. Similarly, gene expression changes suggesting reduction of proliferative activity and transition to a biosynthetic phenotype were also hastened in cows with larger ovulatory follicles. Larger follicles led to increased estradiol concentrations during proestrus and greater progesterone concentrations during early diestrus [8, 14]. Functional enrichment analysis using DAVID identified gene ontology terms associated with regulation of cell proliferation such as pathways in cancer and small cell lung cancer. Similarly, endometrial gene expression at D 7 in one estrous cycle prior to embryo transfer revealed enrichment of GO-terms cell cycle and anti-apoptosis in cows that successfully established

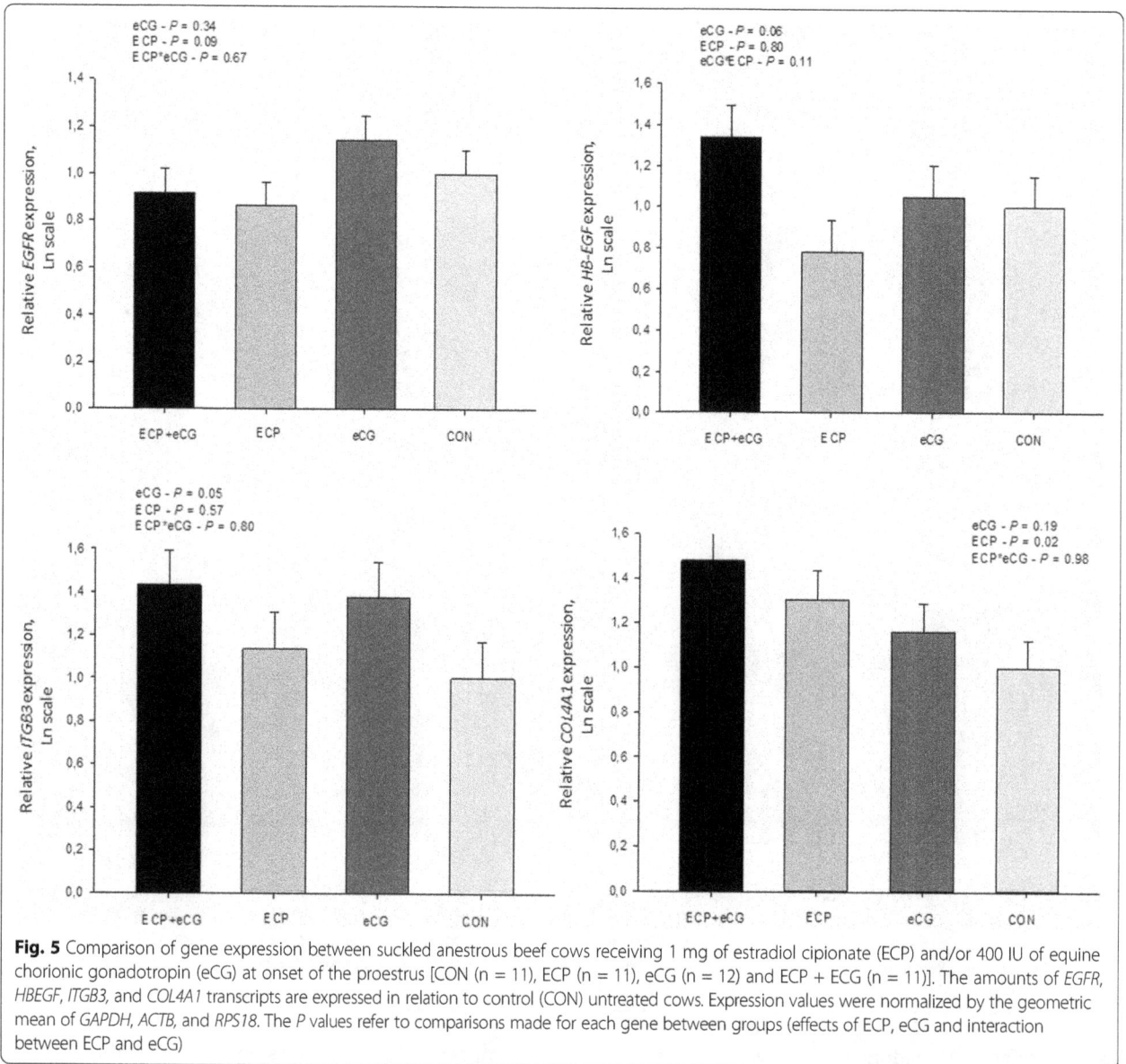

Fig. 5 Comparison of gene expression between suckled anestrous beef cows receiving 1 mg of estradiol cipionate (ECP) and/or 400 IU of equine chorionic gonadotropin (eCG) at onset of the proestrus [CON (n = 11), ECP (n = 11), eCG (n = 12) and ECP + ECG (n = 11)]. The amounts of *EGFR*, *HBEGF*, *ITGB3*, and *COL4A1* transcripts are expressed in relation to control (CON) untreated cows. Expression values were normalized by the geometric mean of *GAPDH*, *ACTB*, and *RPS18*. The *P* values refer to comparisons made for each gene between groups (effects of ECP, eCG and interaction between ECP and eCG)

pregnancy [50]. It is noteworthy that the above mentioned studies obtained samples from non-lactating cyclic cows [8] or heifers [50], whereas in the present study all cows were lactating and in anestrus. Additionally, assessment of the expression of proliferation-related candidate genes showed that *PTCH2* and *COL4A1* were induced by ECP treatment, whereas *EGFR* expression was suppressed. *PTCH2* is a membrane receptor, member of the Hedgehog signaling pathway [51], and has been associated with proliferation-related disorders such as endometriosis and ovarian carcinoma [52], playing a role as a tumor suppressor gene [53]. *COL4A1* encodes a type IV collagen protein that is an integral component of basement membranes [54]. In the endometrium, the breakdown of the basement membrane as well as increased expression of *COL4A1*

have been related with inhibition of angiogenesis and reduced tumor growth [55]. In addition, the oncogene *EGFR*, which is associated with growth of placental tissue [56], was downregulated by ECP-treatment suggesting a suppression of the endometrial ability to respond to mitogenic stimuli. The collective interpretation of these data is that estrogenic stimulus given by ECP induced a non-proliferative status on D6 endometrium. Such findings are consistent with our previous report, in which ovulation of a larger follicle (associated with greater proestrus and estrus plasma concentrations of estradiol) inhibited proliferation in both luminal and glandular epithelial cells on D 7 endometrium [8]. Importantly, such regulation occurred despite similar plasma concentrations of P4 between animals that received an did not receive ECP.

The most remarkable eCG-induced changes in gene expression were associated to E2 signaling. Indeed, transcript abundance was greater for *ESR2* and lesser for *ESR1* in eCG-treated cows than No eCG-treated cows, suggesting the establishment of a transition phase, from proliferative to secretory. The recognized proliferative role of estrogens in the female reproductive tract appears to be mediated by *ESR1* [57]. After estrus, *ESR1* abundance decreases and reaches nadir endometrium concentrations during the mid-luteal phase of the estrous cycle [58]. In contrast, uterine *ESR2* expression is positively associated with the increasing P4 concentration observed from early to mid diestrus. The greater abundance of *ESR2* expression found in eCG treated cows could be justified by the positive effect of eCG on P4 concentration during early diestrus. Altogether, these results suggest that the endometrium of suckled anestrous cows at Day 6 receiving either ECP or eCG is still transitioning from a proliferative to a secretory state, as previously reported [8].

Conclusions

Supplementation with ECP or eCG at onset of the synchronized proestrus of suckled anestrous beef cows significantly influence the ovarian responses; however, the impact on global uterine gene expression is discrete, presenting few DEG that are associated with ceasing cell proliferation. Such phenotype is consistent with the beginning of the secretory phase of the endometrium, required to support conceptus growth and survival.

Additional files

Additional file 1: Table S1. Number of reads from all samples from suckled cows receiving (ECP) or not (CON) 1 mg of ECP at the onset of the proestrous

Additional file 2: Table S2. Differential gene expression results. BaseMean is the average of all samples expression profile after normalization; lfcSE – standard error from log2FoldChange; padj – *P*-value adjusted after correction of BH-FDR for multiple tests

Additional file 3: Table S3. Bio-samples and Experiment accession numbers of the Raw reads resulted from the RNAseq of endometrial biopsis in the SRA data base

Abbreviations

ACTB: actin, Beta; ADAM9: ADAM metallopeptidase domain 9; ANOVA: Analysis of variance; ATP5F1: ATP Synthase, H+ Transporting, Mitochondrial Fo Complex Subunit B1; ATP5J: ATP Synthase, H+ Transporting, Mitochondrial Fo Complex Subunit F6; BCS: Body contition score; CEACAM1: Carcinoembryonic antigen-related cell adhesion molecule 1; CL: Corpus luteum; COL4A1: Collagen, type IV, alpha 1; CON: Control; Cq: Quantification cycle; DAVID: Database for Annotation, Visualization, and Integrated Discovery; DEG: Differentially expressed genes; DF: Dominant follicle; E2: Estradiol; eCG: Equine chorionic gonadotropin; ECP: Estradiol cypionate; EDTA: Ethylenediaminetetraacetic acid; EEF1D: Eukaryotic translation elongation factor 1 delta; EGFR: Epidermal growth factor receptor; ESR1: Estrogen receptor alpha; ESR2: Estrogen receptor beta; GABARAP: GABA (A) receptor-associated protein; GAPDH: Glyceraldehyde-3-Phosphate Dehydrogenase; GEMIN7: Gem nuclear organelle associated protein 7; GEO: NCBI's gene expression omnibus; GnRH: Gonadotropin-releasing hormone; GO: Gene ontology; GPI: Glucose-6-Phosphate Isomerase; HB-EGF: Heparin-binding EGF-like growth factor; IDH3B: Isocitrate dehydrogenase 3 (NAD(+)) Beta; IM: Intramuscular; ITGB3: Integrin, beta 3; LAMC3: Laminin Subunit Gamma 3; LN: Natural logarithm; NDUFB3: NADH:Ubiquinone Oxidoreductase Subunit B3; P4: Progesterone; PEPD: Peptidase D; PGR: Progesterone receptor; PIAS1: Protein Inhibitor Of Activated STAT 1; PIK3R3: Phosphoinositide-3-Kinase regulatory subunit 3; PTCH1: Patched 1; PTCH2: Patched homolog 2; qPCR: Quantitative real time PCR; RIN: RNA integrity number; RNAseq: RNA sequencing; RPS18: Ribosomal Protein S18; RPS2: Ribosomal protein S2; SEM: Standard error of the mean; SG100G: Calcium binding protein G; SNRPD2: Small nuclear ribonucleoprotein D2 polypeptide; SRA: Sequence read archive; STRAP: Serine/Threonine Kinase Receptor Associated Protein; TAI: Timed artificial insemination

Acknowledgements
We would like to thank the Empyreo Farm (Jacarezinho – PR) for allowing the use of their animals and facilities during the trial. These experiments were supported by Firmasa-Pecuária com Tecnologia. This research was funded, in part, by São Paulo Research Foundation (FAPESP; 2012/14731-4). We are thankful to Professor Marcos Roberto Chiaratti from Departamento de Genética e Evolução, CCBS, Universidade Federal de São Carlos - UFSCar for the statistical analysis of qPCR data. We also want to thank Bruno Moura Monteiro, Julia G. Soares and Milena L. Oliveira for technical support.

Funding
FAPESP (2012/14731–4) to MFSF.
CAPES PEC-PG 15068–12-9 to AMGD.
CNPq 142,387–2015-0 to MS.
CNPq 481,199/2012–8 and FAPESP- 2011/03226–4 to MB.
The funding bodies had no participation on the study, collection, analysis, interpretation of data or in writing the manuscript.

Authors' contributions
MFSF contributed to experimental design, animal management, samples preparation for RNAseq and qPCR analysis, statistical analysis of transcripts and was a major contributor in writing the manuscript. RSR, and MFM contributed to animal management. AMGD, GP, and MS performed samples preparation for RNAseq and qPCR analysis, performed qPCR analysis, and contributed in writing the manuscript. SSCA, GG, and LLC performed the RNAseq and the bioinformatics analysis. MDG, FSM, and PSB contributed to experiment design and writing the manuscript. MB was the PI and contributed to experiment design and writing the manuscript. All authors read and approved the final manuscript.

Competing interests
The authors declare that they have no competing interests.

Author details
[1]Departamento de Reprodução Animal, FMVZ-USP, São Paulo, SP, Brazil. [2]Departamento de Genética e Biologia Evolutiva, IB-USP-, São Paulo, SP, Brazil. [3]Laboratório de Biotecnologia Animal, ESALQ-USP, Av Pádua Dias, Piracicaba, SP 11, Brazil. [4]Universidade Federal do Pampa, Uruguaiana, RS, Brazil. [5]Universidade de São Paulo, Faculdade de Medicina Veterinária e Zootecnia, Departamento de Reprodução Animal, Avenida Duque de Caxias Norte, 225, Pirassununga, SP Zip Code 13635900, Brazil.

References
1. Baruselli PS, Reis EL, Marques MO, Nasser LF, Bó GA. The use of hormonal treatments to improve reproductive performance of anestrous beef cattle in tropical climates. Anim Reprod Sci. 2004;82–83:479–86.
2. Baruselli PS, Sales JNS, Sala RV, Vieira LM, Sa Filho MF. History, evolution and perspectives of timed artificial insemination programs in Brazil. Anim Reprod. 2012;9(3):139–52.
3. Sa Filho MF, Penteado L, Reis EL, Reis TANPS, Galvao KN, Baruselli PS. Timed artificial insemination early in the breeding season improves the reproductive performance of suckled beef cows. Theriogenology. 2013;79(4):625–32.
4. Meneghetti M, Sa Filho OG, Peres RFG, Lamb GC, Vasconcelos JLM. Fixed-time artificial insemination with estradiol and progesterone for Bos Indicus

cows I: basis for development of protocols. Theriogenology. 2009;72(2):179–89.

5. Sa Filho OG, Meneghetti M, Peres RFG, Lamb GC, Vasconcelos JLM. Fixed-time artificial insemination with estradiol and progesterone for Bos Indicus cows II: strategies and factors affecting fertility. Theriogenology. 2009;72(2):210–8.

6. Sa Filho MF, Crespilho AM, Santos JEP, Perry GA, Baruselli PS. Ovarian follicle diameter at timed insemination and estrous response influence likelihood of ovulation and pregnancy after estrous synchronization with progesterone or progestin-based protocols in suckled Bos Indicus cows. Anim Reprod Sci. 2010;120(1–4):23–30.

7. Peres RFG, Claro Junior I, Sa Filho OG, Nogueira GP, Vasconcelos JLM. Strategies to improve fertility in Bos Indicus postpubertal heifers and nonlactating cows submitted to fixed-time artificial insemination. Theriogenology. 2009;72(5):681–9.

8. Mesquita FS, Ramos RS, Pugliesi G, Andrade SC, Van Hoeck V, Langbeen A, et al. The receptive endometrial Transcriptomic signature indicates an earlier shift from proliferation to metabolism at early Diestrus in the cow. Biol Reprod. 2015;93(2):52.

9. Satterfield MC, Song G, Kochan KJ, Riggs PK, Simmons RM, Elsik CG, et al. Discovery of candidate genes and pathways in the endometrium regulating ovine blastocyst growth and conceptus elongation. Physiol Genomics. 2009;39(2):85–99.

10. Bauersachs S, Ulbrich SE, Gross K, Schmidt SEM, Meyer HHD, Einspanier R, et al. Gene expression profiling of bovine endometrium during the oestrous cycle: detection of molecular pathways involved in functional changes. J Mol Endocrinol. 2005;34(3):889–908.

11. Miller BG, Moore NW. Effects of progesterone and oestradiol on RNA and protein metabolism in the genital tract and on survival of embryos in the ovariectomized ewe. Aust J Biol Sci. 1976;29(5–6):565–73.

12. Moore NW. The use of embryo transfer and steroid-hormone replacement therapy in the study of prenatal mortality. Theriogenology. 1985;23(1):121–8.

13. Wilmut I, Sales DI, Ashworth CJ. Maternal and embryonic factors associated with prenatal loss in mammals. J Reprod Fertil. 1986;76(2):851–64.

14. Mesquita FS, Pugliesi G, Scolari SC, Franca MR, Ramos RS, Oliveira M, et al. Manipulation of the periovulatory sex steroidal milieu affects endometrial but not luteal gene expression in early diestrus Nelore cows. Theriogenology. 2014;81(6):861–9.

15. Forde N, Beltman ME, Duffy GB, Duffy P, Mehta JP, O'Gaora P, et al. Changes in the endometrial Transcriptome during the bovine estrous cycle: effect of low circulating progesterone and consequences for Conceptus elongation. Biol Reprod. 2011;84(2):266–78.

16. Forde N, Carter F, Fair T, Crowe MA, Evans ACO, Spencer TE, et al. Progesterone-regulated changes in endometrial gene expression contribute to advanced Conceptus development in cattle. Biol Reprod. 2009;81(4):784–94.

17. Okumu LA, Forde N, Fahey AG, Fitzpatrick E, Roche JF, Crowe MA, et al. The effect of elevated progesterone and pregnancy status on mRNA expression and localisation of progesterone and oestrogen receptors in the bovine uterus. Reproduction. 2010;140(1):143–53.

18. Mann GE, Lamming GE. The role of sub-optimal preovulatory oestradiol secretion in the aetiology of premature luteolysis during the short oestrous cycle in the cow. Anim Reprod Sci. 2000;64(3–4):171–80.

19. Bridges GA, Mussard ML, Pate JL, Ott TL, Hansen TR, Day ML. Impact of preovulatory estradiol concentrations on conceptus development and uterine gene expression. Anim Reprod Sci. 2012;133(1–2):16–26.

20. Lamming GE, Mann GE. Control of endometrial oxytocin receptors and prostaglandin F2 alpha production in cows by progesterone and oestradiol. J Reprod Fertil. 1995;103(1):69–73.

21. Robinson RS, Mann GE, Lamming GE, Wathes DC. Expression of oxytocin, oestrogen and progesterone receptors in uterine biopsy samples throughout the oestrous cycle and early pregnancy in cows. Reproduction. 2001;122(6):965–79.

22. Kieborz-Loos KR, Garverick HA, Keisler DH, Hamilton SA, Salfen BE, Youngquist RS, et al. Oxytocin-induced secretion of prostaglandin F2alpha in postpartum beef cows: effects of progesterone and estradiol-17beta treatment. J Anim Sci. 2003;81(7):1830–6.

23. Ulbrich SE, Frohlich T, Schulke K, Englberger E, Waldschmitt N, Arnold GJ, et al. Evidence for estrogen-dependent uterine Serpin (SERPINA14) expression during estrus in the bovine endometrial glandular epithelium and lumen. Biol Reprod. 2009;81(4):795–805.

24. Hillegass J, Lima FS, Filho MFS, Santos JEP. Effect of time of artificial insemination and supplemental estradiol on reproduction of lactating dairy cows. J Dairy Sci. 2008;91(11):4226–37.

25. Sa Filho MF, Santos JEP, Ferreira RM, Sales JNS, Baruselli PS. Importance of estrus on pregnancy per insemination in suckled Bos Indicus cows submitted to estradiol/progesterone-based timed insemination protocols. Theriogenology. 2011;76(3):455–63.

26. Pitaluga PCSF, Sa Filho MF, Sales JNS, Baruselli PS, Vincenti L. Manipulation of the proestrous by exogenous gonadotropin and estradiol during a timed artificial insemination protocol in suckled Bos Indicus beef cows. Livest Sci. 2013;154(1–3):229–34.

27. Souza AH, Gumen A, Silva EP, Cunha AP, Guenther JN, Peto CM, et al. Supplementation with estradiol-17beta before the last gonadotropin-releasing hormone injection of the Ovsynch protocol in lactating dairy cows. J Dairy Sci. 2007;90(10):4623–34.

28. Jinks EM, Smith MF, Atkins JA, Pohler KG, Perry GA, MacNeil MD, et al. Preovulatory estradiol and the establishment and maintenance of pregnancy in suckled beef cows. J Anim Sci. 2013;91(3):1176–85.

29. Sa Filho MF, Ayres H, Ferreira RM, Marques MO, Reis EL, Silva RCP, et al. Equine chorionic gonadotropin and gonadotropin-releasing hormone enhance fertility in a norgestomet-based, timed artificial insemination protocol in suckled Nelore (Bos Indicus) cows. Theriogenology. 2010;73(5):651–8.

30. Sales JNS, Crepaldi GA, Girotto RW, Souza AH, Baruselli PS. Fixed-time AI protocols replacing eCG with a single dose of FSH were less effective in stimulating follicular growth, ovulation, and fertility in suckled-anestrus Nelore beef cows. Anim Reprod Sci. 2011;124(1–2):12–8.

31. Ayres H, Ferreira RM, de Souza Torres-Junior JR, Borges Demetrio CG, de Lima CG, Baruselli PS. Validation of body condition score as a predictor of subcutaneous fat in Nelore (Bos Indicus) cows. Livest Sci. 2009;123(2–3):175–9.

32. Pugliesi G, Scolari SC, Mesquita FS, Maturana Filho M, Araujo ER, Cardoso D, et al. Impact of probing the reproductive tract during early pregnancy on fertility of beef cows. Reprod Domestic Anim. 2014;49(4):E35–E9.

33. Garbarino EJ, Hernandez JA, Shearer JK, Risco CA, Thatcher WW. Effect of lameness on ovarian activity in postpartum Holstein cows. J Dairy Sci. 2004;87(12):4123–31.

34. Langmead B, Salzberg SL. Fast gapped-read alignment with Bowtie 2. Nat Methods. 2012;9(4):357–9.

35. Li H, Handsaker B, Wysoker A, Fennell T, Ruan J, Homer N, et al. The Sequence Alignment/Map format and SAMtools. Bioinformatics. 2009;25:2078–9.

36. Love MI, Huber W, Anders S. Moderated estimation of fold change and dispersion for RNA-seq data with DESeq2. Genome Biol. 2014;15(12):550.

37. Gentleman RC, Carey VJ, Bates DM, Bolstad B, Dettling M, Dudoit S, et al. Bioconductor: open software development for computational biology and bioinformatics. Genome Biol. 2004;5(10):R80.

38. Benjamini Y, Hochberg Y. Controlling the false discovery rate - a practical and powerful approach to multiple testing. J R Stat Soc Series B. 1995;57(1):289–300.

39. Dennis G Jr, Sherman BT, Hosack DA, Yang J, Gao W, Lane HC, et al. DAVID: database for annotation, visualization, and integrated discovery. Genome Biol. 2003;4(5):P3.

40. Smit A, Hubley R, Green P. RepeatMasker Open-3.0. http://www.repeatmasker.org/.

41. Vandesompele J, De Preter K, Pattyn F, Poppe B, Van Roy N, De Paepe A, et al. Accurate normalization of real-time quantitative RT-PCR data by geometric averaging of multiple internal control genes. Genome Biol. 2002;3(7)RESEARCH 0034.1-0034.12.

42. Perry GA, Swanson OL, Larimore EL, Perry BL, Djira GD, Cushman RA. Relationship of follicle size and concentrations of estradiol among cows exhibiting or not exhibiting estrus during a fixed-time AI protocol. Domest Anim Endocrinol. 2014;48:15–20.

43. Fields SD, Gebhart KL, Perry BL, Gonda MG, Wright CL, Bott RC, et al. Influence of standing estrus before an injection of GnRH during a beef cattle fixed-time AT protocol on LH release, subsequent concentrations of progesterone, and steriodogenic enzyme expression. Domest Anim Endocrinol. 2012;42(1):11–9.

44. Perry GA, Smith MF, Roberts AJ, MacNeil MD, Geary TW. Relationship between size of the ovulatory follicle and pregnancy success in beef heifers. J Anim Sci. 2007;85(3):684–9.

45. Minten MA, Bilby TR, Bruno RGS, Allen CC, Madsen CA, Wang Z, et al. Effects of Fertility on Gene Expression and Function of the Bovine Endometrium. Plos One. 2013;8(8)E69944.

46. Killeen AP, Morris DG, Kenny DA, Mullen MP, Diskin MG, Waters SM. Global

gene expression in endometrium of high and low fertility heifers during the mid-luteal phase of the estrous cycle. BMC Genomics. 2014;15:234.

47. Perry GA, Perry BL. Effect of preovulatory concentrations of estradiol and initiation of standing estrus on uterine pH in beef cows. Domest Anim Endocrinol. 2008;34(3):333–8.

48. Allrich RD. Endocrine and neural control of estrus in dairy-cows. J Dairy Sci. 1994;77(9):2738–44.

49. Shimizu T, Krebs S, Bauersachs S, Blum H, Wolf E, Miyamoto A. Actions and interactions of progesterone and estrogen on transcriptome profiles of the bovine endometrium. Physiol Genomics. 2010;42A(4):290–300.

50. Salilew-Wondim D, Hoelker M, Rings F, Ghanem N, Ulas-Cinar M, Peippo J, et al. Bovine pretransfer endometrium and embryo transcriptome fingerprints as predictors of pregnancy success after embryo transfer. Physiol Genomics. 2010;42(2):201–18.

51. Toftgard R. Hedgehog signalling in cancer. Cell Mol Life Sci. 2000;57(12):1720–31.

52. Worley MJ Jr, Liu S, Hua Y, Kwok JS-L, Samuel A, Hou L, et al. Molecular changes in endometriosis-associated ovarian clear cell carcinoma. Eur J Cancer. 2015;51(13):1831–42.

53. Zhulyn O, Nieuwenhuis E, Liu YC, Angers S, Hui C-C. Ptch2 Shares overlapping functions with Ptch1 in Smo regulation and limb development. Dev Biol. 2015;397(2):191–202.

54. Kuhn K. Basement membrane (type IV) collagen. Matrix Biol. 1995;14(6):439–45.

55. Rogers PAW, Lederman F, Taylor N. Endometrial microvascular growth in normal and dysfunctional states. Hum Reprod Update. 1998;4(5):503–8.

56. Hu T, Li C. Convergence between Wnt-beta-catenin and EGFR signaling in cancer. Mol Cancer. 2010;9:236.

57. Wang H, Masironi B, Eriksson H, Sahlin L. A comparative study of estrogen receptors alpha and beta in the rat uterus. Biol Reprod. 1999;61(4):955–64.

58. Kurita T, Young P, Brody JR, Lydon JP, O'Malley BW, Cunha GR. Stromal progesterone receptors mediate the inhibitory effects of progesterone on estrogen-induced uterine epithelial cell deoxyribonucleic acid synthesis. Endocrinology. 1998;139(11):4708–13.

18

Breed and adaptive response modulate bovine peripheral blood cells' transcriptome

Nataliya Poščić[1][*][†] (ORCID), Tommaso Montanari[1][†], Mariasilvia D'Andrea[2], Danilo Licastro[3], Fabio Pilla[2], Paolo Ajmone-Marsan[4], Andrea Minuti[4] and Sandy Sgorlon[1]

Abstract

Background: Adaptive response includes a variety of physiological modifications to face changes in external or internal conditions and adapt to a new situation. The acute phase proteins (APPs) are reactants synthesized against environmental stimuli like stress, infection, inflammation.

Methods: To delineate the differences in molecular constituents of adaptive response to the environment we performed the whole-blood transcriptome analysis in Italian Holstein (IH) and Italian Simmental (IS) breeds. For this, 663 IH and IS cows from six commercial farms were clustered according to the blood level of APPs. Ten extreme individuals (five APP+ and APP- variants) from each farm were selected for the RNA-seq using the Illumina sequencing technology. Differentially expressed (DE) genes were analyzed using dynamic impact approach (DIA) and DAVID annotation clustering. Milk production data were statistically elaborated to assess the association of APP+ and APP- gene expression patterns with variations in milk parameters.

Results: The overall de novo assembly of cDNA sequence data generated 13,665 genes expressed in bovine blood cells. Comparative genomic analysis revealed 1,152 DE genes in the comparison of all APP+ vs. all APP- variants; 531 and 217 DE genes specific for IH and IS comparison respectively. In all comparisons overexpressed genes were more represented than underexpressed ones. DAVID analysis revealed 369 DE genes across breeds, 173 and 73 DE genes in IH and IS comparison respectively. Among the most impacted pathways for both breeds were vitamin B6 metabolism, folate biosynthesis, nitrogen metabolism and linoleic acid metabolism.

Conclusions: Both DIA and DAVID approaches produced a high number of significantly impacted genes and pathways with a narrow connection to adaptive response in cows with high level of blood APPs. A similar variation in gene expression and impacted pathways between APP+ and APP- variants was found between two studied breeds. Such similarity was also confirmed by annotation clustering of the DE genes. However, IH breed showed higher and more differentiated impacts compared to IS breed and such particular features in the IH adaptive response could be explained by its higher metabolic activity. Variations of milk production data were significantly associated with APP+ and APP- gene expression patterns.

Keywords: Acute phase proteins, Adaptive response, Dynamic impact approach (DIA), Hypothalamic-pituitary-adrenal (HPA) axis, RNA-Seq, Stress response, Transcriptomics

* Correspondence: poscic.nataliya@spes.uniud.it
[†]Equal contributors
[1]Department of Agriculture, Food, Environment and Animal Science (DI4A), University of Udine, via delle Scienze 206, 33100 Udine, Italy
Full list of author information is available at the end of the article

Background

In the context of adaptation, stress response is an important neurobehavioral and physiological reaction and it is essential for the survival of living organisms. In response to a stressor, the body orchestrates changes in brain activity followed by the secretion of "stress mediators", including cytokines, metabolic hormones and corticosteroids [1].

The body's response during the first stage of stress is known as fight-or-flight response. It includes the activation of sympathetic nervous system and the stimulation of the production of adrenaline and noradrenaline by adrenal glands. These molecules increase the heart rate and the glycemia and modify blood distribution to supply greater levels of glucose to organs where they are needed, like brain and skeletal muscles. Shortly after, the hypothalamic-pituitary-adrenal (HPA) axis is activated and releases corticosteroids (in particular adrenal glucocorticoids). In turn, these produce a negative feedback onto immune cells and suppress further synthesis and release of cytokines, thereby protecting the host from the detrimental consequences of an overactive immune response (e.g., tissue damage, autoimmunity, septic shock) [2].

The long-term activation of the stress-response mechanisms may also cause irreversible damages, like cardiovascular diseases, immunosuppression, dysfunction of digestive and reproductive systems, type-II diabetes mellitus, impairment of thyroid function, weakening and loss of body lean mass [3, 4]. Such pre-pathological or pathological consequences seriously affect not only the efficiency of animal production and the quality of the product, but undoubtedly reduce animal welfare.

During the acute phase reaction (APR), the body mounts a multifactorial response trying to remove or replace damaged tissues and one of the mechanisms involved is the secretion of the so-called acute phase proteins (APPs). The concentration of some APPs increases several fold during the APR, while others, including albumin, decreases as the liver switches the production of proteins towards the synthesis of proteins required to deal with the damage [5, 6].

In ruminants, APPs are very sensitive factors that allow the early and precise detection of inflammation [7]. The most frequently investigated proteins in cattle are: haptoglobin (Hp), serum amyloid A (SAA), fibrinogen (Fb), ceruloplasmin, α 1-antitrypsin and α 1-acid glycoprotein (α1-AGP) [5, 8–10]. It is possible that the synthesis of APPs in cattle is influenced by cortisol [11, 12], which is the key effector molecule of the HPA axis and is recognized as the physiological response to stress [13–16].

Stress response mechanisms in cattle are still not well understood and the research is complicated by individual differences in stress response [17]. Today, the investigation of how dairy cattle adapt to intensive production is particularly important, since the animal welfare is a growing public concern and stressed animals are less efficient, producing less than predicted by their genetic potential mostly due to a higher environmental impact.

Next-generation high-throughput RNA sequencing technology (RNA-seq) is a recently-developed method for discovering, profiling, and quantifying RNA transcripts. Such approach is used to analyze the continually changing cellular transcriptome and might help identifying gene patterns involved in adaptive response. Applicability of RNA-seq for transcriptome analysis of whole blood samples was already confirmed by many research groups [18–20]. Among the most distinct advantages of RNA-seq over prior methods for mapping and quantifying the transcriptome are unbiased whole-transcriptome profiling, higher sensitivity and accurate estimation of lowly expressed transcripts in peripheral whole blood with or without globin depletion [20].

Up to date RNA-seq technique was highly applied for the assessment of changes in blood transcript abundance in response to stress events, pathogenic processes, and specific physiological and metabolic statuses in dairy cattle [19–22]. However, no study has comprehensively evaluated the adaptive response on molecular changes in dairy cattle whole blood cell transcriptome as an indicator of immune activity without the visible environmental perturbations.

In this context, we used a whole-transcriptome analysis to understand if and how differential gene expression contributes to such a complex phenomenon as adaptive response. Therefore, in the present research, the transcriptome of blood cells was analyzed in selected bovines belonging to Italian Holstein (IH) and Italian Simmental (IS) breeds from six commercial farms in Friuli-Venezia Giulia region, Italy. Cows were clustered for blood APPs, plasma Zn, milk cortisol and somatic cell count (SCC) in milk. The analysis included RNA isolation from blood [23], sequencing by RNA-seq with Illumina pipeline [24–27] and the use of the normalized data for the identification of genes expression of which is significantly associated to the adaptive response to the undefined stress conditions. Genes and metabolic pathways were further analyzed using the dynamic impact approach (DIA) and DAVID online software tool [28].

Methods

Animals and management

A total of 663 IH and IS cows from six commercial farms in Friuli-Venezia Giulia region of Italy were included in the experiment. All animals were kept under the same feeding and management conditions and were in the stage of lactation. Farm veterinary practitioner confirmed that all sampled animals passed the preliminary veterinary checkup, were clinically healthy and were not under any treatments for at least 1 month before the collection day. Composition of herds and characteristics of animals included in the analysis are reported in the Tables 1 and 2.

Farmers and farm veterinary practitioners gave an oral informed consent to the study and had a copy of all the data obtained from the laboratory analyses. All farms involved in the present study adhere to a high standard of veterinary care based on best practice manual under the supervision of the official veterinary service. Sample collection was approved by the Bioethics Committee of the University of Udine.

Cows were housed in a free stall barn with cubicle design and automated milking parlour. They were milked twice a day, at approximately 12 h interval. Cows had free access to water and were fed ad libitum twice a day a total mixed ration (TMR) based on corn silage and formulated to cover nutrient requirements [29]. TMR was administered after each milking. To ensure that no dietary variations occurred during the time window of the study, the ration formulation and the offered amount were recorded using registrations of the TMR mixed feeder. In Table 3 are summarized composition, chemical properties and nutritional values of the diet in the six commercial farms.

Milk and blood sampling and assays

Milk was sampled on the day of the official record. Coccygeal vein blood samples were collected just before the morning milking and prior the feeding process. The same collection protocol was used across all farms. Blood was collected in PAXgene Blood RNA Tubes (PreAnalytiX GmbH, Switzerland), frozen 4 h after the collection and stored at -80 °C until the RNA isolation.

Prior to RNA isolation, blood samples were thawed at +4 °C for at least 12 h. RNA was isolated according to PAXgene Blood RNA Kit (PreAnalytiX GmbH, Switzerland) protocol.

Blood biochemical parameters, i.e., total protein, albumin, urea, glucose, creatinine, total bilirubin, cholesterol, AST/GOT (aspartate transaminase/glutamic oxaloacetic transaminase), gGT (gamma-glutamyl transpeptidase), zinc, ceruloplasmin, haptoglobin and paraoxonase were assayed as described in Sgorlon et al. [30]. Milk composition data, i.e., fat and total protein percentage, casein, urea, SCC and milk cortisol, were obtained from milk samples collected the same day of blood sampling and are described in Sgorlon et al. [31].

Clustering of animals

To identify animals differing in their adaptive response to the environment, cows were clustered according to the level of acute phase proteins and molecules (total protein, albumin, zinc, ceruloplasmin, haptoglobin, paraoxonase, milk cortisol and SCC). To control the differences in adaptive response between breeds and to correct it for the potential effect of the environment the clustering was performed separately for each farm. For this the principal component analysis (PCA) using the correlation matrix in SPSS package was applied. According to the first two principal components the ten extreme individuals (five "plus" [APP+] and five "minus" [APP-] variants) from each farm were selected for gene expression analysis.

RNA quality control and sequencing

First RNA was quantified and quality controlled by Nano-Drop ND-1000 Spectrophotometer analysis (Thermo Fisher Scientific Inc., United States). Further RNAs with the highest quality was assigned the RNA integrity number (RIN) score by the Agilent 2100 Bioanalyzer (Agilent Technologies, United States) [32]. Finally samples with RIN $\geq$ 7 were selected for sequencing [33].

Previously was reported that high-throughput sequencing by RNA-Seq is highly reproducible within a large dynamic range of detection and provides an accurate estimation of RNA concentration in peripheral whole blood [20]. Thus, the experimental globin depletion from RNA samples was avoided as it could significantly reduce the amount and quality of isolated RNA and biological samples in our occasion were not possible to replenish.

The 60RNA samples (30 APP+ and 30 APP- variants) were sequenced by RNA-seq technology with the Illumina pipeline [24–27]. Reads obtained from the sequencing were aligned against *Bos taurus* UMD 3.1 reference genome assembly [34].

Table 1 Composition of herds and number of animals included in the analysis

	F1	F2	F3	F4	F5	F6
Breed	IH	IH	IH	IS	IS	IS
Herd size	654	456	442	538	201	270
Dairy animals	347	250	235	280	119	147
First calving	131	85	82	86	41	36
Lactating cows	313	227	195	225	96	123
Cows >50 DIM	279	204	147	185	84	111
Sampled cows	184	112	75	126	78	88
of which:						
1st parity	85	40	36	42	35	23
2nd parity	49	33	18	31	11	19
3rd parity	23	18	11	29	18	13
4th parity	17	10	4	14	9	17
>4th parity	10	11	6	10	5	16

(*F* farm, *IH* Italian Holstein, *IS* Italian Simmental, *DIM* days in milk)

Table 2 Characteristics of the sampled animals within each farm

	F1	F2	F3	F4	F5	F6
BCS	2.4 ± 0.4	2.4 ± 0.5	2.2 ± 0.5	3.0 ± 0.5	3.4 ± 0.4	2.8 ± 0.3
DIM	179 ± 63	179 ± 53	111 ± 62	160 ± 63	162 ± 99	166 ± 66
Milk yield, kg	42.9 ± 9.1	36.9 ± 7.7	33.7 ± 7.7	26.9 ± 7.8	30.2 ± 6.9	27.8 ± 1.5
Milk fat, %	3.3 ± 0.5	4.1 ± 0.6	3.5 ± 0.6	3.7 ± 0.7	4.1 ± 4.5	3.5 ± 0.7
Milk protein, %	3.1 ± 0.3	3.4 ± 0.3	3.1 ± 0.3	3.7 ± 0.3	3.6 ± 0.4	3.5 ± 0.3
Milk casein, %	2.5 ± 0.2	2.6 ± 0.2	2.5 ± 0.2	2.9 ± 0.2	2.8 ± 0.3	2.8 ± 0.3
Milk urea, mg/dL	18.2 ± 3.4	19.4 ± 4.2	20.8 ± 3.4	21.5 ± 5.5	20.5 ± 4.1	20.9 ± 4.2
SCC	369 ± 732	322 ± 594	439 ± 847	659 ± 1,210	181 ± 503	485 ± 1,319
Milk cortisol, pg/mL	492 ± 335	586 ± 840	562 ± 314	636 ± 275	448 ± 174	481 ± 319
Blood parameters:						
Ceruloplasmin, µmol/L	2.8 ± 0.5	2.9 ± 0.9	2.8 ± 0.6	3.2 ± 0.6	2.4 ± 0.7	2.4 ± 0.5
Total proteins, g/L	77.5 ± 7.8	80.0 ± 6.9	80.6 ± 7.2	81.0 ± 5.2	75.6 ± 5.1	79.7 ± 4.8
Albumin, g/L	37.3 ± 3.1	38.2 ± 3.5	35.3 ± 3.1	37.1 ± 2.1	36.7 ± 2.2	38.6 ± 1.6
Haptoglobin, g/L	0.40 ± 0.33	0.42 ± 0.47	0.50 ± 0.46	0.29 ± 0.30	0.33 ± 0.21	0.42 ± 0.42
Paraoxonase, U/mL	112 ± 26	100 ± 24	104 ± 25	88 ± 21	101 ± 16	90 ± 22
Zinc, µmol/L	14.8 ± 4.3	13.5 ± 2.5	12.1 ± 2.5	12.3 ± 2.7	13.0 ± 2.3	12.1 ± 1.8

(*F* farm, *BCS* body condition score, *DIM* days in milk, *SCC* somatic cell count, values are expressed as mean ± SD)

Post-sequencing analysis

Raw counts produced by RNA-seq were normalized with the DeSeq2 software [27, 35]. To identify differentially expressed genes APP+ and APP- cows were compared either ignoring or considering their breed of origin: i) all APP+ vs. all APP-; ii) IH APP+ vs. IH APP-; iii) IS APP+ vs. IS APP-.

For each comparison, normalized RNA-seq data were analyzed with DeSeq2 software to calculate differential expression values (as $\log_2$ of the fold change) and raw *P*-values. To identify the significant genes raw *P*-values were corrected with the false discovery rate (FDR) method [36], using the cutoff of 0.05.

Dynamic impact approach analysis

Gene expression data were also analyzed by the "dynamic impact approach" (DIA) developed by Bionaz and colleagues [28] for the transcriptome analysis.

DIA produces the list of the most impacted pathways integrating information coming from the dataset of the whole list of genes (regardless of their significance), differential expression values, FDR correction factor and raw p-value calculated by the DeSeq2 software. Graphically, the output is well demonstrated through two types of bars: the Impact bar indicating entity of the impact (colored in blue), and the Flux bar showing direction of the impact (red color represents the overexpression of the pathway, green color represents the under-expression of the pathway and yellow color indicates the absence in expression differences).

Annotation clustering

Significant genes were submitted to the Database for Annotation, Visualization and Integrated Discovery (DAVID) to perform a serial annotation clustering [37]. This pipeline allowed us to form series of clusters with genes grouped according to their biological function. *P*-values automatically associated to each cluster were corrected by the Benjamini-Hochberg method. Clusters were considered significant if corrected *P*-values were lower than 0.05. Thereafter, significant genes from different clusters were grouped in a single list and further checked across comparisons to find out genes in common. The clustering procedure was applied separately for each comparison.

Statistical analysis of milk production data

To investigate differences in milk composition across farms and APP groups, a mixed model analysis of variance (ANOVA) with nested design and Fisher's least significant difference (LSD) test was applied for each breed separately. Data were analyzed with the SPSS package using the following statistical model:

$$Y_{ijk} = m + Farm_i + APP(Farm)_{ji} + e_{ijk}$$

Where:

- Y_{ijk}: dependent variable;
- m: general mean;
- $Farm_i$: fixed effect for the farm, with i ranging from 1 to 3;

Table 3 Diet composition (kg/d) in the selected commercial farms, chemical properties and nutritional values of the rations

Item	Farms					
	F1	F2	F3	F4	F5	F6
Lucerne	8.0	4.4	5.0	4.5	3.5	4.5
Pasture hay			1.5	1.5	3.3	
Wheat silage		2.2				
Corn silage	22.5	20.0	15.0	20.0	32.0	20.0
Ryegrass silage		8.0				5.0
Cotton seeds		1.2	1.0	1.0		
Corn meal	5.2	6.0	6.0	4.5		6.0
Soybean meal	2.0	2.1	2.0	1.0		1.0
Rapeseed meal		1.5		1.4		
Linseed			0.3			
Straw	0.5					
Wheat bran				1.5		
Supplements	3.4	0.2	0.6	0.5	9.5	2.5
Total	41.6	45.6	31.4	35.9	48.3	39.0
DMI	23.3	22.6	19.5	20.7	23.4	19.8
Starch, % DM	25.2	27.8	28.6	26.8	25.5	30.1
CP, % DM	14.7	14.0	14.8	13.2	13.1	13.2
EE, % DM	2.9	3.9	4.1	4.1	2.8	3.0
Ashes, % DM	6.6	6.7	7.5	7.4	7.8	6.4
NDF, % DM	38.7	37.4	35.4	36.9	38.5	38.8
MFU	20.0	20.7	17.9	17.9	21.3	17.5
PDIN, g/d	2.222	2.058	1.972	1.776	1.996	1.726
PDIE, g/d	2.154	2.043	1.940	1.784	1.960	1.785

(*F* farm, *DMI* dry matter intake, *DM* dry matter, *CP* crude protein, *EE* ether extract, *NDF* neutral detergent fiber, *MFU* milk fodder units, *PDIN* protein digested in small intestine when rumen-fermentable nitrogen is limiting, *PDIE* protein digested in small intestine when rumen-fermentable energy is limiting)

- $APP(Farm)_{ji}$: nested effect for the APP group of animals, with j ranging from 1 to 2 within the ith farm;
- e_{ijk}: residual error.

Results

Principal component analysis

The result of PCA is plotted in Fig. 1. The total variability explained by the first two components for each PCA separately was on average 50% (with a range from 43 to 54%) for the 6 commercial farms. The first component explained on average 31% of variability (range of 27-36%) and the second component explained on average 18% of variability (range 15-25%). More accurate information about % of variability explained by PCA is reported in Table 4. Variables related to ceruloplasmin, haptoglobin and SCC were among the most important as they were highly correlated with the 1st PC within each farm (PCA). Their loadings correlation values ranged from 0.4-0.8 for

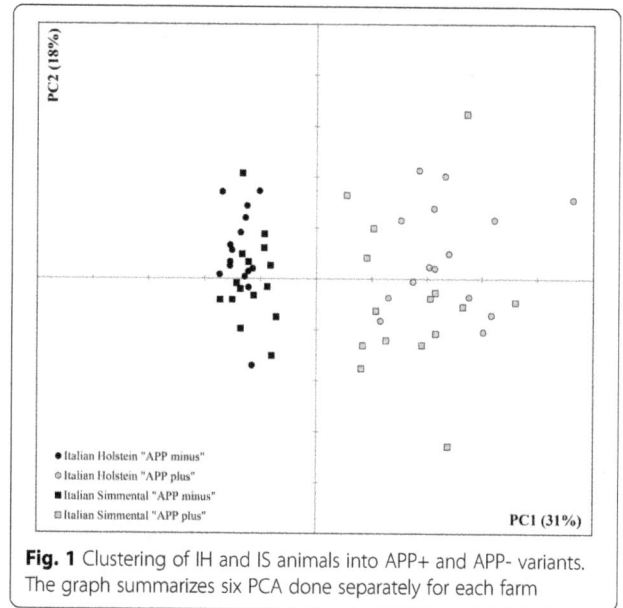

Fig. 1 Clustering of IH and IS animals into APP+ and APP- variants. The graph summarizes six PCA done separately for each farm

each farm. Variable related to the total proteins was also very important as it was highly correlated (around 0.7) to the PC1 in the 4 out of 6 tested farms. Characteristics of groups of animals chosen for the final transcriptome analysis in the selected commercial farms are reported in Additional file 1.

Statistical analysis of milk production data

Significant differences between plus and minus APP groups were observed for milk urea in IS ($P \leq 0.001$). Other parameters did not show significant differences (Table 5).

Between farms, IS cows showed significant differences in milk protein percentage ($P \leq 0.001$), milk yield ($P \leq 0.01$) and percentage of caseins ($P \leq 0.01$). For IH animals the statistically significant differences between farms were observed only for milk fat percentage ($P < 0.05$).

Unlike IS animals, APP+ IH animals demonstrated a marked, even if not significant decrease in milk yield. Milk urea in APP+ animals showed a marked decrease in absolute values in both breeds; however in IS breed the decrease reached the significant level ($P < 0.001$).

Table 4 The total variability explained by the two first components for each PCA separately

% Explained variance						
	F1	F2	F3	F4	F5	F6
	IH	IH	IH	IS	IS	IS
PC1	27.5	33.9	27.2	29.1	35.4	35.6
PC2	19.2	18.0	15.8	24.6	16.7	14.6
Total	46.7	51.9	43.0	53.7	52.1	50.2

(*PC* principal component, *F* farm, *IH* Italian Holstein, *IS* Italian Simmental)

Table 5 Differences in milk parameters among commercial farms and APP groups of animals with p-values and mean standard errors of the statistical analysis performed with SPSS package

Italian Holstein

Milk production data	F1	F2	F3	APP-	APP+	P_{Farm}	$P_{APP(Farm)}$	MSE, %
Milk yield, kg	39.5	37.3	33.0	40.7	32.6	0.231	0.086	4.90
Fat, %	3.08	3.99	3.52	3.37	3.70	*	0.351	2.98
Proteins, %	3.15	3.28	3.17	3.18	3.21	0.471	0.630	0.65
Casein, %	2.50	2.55	2.53	2.53	2.52	0.837	0.601	0.57
Urea, mg/dL	15.75	19.77	19.27	20.41	16.12	0.103	0.092	5.38

Italian Simmental

Milk production data	F4	F5	F6	APP-	APP+	P_{Farm}	$P_{APP(Farm)}$	MSE, %
Milk yield, kg	22.4	33.8	28.3	29.5	26.7	**	0.219	4.43
Fat, %	3.80	3.35	3.27	3.34	3.61	0.208	0.503	3.94
Proteins, %	3.77	3.32	3.31	3.47	3.47	***	0.287	0.58
Casein, %	2.88	2.62	2.56	2.70	2.66	**	0.459	0.58
Urea, mg/dL	21.62	19.57	20.68	23.12	18.12	0.453	***	2.87

(*: low significance [$P \leq 0.05$]; **: high significance [$P \leq 0.01$]; ***: very high significance [$P \leq 0.001$])

Post-sequencing analysis

Alignment of RNA-seq data to the UMD 3.1 bovine reference genome identified 13,665 genes expressed in bovine blood cells. A total of 1,152 significant differentially expressed genes ($P < 0.05$) were identified in the comparison of all APP+ vs. all APP- animals across breeds; 531 in comparison of IH APP+ vs. APP- and 217 in comparison of IS APP+ vs. APP-. The number of shared and unique transcripts within each comparison is indicated on the Venn diagram (Fig. 2).

The higher number of significant genes obtained in the global comparison among APP+ and APP- cows may be explained by the procession of data from all

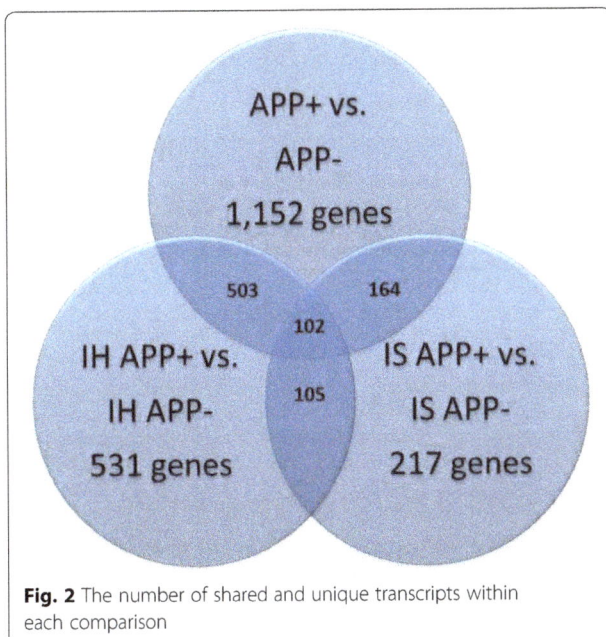

Fig. 2 The number of shared and unique transcripts within each comparison

sampled animals, hence each gene had expression data from both IF and IS cows. This fact increased the level of significance of number of genes in the comparison of all APP+ vs. all APP- variants and showed a less robust significance within intra-breed comparisons.

This analysis also allowed to evaluate the number of over- and underexpressed genes in the list of significant DE genes. Since the expression rate was indicated as the $\log_2$ of the fold change, we assumed that genes with an expression rate greater than 1 were overexpressed and those with the expression rate lower than 1 were underexpressed. The three comparisons, despite the great diversity in the number of significant genes, showed similar ratios between over- and underexpressed genes: in each case, the number of overexpressed genes was far greater than the number of underexpressed ones (Fig. 3). In the global comparison between APP+ and APP- cows the upregulated genes were about 2-fold higher than the downregulated: 763 overexpressed and 389 underexpressed genes. In IF comparison overexpressed genes were about 3-fold higher than underexpressed ones: 396 overexpressed and 135 underexpressed genes. In IS comparison overexpressed genes were about 4-fold more numerous than underexpressed ones: 174 overexpressed and 43 underexpressed genes.

These data highlights that the adaptive response affects blood transcriptome principally by increasing the expression of a high number of genes, while the downregulation is a mechanism with much lower extent.

Dynamic impact approach

The ten most impacted KEGG pathways were identified by DIA for the comparisons with or without breed consideration (Fig. 4).

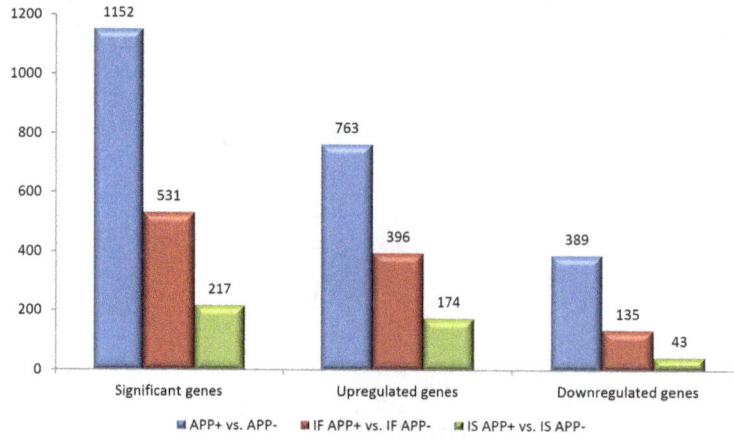

Fig. 3 The number of upregulated and downregulated genes within each comparison

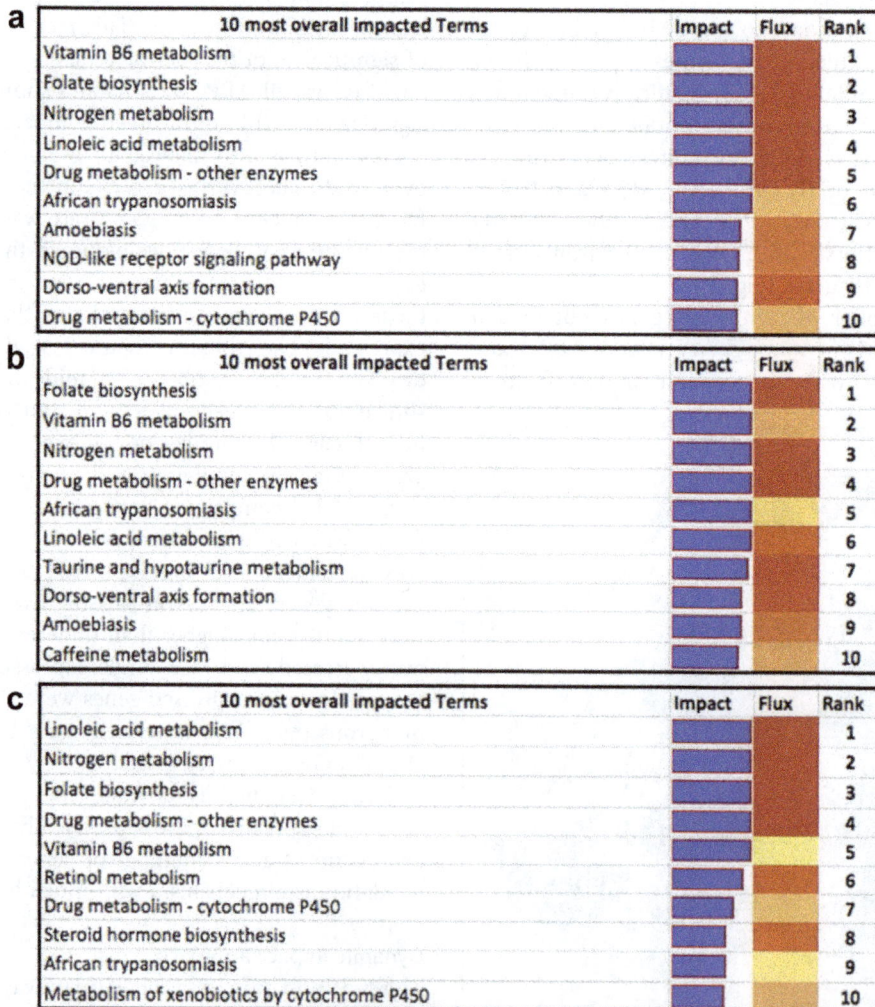

Fig. 4 DIA outputs for each comparison (**a**: APP+ vs. APP-, **b**: IH APP+ vs. IH APP-, **c**: IS APP+ vs. IS APP-)

The three most impacted pathways in the comparison APP+ vs. APP- across breeds and within IH (Figs. 4a and b) are vitamin B6 metabolism, folate biosynthesis and nitrogen metabolism. In IS (Fig. 4c) these pathways are within the first five, in particular nitrogen metabolism is the second, folate biosynthesis the third and vitamin B6 metabolism the fifth most impacted pathway. Other pathways found in all three comparisons are linoleic acid metabolism, drug metabolism-other enzymes and African trypanosomiasis. Some other pathways are present in one comparison or in two out of three comparisons. Amoebiasis and dorso-ventral axis formation are present in the comparison across breeds and in IH; drug metabolism-cytochrome P450 is present in the comparison across breeds and in IS; NOD-like receptor signaling pathway is present only in the comparison

across breeds; taurine and hypotaurine metabolism and caffeine metabolism pathways are present only in the IH comparison; retinol metabolism, steroid hormone biosynthesis and metabolism of xenobiotics by cytochrome P450 are present only in the IS comparison.

The complete list of all impacted pathways in each comparison is presented in Additional file 2.

Gene annotation clusters

Function terms of each cluster identified by DAVID within each comparison were predicted by Gene ontology (GO) (Table 6). After elimination of repeated gene terms, 369 significant differentially expressed genes across breeds, 173 in IH and 73 in IS remained included in significant clusters.

Table 6 Groups of significant clusters in each comparison with indication of the Benjamini-Hochberg-corrected P-values (significance threshold: $P < 0.05$) and the number of genes in common among the three comparisons (see Table 7). Clusters were produced using DAVID database

Annotation Cluster	# Genes	Function terms	Corrected P-value	# Common genes
APP+ vs. APP-				
C1	162	Purine nucleotide binding	0.001	2
C2	44	Cytoplasmic vesicle	0.003	1
C3	37	Cell fraction	0.009	2
C4	15	Positive regulation of cytokine production	0.009	N.A.
C5	150	Glycoprotein	0.016	23
C6	32	Response to wounding	0.018	1
C7	21	Regulation of cytokine production	0.026	1
C8	25	Leukocyte activation	0.029	N.A.
IH APP+ vs. IH APP-				
CH1	11	Tyrosine protein kinase	0.003	N.A.
CH2	72	Purine nucleotide binding	0.005	2
CH3	21	Protein dimerization activity	0.006	N.A.
CH4	83	Glycoprotein	0.006	23
CH5	23	Cytoplasmic vesicle	0.035	1
IS APP+ vs. IS APP-				
CS1	6	Calcium-binding region	0.004	N.A.
CS2	5	Anchored to membrane	0.016	3
CS3	7	Enzyme inhibitor activity	0.017	1
CS4	21	Extracellular space	0.018	12
CS5	5	Negative regulation of molecular function	0.018	N.A.
CS6	3	Cytokine biosynthetic process	0.019	1
CS7	27	Glycoprotein	0.024	19
CS8	42	Glycoprotein	0.025	23
CS9	10	Positive regulation of molecular function	0.030	4
CS10	15	Organelle lumen	0.033	1
CS11	4	Response to steroid hormone stimulus	0.034	2
CS12	5	Nuclear membrane	0.039	1

A total of 24 genes were differentially expressed in all 3 comparisons. Gene names and differential expression values for each of the 24 genes within each group are listed in Table 7.

Discussion

The aim of the present study was to investigate the impact of stress response on gene expression patterns in peripheral blood cells of lactating cows. The analysis of transcriptome variation was performed after the peak of lactation as the transition period is the most challenging in dairy cows and can interfere with the metabolic imbalance of animals [38]. Considering that animals on each commercial farm were kept under the same environmental conditions and that the farm factor was considered in the statistical model, the influence of management on animal adaptive response should have been minimized. Hence, the different levels of plasma APPs are likely to result from individual animal response to subclinical inflammatory/infective events or other stresses, since cows

did not show visible clinical signs or symptoms of the presence of functional disorders.

Stress response is a very complex phenomenon as it can affect overall physiology through different mechanisms, like activation of sympathetic nervous system with the release of catecholamines, activation of HPA axis and non-circadian production of glucocorticoids [1, 4, 39], onset of an acute phase response [14]. Activation of these mechanisms may cause harmful and sometimes irreversible effects on many body systems. Stress may affect circulating glucocorticoids [4, 40, 41] with consequences on female reproductive system [42], immune system, osteoblastogenesis and bone metabolism [43–45], muscle production [46], metabolism of nutrients [47–49], functioning of the thyroid gland [50] and growth hormone axis [51]. Stress research is therefore complicated by these complex and diverse mechanisms and by individual and interspecies differences in stress response [17]. Understanding the biological basis of stress response in livestock is important for improving animal welfare in intensive production systems. In addition of being a growing public concern, animal welfare is important for production efficiency and influence both farm economy and environmental footprint. Whole-transcriptome analysis is crucial to understand how stress influences gene expression to elicit the complex phenomenon of adaptive response. Here we investigated differential expression in blood samples obtained from cows with high and low levels of positive APPs as proxy of stress status and identified stress response-related genes pathways in white blood cells.

Table 7 Common genes for the Annotation Clusters (see Table 6) and the relative differential expression values within each comparison

Gene name	Differential expression (n-fold)		
	APP+ vs. APP-	IH APP+ vs. IH APP-	IS APP+ vs. IS APP-
CA4	3.32	3.68	3.00
ALPL	3.81	4.89	2.97
IGF2	0.33	0.26	0.42
IL10	1.67	1.90	1.47
IGF2R	2.50	1.91	1.36
IL2RA	2.18	2.23	2.14
SCARB1	1.67	1.97	1.42
SLC6A2	2.56	2.79	2.35
MMP9	1.84	2.03	1.66
CHI3L1	2.25	2.87	1.76
PROK2	2.27	2.42	2.13
NMUR2	2.76	2.92	2.60
IL34	0.50	0.46	0.54
ACE2	2.19	2.51	1.91
HEPACAM2	1.97	2.21	1.75
A2M	1.85	2.18	1.57
TMEM120A	1.54	1.83	1.29
CD163	2.61	3.85	1.77
GPR84	2.17	2.57	1.83
PTX3	5.50	6.71	4.51
LYPD8	0.44	0.46	0.43
DEFB7	1.94	2.00	1.88
GNAL	1.60	1.79	1.43
DEFB10	2.19	2.10	2.28

Impacted pathways by DIA

Once fed a list of differentially expressed genes, DIA exploits an online sheet of the Kyoto Encyclopedia of Genes and Genomes (KEGG) database [52] to detect significantly impacted pathways. It calculates the entity and direction of the impact and whether the pathway is entirely overexpressed, underexpressed or if the expression is not altered at all.

The most impacted pathways in APP+ cows in across and within breed analyses are presented in the Fig. 4. Significant genes in significant KEGG pathways have been analyzed in detail to understand gene and pathway function, since the names of KEGG pathways reported in the DIA output files are sometimes misleading.

The pathway of vitamin B6 metabolism (KEGG bta00750), was among the three most significant ones (rank 1 across breeds and in IH and rank 5 in IS). In this pathway we found two significant genes, *PDXK* (pyridoxal kinase) and *AOX1* (aldehyde oxidase 1). The latter was not significant in IS comparison, but it shows xa similar expression pattern. Pyridoxal kinase is involved in the ATP-dependent phosphorylation of pyridoxal, pyridoxamine and pyridoxine to pyridoxal-5-phosphate (PLP), pyridoxamine-5-phosphate and pyridoxine-5-phosphate, respectively. There is a requirement for ubiquitous expression of piridoxal kinase in mammalian tissues as PLP, before entering a cell, must be dephosphorylated and after diffusing through cell membranes it is converted back to the active cofactor by cytosolic pyridoxal kinase [53]. PLP is a very important enzymatic cofactor, as it participates to all transamination reactions and, in some cases, to decarboxylation, deamination and racemization of amino acids [54], catalyzes the rate-limiting step in glycogenolysis [55]. The impact of PLP on amino acid metabolism has direct consequences also on protein synthesis, whose physiological level is altered during an adaptive response. This evidence suggests that overexpression of *PDXK* gene is fundamental for metabolism regulation during adaptive phenomenon, as the cofactor affects protein and energy metabolism, that are likely increased during adaptation, and this is true not only for peripheral white blood cells, but for the vast majority of tissues in an organism. Aldehyde oxidase 1 is an enzyme mainly found in liver, shows broad substrate specificity, including pyridoxal, and catalyzes the oxidation of several endogenous and exogenous aldehydes, with the production of hydrogen peroxide and superoxide ion. Aldehyde oxidase isolated from polymorphonuclear leukocytes showed a more narrow substrate specificity: for example, the enzyme found in leukocytes is inactive on xanthine [56]. This enzyme has a role in oxidative stress and regulation of reactive oxygen species (ROS) homeostasis [57]. The catalysis requires the presence of flavin adenine dinucleotide (FAD) and a molybdopterin cofactor (MoCo) [57–59]. The enzyme also has a role in nitric oxide (NO) biosynthesis [60]. NO has various functions in the organism; white blood cells, mainly macrophages, secrete it as a chemical defense against bacteria and to induce vasodilation [61]. The involvement of aldehyde oxidase 1 in several oxidative metabolisms, in oxidative stress and in NO signaling may explain the overexpression of *AOX1* gene in APP+ cows.

Interesting significant genes emerged from the analysis of folate biosynthesis pathway (KEGG bta00790; rank 2 across breeds and in IH, rank 3 in IS). These are *ALPL* (alkaline phosphatase liver/bone/kidney) and *MOCS1* (molybdenum cofactor synthesis 1). The *ALPL* gene is mainly expressed in neutrophils and monocytes [62]. The role of alkaline phosphatase is fundamental as a high number of signal transduction cascades are involved in adaptive processes. Phosphorylation and dephosphorylation of signal proteins is the key determinant in the phenomenon that regulates the transduction and the amplification of a stimulus from the cell membrane receptor to the nucleus, where the modulation of gene expression occurs. Elevations in plasma alkaline phosphatase, whose sources include neutrophils and monocytes, can be also

related to pathological conditions [62]. *MOCS1* encodes for a protein involved in the biological activation of molybdenum and it is highly expressed by peripheral white blood cells. Participating in the production of MoCo, it is indirectly involved in the catalytic activity of several enzymes, including aldehyde oxidase and xanthine oxidase [63]. *MOCS1* was significant only in APP+ vs. APP- comparison across breeds and the involvement of the gene in a number of metabolic oxidative pathways is likely the reason for its significant overexpression in the APP+ bovines.

In nitrogen metabolism pathway (KEGG bta00910; rank 3 across breed and in IH, rank 2 in IS), *GLUL* (glutamate-ammonia ligase) and *CA4* (carbonic anhydrase IV) genes were significantly differentially expressed. Glutamate-ammonia ligase is a PLP-dependent enzyme that produces glutamine from glutamate and free NH_3. Glutamine is a common metabolite in many amino acid, purine and pyrimidine biosynthetic pathways, so this enzyme has a major role in protein and nucleic acid metabolism. It is also involved in acid-base homeostasis, cell signaling, cell proliferation and biosynthesis of γ-aminobutyric acid (GABA) [64]. Carbonic anhydrase IV is an important Zn-dependent enzyme present in several tissues and, among leukocytes, it is expressed principally by eosinophils and neutrophils. It has a main role in the control of acid-base balance in blood and other tissues [65]. Particularly, this isoform exists in the form of a glycophosphatidylinositol (GPI)-anchored protein and plays an important role in maintaining an appropriate cellular environment for the reactions that occur during adaptive responses. Both these genes are overexpressed in APP+ cows.

The pathway for linoleic acid metabolism (KEGG bta00591; rank 4 across breeds, rank 6 in IH, rank 1 in IS) includes a number of significant genes involved in inflammatory response and metabolism of drugs and xenobiotics. The phospholipase A2 genes, *PLA2G4F* and *PLA2G4A*, selectively hydrolyze membrane phospholipids. The first one has high selectivity for phosphatidylethanolamine, hydrolyzing the ester bond in sn-2 position, and has a role in mitogen-associated protein kinase (MAPK) and Ras signaling pathways [66]. The latter pathway leads to the production of free arachidonic acid, which is further converted in eicosanoids, involved in inflammatory response, and lysophospholipids, that are precursors of platelet-activating factor (PAF). Hence, this enzyme has a role in inflammatory response and hemodynamics, and is also involved in MAPK and G protein-coupled receptor (GPCR) signaling pathways [67]. Some significant genes in this pathway belong to cytochrome P450 superfamily (in detail: *CYP2E1*, *CYP3A4*, *CYP3A5*). These genes are also involved in metabolism of steroid hormones, drugs and carcinogens, playing a role in steroid-mediated physiological responses, activation and metabolic drug clearance and carcinogenesis [68–70].

In the pathway of drug-metabolizing enzymes (KEGG bta00983; rank 5 across breeds, rank 4 in IH and IS) there are several significant genes with a role in the metabolism of nucleotides, suggesting their role in an adaptive response. Xanthine dehydrogenase (*XDH*) is a paralog of *AOX1*, which can operate either as a dehydrogenase or as an oxidase. Xanthine dehydrogenase is involved in metabolism of hypoxanthine and xanthine and in the generation of ROS [71]. Recently, its role in recruiting macrophages through inflammasome activation has been investigated [72]. Cytidine deaminase (*CDA*) preserves pyrimidine pool by irreversibly deaminating cytidine and deoxycytidine to uridine and deoxyuridine, respectively. It is also involved in antibody diversification [73]. Uridine phosphorylase (*UPP1*) reversibly cleaves ribose-1-phosphate and deoxyribose-1-phosphate from uridine and deoxyuridine, releasing free uracil [74]. Another significant enzyme involved in pyrimidine metabolism is dihydropyrimidine dehydrogenase (*DPYD*) [75].

In African trypanosomiasis (KEGG bta05143; rank 6 across breeds, rank 5 in IH, rank 9 in IS) and amoebiasis (KEGG bta05146; rank 7 across breeds, rank 9 in IH) pathways we found a large number of significant genes directly involved in the onset of a stress response or inflammation. Among these, we found several proinflammatory cytokine genes, as *IL12B*, *IL18*, *IL1B* and *TNFα*. These cytokines are produced by different types of immune cells involved in growth, differentiation, chemotaxis and proliferation of white cells during an inflammatory event [76–80]. *NFkB1* and *RELA*, the nuclear transcription factor genes forming the same protein complex, were also among affected ones. *NFkB1* is activated by cytokines, free radicals, UV ray and pathogens' products and it is involved in regulation of inflammation-mediated pathways [81] and in regulation of the expression of number of genes involved in cell adhesion and migration across vascular endothelium, like vascular cell adhesion molecule (*VCAM1*), laminin genes (*LAMA4*, *LAMC1*), fibronectin (*FN1*) and integrin genes (*ITGB2*, *ITGAM*). Some of the latter genes are strongly induced by cytokine signaling, so they have a primary importance in leukocyte adhesion and in cell signal transduction [82–86]. Significant genes in the African trypanosomiasis and amoebiasis pathways include also a number of cell surface receptors involved in inflammation signaling cascades, regulation of cell physiology during these events and in functionality of activated leukocytes. Among these genes there are Fas cell surface death receptor (*FAS*), toll-like receptors (*TLR2*, *TLR4*), *CD14* molecule and complement protein genes (*C8A*) [87–90]. These receptors

transducer signals through different molecules, including myeloid differentiation factor (*MYD88*) [91]. Moreover, some significant genes in these pathways include enzymes that regulate the production of second messengers, like phosphatidylinositide-3-kinases (*PIK3R2*, *PIK3CD*) [92] and phospholipases C (*PLCB3*, *PLCB4*) [93], and proteases that regulate cell protein homeostasis, limiting tissue damage produced by overexpressed proteolytic enzymes, like serpin peptidases (*SERPINB1*, *SERPINB3*) [94]. In IS breed amoebiasis pathway was found to be not significantly impacted between APP+ and APP- animals.

NOD-like receptor signaling pathway (KEGG bta04621) was found to be significant only in the APP+ vs. APP- across breed comparison (rank 8). Significant genes belonging to the pathway are mostly involved in regulation of cell cycle and pro-apoptotic signaling, interacting with the nuclear factor NFkB1 to activate it. Pro-apoptotic genes include the nucleotide-binding oligomerization domain containing receptors (*NOD1*, *NOD2*) [95] and genes of the caspase recruitment domain family (*CARD6*, *PYCARD*, *NLRP3*), which also participate to the formation of inflammasomes [96–98]. In this pathway there are also two significantly overexpressed molecular chaperones, *HSP90AA1* and *HSP90B1*.

Dorso-ventral axis formation (KEGG bta04320) pathway was significantly impacted in APP+ vs. APP- across breeds (rank 9) and in IH (rank 8). However, some genes belonging to this pathway were significant also in IS breed. In this pathway there are two significant transcription factors, *ETV6* and *ETS2*, which are oncogenes with a role in hematopoiesis and apoptosis [99, 100].

The cytochrome P450-mediated metabolism of drugs (KEGG bta00982) is among the significantly impacted pathways across breeds (rank 10) and in IS (rank 7). Important genes involved in oxidative stress and detoxification of oxidation by-products are included in this pathway, such as membrane-bound microsomal glutathione S-transferase (*MGST1*), with a role in the development of inflammation and in cellular defense [101], aldehyde dehydrogenase (*ALDH3B1*) [102] and monoamine oxidase A (*MAOA*) [103]. Monoamine oxidase A has a role in the metabolism of serotonin: this molecule has been shown to be synthesized, released and degraded also by T lymphocytes [104]). Among the significant genes, we found also some terms significantly overexpressed in other pathways, as *AOX1* and *CYP2E1*.

Metabolism of taurine and hypotaurine (KEGG bta00430) was strongly impacted only in IH (rank 7) and the only significant gene detected was a member of γ-glutamyltransferase family (*GGT5*). This gene is involved in metabolism of glutathione and leukotrienes and plays a role in oxidative stress and inflammatory response [105].

The pathway of caffeine metabolism (KEGG bta00232) showed a high impact only in IH (rank 10). It includes the gene *XDH*, the function of which has been previously discussed.

Retinol metabolism (KEGG bta00830) was highly impacted only in IS (rank 6). Important differentially expressed genes in this pathway are *AOX1*, *CYP3A4* and *CYP3A5* and their role in adaptive response was described previously.

Another pathway impacted only in IS included steroid hormone biosynthesis pathway (KEGG bta00140; rank 8). It included *CYP3A4*, *CYP3A5* and *CYP2E1* genes, involved in the biosynthesis of different types of steroids, including corticosteroids, which have a relevant role in adaptive response.

The same scenario occurred also in the pathway relative to metabolism of xenobiotics mediated by cytochrome P450 (KEGG bta00980; rank 10 in IS). It included genes *MGST1*, *CYP2E1* and *ALDH3B1*, which are significant also in other pathways.

The same significant gene plays different roles in different pathways, for example in the metabolism of different compounds or in the signaling of a number of signal transduction cascades. Thus, to obtain a comprehensive analysis and to confirm the significance of the relevant genes, the most impacted pathways were explored by DAVID.

DAVID annotation clusters

Annotation clusters produced by DAVID online tool confirmed the significance of a large number of genes involved in adaptive response. Genes were clustered according to their common structural characteristics or molecular function. Table 6 summarizes the significant clusters, the number of total genes per cluster and the number of genes in common with at least another cluster. In Table 7 are listed common genes between all three comparisons. A comprehensive list of genes included in each cluster is available in Additional file 3.

The highest number of common genes among clusters was included in C5, CH4 and CS8, which are clusters of glycoproteins, based on the GO terms. Twenty-three out of 24 shared genes are included in these clusters. The only one excluded is the *GNAL* gene. We can assume that genes present in these clusters are grouped only according to structural features, as they have various molecular functions, but are all classified as glycoproteins. *GNAL* is included in the cluster that reports purine nucleotide-interacting proteins.

In these clusters we can find some genes that were significant also in DIA output, i.e., *CA4*, *ALPL* and *IL10*. Other shared genes were not included in the most impacted pathways, but they have an important role in the regulation of adaptive response. For example, the receptor

of insulin-like growing factor II (*IGF2R*) is overexpressed in all three comparisons, possibly because of its role in intracellular trafficking of lysosomal enzymes and degradation of IGF-II [106].

Angiotensin I-converting enzyme 2 (*ACE2*) has a direct effect on cardiac and renal functions [107]. Its overexpression may be another consequence of hyperactivation of HPA axis, as glucocorticoids have a direct effect on blood pressure and cardiovascular system, and of an increased uremia, as this condition may trigger the recruitment of pro-atherogenic, ACE2-expressing monocytes [108].

As most of cytokines, present in all three comparisons, interleukin 10 (*IL10*) is significantly overexpressed. Such overexpression can be explained by its important role in lymphocytes differentiation and proliferation and in the production of antibodies by B cells. In the same list of overexpressed genes are present receptors for cytokines, like interleukin 2 receptor alpha (*IL2RA*), involved in intracellular signaling pathways.

The only one underexpressed cytokine revealed among three comparisons is the pro inflammatory cytokine IL34.

Defensins are small antimicrobial, cytotoxic peptides produced by neutrophils [109]. β-defensins (*DEFB7* and *DEFB10* in Table 7), belonging to one of the three existing groups of defensins, were overexpressed in all comparisons. Their overexpression could be associated to the presence of a bacterial infection in mammary gland [110] and this result is in agreement with the highly impacted pathway relative to *S. aureus* infection (see Additional file 2).

Another interesting overexpressed gene is the gene encoding matrix metalloproteinase 9 (*MMP9*). Its overexpression can be associated with an augmented production of hematopoietic stem cells [111] that give rise to all red and white blood cells and platelets. *MMP9* has recently been investigated for its role in promoting the secretion of pro-inflammatory cytokines and the migration of T cells towards inflammation sites. Moreover, the protein increases the permeability of brain-blood barrier to cytokines, showing an important involvement in neuroinflammation [112].

Summarizing the results, a lot of highly impacted pathways were found in common between IH and IS breeds, indicating a similar variation in the gene expression under environmental adaptation. This observation was also confirmed by annotation clustering of the significantly expressed genes: in fact, some of the significantly clustered genes are present also in the most impacted pathways, with similar levels of expression.

Differences between breeds were observed only on the level of individual genes. Therefore, considering the overall variation in stress-inducing factors and stress-related gene expression, the general patterns can be considered very similar in the two breeds investigated.

The obtained data can be considered reliable even if the further validation analysis of the differential gene expression in the population could improve the relevance of the conclusions. This point can be developed in a future research.

Analysis of milk production data

The differences in milk production data between APP- and APP+ cows are reported in Table 5. Significant association with high APP level was observed only for milk urea in IS, which showed a marked decrease in APP+ animals. A marked decrease, close to being statistically significant, was observed also in IH milk yield and milk urea.

The decrease in milk yield can be associated to the onset of a stress response as energy resources are driven towards other organs and other more important physiological processes to guarantee the animal's survival. The decrease in milk urea can be linked to the overall increased protein synthesis. In dairy cows milk urea reflects the catabolism of protein by the ruminant tissues and within the rumen by bacteria. The decrease of rumen ammonia may indicate that animals increase the protein synthesis during an adaptive response [113]. Indeed, high levels of APPs and at the same time high expression of gene products of several pathways were observed in APP+ animals.

Conclusions

This transcriptomic study in lactating dairy cows from IH and IS breeds allowed to assess that the onset of an adaptive response to the environment involves a large number of pathways that are regulated in a similar way in the two breeds, with marginal differences in significantly over- or underexpressed single genes. The altered expression of these genes is statistically associated with the variations of only certain milk production parameters between groups of cows with activated and not activated stress response mechanisms.

Considering all the results we can conclude that the two studied breeds have similar patterns but vary in the degree of activation of metabolic and physiological mechanisms of adaptation to the environment. IH showed a higher rate of significant genes and impacted pathways, demonstrating a higher metabolic activity, respect to the IS breed.

Additional files

Additional file 1: Characteristics of animals chosen for the transcriptome analysis in the selected commercial farms. F: farm, IH: Italian Holstein, IS: Italian Simmental, DIM: days in milk, BCS: body condition score, SCC: somatic cell count, values are expressed as mean ± SD.

Additional file 2: Complete set of 10% most impacted pathways produced by DIA analysis in each comparison. Three images are presented. The first image (named A) collects the 10% most impacted pathways for

APP+ vs. APP- comparison. The second image (named B) lists the 10% most impacted pathways for IH APP+ vs. IH APP- comparison. The third image (named C) lists the 10% most impacted pathways for IS APP+ vs. IS APP- comparison.

Additional file 3: Complete list of genes included in DAVID Annotation Clusters. Each sheet of the.xlsx file includes one of the Annotation Clusters listed in Table 6 of the present article. In each Annotation Cluster the included genes are listed. The genes found to be common in each comparison are bolded.

Abbreviations

AP: Acute phase reaction; APP: Acute phase protein; DIA: Dynamic impact approach; F1: Farm 1; F2: Farm 2; F3: Farm 3; F4: Farm 4; F5: Farm 5; F6: Farm 6; HPA: Hypothalamic-pituitary-adrenal; IH: Italian Holstein; IS: Italian simmental; PCA: Principal component analysis; RIN: RNA integrity number; SCC: Somatic cell count

Acknowledgements

This work was conceived within the EU COST Action FA1308 "DairyCare". We are grateful to Prof. Bruno Stefanon for the helpful comments and discussion points leading to refinement of our concepts.

Funding

Project was approved by independent peer reviewers and on the basis of these evaluations the research was funded by the Italian Ministry of Education, University and Research (PRIN GEN2PHEN). The grant covered all the expensiveness of the study.

Authors' contribution

NP and TM performed RNA extraction from blood, elaboration of sequencing data applying DIA and DAVID tools, statistical elaboration of results and drafted the manuscript. MD and FP participated in the design of the study, in animal sampling and helped to draft the manuscript. DL elaborated raw RNA-Seq data using DeSeq2. PAM contributed to the final revision of the manuscript. AM participated in sample collection and helped in manuscript revision. SS participated in the design of the study and its coordination and also in animal sampling and biochemical analysis of blood and milk samples. All authors read and approved the final manuscript.

Competing interests

The authors declare that they have no competing interests.

Author details

[1]Department of Agriculture, Food, Environment and Animal Science (DI4A), University of Udine, via delle Scienze 206, 33100 Udine, Italy. [2]Department of Agricultural, Environmental and Food Sciences, University of Molise, via F. De Sanctis snc, 86100 Campobasso, Italy. [3]CBM S.c.r.l, SS 14 – km 163.5 AREA Science Park, 34149 Basovizza, TS, Italy. [4]Institute of Zootechnics, Catholic University of the Sacred Heart, via Emilia Parmense 84, 29133 Piacenza, Italy.

References

1. Kinlein SA, Wilson CD, Karatsoreos IN. Dysregulated hypothalamic-pituitary-adrenal axis function contributes to altered endocrine and neurobehavioral responses to acute stress. Front Psychiatry. 2015;6:31.
2. Silverman MN, Pearce BD, Biron CA, Miller AH. Immune modulation of the hypothalamic-pituitary-adrenal (HPA) axis during viral infection. Viral Immunol. 2005;18(1):41–78.
3. Selye HA. Syndrome produced by diverse nocuous agents. Nature. 1936;138:132.

4. Elenkov IJ, Webster EL, Torpy DJ, Chrousos GP. Stress, corticotropin-releasing hormone, glucocorticoids, and the immune/inflammatory response: acute and chronic effects. Ann NY Acad Sci. 1999;876:1–11.

5. Eckersall PD, Conner JG. Bovine and canine acute phase proteins. Vet Res Commun. 1988;12(2-3):169–78.

6. Haeryfar SM, Berczi I. The thymus and the acute phase response. Cell Mol Biol (Noisy-le-grand). 2001;47(1):145–56.

7. Kent J. Acute phase proteins: their use in veterinary diagnosis. Br Vet J. 1992; 148(4):279–82.

8. Conner JG, Eckersall PD, Doherty M, Douglas TA. Acute phase response and mastitis in the cow. Res Vet Sci. 1986;41(1):126–8.

9. Horadagoda A, Eckersall PD, Alsemgeest SP, Gibbs HA. Purification and quantitative measurement of bovine serum amyloid-A. Res Vet Sci. 1993; 55(3):317–25.

10. Dowling A, Hodgson JC, Schock A, Donachie W, Eckersall PD, McKendrick IJ. Experimental induction of pneumonic pasteurellosis in calves by intratracheal infection with Pasteurella multocida biotype A:3. Res Vet Sci. 2002;73(1):37–44.

11. Alsemgeest SP, Taverne MA, Boosman R, van der Weyden BC, Gruys E. Peripartum acute-phase protein serum amyloid-A concentration in plasma of cows and fetuses. Am J Vet Res. 1993;54(1):164–7.

12. Jawor P, Stefaniak T. Acute phase proteins in cattle. In: Veas F, editor. Acute phase proteins as early non-specific biomarkers of human and veterinary diseases. Rijeka: InTech d.o.o; 2011. p. 381–408.

13. Amadori M, Stefanon B, Sgorlon S, Farinacci M. Immune system response to stress factors. Ital J Anim Sci. 2009;8(1):287–99.

14. Cray C, Zaias J, Altman NH. Acute phase response in animals: a review. Comp Med. 2009;59(6):517–26.

15. Ranabir S, Reetu K. Stress and hormones. Indian J Endocrinol Metab. 2011; 15(1):18–22.

16. DairyCare COST Action FA1308. 2014. http://www.dairycareaction.org. Accessed 21 June 2016

17. Hing S, Narayan E, Thompson RCA, Godfrey S. A review of factors influencing the stress response in Australian marsupials. Conserv Physiol. 2014;2(1):cou027.

18. Solano-Aguilar G, Molokin A, Botelho C, Fiorino A-M, Vinyard B, Li R, et al. Transcriptomic Profile of Whole Blood Cells from Elderly Subjects Fed Probiotic Bacteria Lactobacillus rhamnosus GG ATCC 53103 (LGG) in a Phase I Open Label Study. PLoS One. 2016;11:2.

19. McLoughlin KE, Nalpas NC, Rue-Albrecht K, Browne JA, Magee DA, Killick KE, et al. RNA-seq transcriptional profiling of peripheral blood leukocytes from cattle infected with mycobacterium bovis. Front Immunol. 2014;5:396.

20. Shin H, Shannon CP, Fishbane N, Ruan J, Zhou M, Balshaw R, et al. Variation in RNA-Seq transcriptome profiles of peripheral whole blood from healthy individuals with and without globin depletion. PLoS One. 2014;9:3.

21. Weikard R, Demasius W, Hadlich F, Kühn C. Different blood cell-derived transcriptome signatures in cows exposed to vaccination pre- or postpartum. PLoS One. 2015;10:8.

22. O'Loughlin A, Lynn DJ, McGee M, Doyle S, McCabe M, Earley B. Transcriptomic analysis of the stress response to weaning at housing in bovine leukocytes using RNA-seq technology. BMC Genomics. 2012;13:250.

23. Chai V, Vassilakos A, Lee Y, Wright JA, Young AH. Optimization of the PAXgene blood RNA extraction system for gene expression analysis of clinical samples. J Clin Lab Anal. 2005;19(5):182–8.

24. Mardis ER. Next-generation DNA, sequencing methods. Annu Rev Genomics Hum Genet. 2008;9:387–402.

25. Wang Z, Gerstein M, Snyder M. RNA-Seq: a revolutionary tool for transcriptomics. Nat Rev Genet. 2009;10(1):57–63.

26. Wickramasinghe S, Cánovas A, Rincón G, Medrano JF. RNA-sequencing: a tool to explore new frontiers in animal genetics. Livest Sci. 2014;166:206–16.

27. Han Y, Gao S, Muegge K, Zhang W, Zhou B. Advanced applications of RNA sequencing and challenges. Bioinform Biol Insights. 2015;9 Suppl 1:29–46.

28. Bionaz M, Periasamy K, Rodriguez-Zas SL, Hurley WL, Loor JJ. A novel dynamic impact approach (DIA) for functional analysis of time-course omics studies: validation using the bovine mammary transcriptome. PloS One. 2012;7(3):e32455.

29. Institut National de la Recherche Agronomique. Introduction. Feeding standards for ruminants. In: Jarrige R, editor. Ruminant Nutrition. Recommended Allowances and Feed Tables. London: Eurotext; 1989. p. 15–22.

30. Sgorlon S, Fanzago M, Sandri M, Gaspardo B, Stefanon B. Association of index of welfare and metabolism with the genetic merit of Holstein and Simmental cows after the peak of lactation. Ital J Anim Sci. 2015;14(3):368–73.

31. Sgorlon S, Fanzago M, Guiatti D, Gabai G, Stradaioli G, Stefanon B. Factors affecting milk cortisol in mid lactating dairy cows. BMC Vet Res. 2015;11:259.

32. Schroeder A, Mueller O, Stocker S, Salowski R, Leiber M, Gassmann M, et al. The RIN: an RNA integrity number for assigning integrity values to RNA measurements. BMC Mol Biol. 2006;7:3.

33. Sgorlon S, Colitti M, Asquini E, Ferrarini A, Pallavicini A, Stefanon B. Administration of botanicals with the diet regulates gene expression in peripheral blood cells of Sarda sheep during ACTH challenge. Domest Anim Endocrinol. 2012;43(3):213–26.

34. Elsik GC, Unni DR, Diesh CM, Tayal A, Emery ML, Nguyen HN, et al. Bovine Genome Database: new tools for gleaning function from the Bos taurus genome. Nucleic Acids Res. 2015;44(D1):D834–9.

35. Love MI, Huber W, Anders S. Moderated estimation of fold change and dispersion for RNA-seq data with DESeq2. Genome Biol. 2014;15(12):550.

36. Benjamini Y, Hochberg Y. Controlling the false discovery rate: a practical and powerful approach to multiple testing. J R Stat Soc Series B Stat Methodol. 1995;57(1):289–300.

37. Huang DW, Sherman BT, Tan Q, Collins JR, Alvord WG, Roayaei J, et al. The DAVID gene functional classification tool: a novel biological module-centric algorithm to functionally analyze large gene lists. Genome Biol. 2007;8(9):R183.

38. Sundrum A. Metabolic disorders in the transition period indicate that the dairy cows' ability to adapt is overstressed. Animals (Basel). 2015;5(4):978–1020.

39. Papadimitriou A, Priftis KN. Regulation of the hypothalamic-pituitary-adrenal axis. Neuroimmunomodulation. 2009;16(5):265–71.

40. Buckingham JC, Loxley HD, Christian HC, Philip JG. Activation of HPA axis by immune insults: roles and interactions of cytokines, eicosanoids, glucocorticoids. Pharmacol Biochem Behav. 1996;54(1):285–98.

41. Kitajima T, Ariizumi K, Bergstresser PR, Takashima A. A novel mechanism of glucocorticoid-induced immune suppression: the inhibition of T cell-mediated terminal maturation of a murine dendritic cell line. J Clin Invest. 1996;98(1):142–7.

42. Chrousos GP, Torpy DJ, Gold PW. Interactions between the hypothalamic-pituitary-adrenal axis and the female reproductive system: clinical implications. Ann Intern Med. 1998;129(3):229–40.

43. Weinstein RS, Jilka RL, Parfitt AM, Manolagas SC. Inhibition of osteoblastogenesis and promotion of apoptosis of osteoblasts and osteocytes by glucocorticoids. Potential mechanism of their deleterious effects on bone. J Clin Invest. 1998;102(2):274–82.

44. Doga M, Bonadonna S, Giustina A. Glucocorticoids and bone: cellular, metabolic and endocrine effects. Hormones (Athens). 2004;3(3):184–90.

45. Tamura Y, Okinaga H, Takami H. Glucocorticoid-induced osteoporosis. Biomed Pharmacother. 2004;58(9):500–4.

46. Schakman O, Gilson H, Kalista S, Thissen JP. Mechanism of muscle atrophy induced by glucocorticoids. Horm Res. 2009;72(1):36–41.

47. Tsigos C, Young RJ, White A. Diabetic neuropathy is associated with increased activity of the hypothalamic-pituitary-adrenal axis. J Clin Endocrinol Metab. 1993;76(3):554–8.

48. Anagnostis P, Athyros VG, Tziomalos K, Karagiannis A, Mikhailidis DP. Clinical review: The pathogenetic role of cortisol in the metabolic syndrome: a hypothesis. J Clin Endocrinol Metab. 2009;94(8):2692–701.

49. Rose AJ, Vegiopoulos A, Herzig S. Role of glucocorticoids and the glucocorticoid receptor in metabolism: insights from genetic manipulations. J Steroid Biochem Mol Biol. 2010;122(1-3):10–20.

50. Joseph-Bravo P, Jaimes-Hoy L, Charli JL. Regulation of TRH neurons and energy homeostasis-related signals under stress. J Endocrinol. 2015;224(3):R139–59.

51. Nicolaides NC, Kyratzi E, Lamprokostopoulou A, Chrousos GP, Charmandari E. Stress, the stress system and the role of glucocorticoids. Neuroimmunomodulation. 2015;22(1-2):6–19.

52. Kaneisha M, Goto S. KEGG: Kyoto Encyclopedia of Genes and Genomes. Nucleic Acids Res. 2000;28(1):27–30.

53. Hanna MC, Turner AJ, Kirkness EF. Human pyridoxal kinase. cDNA cloning, expression, and modulation by ligands of the benzodiazepine receptor. J Biol Chem. 1997;272(16):10756–60.

54. Toney MD. Reaction specificity in pyridoxal phosphate enzymes. Arch Biochem Biophys. 2005;433(1):279–87.

55. Livanova NB, Chebotareva NA, Eronina TB, Kurganov BI. Pyridoxal 5′-phosphate as a catalytic and conformational cofactor of muscle glycogen phosphorylase b. Biochemistry Mosc. 2002;67(10):1089–98.

56. Beedham C. Similarity in the aldehyde oxidases from guinea-pig liver and polymorphonuclear leucocytes. J Pharm Pharmacol. 1986;38:57–8.

57. Kundu TK, Hille R, Velayutham M, Zweier JL. Characterization of superoxide production from aldehyde oxidase: an important source of oxidants in biological tissues. Arch Biochem Biophys. 2007;460(1):113–21.

58. Sahi J, Khan KK, Black CB. Aldehyde oxidase activity and inhibition in hepatocytes and cytosolic fractions from mouse, rat, monkey and human. Drug Metab Lett. 2008;2(3):176–83.

59. Garattini E, Fratelli M, Terao M. The mammalian aldehyde oxidase gene family. Hum Genomics. 2009;4(2):119–30.

60. Li H, Kundu TK, Zweier JL. Characterization of the magnitude and mechanism of aldehyde oxidase-mediated nitric oxide production from nitrite. J Biol Chem. 2009;284(49):33850–8.

61. Lundberg JO, Weitzberg E, Gladwin MT. The nitrate-nitrite-nitric oxide pathway in physiology and therapeutics. Nat Rev Drug Discov. 2008;7:156–67.

62. Izumi M, Ishikawa J, Takeshita A, Maekawa M. Increased serum alkaline phosphatase activity originating from neutrophilic leukocytes. Clin Chem. 2005;51(9):1751–2.

63. Hänzelmann P, Hernández HL, Menzel C, García-Serres R, Huynh BH, Johnson MK, et al. Characterization of MOCS1A, an oxygen-sensitive iron-sulfur protein involved in human molybdenum cofactor biosynthesis. J Biol Chem. 2004;279(33):34721–32.

64. Labow BI, Souba WW, Abcouwer SF. Mechanisms governing the expression of the enzymes of glutamine metabolism – glutaminase and glutamine synthetase. J Nutr. 2001;131Suppl 9:2467S–74S. discussion 2486S-7S.

65. Wen T, Mingler MK, Wahl B, Khorki ME, Pabst O, Zimmermann N, et al. Carbonic anhydrase IV is expressed on IL-5-activated murine eosinophils. J Immunol. 2014;192(12):5481–9.

66. Ohto T, Uozumi N, Hirabashi T, Shimizu T. Identification of novel cytosolic phospholipase A2s, murine cPLA2δ, ε, and ζ, which form a gene cluster with cPLA2β. J Biol Chem. 2005;280(26):247576–83.

67. Burke JE, Dennis EA. Phospholipase A2 structure/function, mechanism, and signaling. J Lipid Res. 2009;50(Suppl):S237–42.

68. Schulz-Utermoehl T, Mountfield RJ, Bywater RP, Madsen K, Jørgensen PN, Hansen KT. Structure-function analysis of human CYP3A4 using a specific proinhibitory antipeptide antibody. Drug Metab Dispos. 2000;28(7):718–25.

69. Patki KC, Von Moltke LL, Greenblatt DJ. In vitro metabolism of midazolam, triazolam, nifedipine, and testosterone by human liver microsomes and recombinant cytochromes p450: role of cyp3a4 and cyp3a5. Drug Metab Dispos. 2003;31(7):938–44.

70. Hanioka N, Yamamoto M, Iwabu H, Jinno H, Tanaka-Kagawa T, Naito S, et al. Functional characterization of human and cynomolgus monkey cytochrome P450 2E1 enzymes. Life Sci. 2007;81(19-20):1436–45.

71. Harrison R. Structure and function of xanthine oxidoreductase: where are we now? Free Radic Biol Med. 2002;33(6):774–97.

72. Ives A, Nomura J, Martinon F, Roger T, LeRoy D, Miner JN, et al. Xanthine oxidoreductase regulates macrophage IL1β secretion upon NLRP3 inflammasome activation. Nat Commun. 2015;6:6555.

73. Chung SJ, Fromme JC, Verdine GL. Structure of human cytidine deaminase bound to a potent inhibitor. J Med Chem. 2005;48(3):658–60.

74. Temmink OH, de Bruin M, Laan AC, Turksma AW, Cricca S, Masterson AJ, et al. The role of thymidine phosphorylase and uridine phosphorylase in (fluoro) pyrimidine metabolism in peripheral blood mononuclear cells. Int J Biochem Cell Biol. 2006;38(10):1759–65.

75. Van Kuilenburg AB, Meinsma R, Beke E, Bobba B, Boffi P, Enns GM, et al. Identification of three novel mutations in the dihydropyrimidine dehydrogenase gene associated with altered pre-mRNA splicing or protein function. Biol Chem. 2005;386(4):319–24.

76. Boraschi D, Tagliabue A. Human interleukin 1: structure-function relationship. Ann Ist Super Sanita. 1990;26(3-4):273–82.

77. Oshimi K, Hoshino S. Function and molecular structure of IL-12. Nihon Rinsho. 1992;50(8):1840–4.

78. Burdin N, Rousset F, Banchereau J. B-cell-derived IL-10: production and function. Methods. 1997;11(1):98–111.

79. Biet F, Locht C, Kremer L. Immunoregulatory functions of interleukin 18 and its role in defense against bacterial pathogens. J Mol Med. 2002;80(3):147–62.

80. Mukai Y, Shibata H, Nakamura T, Yoshioka Y, Abe Y, Nomura T, et al. Structure-function relationship of tumor necrosis factor (TNF) and its

81. receptor interaction based on 3D structural analysis of a fully active TNFR1-selective TNF mutant. J Mol Biol. 2009;385(4):1221–9.

81. Pereira SG, Oakley F. Nuclear factor-kappaB1: regulation and function. Int J Biochem Cell Biol. 2008;40(8):1425–30.

82. Ponce ML, Nomizu M, Delgado MC, Kuratomi Y, Hoffmann MP, Powell S, et al. Identification of endothelial cell binding sites on the laminin γ1 chain. Circ Res. 1999;84(6):688–94.

83. Dib K, Andersson T. β2 integrin signaling in leukocytes. Front Biosci. 2000;5:D438–51.

84. Cook-Mills JM. VCAM-1 signals during lymphocyte migration: role of reactive oxygen species. Mol Immunol. 2002;39(4):499–508.

85. Sottile J, Hocking DC. Fibronectin polymerization regulates the composition and stability of extracellular matrix fibrils and cell-matrix adhesions. Mol Biol Cell. 2002;13(10):3546–59.

86. DeHahn KC, Gonzales M, Gonzalez AM, Hopkinson SB, Chandel NS, Brunelle JK, et al. The α4 laminin subunit regulates endothelial cell survival. Exp Cell Res. 2004;294(1):281–9.

87. Kawakami A, Eguchi K, Matsuoka N, Tsuboi M, Koji T, Urayama S, et al. Expression and function of Fas and Fas ligand on peripheral blood lymphocytes in normal subjects. J Lab Clin Med. 1998;132(5):404–13.

88. Viriyakosol S, Mathison JC, Tobias PS, Kirkland TN. Structure-function analysis of CD14 as a soluble receptor for lipopolysaccharide. J Biol Chem. 2000; 275(5):3144–9.

89. Sabroe I, Dower SK, Whyte MK. The role of Toll-like receptors in the regulation of neutrophil migration, activation, and apoptosis. Clin Infect Dis. 2005;41 Suppl 7:S421–6.

90. Hadders MA, Beringer DX, Gros P. Structure of C8α-MACPFF reveals mechanism of membrane attack in complement immune defense. Science. 2007;317(5844):1552–4.

91. Qiu Y, Shen Y, Li X, Ding C, Ma Z. Molecular cloning and functional characterization of a novel isoform of chicken myeloid differentiation factor 88 (MyD88). Dev Comp Immunol. 2008;32(12):1522–30.

92. Koyasu S. The role of PI3K in immune cells. Nat Immunol. 2003;4(4):313–9.

93. Ting AE, Pagano RE. Detection of a phosphatidylinositol-specific phospholipase C at the surface of Swiss 3 T3 cells and its potential role in the regulation of cell growth. J Biol Chem. 1990;265(10):5337–40.

94. Christensen S, Sottrup-Jensen L. Characterization of two serpins from bovine plasma and milk. Biochem J. 1994;303 Pt.2:383–90.

95. Ekman AK, Cardell LO. The expression and function of Nod-like receptors in neutrophils. Immunology. 2010;130(1):55–63.

96. Dufner A, Pownall S, Mak TW. Caspase recruitment domain protein 6 is a microtubule-interacting protein that positively modulates NF-κB activation. Proc Natl Acad Sci USA. 2006;103(4):988–93.

97. Zhou R, Yazdi AS, Menu P, Tschopp J. A role for mitochondria in NLRP3inflammasome activation. Nature. 2011;469(7329):221–5.

98. Proell M, Gerlic M, Mace PD, Reed JC, Riedi SJ. The CARD plays a critical role in ASC foci formation and inflammasome signaling. Biochem J. 2013;449(3):613–21.

99. Wang LC, Swat W, Fujiwara Y, Davidson L, Visvader J, Kuo F, et al. The TEL/ETV6 gene is required specifically for hematopoiesis in the bone marrow. Genes Dev. 1998;12(15):2392–402.

100. Dwyer J, Li H, Xu D, Liu JP. Transcriptional regulation of telomerase activity: roles of the Ets transcription factor family. Ann NY Acad Sci. 2007;1114:36–47.

101. Siritantikorn A, Johansson K, Ahlen K, Rinaldi R, Suthiphongchai T, Wilairat P, et al. Protection of cells from oxidative stress by microsomal glutathione transferase 1. Biochem Biophys Res Commun. 2007;355(2):592–6.

102. Marchitti SA, Orlicky DJ, Vasiliou V. Expression and initial characterization of human ALDH3B1. Biochem Biophys Res Commun. 2007;356(3):792–8.

103. Medvedev AE, Gorkin VZ. The role of monoamine oxidase in the regulation of mitochondrial energy functions. Vopr Med Khim. 1991;37(5):2–6.

104. Chen Y, Leon-Ponte M, Pingle SC, O'Connell PJ, Ahern GP. T lymphocytes possess the machinery for 5-HT synthesis, storage, degradation and release. Acta Physiol (Oxf). 2015;213(4):860–7.

105. Heisterkamp N, Groffen J, Warburton D, Sneddon TP. The human gamma-glutamyltransferase gene family. Hum Genet. 2008;123(4):321–32.

106. Morgan DO, Edman JC, Standring DN, Fried VA, Smith MC, Roth RA, et al. Insulin-like growth factor II receptor as a multifunctional binding protein. Nature. 1987;329(6137):301–7.

107. Ingelfinger JR. Angiotensin-converting enzyme 2: implications for blood pressure and kidney disease. Curr Opin Nephrol Hypertens. 2009;18(1):79–84.

108. Trojanowicz B, Ulrich C, Kohler F, Bode V, Seibert E, Fiedler R, et al. Monocytic angiotensin-converting enzyme 2 relates to atherosclerosis in patients with chronic kidney disease. Nephrol Dial Transplant. 2016;0:1–12.

109. Selsted ME, Tang YQ, Morris WL, McGuire PA, Novotny MJ, Smith W, et al. Purification, primary structures, and antibacterial activities of β-defensins, a new family of antimicrobial peptides from bovine neutrophils. J Biol Chem. 1993;268(9):6641–8.

110. Kościuczuk EM, Lisowski P, Jarczak J, Krżyzewski J, Zwierzchowski L, Bagnicka E. Expression patterns of β-defensin and cathelicidin genes in parenchyma of bovine mammary gland infected with coagulase-positive or coagulase-negative Staphylococci. BMC Vet Res. 2014;10:246.

111. Janowska-Wieczorek A, Marquez LA, Nabholtz JM, Cabuhat ML, Montaño J, Chang H, et al. Growth factors and cytokines upregulate gelatinase expression in bone marrow CD34+ cells and their transmigration through reconstituted basement membrane. Blood. 1999;93(10):3379–90.

112. Vafadari B, Salamian A, Kaczmarek L. MMP-9 in translation: from molecule to brain physiology, pathology, and therapy. J Neurochem. 2016;139 Suppl 2:91–114.

113. Schrödl W, Büchler R, Wendler S, Reinhold P, Muckova P, Reindl J, et al. Acute phase proteins as promising biomarkers: perspectives and limitations for human and veterinary medicine. Proteomics Clin Appl. 2016;00:1–16.

19

Effects of total replacement of corn silage with sorghum silage on milk yield, composition, and quality

M. Cattani[1], N. Guzzo[2*], R. Mantovani[2] and L. Bailoni[1]

Abstract

Background: In the last years, difficulties occurring in corn cultivation (i.e., groundwater shortages, mycotoxin contamination) have been forcing dairy farmers to consider alternative silages. Some experiments conducted on lactating cows have proven that the total replacement of corn silage with sorghum silage did not reduce milk yield. However, this kind of substitution involves supplementing sorghum-based diets with grains, to compensate for the lower starch content of sorghum silage compared to corn silage. Change of silage type and inclusion of starch sources in the diet would influence rumen fermentations, with possible effects on milk composition (i.e., fatty acid profile) and coagulation properties. A worsening of milk coagulation properties would have a negative economic impact in Italy, where most of the milk produced is processed into cheese.

This study was designed to compare milk composition and quality, with emphasis on fatty acid profile and coagulation properties, in dairy cows fed two diets based on corn or sorghum silage.

Results: The sorghum diet reduced milk yield ($P = 0.043$) but not 4% fat corrected milk ($P = 0.85$). Feeding sorghum silage did not influence milk contents of protein ($P = 0.07$) and lactose ($P = 0.65$), and increased fat content ($P = 0.024$). No differences emerged for milk concentrations of saturated ($P = 0.61$) and monounsaturated fatty acids ($P = 0.50$), whereas polyunsaturated fatty acids were lower ($P < 0.001$) for the sorghum diet. Concentrations of n-6 ($P < 0.001$) and n-3 fatty acids ($P = 0.017$) were lower in milk of cows fed the sorghum diet. Milk coagulation properties did not differ between the two diets, except the "a30" (the curd firmness, expressed in mm, 30 min after rennet addition), that was lower ($P = 0.042$) for the sorghum diet.

Conclusions: Feeding a forage sorghum silage, properly supplemented with corn meal, as total replacement of corn silage maintained milk composition and did not influence negatively milk coagulation properties, which have a great economic relevance for the Italian dairy industry. Thus, silages obtained from forage sorghums could have a potential as substitute of corn silages in dairy cow diets.

Keywords: Dairy cows, Forage sorghum silage, Mean particle size, Milk coagulation properties, Milk fatty acid profile

Background

Corn silage is the main ingredient of diets fed to lactating cows in the farms of the Po Valley (North-Italy), with exception of those producing milk processed into some Protected Denomination of Origin cheeses (i.e., Parmigiano-Reggiano). However, because of some difficulties occurring in corn cultivation (i.e., groundwater shortages, plant

attack by specific parasites, mycotoxin contamination), in the last few years dairy farmers have been considering the use of alternative forages for ensiling. Experiments conducted on lactating cows have proven that the total replacement of corn silage with sorghum silage did not affect milk yield [1–4]. However, in some studies such results were obtained through the supplementation of sorghum diets with starch sources (i.e., corn meal), to compensate for the lower starch content of sorghum silage compared to corn silage [3, 4]. There is evidence from literature that the change of silage type and the inclusion of starch sources in the diet would influence

* Correspondence: nadia.guzzo@studenti.unipd.it
[2]Department of Agronomy Food Natural resources Animals and Environment (DAFNAE), University of Padova, Viale dell'Università 16, 35020 Legnaro, PD, Italy
Full list of author information is available at the end of the article

rumen fermentation patterns, with effects on composition of milk fat [5] and milk protein [6]. In turn, a modification of protein fractions would affect some milk coagulation properties (MCP) as renneting time and curd firmness [7]. To our knowledge, no attempts have been made to explore whether the total replacement of corn silage with sorghum silage could influence milk fatty acids (FA) profile and MCP. These milk properties have a great economic relevance in Italy, where about 50% of cow milk is processed into Protected Denomination of Origin (PDO) cheeses, even if some specifications (i.e. that of Parmigiano-Reggiano cheese) do not allow the use of silages in dairy cow diets.

In this framework, this study was designed to compare milk yield, composition, and quality, with particular emphasis on milk FA profile and MCP, in dairy cows fed two experimental diets containing silages obtained from corn or forage sorghum.

Methods

Corn and forage sorghum cultivation, harvest, and ensiling

Corn plants (hybrid Kayras, FAO class 600; KWS Italia Spa, Monselice, Italy) were sown on 29 May, 2014 on a soil that previously hosted alfalfa (*Medicago sativa*). The soil belonged to a farmhouse located in San Martino di Lupari, Province of Padova, latitude 45.3°N, longitude 11.5°E; elevation: 40 m above sea level. The corn was fertilized with 200 units of Nitrogen, 140 units of Phosphorous, and 180 units of Potassium/hectare, and weed control was applied with 1.5 L/hectare of herbicide Adengo (Bayer CropScience S.r.l. Milano, Italy). Corn was harvested at two-thirds milk line stage of maturity on 30 September, 2014. Plants of forage sorghum (hybrid Hannibal; KWS Italia Spa, Monselice, Italy) were sown on an adjacent field in the same farmhouse as above on 17 June, 2014 on a soil that previously hosted barley. The sorghum was fertilized with 160 units of Nitrogen, 80 units of Phosphorous, and 100 units of Potassium/hectare, and weed control was applied with 0.3 kg/hectare of herbicide Casper (Syngenta Italia S.p.A., Milano, Italy). Sorghum was harvested at the beginning of the early bloom stage, on 30 September, 2014. No irrigation was needed for either forages due to the favorable climate situation. For both corn and sorghum a forage harvester with 8 line-head for chopping (New Holland Agriculture, Turin Italy) was used. Corn plants were harvested with knives adjusted to a 13-mm theoretical length of cut and processed with a 3-mm roller clearance. Sorghum plants were harvested with knives adjusted to a 20-mm theoretical length of cut, without any processing. After harvest, corn and forage sorghum were ensiled into bunker silos without inoculation and covered with nylon film for 60 d. Sample of silages (n = 4 per corn and sorghum) were obtained collecting about 3 kg of silage from 4 different points in the front of each bunker silos. The chemical composition of the two silages is given in Table 1.

Table 1 Descriptive statistics for nutrient composition, fatty acid profile, and particle size of the two silages used in the study

Component (g/kg DM unless noted)	Corn silage		Sorghum silage	
	Mean	SEM	Mean	SEM
Nutrients				
DM as fed	331.9	4.8	222.7	0.8
CP	76.8	0.9	73.7	1.9
Fat	29.0	1.2	24.7	0.7
Ash	43.2	6.0	79.3	0.4
NDF	348.7	10.0	711.3	14.8
ADF	197.4	5.7	459.9	11.5
ADL	29.5	1.2	73.3	1.9
Starch	351.9	20.2	25.7	2.4
NSC[a]	502.4	10.6	111.9	15.8
NE$_L^b$, Mcal/kg DM	1.82	0.02	0.92	0.04
pH	3.68	0.02	4.08	0.07
Fatty acid profile				
SFA[c], % of total FA[d]				
C16:0	14.4	0.1	22.6	0.5
C18:0	2.31	0.03	2.77	0.21
Total SFA	19.7	0.4	34.2	0.9
MUFA[e], % of total FA				
C18:1 n-9	15.3	2.0	5.2	1.1
Total MUFA	17.9	2.0	9.1	1.0
PUFA[f], % of total FA				
C18:2 n-6	54.4	1.5	22.6	1.2
C18:3 n-3	7.0	0.3	32.5	1.9
Total PUFA	62.4	1.7	56.7	0.8
n-6, % of total FA	54.5	1.5	22.9	1.2
n-3, % of total FA	7.2	0.4	32.6	1.9
Particle size fraction, % retained (as-fed basis)				
> 19 mm	4.3	0.8	4.6	0.9
> 8 to 19 mm	68.0	1.4	80.7	0.9
< 8 mm	27.7	1.2	14.7	0.3
pef[g]	0.72	0.01	0.85	0.01
peNDF[h], % of DM	24.6	0.6	62.1	0.4
Mean particle size[i], mm	6.2	0.2	8.7	0.1

[a]*NSC* non-structural carbohydrates, calculated as: 1000 – (CP + Fat + Ash + NDF), [b]*NE$_L$* calculated according to NRC (2001 [8]), [c]*SFA* saturated fatty acids, [d]*FA* fatty acids, [e]*MUFA* monounsaturated fatty acids, [f]*PUFA* polyunsaturated fatty acids, [g]*pef* physical effectiveness factor, calculated as sum of the proportion of feed particles retained on sieves with openings of 19 and 8 mm, [h]*peNDF* physically effective NDF, calculated as pef multiplied by the corresponding NDF content, [i]Mean particle size = calculated according to ASABE (2007 [17])

Treatments, animals and experimental design

All experimental procedures were carried out according to Italian law on animal care (Legislative Decree No. 26 of March 14, 2014). Eighteen Holstein-Friesian dairy cows (DIM: 146 ± 96 d; parity: 2.0 ± 1.2; milk yield: 28.4

± 6.3 kg/d), housed in a tie-stall commercial dairy farm belonging to the farmhouse described above (San Martino di Lupari, Padova province), were used. Animals were randomly assigned to 2 groups subjected to 2 treatment sequences in a cross-over experimental design of 28-d periods. Each experimental period was preceded by a 10-d preliminary period, thus the whole experiment lasted 76 d. Each group received a total mixed ration (TMR) based on corn silage (CS diet) or forage sorghum silage (SS diet). Two silages were included in different proportions (295 and 195 g/kg DM for CS and SS, respectively), according to their chemical composition. The SS diet contained an amount of corn meal that was almost double compared to the CS diet (254 vs. 142 g/kg DM), in order to compensate for the lower starch content of forage sorghum silage compared to corn silage. The two TMR were prepared using a total mixer wagon. During preparation of TMR, the two experimental diets were only mixed, without a further chopping of silages. The diets were formulated to be isoenergetic and isonitrogenous using the Plurimix system® (Fabermatica Sas, Ostiano, Cremona, Italy), and according to NRC system [8] considering a predicted milk yield of about 30 kg/d and a feed intake of about 25 kg/d per cow. The chemical composition of the experimental diets is given in Table 2. The forage NDF contents resulted 182 and 218 g/kg DM for CS and SS, respectively [8] (data not shown). During the trial the cows had free access to water and they were fed once daily. Cows were milked twice per day (at 0700 h and 1900 h). Milking procedure occurred at stall using a milking pipeline system.

Data and sample collection

The amount of diet distributed in the manger of each experimental group was measured daily and recorded by the weighing station of the mixer wagon. The orts were daily collected and analyzed for DM in order to obtain the daily DM intake (DMI), on a group basis, as difference between consumed and residual DM. Individual milk yield was recorded and individual milk samples (50 mL/each) from the morning milking only were collected from each cow to be analyzed for composition, FA profile, and MCP (n = 2 at about d 15 and d 28 of each period for a total of 4 samples per cow). Representative samples (about 3 kg as fed) of the two silages and of the two diets were collected to be analyzed for chemical composition and particle size distribution. Feed and diet samples were collected at the beginning and at the end of each period (n = 4 samples for each silage and diet).

Sample analysis

Samples of silages and diets were analyzed in duplicate for DM (DM; # 934.01; [9]), N (# 976.05; [9]), lipids (# 920.29; [9]) and ash (# 942.05; [9]). The NDF content was measured using α-amylase and sodium sulphite, with the Ankom[220]

Fiber Analyzer (Ankom Technology®, Macedon, NY, USA). The ADF, inclusive of residual ash, and ADL contents were sequentially determined according to [10]. Starch was analyzed by high-performance liquid chromatography, following the quickly and more precise to conventional assay method suggested by [11]. The FA profile of silages and diets was determined according to procedure suggested by [12].

Fat, protein, and lactose contents of the milk were analyzed using the FIL-IDF procedure [13] by MilkoScan™ FT1 apparatus (Foss Electric, DK-3400, Hillerød, Denmark). Milk urea nitrogen was measured automatically by the conduct metric-enzymatic method (CL 10 micro analyser, Eurochem, Roma, Italy). Somatic cell count was performed using a Fossomatic™ 5000 (Foss Electric, DK-3400, Hillerød, Denmark) according to the standard FIL-IDF148a [14], and transformed in logarithmic terms using the following equation: $SCS = 3 + ln_2$ (somatic cell count × 10^{-5}). Milk lipids were extracted according to the procedures described by [15]. Fatty acid methyl esters were analyzed using a two-dimensional gas-chromatography instrument (Agilent 7890A, Agilent Technologies, Milan, Italy) equipped with a modulator (Agilent G3486 A CFT), an automatic sampler (Agilent 7693), a flame-ionization detector connected to a chromatography software (Agilent Chem Station), and two columns in series, to separate and identify each FA on a 2-dimensional basis [15]. The first column was a 75 m × 180 μm (internal diameter) × 0.14 μm film thickness column (23348U, Supelco, Bellefonte, PA) and used H_2 as carrier gas (flow of 0.22 mL/min). The second was a 3.8 m × 250 μm (internal diameter) × 0.25 μm film thickness column (J&W 19091-L431, Agilent Technologies) and used H_2 as carrier gas (flow of 22 mL/min). A Computerized Renneting Meter (CRM-48, Polo Trade, Monselice, Italy) was used to determine MCP within 5 h after collection of samples. Milk samples (10 mL) were preheated at 35°C, and 200 μL of rennet (NATUREN TM STANDARD 215, Hansen 215 IMCU/mL, Pacovis Amrein AG, Bern, Switzerland) were diluted to 1.2% (v/v) in distilled water and then added to the milk. This analysis provided measurements of rennet coagulation time (RCT; the time occurring from addition of rennet to the beginning of coagulation), k20 (the time to observe a curd firmness of 20 mm after the rennet was added to the sample), and a30 (curd firmness 30 min after rennet addition). Measures of pH (pH-Burette 24, Crison) were conducted before measuring MCP.

The particle size distribution of silages and diets was determined by dry sieving using a Penn State Particle Separator equipped with three screens (diameter openings = 19 mm, 8 mm, plus the bottom pan). Approximately 250 g of feed sample was placed on the upper screen of the separator and sieved according to the procedure described by [16].

Table 2 Ingredients ad descriptive statistics for nutrient composition, fatty acid profile, and particle size of the two experimental diets

Component (g/kg DM unless noted)	Corn silage diet		Sorghum silage diet	
	Mean	SEM	Mean	SEM
Ingredients				
Corn silage	295	-	-	-
Sorghum silage	-	-	195	-
Alfalfa hay	172	-	170	-
Corn meal	142	-	254	-
Soybean meal	87	-	85	-
Barley meal	65	-	63	-
Cottonseeds	65	-	63	-
Dry sugar beet pulp	53	-	52	-
Distillers dried grains with solubles	44	-	43	-
Wheat straw	44	-	43	-
Soybean hulls	30	-	30	-
Urea	2	-	2	-
Nutrients				
DM, % fresh matter	539.2	2.1	158.9	0.2
CP	141.6	3.3	142.6	2.7
Fat	38.3	2.4	33.1	1.1
Ash	67.7	0.8	69.4	2.4
NDF	365.1	2.9	397.0	9.1
ADF	218.8	2.5	238.4	3.9
ADL	41.9	1.5	47.7	1.0
Starch	229.5	8.2	202.3	10.7
NSC[a]	387.4	0.6	357.9	9.3
NE$_L$[b], Mcal/kg DM	1.61	0.02	1.59	0.02
Fatty acid profile				
SFA[c], % of total FA[d]				
C16:0	18.7	0.2	19.3	0.1
C18:0	2.48	0.03	2.28	0.02
Total SFA	24.6	0.3	24.8	0.1
MUFA[e], % of total FA				
C18:1 n-9	18.8	0.3	19.3	0.3
Total MUFA	21.6	0.3	21.7	0.2
PUFA[f], % of total FA				
C18:2 n-6	49.3	0.2	48.6	0.7
C18:3 n-3	4.0	0.2	4.2	0.4
Total PUFA	53.8	0.1	53.4	0.3
n-6, % of total FA	49.4	0.2	48.6	0.7
n-3, % of total FA	4.0	0.2	4.2	0.4

Table 2 Ingredients ad descriptive statistics for nutrient composition, fatty acid profile, and particle size of the two experimental diets (Continued)

Particle size fraction, % retained (as-fed basis)				
> 19 mm	5.7	1.4	3.8	0.6
> 8 to 19 mm	27.8	0.7	33.7	0.9
< 8 mm	66.5	1.4	62.5	0.8
pef[g]	0.34	0.01	0.37	0.01
peNDF[h], % of DM	12.3	0.5	15.2	0.5
Mean particle size[i], mm	2.2	0.1	2.4	0.1

[a]NSC non-structural carbohydrates, calculated as: 1000 − (CP + Fat + Ash + NDF), [b]NE$_L$ calculated according to NRC (2001 [8]), [c]SFA saturated fatty acids, [d]FA fatty acids, [e]MUFA monounsaturated fatty acids, [f]PUFA polyunsaturated fatty acids, [g]pef physical effectiveness factor, calculated as sum of the proportion of feed particles retained on sieves with openings of 19 and 8 mm, [h]peNDF physically effective NDF, calculated as pef multiplied by the corresponding NDF content, [i]Mean particle size = calculated according to ASABE (2007 [17])

Calculations and statistical analysis

Mean particle size of silages and diets was calculated according to [17]. The physical effectiveness factor (pef; ranging from 0 to 1) of silages and diets was computed as the sum of the proportion of feed particles retained on sieves with openings of 19 and 8 mm [18]. The content of physically effective NDF (peNDF; % of DM) of silages and diets was calculated by multiplying the pef by the corresponding NDF content [18]. Yield of solids-corrected milk (SCM) was computed according to [19]. Feed efficiency of cows, intended as the ability to convert feed to milk, was calculated as the ratio between the SCM and the DMI.

All individual data were preliminarily analyzed using the TTSET procedure of [20] to investigate possible carryover effects due to the sequence. Because of the washout interval of 10 d between the two experimental periods, no carryover was detected. Data were then analyzed by a MIXED procedure of [20] accounting for the fixed effects of diet (CS and SS), period (first and second period), and sequence of treatment (CS-SS or SS-CS sequence), and the random effect of cow. A simple one way ANOVA by a GLM procedure of [20] was carried out for the dry matter intake accounting for the diet effect only.

Results
Chemical composition and mean particle size of silages and diets

Compared to the corn silage, the sorghum silage showed, as expected, a lower DM content (223 vs. 332 g/kg; Table 1). On the contrary, the protein and lipid contents were similar for the two silages. As expected, levels of fibrous fractions (NDF, ADF, and ADL) resulted greater for the forage sorghum silage compared to the corn silage. Consequently, the sorghum silage showed a lower content of NSC (non-structural carbohydrates; 112 vs. 502 g/kg

DM) and starch (26 vs. 352 g/kg DM) with respect to the corn silage. Compared to the corn silage, the sorghum silage had a greater proportion of saturated fatty acids (SFA) and n-3 but lower proportions of MUFA and PUFA (Table 1). As regards to the main FA, the sorghum silage showed higher proportions of palmitic and α-linolenic acid and lower proportions of oleic and linoleic acid. Considering the particle size distribution, the sorghum silage had a greater percentage of feed material retained on the screen of 8 mm and a lower percentage retained on the bottom pan in comparison with the corn silage. The sorghum silage had a greater pef (0.85 vs. 0.72) and peNDF (62.1 vs. 24.6% of DM) compared to the corn silage, as a consequence of particle size distribution and greater NDF content, and a greater mean particle size. Compared to the CS diet, the SS diet had a greater NDF content (397 vs. 365 g/kg DM; Table 2) and a lower starch content (202 vs. 229 g/kg DM). The diets did not differ for FA composition, showing comparable contents of SFA, MUFA, and PUFA. As regards to the main FA, the only difference emerged for stearic acid that was lower for the SS diet compared to the CS diet. The SS diet showed a greater percentage of feed particles retained on the screen of 8 mm and a greater value of peNDF with respect to the CS diet (Table 2). However, the two diets did not differ in terms of pef and mean particle size.

Dry matter intake, milk yield, milk composition and coagulation properties, and feed efficiency

The DMI did not differ between the two diets ($P = 0.88$; Table 3). Milk yield (kg/d) was lower (29.8 vs. 31.6 kg/d; $P = 0.043$) for the SS diet, whereas yields of 4% fat corrected milk ($P = 0.85$), fat ($P = 0.78$), protein ($P = 0.23$), and lactose ($P = 0.07$) did not differ between the two diets. The yield of SCM was comparable for the two diets ($P = 0.59$) and resulted 31.6 and 31.0 kg/d, for the CS and the SS diet, respectively. In terms of milk composition, the SS diet resulted in a greater fat content (4.26 vs. 3.98%; $P = 0.02$) and a nominally greater protein content (3.66 vs. 3.55%; $P = 0.07$) compared to the CS diet. Lactose, urea and SCS in milk were not influenced by the diet. Except for the value of a30, that was lower (31.9 vs. 35.5 mm; $P = 0.04$) in the SS diet, milk coagulation properties did not differ between diets (Table 3). Milk samples of cows fed the SS diet showed, on average, a small but significantly greater pH compared to those fed the CS diet (6.84 vs. 6.86, for the CS and SS diet, respectively; $P = 0.008$).

Feed efficiency, expressed as the ratio between the SCM and the DMI, was similar for the two groups of cows ($P = 0.97$).

Total concentrations of saturated FA and monounsaturated FA in milk did not differ between the two diets ($P = 0.61$ and $P = 0.50$, respectively; Table 4), whereas that of polyunsaturated FA resulted lower (4.07 vs. 4.58% of total FA; $P < 0.001$)

Table 3 Least square means and pooled standard error (SE) for DMI, milk yield and composition of cows fed the two experimental diets

Item	Diet		Pooled SE	P value
	Corn silage	Sorghum silage		
DMI, kg/d	24.88	24.52	1.44	0.878
Yield, kg/d				
Milk	31.63	29.79	1.97	0.043
4% FCM[a]	31.83	31.54	1.92	0.848
Fat	1.23	1.25	0.07	0.781
Protein	1.11	1.07	0.06	0.229
Lactose	1.54	1.45	0.10	0.067
SCM[b]	31.63	31.02	1.73	0.589
Milk composition, %				
Fat	3.98	4.26	0.16	0.024
Protein	3.55	3.66	0.08	0.065
Lactose	4.85	4.84	0.04	0.634
Urea, mg/L	24.01	25.42	0.52	0.510
SCS[c], units	3.59	3.65	0.44	0.897
Milk coagulation properties				
RCT[d], min	19.75	19.76	1.26	0.986
k20[e], min	6.66	6.50	0.35	0.732
a30[f], mm	35.45	31.90	2.25	0.042
pH	6.84	6.86	0.02	0.008
Feed efficiency, kg/kg				
SCM:DMI	1.27	1.27	0.07	0.969

[a]FCM milk corrected milk, [b]SCM solids-corrected milk, calculated according to Tyrrell and Reid (1965 [19]), [c]SCS somatic cell score, calculated as: SCS = 3 + $\ln_2$(somatic cell count/100, 000), [d]RCT rennet coagulation time, [e]k20 time required to reach a curd firmness of 20 mm, [f]a30 curd firmness 30 min after the addition of rennet

for the SS diet in comparison to the CS diet (Table 4). As a consequence, the ratio between monounsaturated FA and polyunsaturated FA was greater (6.27 vs. 5.51% of total FA; $P < 0.001$) for the SS diet compared to the CS diet. Among saturated FA, concentrations of butyric (C4:0; $P = 0.003$), caprylic (C8:0; $P = 0.003$), palmitic (C16:0; $P = 0.010$), and stearic acid (C18:0; $P = 0.019$) were greater in the milk of cows fed the SS diet; opposite trends were observed for capric (C10:0), and lauric acid (C12:0). Among polyunsaturated FA, concentrations of n-6 and n-3 were lower in milk of cows fed the SS diet ($P < 0.001$ and $P = 0.017$, respectively). The ratio n-6/n-3 also resulted lower ($P < 0.001$) for the SS diet compared to the CS diet. Conjugated linoleic acid isomers (CLA) content, expressed as the sum of the single isomers, was greater ($P = 0.004$) in the milk of cows fed the SS diet.

Discussion

Effects on dairy cows performance

The presence of a sorghum silage in the diet, as total replacement of corn silage, has often been associated with

Table 4 Least square means and pooled standard error (SE) for milk fatty acids from cows fed the two experimental diets

Item	Diet		Pooled SE	P value
	Corn silage	Sorghum silage		
SFA[a], % of total FA[b]				
C4:0	3.02	3.24	0.06	0.003
C6:0	2.09	2.08	0.02	0.507
C8:0	1.31	1.21	0.02	0.003
C10:0	3.11	2.71	0.09	<0.001
C12:0	3.69	3.13	0.12	0.080
C14:0	11.90	11.00	0.36	0.062
C16:0	30.76	32.05	0.56	0.010
C18:0	9.93	10.94	0.35	0.019
Others	4.57	4.26	0.11	0.037
Total SFA	70.38	70.60	0.59	0.607
MUFA[c], % of total FA				
C14:1	0.92	0.87	0.05	0.116
C16:1	1.28	1.26	0.05	0.563
C18:1 n-7	1.87	2.03	0.06	0.018
C18:1 n-9	17.75	18.39	0.50	0.124
Others	3.22	2.77	0.06	<0.001
Total MUFA	25.04	25.32	0.51	0.496
PUFA[d], % of total FA				
C18:2 n-6	2.48	2.02	0.07	<0.001
C18:2 c9, t11 CLA[e]	0.44	0.52	0.02	0.006
C18:2 t10, c12 CLA	0.01	0.01	0.01	0.396
C18:3 n-3	0.30	0.27	0.01	0.001
C18:3 n-6	0.06	0.05	0.01	0.044
C20:3 n-6	0.15	0.13	0.01	<0.001
C20:4 n-6	0.19	0.16	0.01	<0.001
Others	0.97	0.91	0.03	0.031
Total PUFA	4.58	4.07	0.12	<0.001
SFA/(MUFA + PUFA)	2.40	2.42	0.07	0.714
MUFA/PUFA	5.51	6.27	0.14	<0.001
n-6, % of total FA	2.91	2.39	0.08	<0.001
n-3, % of total FA	0.39	0.36	0.01	0.017
n-6/n-3	7.49	6.73	0.16	<0.001
CLA, % of total FA	0.49	0.58	0.02	0.004

[a]SFA saturated fatty acids, [b]FA fatty acids, [c]MUFA monounsaturated fatty acids, [d]PUFA polyunsaturated fatty acids, [e] CLA conjugated linoleic acid isomers

a decreased DMI [2, 4]. However, other authors [1, 21] did not find differences in DMI for diets based on sorghum or corn silages. The comparable DMI observed in this experiment can be partially related to the similar mean particle size of the two diets. The sorghum silage had a greater pef, peNDF, and particle size compared to the corn silage, because it was cut longer at harvesting.

However, such differences were not reflected in the physical characteristics of the final diets, as the SS diet was supplemented with a greater amount of finely ground corn, allowing a reduction of particle size in the complete diet. Such a supplementation was necessary because of low starch content of sorghum, which is confirmed by others [22]. The general high DMI observed in this experiment might be related to the proper particle size of the two diets that were fractioned by sieving according to guidelines: 2 to 8% of feed material was retained on the upper screen, 30 to 50% on the middle one, and 30 to 70% on the bottom pan [23].

In this experiment milk yield was greater (+1.8 kg/d) when cows received the diet based on corn silage. Such a result is in accordance with findings of [4] and it could be due to the greater daily intake of net energy for lactation (NE_L) when cows received the CS diet (40.1 vs. 38.9 Mcal/d, for the CS and the SS diet, respectively). However, differences disappeared when milk production was corrected for milk fat content, as the SS diet favored a + 7% increase of fat percentage compared to the CS diet. Positive effects of sorghum silage on milk fat percentage have been observed by others [2]. In the case of our study, this result could be related to the greater intake of NDF when cows received the SS diet (9.7 vs. 9.1 kg NDF per day, for cows fed the SS and the CS diet, respectively) and to the greater amount of peNDF provided by the SS diet. Performances of dairy cows were identical in terms of yield of SCM and in terms of SCM:DMI. Values of SCM and SCM:DMI found in this study were, on average, respectively 15% greater and 5% lower than those reported by [2] for cows fed diets based on brown midrib (bmr) forage sorghum or corn silages. It could be speculated that this result was mostly related to the high DMI rather than to the production level of cows that was in line with farm data.

The absence of effects on milk protein, in terms of percentage and production (kg/d), is confirmed by [1]. In the present study milk produced by cows fed the SS diet showed a protein content that was nominally greater than that produced by cows fed the CS diet. However, this effect did not have any consequence on milk urea content. Values of SCS were similar for the two diets, according to other studies [2, 3], and within the physiological range.

Effects on milk fatty acid profile

Results of this study highlighted that milk FA profile was clearly influenced by the diet, with notable changes in proportion of the individual fatty acids. Milk was richer in PUFA when the cows were fed the CS diet, whereas the total contents of SFA and MUFA were not influenced by the dietary treatment.

The current literature encompasses several studies that evaluated effects on milk FA composition due to the replacement of corn silage with grass silages [5]. However, to our knowledge, no trial was focused on sorghum and corn silages, thus a direct comparison with literature is not possible.

Effects of corn silage on milk FA profile are well documented, but their interpretation is often complicated by the inference of chemical characteristics of corn silage and other feed ingredients included in the ration [24].

In this study, the two experimental diets differed for NDF and starch concentrations. Such a result was mainly related to the use of a forage sorghum that, as expected, had contents of NDF and starch, respectively, much greater (711 vs. 349 g NDF/kg DM) and much smaller (26 vs. 352 g starch/kg DM) compared to corn silage. In this context, inclusion of a greater amount of corn meal in the SS diet was not sufficient to compensate chemical differences between the two diets.

The addition of starch sources (i.e., cereal grains) to the ration is often associated with a decrease in milk FA with 6 to 16 carbons, and to an increase in unsaturated FA with 18 carbons [25]. Accordingly, the SS diet, being supplemented with corn meal, determined a significant decrease of C8:0, C10:0, and a numerical decrease of C12:0 and C14:0. On the contrary, the effect on 18-carbon unsaturated FA was more variable.

In addition to starch and NSC, also the dietary fat content and composition exert an influence on milk FA profile [25]. In this study, the SS diet had a lower fat content compared to the CS diet, likely because of significant supplementation with corn meal. However, both fat contents were included in ranges that are typical of rations used in dairy farms of North Italy [26]. The two diets did not differ for FA composition, with the exception of caproic acid, that was greater for the CS diet (data not shown), and stearic acid (2.48 and 2.28% on total FA, for the CS and the SS diet, respectively). Based on these data, we can speculate that differences in terms of milk FA profile between the two groups of cows were mostly related to the carbohydrate fraction rather than to the fat fraction of the diets.

Possible effects on milk FA profile due to the lactation stage, parity, or farming conditions can be excluded, because the two groups of animals were balanced for the first two parameters and farming conditions were the same for all the cows.

Despite the different effects on milk FA profile, the SS diet produced some improvements on the nutritional value of milk in comparison to CS diet. Particularly, the SS diet increased concentration of some FA with anticarcinogenic properties, such as butyric (C4:0), which is mostly protective against colon cancer, and CLA [27]. On the other hand, the CS diet increased the milk concentration of n-3 FA that also minimize the risk of cancer and cardiovascular diseases [28].

Effects on milk coagulation properties

The role of feeding on influencing MCP is still uncertain, even if there is evidence that these properties are more influenced by genotype of animals than by phenotypic factors such as feeding [29]. However, it could be speculated that feeding and MCP are indirectly related. In this study, the protein content of milk produced by cows fed the SS diet tended to be greater than that produced by cows fed the CS diet. A recent review on modeling of MCP [30] reported that mean values (33 estimates obtained from 26 experiments) of RCT and k20 were on average 14.1 ± 4.9 min and 9.2 ± 3.1 min, respectively, for milk from Holstein-Friesian cows. Milk samples analyzed in the present study showed, on average, values of RCT and k20 included in these ranges. The same review indicated that milk produced by Holstein-Friesian cows had a mean value of a30 equal to 29.9 ± 8.0 mm. In this study, milk samples collected from cows fed the CS or SS diet revealed a mean value of a30 (35.5 mm and 31.9 mm, respectively) included in the above-mentioned range. However, the measures of a30 obtained with CRM, which was used in this study to evaluate MCP, should be considered with caution, as they were found to be less repeatable and reproducible compared to those of RCT [31].

Conclusions

Feeding a forage sorghum silage, properly supplemented with corn meal, as total replacement of corn silage, maintained DMI, milk yield, and feed efficiency intended as the ability of cows to convert feed to milk. The substitution of corn silage with sorghum silage did not change the concentration of saturated and monounsaturated fatty acids, but reduced the concentration of polyunsaturated fatty acids in milk. In addition, n-6 and n-3 fatty acids resulted lower in milk of cows fed the sorghum diet as compare to cows fed the corn diet. Milk coagulation properties, which have a great economic relevance for the Italian dairy industry, were not altered by the substitution of the corn silage with the sorghum silage in dairy cows. These preliminary results suggest that forage sorghum silages could have a potential as substitute of corn silages in dairy cow diets.

Abbreviations

a30: Curd firmness 30 min after the addition of rennet; ADF: Acid detergent fibre expressed inclusive of residual ash; ADL: Acid detergent lignin; CLA: Conjugated linoleic acid isomers; CP: Crude protein; CS: Corn silage; DM: Dry matter; DMI: Dry matter intake; FA: Fatty acids; k20: Time required to reach a curd firmness of 20 mm; MCP: Milk coagulation properties; MUFA: Monounsaturated fatty acids; NDF: Neutral detergent fibre assayed with a heat stable amylase and expressed inclusive of residual ash; NE_L: Net energy per lactation; NRC: National Research Council; NSC: Non-structural carbohydrates; pef: Physical effectiveness factor; pefNDF: Physically effective NDF; PUFA: Polyunsaturated fatty acids; RCT: Rennet coagulation time; SCM: Solids corrected milk; SCS: Somatic cell score; SE: Standard error; SEM: Standard error of the mean; SFA: Saturated fatty acids; SS: Sorghum silage; TMR: Total mixed ration

Acknowledgements
This trial was financed by the KWS Italia Spa (Monselice, PD, Italy). The authors would like to thank Andretta Francesco for the availability of the dairy cow farm and his help in collecting the samples. Many thanks to Roberto Cecchinato (KWS Italia Spa) for his support during the cultivation, harvesting and ensiling of sorghum and corn, to Paolo Cracco (Breeder Association of Padova province) for the diet formulation, milk collection and analysis, and to Valentina Bonfatti and Roberta Rostellato for cheese-making activities. A special thanks to the anonymous reviewers for their help in improving the quality of the manuscript.

Funding
This trial was financed by a private company (KWS Italia Spa) located in Monselice, Italy. The funding allowed to cover all costs related to collection, transportation and analysis of feed and milk samples.

Authors' contribution
NG gave a substantial contribution to collection and analysis of samples. NG and RM curated all aspects of the statistical analysis. MC gave contribution to writing of the manuscript. RM and LB contributed in revising critically the manuscript. All authors read and approved the final version of the manuscript.

Competing interests
The authors declare that they have no competing interests.

Author details
[1]Department of Comparative Biomedicine and Food Science (BCA), University of Padova, Viale dell'Università 16, 35020 Legnaro, PD, Italy. [2]Department of Agronomy Food Natural resources Animals and Environment (DAFNAE), University of Padova, Viale dell'Università 16, 35020 Legnaro, PD, Italy.

References
1. Oliver AL, Grant RJ, Pendersen JF, O'Rear J. Comparison of brown midrib-6 and -18 forage sorghum with conventional sorghum and corn silage in diets of lactating dairy cows. J Dairy Sci. 2004;87(3):637–44.
2. Dann HM, Grant RJ, Cotanch KW, Thomas ED, Ballard CS, Rice R. Comparison of brown midrib sorghum-sudangrass with corn silage on lactational performance and nutrient digestibility in Holstein dairy cows. J Dairy Sci. 2008;91(2):663–72.
3. Colombini S, Rapetti L, Colombo D, Galassi G, Crovetto GM. Brown midrib forage sorghum silage for the dairy cow: nutritive value and comparison with corn silage in the diet. Ital J Anim Sci. 2010;9(3):273–77.
4. Colombini S, Galassi G, Crovetto GM, Rapetti L. Milk production, nitrogen balance, and fiber digestibility prediction of corn, whole plant grain sorghum, and forage sorghum silages in the dairy cow. J Dairy Sci. 2012; 95(8):4457–67.
5. Kalač P, Samková E. The effects of feeding various forages on fatty acid composition of bovine milk fat: a review. Czech J Anim Sci. 2010;55(12):521–37.
6. Walker GP, Dunshea FR, Doyle PT. Effects of nutrition and management on the production and composition of milk fat and protein: a review. Aust J Agric Res. 2004;55(10):1009–28.
7. Wedholm A, Larsen LB, Lindmark-Månsson H, Karlsson AH, Andrén A. Effect of protein composition on the cheesemaking properties of milk from individual dairy cows. J Dairy Sci. 2006;89(9):3296–305.
8. National Research Council (NRC). Nutrient requirements of dairy Cattle. 7 revth ed. Washington: National Academy Press; 2001.
9. AOAC International. Official Methods of Analysis. 19th ed. Gaithersburg: AOAC International; 2012.
10. Robertson JB, Van Soest PJ. The detergent system of analysis and its application to human foods. In: James WPT, Theander O, editors. The Analysis of Dietary Fiber in Food. New York: Marcel Dekker Inc; 1981. p. 123–58.
11. Bouchard J, Chornet E, Overend RP. High-performance liquid chromatographic monitoring carbohydrate fractions in partially hydrolyzed corn starch. J Agric Food Chem. 1988;36(6):1188–92.
12. Sukhija PS, Palmquist DL. Rapid method for determination of total fatty acid content and composition of feedstuffs and feces. J Agric Food Chem. 1988; 36(6):1202–06.
13. International Dairy Federation. International IDF Standard 141C. Determination of milk fat, protein and lactose content. Guidance on the operation of mid-infrared instruments. Brussels: Int. Dairy Fed; 2000.
14. International Dairy Federation. International IDF Standard 148A. Milk enumeration of somatic cells. Brussels: Int. Dairy Fed; 1995.
15. Pellattiero E, Cecchinato A, Tagliapietra F, Schiavon S, Bittante G. The use of 2-dimensional gas chromatography to investigate the effect of rumen-protected conjugated linoleic acid, breed, and lactation stage on the fatty acid profile of sheep milk. J Dairy Sci. 2015;98(4):2088–102.
16. Lammers BP, Buckmaster DR, Heinrichs AJ. A simple method for the analysis of particle sizes of forage and total mixed rations. J Dairy Sci. 1996;79(5):922–28.
17. American Society of Agricultural and Biological Engineers (ASABE). Method of determining and expressing particle size of chopped forage materials by screening. ANSI/ASAE S424.1, 663-665. St. Joseph: ASABE; 2007.
18. Beauchemin KA, Yang WZ. Effects of physically effective fiber on intake, chewing activity, and ruminal acidosis for dairy cows fed diets based on corn silage. J Dairy Sci. 2005;88(6):2117–29.
19. Tyrrell HF, Reid JT. Prediction of the energy value of cow's milk. J Dairy Sci. 1965;48(9):1215–23.
20. SAS Institute. SAS User's Guide: Basics. Cary: SAS Inst. Inc.; 2009.
21. Miron J, Zuckerman E, Adin G, Solomon R, Shoshani E, Nikbachat M, et al. Comparison of two forage sorghum varieties with corn and the effect of feeding their silages on eating behavior and lactation performance of dairy cows. Anim Feed Sci Technol. 2007;139(1):23–39.
22. Di Marco ON, Ressia MA, Arias S, Aello MS, Arzadún M. Digestibility of forage silages from grain, sweet and bmr sorghum types: Comparison of in vivo, in situ and in vitro data. Anim Feed Sci Technol. 2009;153(3-4):161–68.
23. Heinrichs J, Kononoff P, Heinrichs J, Kononoff P. Evaluating particle size of forages and TMRs using the New Penn State Forage Separator. 2002. DAS 02-42. Pennsylvania State University Park, PA, USA.
24. Khan NA, Yu P, Ali M, Cone JW, Hendriks WH. Nutritive value of maize silage in relation to dairy cow performance and milk quality. J Sci Food Agric. 2015;95(2):238–52.
25. Tripathi MK. Effect of nutrition on production, composition, fatty acids and nutraceutical properties of milk. J Adv Dairy Res. 2014;2(2):115–25.
26. Pirondini M, Malagutti L, Colombini S, Amodeo P, Crovetto GM. Methane yield from dry and lactating cows diets in the Po Plain (Italy) using an in vitro gas production technique. Ital J Anim Sci. 2012;11(3):330–35.
27. Davoodi H, Esmaeili S, Mortazavian AM. Effects of milk and milk products consumption on cancer: a review. Compr Rev Food Sci Food Saf. 2013;12(3):249–64.
28. de Lorgeril M, Salen P. New insights into the health effects of dietary saturated and omega-6 and omega-3 polyunsaturated fatty acids. BMC Med. 2012;10(1):50.
29. Ikonen T, Morri S, Tyriseva AM, Ruottinen O, Ojala M. Genetic and phenotypic correlations between milk coagulation properties, milk production traits, somatic cell count, casein content, and pH of milk. J Dairy Sci. 2004;87(2):458–67.
30. Bittante G, Penasa M, Cecchinato A. Invited review: Genetics and modeling of milk coagulation properties. J Dairy Sci. 2012;95(12):6843–70.
31. Dal Zotto R, De Marchi M, Cecchinato A, Penasa M, Cassandro M, Carnier P, et al. Reproducibility and repeatability of measures of milk coagulation properties and predictive ability of mid-infrared reflectance spectroscopy. J Dairy Sci. 2008;91(10):4103–12.

20

Comparison of rumen bacteria distribution in original rumen digesta, rumen liquid and solid fractions in lactating Holstein cows

Shoukun Ji[1†], Hongtao Zhang[1†], Hui Yan[1], Arash Azarfar[2], Haitao Shi[1], Gibson Alugongo[1], Shengli Li[1], Zhijun Cao[1] and Yajing Wang[1*]

Abstract

Background: Original rumen digesta, rumen liquid and solid fractions have been frequently used to assess the rumen bacterial community. However, bacterial profiles in rumen original digesta, liquid and solid fractions vary from each other and need to be better established.

Methods: To compare bacterial profiles in each fraction, samples of rumen digesta from six cows fed either a high fiber diet (HFD) or a high energy diet (HED) were collected via rumen fistulas. Rumen digesta was then squeezed through four layers of cheesecloth to separate liquid and solid fractions. The bacterial profiles of rumen original digesta, liquid and solid fractions were analyzed with High-throughput sequencing technique.

Results: Rumen bacterial diversity was mainly affected by diet and individual cow ($P > 0.05$) rather than rumen fraction. Bias distributed bacteria were observed in solid and liquid fractions of rumen content using Venn diagram and LEfSe analysis. Fifteen out of 16 detected biomarkers (using LEfSe analysis) were found in liquid fraction, and these 15 biomarkers contributed the most to the bacterial differences among rumen content fractions.

Conclusions: Similar results were found when using samples of original rumen digesta, rumen liquid or solid fractions to assess diversity of rumen bacteria; however, more attention should be draw onto bias distributed bacteria in different ruminal fractions, especially when liquid fraction has been used as a representative sample for rumen bacterial study.

Keywords: Bacteria biomarker, Rumen bacteria diversity, Rumen content fraction

Background

The bovine rumen harbors a diverse population of microorganisms that convert ingested plant biomass into microbial protein and volatile fatty acids, and their fermentation end-products provide the host with essential nutrients for metabolism. Rumen microbes, therefore, play a key role in the productivity and health of ruminants [1].

Collection and sampling of ruminal content are important in both scientific research and diagnosis of diseases in ruminants [2]. Ruminal microbial diversity has been investigated in numbers of studies using different ruminal fractions including original rumen digesta, rumen liquid or solid fractions [3–8]. It has been demonstrated that rumen sampling methods and/or sampling pre-treatments could affect on the results of rumen microbial community [9, 10]. The relationships among microbial communities in different fractions of rumen content have been studied previously [11, 12]; however, it is still a controversial topic and deserves further investigations. In late studies, development of high-throughput sequencing techniques have allowed subtle effects on microbial community components to be detected as changes in relative numbers of bacterial community [1]. The objective of this study was to assess the differences and similarities of bacterial community in original rumen digesta, rumen liquid and solid fractions using a high-throughput sequencing technique.

* Correspondence: yajingwang@cau.edu.cn
†Equal contributors
[1]State Key Laboratory of Animal Nutrition, Beijing Engineering Technology Research Center of Raw Milk Quality and Safety Control, China Agricultural University, Beijing 100193, China
Full list of author information is available at the end of the article

Methods

Animals and sampling

Six ruminally fistulated lactating Holstein cows were housed in a free stall pen at the Zhongdi Dairy Research Center (Beijing, China) and were cared for according to the practices outlined in the Guide for the Care and Use of Agriculture Animals in Agriculture Research and Teaching (FASS, 2010).

Cows were randomly assigned to two groups with three cows in each group and individually fed by Roughage Intake Control System (RIC, Insentec B.V, Netherland). One group of cows were fed with a high fiber containing diet (HFD group) and the other group was fed with a high energy containing diet (HED group) (Table 1). After a 14-days adaptation period to the experimental diets, approximately 500 g of original rumen digesta of each cow was collected 5 h after morning feeding via rumen fistulas from the middle part of the ventral sac. After the original

rumen digesta sampled, the solid and liquid fractions were obtained by squeezing the original digesta through four layers of sterile cheesecloth. All samples were snap-frozen in liquid nitrogen and were stored at −80 °C until DNA extraction.

DNA isolation

Genomic DNA (gDNA) was extracted from 1 g of original or solid fraction, and from 1 mL of ruminal liquid with Qiagen DNA Extraction Kit™ (Qiagen, Hilden, Germany) using a repeated bead beating method followed by phenol-chloroform extraction according to the manufacturer's protocol. The DNA was re-suspended after being precipitated with ethanol. The quality of extracted DNA was assessed based on the absorbance ratios of 260/280 nm and 260/230 nm using a NanoDrop ND-1000 Spectrophotometer (NanoDrop Technologies, Wilmington, DE, USA). The values for A260/A280 ratio in the present study were ranged from 1.8 to 2.0.

PCR amplification and purification

For illumina MiSeq sequencing, bacterial 16S rRNA gene were amplified using primers covering the V3 region (343 F, 5′-GATCCTACGGGAGGCAGCA-3′ and 534R, 5′-GCTTACCGCGGCTGCTGGC-3′) with barcodes. All PCR reactions were carried out in 30 μL reaction mixtures with 15 μL of Phusion High-Fidelity PCR Master Mix (New England Biolabs), 0.2 μL of both forward and reverse primers and 10 ng of template DNA. PCR amplification was carried out according to the following protocol: initial denaturation for 5 min at 95 °C, followed by 25 cycles of denaturation at 95 °C for 1 min, annealing at 50 °C for 1 min, and elongation at 72 °C for 1 min, with a final elongation step at 72 °C for 7 min. To qualify and quantify PCR products, the same volume of 1 × loading buffer (containing SYBR green) mixed with PCR products and electrophoresis on 2% agarose gel. Samples with bright main strip between 200 and 210 bp were then chosen for further analysis. PCR products from samples for sequencing in the same MiSeq run were pooled at equal molality. The pooled mixture was purified with a QIAquick PCR Purification Kit (Qiagen, Hilden, Germany) and re-quantified with Agilent DNA 1000 Kit (Agilent Technologies Inc.).

Sequencing with high-throughput sequencing technique

Sequencing libraries were generated using NEBNext ultra DNA sample preparation kit (NEB, USA), following standard Illumina sample-preparation protocol. The quality of library was assessed on the Qubit 2.0 Fluorometer (Life technologies, Grand Island, NY, USA) and Agilent Bioanalyzer 2100 system (Agilent Technologies, Palo Alto, Calif.). The library was sequenced on illumina

Table 1 Ingredients and chemical composition of the experimental diets (as dry matter basis)

Items[a]	High energy diet (HED)	High fiber diet (HFD)
Ingredients, kg		
Corn silage	4.51	4.86
Alfafa hay	2.38	1.39
Oat hay	—	2.66
Extruded soybean	0.35	0.27
Flaked corn	3.26	1.32
Corn	1.72	2.87
DDGS	1.79	1.28
Salt	0.08	0.07
Sodium bicarbonate	0.29	0.29
Cottonseed	1.36	0.90
Soybean meal	2.12	1.82
Beet pulp	1.00	0.43
Additives	2.04	1.34
Contents, %		
DM as fed	56.4	53.5
Crude protein	17.98	17.00
NE_L, MCal/kg[b]	1.81	1.67
Fat	5.78	4.43
NDF	32.84	37.81
ADF	17.86	20.53
NFC	40.02	38.35
Ca	0.80	0.80
P	0.39	0.36

[a]*DDGS* dried distillers grains with solubles, *DM* dry matter, *NE$_L$* net energy requirement for lactation, *NDF* neutral detergent fiber, *ADF* acid detergent fiber, *NFC* nonfiber carbohydrates, *Ca* calcium, *P* phosphorus
[b]Calculated using equations from NRC (2001)

MiSeq platform and ~250 bp to ~300 bp paired-end reads were generated.

Quality control of raw data and data processing

Controlling the quality of raw data was done by FastQC (version 0.11.3). Reads with quality score higher than 30 were retained for further analysis. Paired-end reads from the original DNA fragments were merged using FLASH (version 1.2.7) [13]. Paired-end reads was assigned to each sample according to the unique barcodes. Concatenated sequences were detected using USEARCH (v6.1), and subsequently filtered out. Sequences analyses were performed using QIIME pipeline (version 1.5.0) [14]. Generated sequences were distributed into different samples based on barcodes, and the OTUs were defined by clustering sequences together with a 97% identity cut-off at UCLUST software [15] after removing the barcode. The RDP classifier [16] was used for taxonomic classification of generated OTUs [17]. To ensure the comparability of the species diversity between the samples, standardized OTU documents were used to analyze the species and diversity indexes. The threshold for the number of standardized sequences was set at 150,000 sequences.

Data analysis

Alpha diversity indices were calculated using QIIME pipeline (version 1.5.0) [14]. Beta diversity indices between samples were determined based on Bray-Curtis metric, relationship network of each fraction was calculated using Pearson correlation. LEfSe (LDA Effect Size) analysis was performed online (https://huttenhower.sph.harvard.edu/galaxy) to find differentially abundant taxa

(biomarkers) with P- value higher than 0.05 and LDA score higher than 2. ANOSIM analysis was performed with R software (version 3.1.2). Comparisons of bacterial abundance in different experimental groups were performed using Wilcoxon or Kruskal-Wallis test with R software (version 3.1.2).

Results

Bacterial diversities and community composition in different rumen fractions

Six ruminally fistulated Holstein dairy cows were grouped by two different diets containing either high level of fiber or high level of energy. Rumen contents were collected and separated to liquid and solid fractions. After sequencing on the Illumina MiSeq platform, 4,650,173 quality reads were generated, and 23,896 OTUs in total in all samples were detected with an average of $258,343 \pm 26,890$ reads and $5,003 \pm 678$ OTUs in each sample.

Within each dietary treatment, the number of OTUs and Chao1 index between different fractions of rumen samples were similar ($P > 0.05$); however, the number of OTUs and Chao1 index in rumen liquid fraction was higher in HFD group than that of HED group ($P < 0.05$; Additional file 1: Figure S1a and b). Simpson index has no difference among fractions within and between dietary treatments ($P > 0.05$; Additional file 1: Figure S1c).

The microbial diversity difference was displayed by heatmap, the similarity index with Pearson correlation, and ANOSIM analysis with Bray-Curtis similarity (Fig. 1; Additional file 1: Figure S2). We found that bacteria

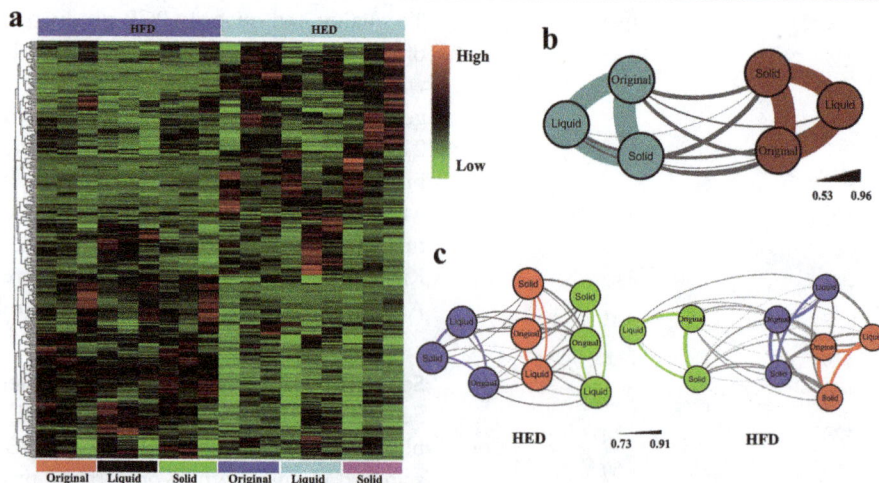

Fig. 1 Comparison of bacterial community in different fractions of rumen content based on OTUs. **a**, Heatmap of top 300 OTUs. **b**, Pearson correlation similarity of fractions, grayblue nodes represent HFD and darkred nodes represent HED, the thickness of lines indicates the Pearson correlation similarity. **c**, Pearson correlation similarity of different fractions of rumen content in HED group (*right*) and HFD group (*left*), nodes in same color represent samples obtained from the same cow, the thickness of lines indicates the Pearson correlation similarity

communities were clustered mainly by diets (Fig. 1a, b; Additional file 1: Figure S2a) and individual cows (Fig. 1c; Additional file 1: Figure S2b).

Venn plot was used to illustrate the distribution of bacteria in different ruminal fractions. After filtering out the rare OTUs (defined as OUTs that only appear in one sample in each group), we identified 2,022 OTUs appeared in ruminal fractions. More unique OTUs (appeared in one fraction but not appeared in others) were detected in HFD group compared with that in HED group (Additional file 1: Figure S3a). Rumen original digesta, solid and liquid fractions shared 53% of detected OTUs, but unique OTUs was also found in different fractions although they were evenly distributed among the rumen fractions (Additional file 1: Figure S3b).

Rumen bacterial taxa change in original and fractional rumen digesta

In total, 22 phyla, 38 classes, 62 orders, 96 families, 127 genera were detected regardless of fractional and dietary treatments (Fig. 2). To identify the taxon distributions in different fractions, LEfSe analysis was performed and biomarkers of liquid and/or solid fractions were found in both HFD and HED groups (Fig. 2). Fifteen taxa were found as biomarker in HFD group (Fig. 2a; Additional file 1: Figure S4a), while only one taxon was detected as biomarker in HED group (Fig. 2b; Additional file 1: Figure S4b).

Within HFD group, 14 taxa increased in liquid fraction and one taxon increased in solid fraction (Fig. 2a; Additional file 1: Figure S5). Among the changed taxa, genus *Coprococcus* and *Oscillospira* were found to be predominant bacteria (appeared in all samples and relative abundance ≥ 1% in at least one sample; Additional file 1: Figure S6) in both rumen solid and liquid fractions in HFD group, respectively. Within HFD group, relative abundance of genus *Succinivibirio* among ruminal fractions were found different (Fig. 2b) with the highest abundance found in solid fraction ($P < 0.05$; Additional file 1: Figure S5).

Discussion

Influence of rumen content fractions on bacterial diversity

Rumen, harboring large number of inhabiting microbes (approximately 10^{11} bacterial cells per g of rumen content) which play important roles in providing necessary nutrients (such as proteins and energy yielding substrates) to the host animal [18, 19]. Many factors such as age [4], diet [11, 19], and animal individual [20] affect rumen microbial community. Original rumen digesta, liquid fraction and solid fraction samples have been frequently used to assess rumen microbes [1, 4]; however, the fractions used may also cause biased observation in rumen microbial studies [1, 20]. In present study, we examined the differences and

similarities of bacterial communities in different rumen fractions from lactating dairy cows fed with either HFD or HED using high throughput sequencing technology.

Many researchers proved that diet is one of the main factors that affect rumen microbial diversity, and that the nutritional plane and/or feed ingredients have great impacts on rumen microbial communities [3, 21–25], which may due to bacteria preferring particular metabolic substrates and rumen environment [26]. This was supported by our findings that HFD- and HED-fed cows harbored different bacterial communities (Fig. 1a; Additional file 1: Figure S3a), which was illustrated by different α-diversity indices (Additional file 1: Figure S1) and different clusters (Fig. 1b; Additional file 1: Figure S2a) between dietary treatment groups.

Regardless the dietary effect, bacterial community of each individual cow was quite unique (Fig. 1c). A clear grouping resulted by individual cows (Additional file 1: Figure S2b) indicating that animal individual had its own distinct microbiota [27]. This finding was also supported by a previous study in which cows were switched from a high forage diet to a high concentrate diet (during acidosis and after recovery) and the bacterial populations exhibited a low taxonomic variability as less than 5% of total identified OTUs differed [11].

Between different ruminal content fractions, the diversity difference was not observed (Fig. 1b and c; Additional file 1: Figure S1; Figure S2c), while different bacteria distribution was detected (Fig. 1a; Additional file 1: Figure S3b) indicating that each fraction had some unique bacteria.

Influence of ruminal fractions on bacterial taxon distribution

Different bacteria prefer particular metabolic substrates and rumen environment and thus might be distributed differently in disparate phases [26]. In the current study, results of LEfSe illustrated that distribution difference of rumen bacteria in different fractions can be observed in both HFD group (Fig. 2a) and HED group (Fig. 2b). Biomarkers of liquid and solid fractions were detected (Fig. 2), and bacterial community of original digesta were largely likely displaying an intermediate state of liquid and solid fractions (Additional file 1: Figure S5). This verifies the common theory that bacterial community in original digesta represents the real rumen bacterial community the best.

Crucially, we found that liquid fraction contributed most to the bacteria difference in different fractions of HFD group (Fig. 2; Additional file 1: Figure S5). Usually, for studying rumen microbial diversity, original rumen digesta sampling through a fistula or from slaughtered animals was described the best method to have a representative sample [27]. Alternatively, rumen

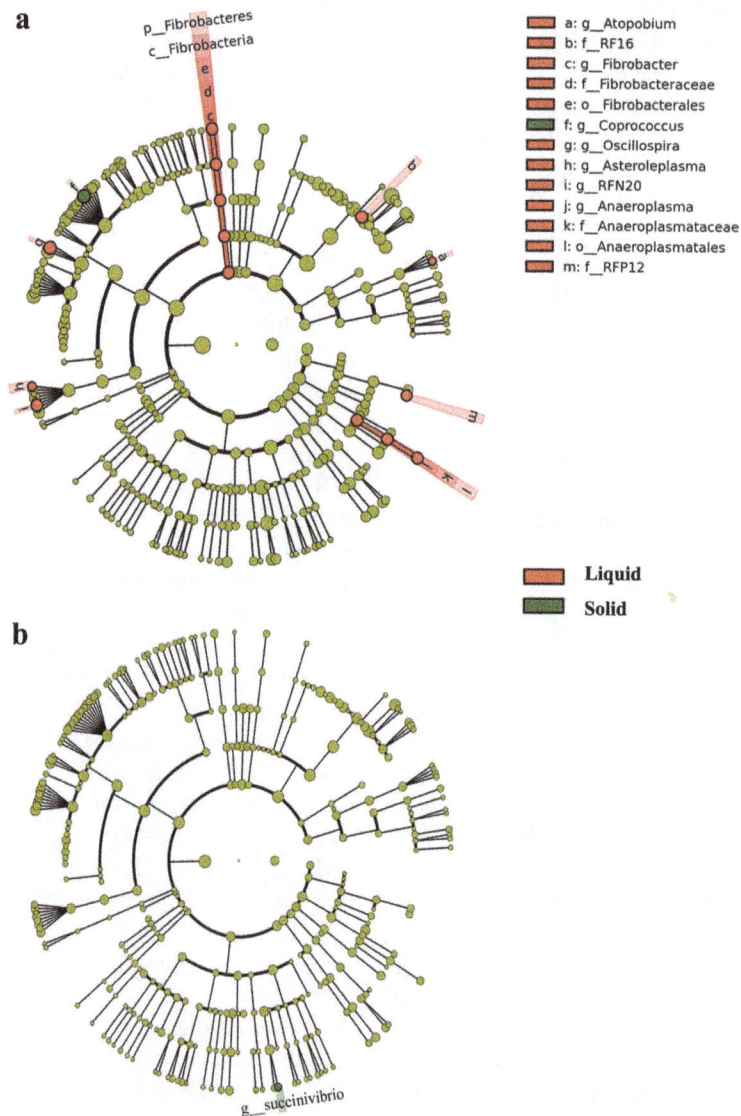

Fig. 2 Ruminal bacteria change in different fractions at genus level. LEfSe cladograms demonstrating taxonomic differences among different fractions in HFD group (**a**) and HED group (**b**) respectively, LDA scores above 2 and *P* value smaller than 0.05 were shown. LEfSe: linear discriminant analysis (LDA) effect size

liquid collected via a stomach tube has become a routinely used method for rumen sample collection because of its easy achievement [1, 28, 29]. A previous study showed significant differences in bacterial communities between rumen solid and liquid content [21]. In current study, liquid fraction had similar bacterial diversity with original rumen digesta and solid fraction, but particular bacteria abundance (including predominant bacteria) in liquid fraction differed with original content and solid fraction (Additional file 1: Figure S5; Figure S6). The biased distribution of bacteria should be taken into consideration when samples of liquid fraction have been used to assess rumen bacterial community.

Conclusions

This study investigated and compared rumen bacteria community of original rumen digesta, liquid and solid fractions from cows fed with HFD and HED. Rumen bacteria diversity was mainly affected by diet and cow individuals rather than by rumen fractions. Bias distributed bacteria were observed in different fractions of rumen content. Liquid fraction contributed most to bacterial differences among rumen content fractions of HFD group. Results indicated that using different fractions to assess rumen bacterial diversity will generate similar results; however, more attention should be paid to bias distributed bacteria in different fractions, especially when liquid fraction is used as a representative sample.

Additional files

Additional file 1: Figure S1. Microbial diversity in different fractions of rumen content. a, the OTU numbers in original, solid or liquid fraction samples. b, Chao1 index in original, solid or liquid fraction samples. c, Simpson index based on OTUs in original, solid, and liquid fraction samples. HFD: High fiber diet; HED: High energy diet. Data are presented as Mean ± SD. **Figure S2.** Analysis of similarity (ANOSIM) in different groups. ANOSIM results are presented with box plot when bacteria communities are grouped by diet (a), cows (b), and ruminal content fractions (c) using Bray-Curtis metric based on OTUs. **Figure S3.** Venn plot for shared OTUs. a, OTUs in HFD and HED. b, OTUs in original, solid and liquid fractions. **Figure S4.** Ruminal bacteria change in different fractions of rumen content at genera level. LEfSe histogram demonstrating taxonomic differences among different fractions in HFD group (a) and HED group (b) respectively, LDA scores above 2 and P value smaller than 0.05 were shown. LEfSe: linear discriminant analysis (LDA) effect size. **Figure S5.** Influence of rumen fractions on biomarker taxa abundance. p_: phylum; c_: class; o_: order; f_: family; g_: genus. Data was presented as Mean ± SD. **Figure S6.** Predominant rumen bacteria at genera level. a, predominant genera higher than 1% in proportion in all samples. b, distribution of predominant genera in each fractions.

Additional file 2: OTUs distribution in each sample.

Abbreviations
ADF: Acid detergent fiber; Ca: Calcium; DDGS: Dried distillers grains with soluble; DM: Dry matter; HED: High energy diet; HFD: High fiber diet; NDF: Neutral detergent fiber; NE$_L$: Net energy requirement for lactation; NFC: Nonfiber carbohydrates; OTUs: Operational Taxonomic Units; P: Phosphorus

Acknowledgements
The authors thank L.F. Zhang for analyses of diet nutrition. The receipt of help in samples collection from Y.Q. Wu, J. Mao, and Y. Du, is gratefully acknowledged.

Funding
This research was supported by National Dairy Industry and Technology System (CARS-37) and National Natural Science Foundation of China (31402099).

Authors' contributions
SKJ, SLL, YJW, and ZJC designed the research; SKJ and HTS collected the samples; SKJ and HTZ performed the research, analyzed data and prepared the manuscript; HY, AA and GMA helped improving the data analysis and manuscript. All authors read and approved the final manuscript.

Competing interests
The authors declare that they have no competing interests.

Author details
[1]State Key Laboratory of Animal Nutrition, Beijing Engineering Technology Research Center of Raw Milk Quality and Safety Control, China Agricultural University, Beijing 100193, China. [2]Faculty of Agriculture, Lorestan University, PO Box 465, Khorramabad, Iran.

References
1. Henderson G, Cox F, Kittelmann S, Miri VH, Zethof M, Noel SJ, et al. Effect of DNA extraction methods and sampling techniques on the apparent structure of cow and sheep rumen microbial communities. PLoS One. 2013; 8:e74787.
2. Shen JS, Chai Z, Song LJ, Liu JX, Wu YM. Insertion depth of oral stomach tubes may affect the fermentation parameters of ruminal fluid collected in dairy cows1. J Dairy Sci. 2012;95:5978–84.
3. Golder HM, Denman SE, McSweeney C, Celi P, Lean IJ. Ruminal bacterial community shifts in grain-, sugar-, and histidine-challenged dairy heifers. J Dairy Sci. 2014;97:5131–50.
4. Jami E, Israel A, Kotser A, Mizrahi I. Exploring the bovine rumen bacterial community from birth to adulthood. ISME J. 2013;7:1069–79.
5. Li RW, Connor EE, Li C, Baldwin RL, Sparks ME. Characterization of the rumen microbiota of pre-ruminant calves using metagenomic tools. Environ Microbiol. 2012;14:129–39.
6. Lima FS, Oikonomou G, Lima SF, Bicalho MLS, Ganda EK, de Oliveira Filho JC, et al. Prepartum and postpartum rumen fluid microbiomes: characterization and correlation with production traits in dairy cows. Appl Environ Microbiol. 2015;81:1327–37.
7. Nieman CC, Steensma KM, Rowntree JE, Beede DK, Utsumi SA. Differential response to stocking rates and feeding by two genotypes of Holstein-Friesian cows in a pasture-based automatic milking system. Animal. 2015;9: 2039–49.
8. Sandri M, Manfrin C, Pallavicini A, Stefanon B. Microbial biodiversity of the liquid fraction of rumen content from lactating cows. Animal. 2014;8:572–9.
9. Fliegerova K, Tapio I, Bonin A, Mrazek J, Callegari ML, Bani P, et al. Effect of DNA extraction and sample preservation method on rumen bacterial population. Anaerobe. 2014;29:80–4.
10. McKain N, Genc B, Snelling TJ, Wallace RJ. Differential recovery of bacterial and archaeal 16S rRNA genes from ruminal digesta in response to glycerol as cryoprotectant. J Microbiol Methods. 2013;95:381–3.
11. Petri RM, Schwaiger T, Penner GB, Beauchemin KA, Forster RJ, McKinnon JJ, et al. Changes in the rumen epimural bacterial diversity of beef cattle as affected by diet and induced ruminal acidosis. Appl Environ Microbiol. 2013; 79:3744–55.
12. Singh KM, Jisha TK, Reddy B, Parmar N, Patel A, Patel AK, et al. Microbial profiles of liquid and solid fraction associated biomaterial in buffalo rumen fed green and dry roughage diets by tagged 16S rRNA gene pyrosequencing. Mol Biol Rep. 2015;42:95–103.
13. Magoc T, Salzberg SL. FLASH: fast length adjustment of short reads to improve genome assemblies. Bioinformatics. 2011;27:2957–63.
14. Caporaso JG, Lauber CL, Walters WA, Berg-Lyons D, Lozupone CA, Turnbaugh PJ, et al. Global patterns of 16S rRNA diversity at a depth of millions of sequences per sample. Proc Natl Acad Sci. 2011;108:4516–22.
15. Edgar RC. Search and clustering orders of magnitude faster than BLAST. Bioinformatics. 2010;26:2460–1.
16. Wang Q, Garrity GM, Tiedje JM, Cole JR. Naive bayesian classifier for rapid assignment of rRNA sequences into the new bacterial taxonomy. Appl Environ Microbiol. 2007;73:5261–7.
17. DeSantis TZ, Hugenholtz P, Larsen N, Rojas M, Brodie EL, Keller K, et al. Greengenes, a chimera-checked 16S rRNA gene database and workbench compatible with arb. Appl Environ Microbiol. 2006;72:5069–72.
18. Russell JB. Factors that alter rumen microbial ecology. Science. 2001;292: 1119–22.
19. Henderson G, Cox F, Ganesh S, Jonker A, Young W, Abecia L, et al. Rumen microbial community composition varies with diet and host, but a core microbiome is found across a wide geographical range. Sci Rep. 2015;5: 14567.
20. Brulc JM, Antonopoulos DA, Miller ME, Wilson MK, Yannarell AC, Dinsdale EA, et al. Gene-centric metagenomics of the fiber-adherent bovine rumen microbiome reveals forage specific glycoside hydrolases. Proc Natl Acad Sci U S A. 2009;106:1948–53.
21. de Menezes AB, Lewis E, O'Donovan M, O'Neill BF, Clipson N, Doyle EM. Microbiome analysis of dairy cows fed pasture or total mixed ration diets. FEMS Microbiol Ecol. 2011;78:256–65.
22. Thoetkiattikul H, Mhuantong W, Laothanachareon T, Tangphatsornruang S, Pattarajinda V, Eurwilaichitr L, et al. Comparative analysis of microbial profiles in cow rumen fed with different dietary fiber by tagged 16S rRNA gene pyrosequencing. Curr Microbiol. 2013;67:130–7.

23. Tajima K, Aminov RI, Nagamine T, Matsui H, Nakamura M, Benno Y. Diet-dependent shifts in the bacterial population of the rumen revealed with real-time PCR. Appl Environ Microbiol. 2001;67:2766–74.

24. Hook SE, Steele MA, Northwood KS, Dijkstra J, France J, Wright AG, et al. Impact of subacute ruminal acidosis (SARA) adaptation and recovery on the density and diversity of bacteria in the rumen of dairy cows. FEMS Microbiol Ecol. 2011;78:275–84.

25. Fernando SC, Purvis HT, Najar FZ, Sukharnikov LO, Krehbiel CR, Nagaraja TG, et al. Rumen microbial population dynamics during adaptation to a high-grain diet. Appl Environ Microbiol. 2010;76:7482–90.

26. Zebeli Q, Tafaj M, Weber I, Steingass H, Drochner W. Effects of dietary forage particle size and concentrate level on fermentation profile, in vitro degradation characteristics and concentration of liquid- or solid-associated bacterial mass in the rumen of dairy cows. Anim Feed Sci Technol. 2008; 140:307–25.

27. Castro-Carrera T, Toral PG, Frutos P, McEwan NR, Hervás G, Abecia L, et al. Rumen bacterial community evaluated by 454 pyrosequencing and terminal restriction fragment length polymorphism analyses in dairy sheep fed marine algae. J Dairy Sci. 2014;97:1661–9.

28. Ramos-Morales E, Arco-Pérez A, Martín-García AI, Yáñez-Ruiz DR, Frutos P, Hervás G. Use of stomach tubing as an alternative to rumen cannulation to study ruminal fermentation and microbiota in sheep and goats. Anim Feed Sci Technol. 2014;198:57–66.

29. Lodge-Ivey SL, Browne-Silva J, Horvath MB. Technical note: Bacterial diversity and fermentation end products in rumen fluid samples collected via oral lavage or rumen cannula. 2009. 87: 2333-7.

PERMISSIONS

All chapters in this book were first published in JASB, by BioMed Central; hereby published with permission under the Creative Commons Attribution License or equivalent. Every chapter published in this book has been scrutinized by our experts. Their significance has been extensively debated. The topics covered herein carry significant findings which will fuel the growth of the discipline. They may even be implemented as practical applications or may be referred to as a beginning point for another development.

The contributors of this book come from diverse backgrounds, making this book a truly international effort. This book will bring forth new frontiers with its revolutionizing research information and detailed analysis of the nascent developments around the world.

We would like to thank all the contributing authors for lending their expertise to make the book truly unique. They have played a crucial role in the development of this book. Without their invaluable contributions this book wouldn't have been possible. They have made vital efforts to compile up to date information on the varied aspects of this subject to make this book a valuable addition to the collection of many professionals and students.

This book was conceptualized with the vision of imparting up-to-date information and advanced data in this field. To ensure the same, a matchless editorial board was set up. Every individual on the board went through rigorous rounds of assessment to prove their worth. After which they invested a large part of their time researching and compiling the most relevant data for our readers.

The editorial board has been involved in producing this book since its inception. They have spent rigorous hours researching and exploring the diverse topics which have resulted in the successful publishing of this book. They have passed on their knowledge of decades through this book. To expedite this challenging task, the publisher supported the team at every step. A small team of assistant editors was also appointed to further simplify the editing procedure and attain best results for the readers.

Apart from the editorial board, the designing team has also invested a significant amount of their time in understanding the subject and creating the most relevant covers. They scrutinized every image to scout for the most suitable representation of the subject and create an appropriate cover for the book.

The publishing team has been an ardent support to the editorial, designing and production team. Their endless efforts to recruit the best for this project, has resulted in the accomplishment of this book. They are a veteran in the field of academics and their pool of knowledge is as vast as their experience in printing. Their expertise and guidance has proved useful at every step. Their uncompromising quality standards have made this book an exceptional effort. Their encouragement from time to time has been an inspiration for everyone.

The publisher and the editorial board hope that this book will prove to be a valuable piece of knowledge for researchers, students, practitioners and scholars across the globe.

LIST OF CONTRIBUTORS

Tahir Usman
Key Laboratory of Animal Genetics, Breeding and Reproduction, Ministry of Agriculture of China, National Engineering Laboratory for Animal Breeding, College of Animal Science and Technology, China Agricultural University, Beijing 100193, People's Republic of China
College of Veterinary Sciences and Animal Husbandry, Abdul Wali Khan University Mardan, Mardan 23200, Pakistan

Yachun Wang, Chao Liu, Yanghua He, Xiao Wang, Yichun Dong and Ying Yu
Key Laboratory of Animal Genetics, Breeding and Reproduction, Ministry of Agriculture of China, National Engineering Laboratory for Animal Breeding, College of Animal Science and Technology, China Agricultural University, Beijing 100193, People's Republic of China

Hongjun Wu
Xieerltala Breeding Farm, Hailaer 021012, Inner Mongolia, China

Airong Liu
Agricultural and Animal Husbandry Administration Bureau, Hailaer 021000, Inner Mongolia, China

Mario Vailati-Riboni and Juan J. Loor
Mammalian NutriPhysioGenomics, Department of Animal Sciences and Division of Nutritional Sciences, University of Illinois, Urbana, IL 61801, USA

Johan S. Osorio
Mammalian NutriPhysioGenomics, Department of Animal Sciences and Division of Nutritional Sciences, University of Illinois, Urbana, IL 61801, USA
Dairy and Food Science Department, South Dakota State University, 1111 College Ave, 113H Alfred DairyScience Hall, Brookings SD 57007, USA

Erminio Trevisi
Istituto di Zootecnica Facoltà di Scienze Agrarie, Alimentari e Ambientali, Università Cattolica del Sacro Cuore, 29122 Piacenza, Italy

Daniel Luchini
Adisseo NA, Alpharetta, GA 30022, USA

Alea Agrawal, Abdulrahman Alharthi, Mario Vailati-Riboni, Zheng Zhou and Juan J. Loor
Mammalian NutriPhysioGenomics, Department of Animal Sciences and Division of Nutritional Sciences, University of Illinois, 1207 West Gregory Drive, Urbana, IL 61801, USA

Lefei Jiao, Fanghui Lin, Shuting Cao, Chunchun Wang, Huan Wu and Miaoan Shu
Animal Science College, Zhejiang University, Key Laboratory of Animal Feed and Nutrition of Zhejiang Province, No.866, Yuhangtang Road, Hangzhou 310058, People's Republic of China

Caihong Hu
Animal Science College, Zhejiang University, Key Laboratory of Animal Feed and Nutrition of Zhejiang Province, No.866, Yuhangtang Road, Hangzhou 310058, People's Republic of China
Key Laboratory of Animal Nutrition and Feed in East China, Ministry of Agriculture, No.866, Yuhangtang Road, Hangzhou 310058, People's Republic of China

Wen Zhu, Zihai Wei, Ningning Xu, Fan Yang, Jianxin Liu and Jiakun Wang
Institute of Dairy Science, College of Animal Sciences, Zhejiang University, 866 Yuhangtang Road, Hangzhou 310058, People's Republic of China

Ilkyu Yoon and Yihua Chung
Diamond V, Cedar Rapids, IA 52405, USA

Lex Ee Xiang Leong and Carl K. Davis
School of Chemistry and Molecular Bioscience, University of Queensland, St Lucia 4072, QLD, Australia

Shahjalal Khan
School of Agriculture and Food Sciences, University of Queensland, St Lucia 4072, QLD, Australia

Stuart E. Denman and Chris S. McSweeney
CSIRO Agriculture and Food, Queensland Bioscience Precinct, St Lucia 4072, QLD, Australia

Yang Zou
State Key Laboratory of Animal Nutrition, Beijing Engineering Technology Research Center of Raw Milk Quality and Safety Control, College of Animal Science and Technology, China Agricultural University, Beijing 100193, China
Beijing Dairy Cattle Center, Beijing 100192, China

Yajing Wang, Youfei Deng, Zhijun Cao and Shengli Li
State Key Laboratory of Animal Nutrition, Beijing Engineering Technology Research Center of Raw Milk Quality and Safety Control, College of Animal Science and Technology, China Agricultural University, Beijing 100193, China

Jiufeng Wang
College of Veterinary Medicine, China Agricultural University, Beijing 100193, China

Aoxing Liu
Laboratory of Animal Genetics, Breeding and Reproduction, Ministry of Agriculture of China, National Engineering Laboratory of Animal Breeding, College of Animal Science and Technology, China Agricultural University, Beijing 100193, China
Center for Quantitative Genetics and Genomics, Department of Molecular Biology and Genetics, Aarhus University, 8830 Tjele, Denmark

Mogens Sandø Lund, Per Madsen and Guosheng Su
Center for Quantitative Genetics and Genomics, Department of Molecular Biology and Genetics, Aarhus University, 8830 Tjele, Denmark

Yachun Wang
Laboratory of Animal Genetics, Breeding and Reproduction, Ministry of Agriculture of China, National Engineering Laboratory of Animal Breeding, College of Animal Science and Technology, China Agricultural University, Beijing 100193, China

Gang Guo and Ganghui Dong
Beijing Sunlon Livestock Development Co., Ltd, Beijing 100176, China

Mariana Piatto Berton, Rafael Medeiros de Oliveira Silva, Elisa Peripolli, Nedenia Bonvino Stafuzza, Fernando Baldi
Departamento de Zootecnia, Faculdade de Ciências Agrárias e Veterinárias, Universidade Estadual Paulista, Via de acesso Prof. Paulo Donato Castellane, s/no, Jaboticabal, SP CEP 14884-900, Brazil

Jesús Fernández Martin, Maria Saura Álvarez and Beatriz Villanueva Gavinã
Instituto Nacional de Investigación y Tecnología Agraria y Alimentaria INIA, Crta. de la Coruña, km 7,5 -, 28040 Madrid, Spain

Miguel Angel Toro
Departamento de Producción Agraria, School of Agricultural, Food and Byosystems Engineering, Universisdad Politécnica de Madrid, Campus Ciudad Universitaria Avda. Complutense 3 - Avda. Puerta Hierro, 28040 Madrid, Spain

Georgget Banchero
Instituto Nacional de Investigación Agropecuária (INIA), Ruta 50 Km. 12, Colonia, Uruguay

Priscila Silva Oliveira, Joanir Pereira Eler and José Bento Sterman Ferraz
Faculdade de Zootecnia e Engenharia de Alimentos, Nucleo de Apoio à Pesquisa em Melhoramento Animal, Biotecnologia e Transgenia, Universidade de São Paulo, Rua Duque de Caxias Norte, 225, Pirassununga, SP CEP 13635-900, Brazil

Afshin Hosseini, Zheng Zhou, James K. Drackley and Juan J. Loor
Department of Animal Sciences and Division of Nutritional Sciences, University of Illinois, 1207 West Gregory Drive, Urbana, IL 61801, USA

Mustafa Salman
Department of Animal Nutrition and Nutritional Diseases, University of Ondokuz Mayıs, 55139 Samsun, Turkey

Marcos Eli Buzanskas
Departamento de Zootecnia, Universidade Federal da Paraíba (UFPB), Areia, Paraíba 58397-000, Brazil

Daniela do Amaral Grossi
Fast Genetics, Saskatoon, SK S7K 2K6, Canada

Ricardo Vieira Ventura
Beef Improvement Opportunities (BIO), Guelph, ON N1K 1E5, Canada

Flavio Schramm Schenkel
Department of Animal and Poultry Science, University of Guelph, Centre for Genetic Improvement of Livestock (CGIL), Guelph, ON N1G 2W1, Canada

Tatiane Cristina Seleguim Chud, Nedenia Bonvino Stafuzza and Danísio Prado Munari
Departamento de Ciências Exatas, Faculdade de Ciências Agrárias e Veterinárias, Universidade Estadual Paulista (Unesp), Jaboticabal, São Paulo 14884-900, Brazil

Luciana Diniz Rola
Departamento de Zootecnia, Núcleo de Pesquisa e Conservação de Cervídeos, Faculdade de Ciências Agrárias e Veterinárias, Universidade Estadual Paulista (Unesp), Jaboticabal, São Paulo 14884-900, Brazil

Sarah Laguna Conceição Meirelles
Department of Animal Science, Federal University of Lavras (UFLA), Lavras, Minas Gerais 37200-000, Brazil.

Fabiana Barichello Mokry
Department of Genetics and Evolution, Federal University of São Carlos (UFSCar), São Carlos, São Paulo 13565-905, Brazil

Maurício de Alvarenga Mudadu and Roberto Hiroshi Higa
Embrapa Agricultural Informatics, Campinas, São Paulo 13083-886, Brazil

Marcos Vinícius Gualberto Barbosa da Silva
Embrapa Dairy Cattle, Juiz de Fora, Minas Gerais 36038-330, Brazil

Maurício Mello de Alencar and Luciana Correia de Almeida Regitano
Embrapa Southeast Livestock, São Carlos, São Paulo 13560-970, Brazil

Nicky-Lee Willson
School of Animal and Veterinary Sciences, The University of Adelaide, Roseworthy, SA 5371, Australia
The Australian Poultry and Cooperative Research Centre, University of New England, Armidale, NSW 2351, Australia

Rebecca E. A. Forder and Philip I. Hynd
School of Animal and Veterinary Sciences, The University of Adelaide, Roseworthy, SA 5371, Australia

Rick G. Tearle
Davies Research Centre, School of Animal and Veterinary Sciences, The University of Adelaide, Roseworthy, SA 5371, Australia

Greg S. Nattrass
South Australian Research and Development Institute (SARDI), Livestock and Farming Systems, Roseworthy, SA 5371, Australia

Robert J. Hughes
School of Animal and Veterinary Sciences, The University of Adelaide, Roseworthy, SA 5371, Australia
South Australian Research and Development Institute (SARDI), Pig and Poultry Production Institute, Roseworthy, SA 5371, Australia

Lúcio Flávio Macedo Mota
Departmento de Zootecnia, Universidade Federal dos Vales do Jequitinhonha e Mucuri, Diamantina, MG 39100-000, Brazil
Faculdade de Ciências Agrárias e Veterinárias, Universidade Estadual Paulista, Jaboticabal, SP 14884-900, Brazil São Paulo, Av. Duque de Caxias Norte nº225, Pirassununga 13635-900, SP, Brazil

Cristina Moreira Bonafé and Aldrin Vieira Pires
Departmento de Zootecnia, Universidade Federal dos Vales do Jequitinhonha e Mucuri, Diamantina, MG 39100-000, Brazil

Pâmela Almeida Alexandre, Miguel Henrique Santana, Francisco José Novais, José Bento Sterman Ferraz and Heidge Fukumasu
Departamento de Medicina Veterinária, Faculdade de Zootecnia e Engenharia de Alimentos, Universidade de

Erika Toriyama
Departamento de Zootecnia e Desenvolvimento Agrossocioambiental Sustentável, Faculdade de Veterinária, Universidade Federal Fluminense, Niteroi, RJ 24230-340, Brazil

Saulo da Luz Silva and Paulo Roberto Leme
Departamento de Zootecnia, Faculdade de Zootecnia e Engenharia de Alimentos, Universidade de São Paulo, Pirassununga 13635-900, SP, Brazil

Janina Skipor, Marta Kowalewska, Aleksandra Szczepkowska and Anna Majewska
Institute of Animal Reproduction and Food Research, Polish Academy of Sciences, Olsztyn, Poland

Tomasz Misztal and Andrzej P. Herman
The Kielanowski Institute of Animal Physiology and Nutrition, Polish Academy of Sciences, Jablonna n/ Warsaw, Olsztyn Poland

Marek Jalynski
Veterinary Medicine Faculty, University of Warmia and Mazury, Olsztyn, Poland

Katarzyna Zabek
Department of Sheep and Goat Breeding, Animal Bioengineering Faculty, University of Warmia and Mazury in Olsztyn, Olsztyn, Poland.

Aline Gomes da Silva
Universidade Federal de Viçosa, Viçosa, Minas Gerais 36570–000, Brazil
Universidade Federal de Mato Grosso do Sul, Campo Grande, Mato Grosso do Sul 79074–460, Brazil

Mário Fonseca Paulino, Edenio Detmann, Román Enrique Maza Ortega, Victor Valério de Carvalho, Josilaine Aparecida da Costa Lima, Felipe Henrique de Moura and Jéssika Almeida Bitencourt
Universidade Federal de Viçosa, Viçosa, Minas Gerais 36570–000, Brazil

Henrique Jorge Fernandes
Universidade Estadual de Mato Grosso do Sul, Aquidauana, Mato Grosso do Sul 79200–000, Brazil

Lincoln da Silva Amorim
Universidade Federal de Viçosa, Viçosa, Minas Gerais 36570–000, Brazil
Biotran - Biotecnologia e Treinamento em Reprodução Animal, Alfenas, Minas Gerais 37130–000, Brazil.

Mariana Benevides Monteiro
Universidade Federal do Acre, Rio Branco, Acre 69920–900, Brazil

Linghong Jin, Wei Wang, Jeroen Degroote, Noémie Van Noten, Honglin Yan and Maryam Majdeddin
Department of Applied Biosciences, Ghent University, Valentin Vaerwyckweg 1, 9000 Ghent, Belgium
Laboratory for Animal Nutrition and Animal Product Quality, Department of Animal Production, Ghent University, Coupure Links 653, 9000 Ghent, Belgium

Mario Van Poucke and Luc Peelman
Department of Nutrition, Genetics and Ethology, Faculty of Veterinary Medicine, Ghent University, Heidestraat 19, 9820 Merelbeke, Belgium

Anne Goderis, Kurt Van De Mierop and Ronny Mombaerts
Nutrex, Achterstenhoek 5, 2275 Lille, Belgium

Stefaan De Smet
Laboratory for Animal Nutrition and Animal Product Quality, Department of Animal Production, Ghent University, Coupure Links 653, 9000 Ghent, Belgium

Joris Michiels
Department of Applied Biosciences, Ghent University, Valentin Vaerwyckweg 1, 9000 Ghent, Belgium

Manoel Francisco de Sá Filho, Angela Maria Gonella-Diaza, Mariana Sponchiado, Marcio Ferreira Mendanha, Guilherme Pugliesi, Roney dos Santos Ramos, Marcelo Demarchi Goissis and Pietro Sampaio Baruselli
Departamento de Reprodução Animal, FMVZ-USP, São Paulo, SP, Brazil

Sónia Cristina da Silva Andrade
Departamento de Genética e Biologia Evolutiva, IB-USP-, São Paulo, SP, Brazil

Gustavo Gasparin and Luiz Lehmann Coutinho
Laboratório de Biotecnologia Animal, ESALQ-USP, Av Pádua Dias, Piracicaba, SP 11, Brazil

Fernando Silveira Mesquita
Universidade Federal do Pampa, Uruguaiana, RS, Brazil

Mario Binelli
Departamento de Reprodução Animal, FMVZ-USP, São Paulo, SP, Brazil
Universidade de São Paulo, Faculdade de Medicina Veterinária e Zootecnia, Departamento de Reprodução Animal, Avenida Duque de Caxias Norte, 225, Pirassununga, SP Zip Code 13635900, Brazil

Nataliya Pošćić, Tommaso Montanari and Sandy Sgorlon
Department of Agriculture, Food, Environment and Animal Science (DI4A), University of Udine, via delle Scienze 206, 33100 Udine, Italy

Mariasilvia D'Andrea and Fabio Pilla
Department of Agricultural, Environmental and Food Sciences, University of Molise, via F. De Sanctis snc, 86100 Campobasso, Italy

Danilo Licastro
CBM S.c.r.l, SS 14 – km 163.5 AREA Science Park, 34149 Basovizza, TS, Italy

Paolo Ajmone-Marsan and Andrea Minuti
Institute of Zootechnics, Catholic University of the Sacred Heart, via Emilia Parmense 84, 29133 Piacenza, Italy

M. Cattani and L. Bailoni
Department of Comparative Biomedicine and Food Science (BCA), University of Padova, Viale dell'Università 16, 35020 Legnaro, PD, Italy

N. Guzzo and R. Mantovani
Department of Agronomy Food Natural resources Animals and Environment (DAFNAE), University of Padova, Viale dell'Università 16, 35020 Legnaro, PD, Italy

Shoukun Ji, Hongtao Zhang, Hui Yan, Haitao Shi, Gibson Alugongo, Shengli Li, Zhijun Cao and Yajing Wang
State Key Laboratory of Animal Nutrition, Beijing Engineering Technology Research Center of Raw Milk Quality and Safety Control, China Agricultural University, Beijing 100193, China

Arash Azarfar
Faculty of Agriculture, Lorestan University, Khorramabad, Iran

Index